ESTUARINE PROCESSES

Volume I

Uses, Stresses, and Adaptation to the Estuary

Produced by the
Estuarine Research Federation
with support from the

Bureau of Land Management, Department of Interior
Fish and Wildlife Service, Department of Interior
U. S. Environmental Protection Agency
National Marine Fisheries Service, National Oceanographic and Atmospheric
    Administration
Marine Ecosystems Analysis Program, NOAA
Office of Applications, National Aeronautics and Space Administration
Council on Environmental Quality
Energy Research and Development Administration

ESTUARINE PROCESSES

Volume I
Uses,
Stresses,
and Adaptation to the Estuary

Edited by

MARTIN WILEY

Chesapeake Biological Laboratory
University of Maryland
Center for Environmental
and Estuarine Studies
Solomons, Maryland

ACADEMIC PRESS   NEW YORK   SAN FRANCISCO   LONDON   1976

A Subsidiary of Harcourt Brace Jovanovich, Publishers

ACADEMIC PRESS, INC.
111 Fifth Avenue, New York, New York 10003

United Kingdom Edition published by
ACADEMIC PRESS, INC. (LONDON) LTD.
24/28 Oval Road. London NW1

ISBN 0–12–751801–0

CONTENTS

POPULATION DYNAMICS

WETLANDS USES

CYCLING OF POLLUTANTS
Convened by Thomas W. Duke   481

LIST OF CONVENERS

Numbers in parentheses indicate the pages on which the conveners' sessions begin.

ARMANDO A. de la CRUZ (217), *Department of Zoology, Mississippi State University, P. O. Drawer Z, Mississippi State, Mississippi 39762*

THOMAS W. DUKE (481), *U.S. Environmental Protection Agency, Environmental Research Laboratory, Gulf Breeze, Florida 32561*

JAMES J. McCARTHY (67), *The Biological Laboratories, Harvard University, Cambridge, Massachusetts 02138*

BORI L. OLLA (277), *National Marine Fisheries Service, Middle Atlantic Coastal Fisheries Center, Sandy Hook Laboratory, Highlands, New Jersey 07732*

SAUL B. SAILA (135), *Graduate School of Oceanography, Narragansett Bay Campus, University of Rhode Island, Kingston, Rhode Island 02881*

J. R. SCHUBEL (1), *Marine Sciences Research Center, State University of New York, Stony Brook, New York 11794*

RICHARD C. SWARTZ (135), *U.S. Environmental Protection Agency, Marine Science Center, Newport, Oregon 97365*

F. JOHN VERNBERG (333), *Belle W. Baruch Institute for Marine Biology and Coastal Research, University of South Carolina, Columbia, South Carolina 29208*

LIST OF CONTRIBUTORS

Numbers in parentheses indicate the pages on which the authors' contributions begin.

D. G. AHEARN (483), *Georgia State University, Atlanta, Georgia 30303*

JELLE ATEMA (302), *Boston University Marine Program, Marine Biological Laboratory, Woods Hole, Massachusetts 02543*

LARRY P. ATKINSON (69), *Skidaway Institute of Oceanography, P.O. Box 13687, Savannah, Georgia 31406*

F. N. AYDIN (137), *Lawler, Matusky & Skelly Engineers, 415 Route 303, Tappan, New York 10983*

LOWELL H. BAHNER (523), *U.S. Environmental Protection Agency, Environmental Research Laboratory, Sabine Island, Gulf Breeze, Florida 32561*

STEVE BARNES (493), *Department of Chemistry, Texas A&I University at Corpus Christi, Corpus Christi, Texas 78411*

B. L. BAYNE (432), *Institute for Marine Environmental Research, 67 Citadel Road, Plymouth PL1 3DH, England*

THOMAS J. BERGGREN (166), *Texas Instruments Incorporated, P.O. Box 237, Buchanan, New York 10511*

N. H. BERNER (483), *Georgia State University, Atlanta, Georgia 30303*

STEPHEN H. BISHOP (414), *Marrs McLean Department of Biochemistry, Baylor College of Medicine, Houston, Texas 77025*

DONALD F. BOESCH (177), *Virginia Institute of Marine Science, Gloucester Point, Virginia 23062*

MALCOLM J. BOWMAN (28), *Marine Sciences Research Center, State University of New York, Stony Brook, New York 11794*

xi

ROBERT H. CHABRECK* (226), *School of Forestry and Wildlife Management, Louisiana State University, Baton Rouge, Louisiana 70803*

JOHN D. COSTLOW, Jr. (279), *Duke University Marine Laboratory, Beaufort, North Carolina 28516,* and *Zoology Department, Duke University, Durham, North Carolina 27706*

CLAUDE R. CRIPE (313), *Department of Biological Science, Florida State University, Tallahassee, Florida 32306*

S. A. CROW (483), *Georgia State University, Atlanta, Georgia 30303*

ARMANDO A. de la CRUZ (267), *Department of Zoology, Mississippi State University, Mississippi State, Mississippi 39762*

WILLIAM M. DUNSTAN (69), *Skidaway Institute of Oceanography, P.O. Box 13687, Savannah, Georgia 31406*

T. L. ENGLERT (137), *Lawler, Matusky & Skelly Engineers, 415 Route 303, Tappan, New York 10983*

CARL W. ERKENBRECHER (381), *Department of Biology and Belle W. Baruch Institute for Marine Biology and Coastal Research, University of South Carolina, Columbia, South Carolina*

MILTON FINGERMAN (449), *Department of Biology, Tulane University, New Orleans, Louisiana 70118*

RICHARD B. FORWARD, Jr. (279), *Duke University Marine Laboratory, Beaufort, North Carolina 28516,* and *Zoology Department, Duke University, Durham, North Carolina 27706*

MILES J. FURNAS (118), *Graduate School of Oceanography, University of Rhode Island, Kingston, Rhode Island 02881*

M. GRANT GROSS (3), *Chesapeake Bay Institute, The Johns Hopkins University, Baltimore, Maryland 21218*

THURMAN L. GROVE** (166), *Texas Instruments Incorporated, P.O. Box 237, Buchanan, New York 10511*

*Present address: U.S. Fish and Wildlife Service, Coastal Ecosystems Team, National Space Technology Laboratories, Bay St. Louis, Mississippi 39520.
**Present address: Beak Consultants, Inc., Cornell Industry Research Park, Brown Road, Ithaca, New York 14850.

LEONARD W. HAAS (90), *Virginia Institute of Marine Science, Gloucester Point, Virginia 23062*

C. S. HAMMEN (347), *Department of Zoology, University of Rhode Island, Kingston, Rhode Island 02881*

ERIC O. HARTWIG (103), *Chesapeake Bay Institute, The Johns Hopkins University, Baltimore, Maryland 21218*

GARY L. HITCHCOCK (118), *Graduate School of Oceanography, University of Rhode Island, Kingston, Rhode Island 02881*

ROBERT L. HORTON (290), *School of Oceanography, Oregon State University, Corvallis, Oregon 97331*

DONALD E. HOSS (335), *National Marine Fisheries Service, Atlantic Estuarine Fisheries Center, Beaufort, North Carolina 28516*

WILLIAM W. KIRBY-SMITH (469), *Duke University Marine Laboratory, Beaufort, North Carolina 28516*

BJÖRN KJERFVE (44), *Belle W. Baruch Institute for Marine Biology and Coastal Research, University of South Carolina, Columbia, South Carolina 29208*

PATRICIA KREMER* (197), *Graduate School of Oceanography, University of Rhode Island, Kingston, Rhode Island 02881*

PIERRE LASSERRE (395), *Université de Bordeaux, Institut de Biologie Marine, 33120 Arcachon, France*

ROGER A. LAUGHLIN (313), *Department of Biological Science, Florida State University, Tallahassee, Florida 32306*

J. P. LAWLER (137), *Lawler, Matusky & Skelly Engineers, 415 Route 303, Tappan, New York 10983*

FRANK G. LEWIS, III (313), *Department of Biological Science, Florida State University, Tallahassee, Florida 32306*

ROBERT J. LIVINGSTON (313, 507), *Department of Biological Science, Florida State University, Tallahassee, Florida 32306*

J. L. McHUGH (15), *Marine Sciences Research Center, State University of New York, Stony Brook, New York 11794*

*Present address: Allan Hancock Foundation, University of Southern California, Los Angeles, California 90007.

CHARLOTTE P. MANGUM (356), *Department of Biology, College of William and Mary, Williamsburg, Virginia 23185*

S. P. MEYERS (483), *Louisiana State University, Baton Rouge, Louisiana 70803*

J. A. MIHURSKY (151), *Chesapeake Biological Laboratory, Center for Environmental and Estuarine Studies, University of Maryland, Solomons, Maryland 20688*

D. HOWARD MILES (267), *Department of Chemistry, Mississippi State University, Mississippi State, Mississippi 39762*

R. P. MORGAN, II (151), *Chesapeake Biological Laboratory, Center for Environmental and Estuarine Studies, University of Maryland, Solomons, Maryland 20688*

JULIA F. MORTON (254), *Director, Morton Collectanea, University of Miami, Coral Gables, Florida 33124*

DEL WAYNE R. NIMMO (523), *U.S. Enviromental Protection Agency, Environmental Research Laboratory, Sabine Island, Gulf Breeze, Florida 32561*

PATRICK PARKER (493), *University of Texas, Marine Science Laboratory, Port Aransas, Texas 78373*

WALTER H. PEARSON* (290), *School of Oceanography, Oregon State University, Corvallis, Oregon 97331*

DAVID S. PETERS (335), *National Marine Fisheries Service, Atlantic Estuarine Fisheries Center, Beaufort, North Carolina 28516*

TIBOR T. POLGAR (151), *Martin Marietta Corporation, Environmental Technology Center, 1450 South Rolling Road, Baltimore, Maryland 21227*

DENNIS A. POWERS (166), *Department of Biology, The Johns Hopkins University, Baltimore, Maryland*

WARREN PULICH (493), *University of Texas Marine Science Laboratory, Port Aransas, Texas 78373*

ROBERT J. REIMOLD (219), *Marine Resources Extension Center, The University of Georgia, P.O. Box 517, Brunswick, Georgia 31520*

J. R. SCHUBEL (57), *Marine Sciences Research Center, State University of New York, Stony Brook, New York 11794*

*Present address: National Marine Fisheries Service, Sandy Hook Marine Laboratory, Highlands, New Jersey 07732.

THEODORE J. SMAYDA (118), *Graduate School of Oceanography, University of Rhode Island, Kingston, Rhode Island 02881*

L. HAROLD STEVENSON (381), *Department of Biology and Belle W. Baruch Institute for Marine Biology and Coastal Research, University of South Carolina, Columbia, South Carolina*

J. L. TAFT (79), *Chesapeake Bay Institute, Johns Hopkins University, Baltimore, Maryland 21218*

W. ROWLAND TAYLOR (79), *Chesapeake Bay Institute, Johns Hopkins University, Baltimore, Maryland 21218*

JOHN M. TEAL (234), *Woods Hole Oceanographic Institution, Woods Hole, Massachusetts 02543*

R. E. ULANOWICZ (151), *Chesapeake Biological Laboratory, Center for Environmental and Estuarine Studies, University of Maryland, Solomons, Maryland 20688*

G. VACHTSEVANOS (137), *Lawler, Matusky & Skelly Engineers, 415 Route 303, Tappan, New York 10983*

IVAN VALIELA (234), *Boston University Marine Program, Marine Biological Laboratory, Woods Hole, Massachusetts 02543*

SUSAN VINCE (234), *Boston University Marine Program, Marine Biological Laboratory, Woods Hole, Massachusetts 02543*

ROBERT W. VIRNSTEIN (177), *Virginia Institute of Marine Science, Gloucester Point, Virginia 23062*

MARVIN L. WASS (177), *Virginia Institute of Marine Science, Gloucester Point, Virginia 23062*

KENNETH L. WEBB (90), *Virginia Institute of Marine Science, Gloucester Point, Virginia 23062*

J. S. WILSON (151), *Chesapeake Biological Laboratory, Center for Environmental and Estuarine Studies, University of Maryland, Solomons, Maryland 20688*

PREFACE

Planning for the Third International Estuarine Research Conference began at Myrtle Beach, while the second conference was still in session. Before moving ahead on details, the Estuarine Research Federation had several questions to answer: "Does the rate of *Recent Advances* in our field warrant another invited symposium two-years hence; if so, is a unifying theme readily apparent; how can we keep expenses down for students and the established investigators of tomorrow; and where should we meet to recognize surging interests of the Federation's new affiliates in the Southern and Gulf states?"

The Governing Board decided that the next *Recent Advances* would emphasize estuarine processes—an attempt to focus on dynamic interactions at several levels of organization. As it worked out, more than 70 papers were presented before 550 registrants in Galveston, Texas, October 7-9, 1975. Most of the papers appear in these two volumes. Success and failure reflect not only the Board's wisdom, but also instabilities that started at a global level with an oil embargo and came to bear on the outlook of all.

On the success side of the balance sheet, special thanks go to the convenors. They were given authority to organize sessions and follow through in the peer review process. The Federation also thanks the eight Federal agencies which provided support for the Conference and associated editorial work. The comments we received on the proposals submitted to the agencies were inevitably coupled with our direct experience, yielding a uniquely comprehensive view on the role of big scientific meetings enriched with opportunities for environmental management.

I believe that our shortcomings—judged from scientific value in the Conference's written record—are intertwined with positive attitudes emanating from the Federation's affiliated societies. Their history is a commitment to excellence in a field that on occasion has been as variable as the very structure of the environment studied. If certain sessions appear less comprehensive and sophisticated than usual, the result may well reflect traditions deeply embedded in these societies. The societies exist to challenge and redirect the results of research, and they rely upon an especially open system of appraisal. It follows that the negative side of the balance sheet is more indicative of needs—either for encouragement or for curtailment in specific areas—than it is an enumeration of wasted effort.

These volumes and the reviews they attract should serve well in the development of critical perspective, the foremost requirement today in shallow-water oceanography.

For the Governing Board, 1973-75

H. Perry Jeffries, President
Estuarine Research Federation

REHABILITATION OF ESTUARIES

Convened by:
J. R. Schubel
Marine Sciences Research Center
State University of New York
Stony Brook, New York 11790

The objectives of this session are to assess the effectiveness of man's past and present strategies to protect and rehabilitate estuaries, and to explore alternative ways of attaining the pervasive goals of pollution abatement and estuarine management. Billions of dollars have been spent over the past half century on pollution control to improve the "quality" of the estuarine environment. These expenditures have been used largely for construction and operation of sewage treatment facilities, and for controlling large industrial point sources. Increasingly, stricter controls and standards have been imposed in an attempt to restore and ensure acceptable environmental quality.

In this session we shall examine historical data from a variety of estuaries to attempt to determine whether past and present pollution control measures have been effective in improving water quality, or whether there has been a continued, slow but perceptible degradation of the quality of the estuarine environment. If accepted strategies have not been effective, will more stringent standards and criteria for industrial and municipal wastes significantly improve water quality, or do we need to explore new and different approaches to estuarine management instead of, or in addition to continuation or acceleration of present methods? Will higher levels of sewage treatment significantly reduce nutrient levels? Will any improvements in the quality of the estuarine environment that might result from stricter controls on disposal of dredged materials outweigh the economic perturbations that will result from such practices? Is "zero discharge" the answer to environmental management, or is it an empty concept? Do we need to take different approaches to estuarine management and rehabilitation, radical approaches, approaches based on sound scientific principles?

1

ESTUARINE CLEANUP—CAN IT WORK?

M. Grant Gross
Chesapeake Bay Institute
The Johns Hopkins University
Baltimore, Maryland 21218

ABSTRACT: Available data show no evidence of major improvement in estuarine water quality in the United States in recent years. Dissolved oxygen (DO) concentrations in New York Harbor have improved only slightly despite decades of building new treatment facilities and upgrading older plants. No improvement in DO levels has yet been documented for the Upper Delaware Estuary. But increased DO values in the Thames Estuary following construction and enlargement of major sewage treatment facilities indicates that estuarine water quality can be improved. Areas in the Thames previously devoid of DO in summer have DO values averaging about 30 percent of saturation. Odor problems have been alleviated and fish now are caught in the estuary.

Successful cleanup of the Thames Estuary required well defined objectives and a regional plan based on a comprehensive scientific study. Capital expenditures exceeded 500 million (1974) dollars, about half of that since 1950. At least 15 years were required to achieve the cleanup objectives; including delays caused by World War II, planning and implementation required several decades.

INTRODUCTION

Present water quality problems in urban estuaries result from two successes in western history: providing a plentiful supply of high quality water to cities beginning in the early 1800s (3); and the subsequent development of a water-borne waste-transport system in the late 1800s (22). Because of these developments, large volumes of domestic and industrial wastes are discharged into estuaries and eventually into the coastal ocean. In small quantities these wastes were diluted by mixing with relatively clean river- and sea-water and carried away by tidal currents causing little or no obvious damage to the estuarine system. When discharged in large quantities, such wastes form deposits in estuaries, deplete dissolved oxygen (DO) concentrations, or cause localized odor problems.

Just as assessing the state of pollution is essentially a matter of subjective judgments, so too is the question of environmental cleanup. In considering the

3

question of whether "estuarine cleanup" is possible it is first necessary to state how we will judge the results. Indeed, one of the basic issues involved in the whole complex of problems involved in estuarine management is: How is "cleanup" of a system to be evaluated?

At least four different criteria could be used:

—Absence of perceived indications of degraded water quality, such as objectionable odors, floatable debris, surface films, or discolored water;

—Achieving some preset level of dissolved oxygen concentration, say 30 percent saturation;

—Absence of estuarine-related public health problems so that estuarine waters can be harvested for shellfish or used for water-contact sports;

—Returning the water body to its original state.

In most cases, it is the disappearance of some perceived indication of pollution that is taken to indicate success in estuarine restoration. Thus the absence of odors and elimination of surface films and floatable debris, and elimination of discolored waters or algal mats are considered indications of successful cleanup efforts.

The next more rigorous level of environmental quality assessment has been attempted in several estuaries. DO saturations are usually the parameter used to evaluate water quality because they can be readily measured and have easily understood effects on aquatic life.

When dissolved oxygen concentrations drop and remain below 3 to 4 parts per million (ppm) resident fish populations die and migrating fish cannot pass the low-oxygen zone, and when dissolved oxygen concentration drops to zero, the system produces objectionable odors, such as that produced by hydrogen sulfide. DO levels are frequently employed as a primary water quality indicator and are the basis for most of this paper. Other indices could also be used in estuarine systems to evaluate water quality. Among these are turbidity or suspended sediment concentrations, temperatures, color, or concentrations of radionuclides.

Levels of bacterial contamination are also used to evaluate water quality for shellfish harvesting or body-contact sports such as swimming or surfing. Little effort seems to have been made on a regional scale to make waters safe for shellfish consumption although in some instances water quality has improved enough for beaches to be reopened (New York Times, August 31, 1957, p. 38).

Despite the obvious appeal of restoring a water body to its original conditions, I found no reports of such an approach having been attempted in an estuarine system in the United States.

THAMES ESTUARY

With the City of London along its shores, the Thames Estuary (Fig. 1) has received human wastes for nearly 2000 years, at least as early as the Romans. Its restoration from a severely deteriorated condition nearly devoid of fish life in the mid 1950s to its condition in the mid 1970s with the return of more than 80 species of fish, including salmon, to the river is one of the success stories of estuarine restoration (10). The history of this undertaking is worth noting.

The first record of pollution of the Thames is the Royal Act of 1388 which prohibited "corruption" of ditches, rivers waters and the air of London and required that all dung, filth, and garbage be carried away, presumably for disposal on nearby vacant land. Householders were ordered not to throw water from their windows but to discharge it to the street; fishmongers had to carry their dirty water to the river.

Until around 1815, primitive sewers drained surface water runoff and domestic sewage went into cesspools. Increased use of the water closet overloaded the cesspools and after 1815 they could legally be connected to the sewers. Thus London gradually changed from disposal of sewage on land to a water-transport sewage system discharging directly into the Thames.

By 1842 odors from the river had become a major problem. Furthermore, some of London's drinking water was taken from the Thames, and there were cholera epidemics in 1831-2, 1848-9, and 1853-4 with up to 20,000 deaths in the worst years. In 1848 the Metropolitan Commission on Sewers was established and began work on problems of waterborne disease and smell.

Planning proceeded for about ten years until construction of a regional sewer system began due, in part, to the "Great Stink of 1858." Committee Rooms in

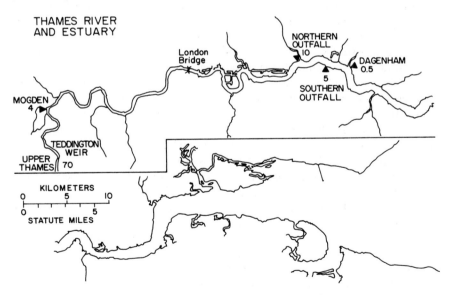

Figure 1. The Thames Estuary showing average discharges (in cubic meters per second) from major sewage treatment plants and the discharge from the Upper Thames (6).

Parliament were untenable because of odors during prolonged hot weather in June, a period of low river flow. Sheets soaked in "disinfectant" were hung in the windows to combat the odors (14).

Construction of London's sewers began in 1859 and was completed by 1865 at a cost of 4.1 million pounds (probably equivalent to more than 100 million [1974] dollars). Many of these sewers are still in use. In 1856 water consumption was 80 to 100 liters (20 to 25 gallons) per person per day. The system was designed to store the flow for six hours. Untreated sewage was discharged on the ebb tide through the outfalls at Beckton and Crossness 18 and 21 km (11 and 13 miles) downstream from London Bridge (Fig. 1). There was immediate improvement of the Thames in central London with a corresponding deterioration near the outfall locations. The problem had simply been moved downstream (15).

Treatment of sewage began in 1889; six ships carried sewage sludges (solids removed during wastewater treatment) to sea. But population increases and more sewage discharges to the Thames caused continued deterioration of conditions in the river, especially near the outfalls.

Sewage treatment and sludge digestion processes were upgraded, regional facilities were constructed during the first half of the twentieth century, but water quality in the river continued to deteriorate. Widespread introduction of synthetic detergents in the 1950s caused severe problems. Sewage treatment plants were buried under 5 meters (15 ft) of foam for weeks and treatment efficiency was reduced by as much as 30 percent (25). In 1955 a 50 kilometer reach of the Thames was devoid of dissolved oxygen and plagued by odor problems. The new plants (or expansion) brought into operation after 1955 caused demonstrable reduction in the extent of the zero-oxygen areas (Fig. 2).

A study of the Thames Estuary was begun in 1949 and completed in 1964. A tidal mixing model was developed and used to predict the effects of different discharges on water quality in the Thames (6). These predictions provide the scientific and technical basis for new corrective measures (9).

Studies of the Thames concluded that freedom from offensive odors was a realistic objective. To achieve this, DO levels of 10 percent saturation or higher at all places and at all times by 1980 were sought.

Water quality in the Thames has improved since the mid-1960s (Fig. 2). Dissolved oxygen concentrations have increased in the most heavily impacted reaches of the River and no odor problems have been reported since the mid-1960s (9, 10).

Achievement of these goals was built on three things:

1. Scientific understanding of the estuary to the degree which permits predictions about future conditions under different discharge conditions.

2. Modest, well-defined goals: more than 10% saturation for DO.

3. Massive capital investments of at least half a billion (1974) dollars, about half of that since 1950.

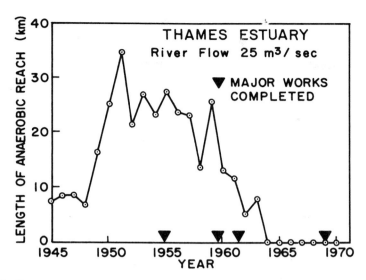

Figure 2. Changes in the length of the anaerobic reach of the Thames Estuary and the approximate completion dates for major sewage treatment works (9).

Construction costs for sewerage and sewage treatment facilities on the Thames can only be approximated as many of the early construction projects were completed more than a century ago and under quite different organizational structures and price levels. In 1974 dollars, the capital construction costs of major works in the Greater London area exceed $500 million dollars; probably a more realistic estimate would be between $600 and $700 million dollars (data supplied by H. Fish, Director Scientific Sciences, Thames Water Authority, written communication 23 June 1975). About $250 million dollars has been spent since 1950 (Table 1) to construct or expand the works that apparently caused the dramatic improvement shown in Fig. 2. Projected capital and operating expenditures were approximately $50 million (current) dollars for 1975 and 1976.

Time is also an important factor. If we take the initiation (1949) of the Thames Estuary Study as the beginning of the most recent phase of Thames "cleanup" and select 1964 as the year in which the Estuary was first completely aerobic (Fig. 2), it required 15 years from the beginning of the study until the goal was achieved. Nine years were required from the time the first major sewage work extension was completed in 1955 until the river was aerobic year-round in 1964. Major improvements made during the 1950s and 1960s had been proposed even earlier and delayed by the second world war (10). So perhaps we should count the time between problem identification and amelioration in terms of decades.

Table 1.   Some capital costs of sewerage and sewerage treatment plants in the Greater London Area, 1865-1976. (Data From H. Fish, Thames Water Authority).

Works and Projects	Completion date	Capital costs in millions of	
		£ (current)	$ (1974)
Intercepting sewers, storm sewage overflows, and Northern and Southern outfalls	1865	4.1	100
Sewers and pumping stations	1914	2.5	60
Regional drainage scheme and sewage treatment plant (Mogden)	1935 1961	5.4 1.5	90 8
Beckton sewage treatment plant and extensions	1954 1975	7.0 26.5	43 53
Hammersmith pumping station	1960+	3.0	17
Regional drainage scheme, Deephams and Ramney March sewage treatment plants and extensions	1965 1971	9.2 2.8	40 9
Flood relief tunnel	1968	1.0	4
Sludge main	1971	1.1	4
Crossness sewage treatment plant	1969	11.0	40
Sludge disposal vessels	1954- 1968	2.4	3.4
Routine construction	1974- 1976	40	29
Approximate Totals		117	500

SEWERAGE AND SEWAGE TREATMENT IN THE UNITED STATES

Sewerage systems in the United States started in 1855 in Chicago, only 12 years after the world's first sanitary sewer system was installed in Hamburg, Germany (7). Sewerage services expanded faster than population growth. By 1932 approximately half the nation's population was served by sanitary sewers. In 1973 the nation's sewered population exceeded slightly the total urban population including large suburban tracts.

Sewage treatment plants in this country were first constructed in 1870. Many sewage treatment plans were constructed between 1910 and 1932 and more were built under the stimulus of "New Deal" programs. By 1957 the sewage from approximately 74 million persons was receiving treatment and in 1973 about 97 percent of the sewered population or 159 million persons was receiving treatment (7).

Between 1855 and 1971, the nation invested an estimated 58 billion (1972) dollars in its public sewerage works. This represents about five percent of the capital expenditures by state and local governments for all purposes since 1915.

In 1971, the replacement value of facilities in place was estimated at 32 billion (1971) dollars (7).

Results of the nation's investment in sewerage and sewage treatment facilities have been surprisingly meager. While the amount of Biochemical Oxygen Demand (BOD) collected by the nation's sewers increased 140 percent between 1957 and 1973, the amount discharged by sewage treatment plants during that same period remained nearly constant (Table 2). These data do not, however, reflect the effects of industrial waste treatment facilities constructed during this period.

Table 2.   Effect of sanitary sewage treatment (7).

Year	Collected by sanitary sewers*	Reduced by treat- ment**	Discharged by treatment plants
	(millions of pounds of BOD_5 per day)		
1957	16.4	7.7	8.7
1962	19.8	10.8	9.0
1968	23.3	15.0	8.3
1973	27.1	18.5	8.6

*Based on 0.167 pounds of BOD_5 per sewered person per day.
**Based on the distribution of treatment facilities and on estimates of removal efficiency.

Three aspects of this situation stand out (7). First, most of the investment of capital in sewerage and treatment works has occurred rather recently, almost 80 percent since 1929 and more than 30 percent since 1961. Second, the amount of capital in place is so large compared to annual investments that replacement and upgrading of facilities has absorbed approximately half of all capital expenditures since 1961. Replacement costs in 1973 were close to one billion dollars a year and rising in proportion to the growth of the capital stock. Third, data on the benefits obtained from this massive capital investment in terms of improved water quality in rivers and estuaries are scarce for rivers (23) and virtually nonexistent for estuaries. Some reaches of the nation's rivers have shown improvements in water quality since the 1930s, such as the Potomac River at Washington, D.C., and the Mississippi at St. Paul, Minn. (24).

NEW YORK HARBOR

In 1850 New York had approximately 700,000 inhabitants, most of them living on Manhattan Island. Many used privies and cesspools that kept their wastes from entering the Harbor. There were, however, about 130 km of sewers, primarily storm drains, serving the city. There is no record of substantial deterioration of water quality affecting large areas of the Harbor at that time; so it has been assumed that dissolved oxygen concentrations were close to saturation. Between 1850 and 1910, the population of the City increased to 4.7 million of which an increasing number discharged their wastes to the Harbor through the steadily expanding sewerage system (17).

Rapid changes in water quality were documented between 1900 and 1916. DO values dropped from near saturation to around 25 percent in the Lower East River (Fig. 3). Sewage treatment plants began operating in the 1920s and were expanded after 1937 when secondary treatment was afforded to more of the region's sewage. The effect of improved sewage treatment can be seen in the slight improvement in DO saturation after 1935.

DO levels in the waters of the Harbor show marked decreases between about 1910 and 1925. After that they remained generally constant until the mid-1960s when an increase in DO values was observed in all parts of the Harbor. But in the early 1970s DO saturation again trended downward for unknown reasons.

The deterioration in water quality virtually eliminated harvesting shellfish and finfish from New York Harbor. Once extensive oyster beds progressively were

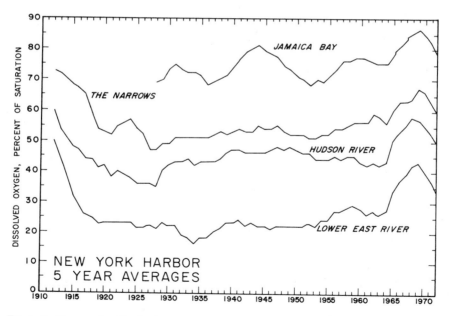

Figure 3. Changes in dissolved oxygen concentrations in parts of New York Harbor, 1909-1973. Data are calculated as five-year running averages. (Unpublished data from New York City, Environmental Protection Administration, Dept. of Water Resources).

depleted or polluted so that the last commercially productive area was in Princess Bay south of Staten Island, a remote part of the Harbor. By 1921, New York Harbor was essentially closed to commercial shellfish production.

Locally produced crabs and lobsters were no longer commercially available after 1887. The commercial fisheries for sturgeon disappeared. Between 1959 and 1965, commercial catches of shad and striped bass from the Hudson River dropped by 75% and 65% respectively (5, p 206).

In addition to the rapidly changing water-quality characteristics, one can also use the type of bottom deposits in New York Harbor as an indication of water quality. Extensive deposits of oily sediments, high in carbon and rich in metals, containing large amounts of sewage-derived solids have accumulated in the Harbor (11). These physically changed the substrate, thereby altering or eliminating benthic communities. Such deposits affect water quality, consuming DO and releasing nutrients (ammonia and phosphates) as well as metals.

Summarizing, it appears that the sewerage and sewage treatment facilities built around New York Harbor have so far only been able to prevent further deterioration of water quality in the Harbor, at least as measured by dissolved oxygen concentrations.

DELAWARE ESTUARY

The Delaware Estuary, especially the upper estuary, has been extensively studied because of pollution problems antedating World War I. In a pioneering study begun in 1962 FWQA, the Federal Water Quality Administration (8), outlined water quality control programs for the estuary, including estimated costs and effects of several alternatives. The control measures were based primarily on a model of steady state CO concentrations.

In the study, FWQA recognized variability among various parts of the estuary and that not all effluents could be treated equally. For instance, 85 percent overall BOD removal was considered necessary but not in all zones. Economies could be achieved by treating sources differently, depending on the materials involved and their locations. Implementation of the least expensive alternative was estimated at about $30 million (1964 dollars) annually and would provide seasonal (summer) average dissolved oxygen levels of 1.0 mg/ℓ in the most heavily impacted reaches of the River. The more ambitious program would provide dissolved oxygen levels of 4.5 mg/ℓ and cost $490 million.

The effect of pollutants on DO concentrations in the Delaware River can be seen in Fig. 4 which shows average DO concentrations during the summer of 1964, a period of severe drought. Note the low DO values just downstream of Philadelphia. The pollution abatement program began in 1968 (12). For 1970 the DRBC, Delaware River Basin Commission (4), reported that "zero oxygen conditions in the river at Philadelphia tend to be diminishing in frequency, intensity and duration;" there were only 20 days of zero oxygen conditions in the river as compared to as many as 123 days in previous years. Part of this

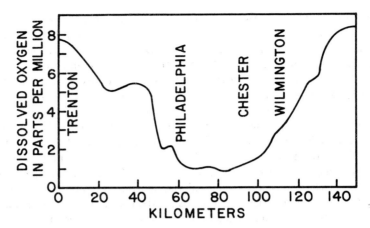

Figure 4. Average dissolved oxygen concentration in Delaware River and Estuary, summer
1964 (16).

improvement in DO levels may be related to the greater river flow during this time.

The model used to predict effects of waste discharges and to allocate permissible loads from specific activities in the Delaware Estuary has been criticized. Among the questions raised by Ackerman and his associates (1, 2) were the steady state assumption that gave inadequate consideration to transient events such as storms, nonpoint sources such as small streams, or oxygen consumption by deposits on the river bottom.

Whipple (20) has argued that planning for the Delaware River is inadequate and does not provide a useful basis for the large expenditures necessary to achieve 90 percent reduction in biochemical oxygen demand or 95 percent reduction in suspended solids. Furthermore, Whipple (20) argues that nonpoint sources such as storm-water runoff and various unrecorded pollution sources were not taken into account.

For the Delaware Estuary, there is as yet no compelling evidence of cleanup of the system. It is too soon to see definitive results; completion of the upgrading of Philadelphia's system is scheduled for 1977 (12). But the model used as the basis for designing the cleanup program has been challenged and the lack of an adequate long-range plan for the region has been noted. In short, the prognosis for effective cleanup of the Delaware Estuary is unclear at this time.

DISCUSSION

If we take the Thames as our only example of successful estuarine cleanup, it is obvious that science has a major role to play in estuarine cleanup. There are, however, some important factors involved.

The science must be first rate and results presented in easily understood forms. All assumptions must be explicit and each step of the analysis must be clearly presented and explained. Limitations of the data must be clearly stated. Failure to do so is likely to result in discrediting the science and diminishing its effectiveness, at least in the public view.

The published literature for U.S. estuaries does not compare with the extensive and comprehensive studies made of the Thames. Criticisms of the model used for the Delaware Estuary planning—apparently the best planned program in the U.S.—have not been answered. A regional plan for the whole of New York Harbor is not yet available.

The data on capital expenditures for sewerage and sewage treatment for cities around U.S. estuaries appear to be comparable to that for the Thames. Massive expenditures have been projected for new facilities and upgrading existing ones.

It seems probable that there has not yet been enough time to evaluate the results of planning and projects begun in the 1960s. But progress to date has been meager.

There is a surprising lack of data on estuarine conditions and the results of cleanup activities. Scientists and engineers have not provided the surveys and analyses that will be required to assess properly the results of projects now underway.

ACKNOWLEDGMENTS

I thank J. R. Schubel, D. W. Pritchard, W. Pressman, C. H. J. Hull, M. G. Wolman, and A. L. H. Gameson for their help in locating literature and for their comments on earlier drafts of the paper.

REFERENCES

1. Ackerman, B. A., and James Sawyer. 1972. The uncertain search for environmental policy: Scientific fact finding and rational decision making along the Delaware River. University of Pennsylvania Law Review 120:419-503.
2. _____, S. R. Ackerman and D.W. Henderson. 1973. The uncertain search for environmental policy: The costs and benefits of controlling pollution along the Delaware River. University of Pennsylvania Law Review 121:1225-1308.
3. Blake, N.M. 1956. Water for the Cities. Syracuse University Press. Syracuse, New York. 34 p.
4. Delaware River Basin Commission. 1971. Annual Report. Philadelphia, Pennsylvania. 24 p.
5. Department of Health, Education, and Welfare. 1965. Conference in the matter of pollution of the interstate waters of the Hudson River and its tributaries—New York and New Jersey. Sept. 28, 29, and 30, 1965. Washington, D.C. 2 volumes.
6. Department of Scientific and Industrial Research. 1964. Effects of Polluting Discharges on the Thames Estuary. Water Pollution Research Tech. Paper 11. H.M. Stationery Office, London. 609 p.

7. Environmental Protection Agency. 1973. The economics of clean water. U. S. Environmental Protection Agency, Washington, D.C. 120 p.
8. Federal Water Quality Administration. 1966. Delaware Estuary Comprehensive Study, Preliminary Report and Findings. Philadelphia, Pennsylvania.
9. Gameson, A.L.H., M.J. Barrett, and J.S. Shewbridge. 1973. The aerobic Thames Estuary, p. 843-850. *In* S.H. Jenkins (ed.) Advances on Water Pollution Research, Sixth International Conference, Jerusalem. June 8-23, 1972. Pergamon Press, New York.
10. _____, and A. Wheeler. In press. Restoration and recovery of the Thames Estuary. *In* Restoration and Recovery of Damaged Ecosystems. Blacksburg, Va., March 22-25, 1972.
11. Gross, M.G. 1972. Geologic aspects of waste solids and marine waste deposits: New York Metropolitan Region. Geol. Soc. Amer. Bull. 83:3163-3176.
12. Hull, C.H.J. 1975. Implementation of interstate water quality plan. Journal of Hydraulics Division, Proceedings American Society of Civil Engineers. 101(HY3):495-509.
13. Kneese, A.V., and C.L. Schultze. 1975. Pollution, prices and public policy. The Brookings Institution, Washington, D.C. 125 p.
14. Mitchell, R.J., and M.D.R. Leys. 1958. A History of London Life. Penguin Books, Baltimore. 348 p.
15. Morrison, Alex. 1974. The restoration of the River Thames. Paper presented at World Successes—Reasons for Hope. Spokane, WA.
16. Schaumberg, Grant. 1967. Water pollution control in the Delaware Estuary. Unpublished thesis, Harvard College, Cambridge, MA.
17. Torpey, W.N. 1967. Response to pollution of New York Harbor and Thames Estuary. J. Water Poll. Control Fed. 39:1797-1809.
18. Vitale, A.M., and P.M. Sprey. 1974. Total water pollution loads: The impact of storm water. Enviro. Control, Inc., Rockville, Maryland. (Distributed by National Technical Information Service, Springfield, Virginia PB-231-730.) 183 p.
19. Wheeler, A. 1970. Fish return to the Thames. Science Journal 6:28-32.
20. Whipple, William, Jr. 1975. Water quality planning for the Delaware Estuary—An example. Water Resources Bulletin 11:300-305.
21. Whipple, W., J.V. Hunter, and S.L. Yu. 1974. Unrecorded pollution from urban runoff. J. Water Poll. Control Fed. 46:873-885.
22. Wolman, Abel. 1956. Disposal of man's wastes, p. 807-16. *In* W.L. Thomas, Jr. (ed.) Man's Role in Changing the Face of the Earth. Univ. of Chicago Press, Chicago. 1193 p.
23. Wolman, M.G. 1971. The nation's rivers. Science 174:905-918.
24. _____. 1974. Crisis and catastrophe in water-resources policy. U.S. Geological Survey Professional Paper 921. p. 17-27.
25. Wood, L.B. 1971. The cleaner Thames. Greater London Council Intelligence Unit Quarterly Bulletin 17. p. 37-48.

ESTUARINE FISHERIES: ARE THEY DOOMED?[1]

J. L. McHugh
Marine Sciences Research Center[2]
State University of New York
Stony Brook, N.Y. 11794

ABSTRACT: In the most recent year for which U.S. commercial and recreational catch statistics are available, total fish and shellfish landings were about 6.5 billion pounds, of which about 4.5 billion were estuarine-dependent. On the Atlantic coast estuarine species were more important in catches from north to south, on the Pacific coast from south to north. In the Gulf of Mexico 98% were estuarine. Effects of human activities therefore should be relatively most severe from Chesapeake Bay south, in the Gulf, and in Alaska. But estuarine resources make important contributions in other areas: New England and Middle Atlantic landed commercial value was $70 million in 1970. In California, estuarine resources were only 3% by weight, but $12 million by landed value.

No one can question the importance of maintaining estuarine environmental quality if fisheries are to continue. Present concerns about the environment probably have had beneficial effects, although cause and effect is difficult to prove. Doom has been averted, but constant vigilance is needed. Public education will be most important, for the issues are much more complicated than most people realize.

INTRODUCTION

It has been generally agreed that wetlands and shallow estuarine waters are essential to fishery production. Over and over again it has been asserted that

[1] Parts of the analysis on which this paper is based are the result of research sponsored by NOAA Office of Sea Grant, Department of Commerce, under Grant No. 04-5-158-34. The U.S. Government is authorized to produce and distribute reprints for governmental purposes notwithstanding any copyright notation that may appear hereon.

[2] Contribution no. 159 from the Marine Sciences Research Center, State University of New York at Stony Brook.

15

destruction of coastal wetlands and deterioration of the estuarine environment will destroy the fisheries. Some ecologists have gone so far as to say that early stages of some ocean-spawning species (e.g. shrimp and menhaden) must get to bay waters or perish. This conclusion is largely intuitive, based on observations of distribution, movements and migrations of the various life-history stages. Yet no one can dispute the high biological productivity of estuaries, especially shallow waters fringed with extensive wetlands, such as those that border most of the coast from New York south and in the Gulf of Mexico.

One thing is certain about the estuarine environment: it is more vulnerable to human influences than is any other part of the ocean. This is because estuaries are partially enclosed by land and are attractive places for people to live, work, and play.

Most estuaries are among the most highly productive biological systems on earth, producing annually from 10,000 to 25,000 kilocalories per square meter gross production (18). Where estuarine systems are well developed around the coasts of the United States they abound in living resources, year round and seasonally. In regions where estuaries are small and few, but rich fisheries exist, as on Georges Bank and along the coast of California, turbulence and upwelling provide the necessary nutrients.

What is not so well established is whether species that spend major parts of their lives in estuarine waters do so simply because the estuaries are there, or whether they are truly estuarine-dependent. Would species like blue crab, menhaden, and other animals found frequently in great abundance in estuaries be less abundant if the expanse of wetlands and shallow, protected waters of intermediate salinity were substantially reduced, and would they perish if the wetlands were destroyed and the estuaries filled? Intuition tell us yes, but solid scientific proof is hard to find. Experimental evidence is suggestive, but it is difficult to measure cause and effect in the wild.

If fishery production of estuaries is greater than from other coastal waters, it might be expected that the states with the greatest fishery resources would be those states in which estuarine species dominate. In 1970 Louisiana was the leading fishing state in the United States, with commercial landings of about 502,250 metric tons of fish and shellfish, about 99 percent of which were estuarine species; but California came second, with about 318,754 metric tons, only about three percent of which were estuarine.[3] Even if tropical tunas are omitted from the California catch, estuarine-dependent species made up less

---

[3] Domestic commercial landings do not necessarily give a very precise index of abundance of marine fishery resources in the coastal waters of a state, for several reasons. Landings represent the point of unloading, not the point of capture. Landings state by state are not necessarily proportional to abundance in local waters. Domestic commercial landings do not include recreational or foreign catches in the area. Making allowance for these things, however, the history of domestic commercial landings can be useful in understanding the condition of the fisheries in a region.

than six percent of the landings in 1970. Virginia ranked third by weight (over 99 percent estuarine), but Alaska was fourth (about two-thirds estuarine). Mississippi was fifth (99 percent estuarine), but Massachusetts was sixth (only 6.5 percent estuarine). Total landings in the marine coastal states excluding Hawaii contained about 73.8 percent estuarine-dependent species. Thus, it does not appear that the states which lack extensive estuarine waters also lack fishery resources.

ESTUARINE-DEPENDENT SPECIES IN THE U.S. CATCH

"Two out of every three species of useful Atlantic fish depend in some way upon tidal lands and the shallowest of our bays for their survival" (1), and "almost two-thirds by value of the United States commercial catch and much of the marine sport catch is composed of species that spend at least a part of their lives within land-bound estuaries" (13). In the Gulf of Mexico in 1961 about 97.5 percent of the total commercial catch was of estuarine species (7). If these authors used the same definition of an estuarine species, then the relative importance of estuarine species must vary widely according to locality.

The distinction between oceanic and estuarine species is fairly easy to make. The definition used here includes resources that live totally within the inshore estuary, use the estuary as a breeding or nursery ground, or move in seasonally for extended periods to feed. The ratio of estuarine to marine species in the commercial catch ranges from slightly over 3 percent in California to 98 percent in the Gulf of Mexico (Table 1). California is scarcely representative, however,

Table 1. Percentage of estuarine-dependent species by weight and value, and total catch of estuarine species, in the United States commercial catch in 1970 by coastal regions. Weights and values in thousands of metric tons and millions of dollars.

Region[1]	Percent estuarine dependent		Total catch of estuarine species	
	By weight	By value	By weight	By value
New England	11.3	12.9	20.3	10.9
Middle Atlantic	44.1	57.5	26.9	16.8
Chesapeake	97.4	95.4	274.8	37.1
South Atlantic	95.2	85.5	116.3	24.1
Gulf of Mexico	98.0	93.4	700.4	150.6
Alaska	68.5	72.8	169.3	69.8
Washington	53.9	80.1	30.9	23.3
Oregon	36.6	55.1	16.2	12.5
California	3.3	11.7	10.5	10.0

[1] New England: Me - Conn incl; Middle Atlantic: NY - Del incl; Chespeake: Md & Va; South Atlantic: NC - east coast of Fla incl; Gulf of Mexico: west coast of Fla - Texas incl; Pacific coast: Hawaii omitted.

Source: Wheeland, Hoyt A. 1973. Fishery Statistics of the United States 1970. U.S. Dept. Commerce, Natl. Marine Fish. Serv., Stat. Digest No. 64:489 p.

Table 2. Percentage of estuarine-dependent species by weight in the United States salt-
water sport catch in 1970 by coastal regions. Weights in thousands of metric tons.

Region[1]	Percent estuarine dependent	Total catch of estuarine species
New England	59.3	72.0
Middle Atlantic	79.1	87.0
South Atlantic	53.3	97.4
Eastern Gulf of Mexico	71.5	108.1
Western Gulf of Mexico	90.5	62.0
North Pacific	71.0	24.9
South Pacific	9.6	4.0

[1] New England: Me - NY incl; Middle Atlantic: NJ - Cape Hatteras; South Atlantic: Cape Hatteras - Southern Fla including Florida Keys; Eastern Gulf: Florida Keys to and including Mississippi R. delta; Western Gulf: Mississippi R. delta to Mexican border; North Pacific: Point Conception, Calif to Washington incl, and Alaska; South Pacific: Mexican border to Point Conception.

Source: Deuel, David G. 1973. 1970 Salt-Water Angling Survey. U.S. Dept. Commerce, Natl. Marine Fish. Serv., Current Fish. Stat: No. 6200:iii+54 p.

for a considerable part of commercial fishery landings in that State is made up of tunas caught off Central and South America. The next lowest region was New England, with about 11 percent estuarine species by weight. Because the coastlines of the Pacific coastal states are so long and so different topographically, regional divisions were made by States. The year 1970 was used because that is the latest year for which a national survey of saltwater sport fishing is available.

The percentages of estuarine-dependent species in the recreational marine fisheries in 1970 are given in Table 2. The results are similar to, although not always very close to the ratios for commercial landings: highest in the Middle Atlantic region and Gulf of Mexico, and on the Pacific coast highest in the north. The differences in relative importance of estuarine species between recreational and commercial catches are not unexpected. Menhaden are presently so dominant in commercial catches in the Chesapeake and Gulf of Mexico regions as to make it inevitable that estuarine species will dominate strongly by weight. Other reasons are that the regional divisions are somewhat different in the two sets of statistics, and that invertebrates, most of which are estuarine and support recreational as well as commercial fisheries, were not included in sport catches in the 1970 survey.

HAZARDS TO ESTUARINE FISHERY RESOURCES

Man-made hazards to estuarine resources or to the harvest of these resources can be placed in four broad categories: physical barriers to migration or distribution; direct effects on human health or quality of life which may require closure of fishing grounds or prohibitions on marketing certain species; catastrophic or chronic effects of human activities on estuarine biota; and social-political phenomena which affect fishery management in various ways.

Any man-made alteration of the estuarine environment can be considered potentially hazardous to living resources, but it is possible that some have been beneficial. One example of possible benefit is the substantial increase in abundance of striped bass over the past 40 years along the Atlantic coast. Mansueti (12) suggested that this may have been caused by nutrient enrichment in the nursery areas although this has not been proven. Another is the continued dominance of Maryland as a striped bass producing State despite the almost complete cessation of spawning in the Susquehanna River (5). Most striped bass spawning in Maryland now is said to take place in or in the vicinity of the Chesapeake and Delaware Canal, a man-made waterway (8). On the other hand, on the Pacific coast, the introduced striped bass stock in the Sacramento River is affected adversely by man-made environmental alterations (20).

A third type of benefit can come from the tendency of marine animals to be attracted to certain structures or other environmental modifications. Examples are aggregations around petroleum drilling rigs, bridges, and warm plumes from power plants. Even dredging, usually regarded as highly detrimental, may produce benefits of a kind. Baymen in New York State are firmly convinced that hard clam production is enhanced at the edges of newly-dredged channels, and this is ecologically reasonable. The power plant at Northport, Long Island, has created a winter sport fishery in Long Island Sound where none existed before. Striped bass, bluefish, white perch, menhaden, and other species remain within the influence of the plume as water temperatures drop in fall. The power company is not always praised for this serendipity, however. When adverse weather conditions or changes in plant operations intervene, fish kills sometimes occur in winter, and fishermen quickly turn on the company for having destroyed a fishery that it had created in the first place (19).

The 10 major estuarine-dependent commercial fishery resources in the Middle Atlantic Bight in 1974, in descending order of rank by weight, were Atlantic menhaden, blue crab, American oyster, alewife[4], scup, summer flounder, hard clam, striped bass, weakfish, and bluefish. Menhaden is an industrial species, alewife is used partly as an industrial resource; five others, scup, summer flounder, striped bass, weakfish, and bluefish, are important recreational resources also; and two of the remaining three, blue crab and hard clam, are also important recreationally but have not been included in the national sport fishing surveys. These 10 species provide examples of the hazards to which living estuarine resources are subject.

The 10 major estuarine-dependent saltwater sport fishes in the Middle Atlantic Bight in 1970 in descending order of rank by weight were bluefish, striped bass, winter flounder, northern puffer, spot, summer flounder, tautog, weakfish, white perch, and black sea bass.

---

[4] This is really two species, alewife (*Alosa pseudoharengus*) and blueback herring (*A. aestivalis*).

Atlantic menhaden (*Brevoortia tyrannus*)

The menhaden fishery of the Middle Atlantic Bight began to flourish when markets for fish meal and oil became available with the collapse of the Pacific coast sardine fishery in the 1940s. Like the sardine fishery, the menhaden industry had a 10-year period of great prosperity, from about 1953 to 1962. During this period, average annual landings from Rhode Island to Virginia inclusive were nearly half a million metric tons. Thereafter, the catch fell off rapidly to a low of about 102,241 metric tons from a high of about 524,000 metric tons in 1956. The principal fishery once was in the New York Bight area, but the growing industry in Virginia so reduced the average life span that few fish now survive to migrate farther north. Despite this intense fishing pressure some unusually successful spawnings occurred in the 1970s and the last five years have included three of the greatest catches on record, and the greatest five-year average catch. Abundance was so great that substantial numbers of menhaden survived to migrate as far north as New England. In the New York Bight area this led to a resumption of the menhaden industry, and the catch in 1974 was more than eight times the 1966 low. In 1974 the catch in the Chesapeake Bay area had dropped by one-third from the 1972 high, and it can be expected that abundance soon will decline sharply north of Chesapeake Bay also.

The future of the menhaden resource will be determined largely by the fishery. There is no evidence that human activities other than fishing have affected the resource adversely. If the present pattern of essentially unregulated fishing continues, the probability is high that the menhaden industry will suffer the fate of the defunct Pacific sardine industry.

Blue Crab (*Callinectes sapidus*)

Blue crab is a major fishery resource in Chesapeake Bay. Until surf clam landings overtook it in 1974 blue crab was the top edible seafood by weight in that area. The resource has been much less important to the north, although it is fished commercially at least as far north as Connecticut whenever it is locally abundant, and in the bays and estuaries it is an important recreational resource. Annual landings in Chesapeake Bay have been trending upward, with major fluctuations, for about half a century. The peak catch in Chesapeake Bay, in 1966, was about 44,000 metric tons. In the New York Bight area catches dropped abruptly from peaks in the 1950s of about 2,990 metric tons to about 180 metric tons in 1968. Since that time catches have risen steadily, to about 2,404 metric tons in 1974. Although there are no comprehensive supporting data, it is likely that recreational catches also have risen in all areas. For the Middle Atlantic Bight as a whole the history of commercial landings contain no hint that the resource has been overfished. It has been suggested that the decline of the fishery north of Chesapeake Bay in the late 1950s and the 1960s may have been caused by water pollution, especially by insecticides, and that the recent recovery may have been caused by the ban on DDT. This is pure speculation, however.

The blue crab resource apparently is highly resistant to fishing, at least over the past range of fishing intensity. As a resource of estuaries and the immediate coastal zone, it might be expected to be susceptible to the effects of water pollution and other man-made environmental changes.

American Oyster (*Crassostrea virginica*)

The decline of oyster production along the Atlantic coast is a prime example of the vulnerability of estuarine resources to man-made hazards. The history of oyster production in the New York Bight area (Fig. 1) illustrates well what has happened all along the coast. Three distinct periods can be seen. The 50 years up to about 1930 were largely occupied with harvesting the resource on natural oyster grounds. The large fluctuations in this period were mainly in response to economic variations. The two decades from the early 1930s to early 1950s embraced a period in which highly mechanized oyster farming was in effect. The industry virtually collapsed in the late 1950s to a much lower level of production. The causes were complex, but they are reasonably well understood.

In the first period the natural oyster grounds were seriously overharvested in many places along the coast. Toward the end of this period water pollution increasingly affected the industry, directly by affecting survival of sensitive larvae and young, and indirectly by forcing closure of shellfish beds because water quality did not meet acceptable standards. The once important oyster industry in Raritan Bay, N.Y. and N.J., collapsed during this period (4). The decline in

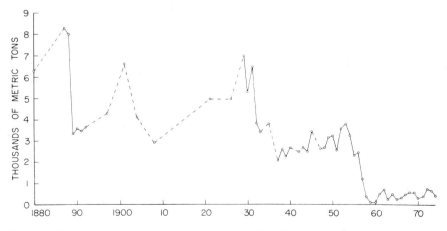

Figure 1. Historic landings of American oyster in New York and New Jersey combined, 1880-1974. Weights of meats only. Broken lines join years between which data are missing.

production in the early 1930s, which led to the second period, was caused primarily by the effects of the economic depression. In New York growing water pollution in the vicinity of New York City forced the oyster industry to move into Long Island Sound and the bays of the south shore of Long Island. The sharp decline in the 1950s to the third period was caused mainly by a massive invasion of predators in Long Island Sound (22), but careless oystering practices also contributed to the decline. In Chesapeake Bay the sequence of events was somewhat the same, although disease, rather than predation, was the principal natural catastrophe. In the Chesapeake area recent increases in production have been caused mainly by a Maryland program of rehabilitation on the public oyster grounds. In the New York Bight area improved production has come about through more sophisticated private oyster farming, which has included hatchery production of young oysters.

Alewife (*Alosa pseudoharengus*)

In the Middle Atlantic Bight area Virginia has been the major producer of anadromous alewife landings. As noted in footnote 4 the catch includes also blueback herring. Landings rose to a peak of about 18,500 metric tons in 1951, and then fluctuated, mainly from natural variations in spawning success, reaching a secondary peak of about 17,500 metric tons in 1965. Subsequently, production declined sharply, to a low of about 5,000 metric tons in 1973. This decline in domestic production was caused by foreign fishing on the high seas (6).

Scup (*Stenotomus chrysops*)

The decline and fall of the commercial scup fishery in the Middle Atlantic Bight is more difficult to explain. The history of domestic commercial scup landings in the New York Bight area (Fig. 2) is illustrative. In the Middle Atlantic Bight as a whole, landings followed a similar trend, rising to a peak of about 22,360 metric tons in 1955, dropping for four years, then rising to a secondary peak almost as high in 1960. Then came a sharp and steady decline to a low of about 3,700 metric tons in 1971, followed by a rise to about 6,500 metric tons in 1974.

Scup is known to vary widely in abundance from natural causes (17). It is not known how much the intensive domestic fishery contributed to the decline in abundance in the 1960s. The decline in domestic landings in the 1960s probably exaggerates the decline in abundance, because catches in the growing recreational fishery and the foreign fisheries are not included. The recent increase in commercial and recreational catches appears to have been caused by a natural increase in abundance. The causes are not known. Widely fluctuating resources like scup are likely to be vulnerable to overfishing at times of scarcity. Incidental catches of scup by foreign fleets are an additional matter of concern at such times.

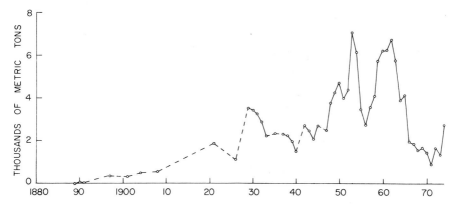

Figure 2. Historic landings of scup (*Stenotomus chrysops*) in New York and New Jersey combined, 1880-1974, live weights. Broken lines join years between which data are missing.

Summer Flounder (*Paralichthys dentatus*)

Commercial landings of summer flounder or fluke in the Middle Atlantic Bight were at a maximum of approximately 6,800 metric tons in the middle 1950s, when the species was especially abundant. The temporary decline in landings of scup at this time was at least partially caused by a shift of the trawl fleet to the higher priced flounder (10). After 1958 summer flounder landings declined steadily to less than 20 percent of the maximum by 1969, and then increased threefold by 1974. The recent increase, which has occurred in recreational catches also, was caused by an increase in abundance. The causes are not known.

Hard Clam (*Mercenaria mercenaria*)

The major producer of hard clam in 1974 was New York State, which accounted for about 68 percent of the total Middle Atlantic Bight catch of about 5,350 metric tons of meats. Peak production of hard clam in the Middle Atlantic Bight was reached in 1947 at about 7,710 metric tons of meats. Many living coastal resources reached peak landings near the end of the second world war and immediately after, as meat rationing increased the demand for other sources of animal protein. Since 1947, although landings in individual states have varied widely (14), hard clam production in the Middle Atlantic Bight has remained fairly stable.

As a non-migratory estuarine resource, hard clam is vulnerable to the effects of water pollution and overfishing, and also to natural environmental fluctuations. No estimates of recreational and subsistence catches are available, but they must be considerable. As just one example, the Town of Islip, on Great South Bay, Long Island, N.Y., in 1974 issued 2,459 permits to residents for limited catches of hard clam, as compared to about 1,700 part-time or full-time commercial clam permits (9). It is also known that considerable quantities are taken illegally.

Striped Bass (*Morone saxatilis*)

Striped bass landings in the Middle Atlantic Bight, like blue crab landings, have followed an upward trend for some 40 years. The greatest commercial catch on record was as recently as 1973, when about 5,000 metric tons were reported. Recreational catches have been growing also, and exceed commercial catches substantially. In 1970 it was reported that about 33,158 metric tons of striped bass were taken by sport fishermen from the Gulf of Maine to Cape Hatteras. There is no reason to doubt that this increase in catches was made possible by an increase in abundance, although part of the rise in catch was related to a substantial increase in numbers of recreational fishermen (3). Possible reasons for the increase have already been given. None of the hypotheses about causes can be confirmed by scientific evidence.

Weakfish (*Cynoscion regalis*)

Like several other important living resources of the coastal zone of the Middle Atlantic Bight, weakfish or gray sea trout produced peak landings at about the end of the second world war (about 16,420 metric tons in 1945) then declined to much lower levels of abundance. In 1967 only about 544 metric tons were landed in the Middle Atlantic Bight but by 1973 the commercial catch had risen almost tenfold, to about 4,500 metric tons. The drop and recent rise in catches were caused by changes in abundance. Recreational catches also have risen sharply recently. Part of the fluctuation almost certainly has been caused by natural environmental changes. To what extent other factors, like fishing and man-made environmental effects, have played a part is not known.

Bluefish (*Pomatomus saltatrix*)

Bluefish also is highly variable in abundance. Commercial landings were reported to be much higher at the turn of the century than they are now, but it is not known how comparable these early records are with those of today. Landings in the Middle Atlantic Bight in the early 1940s reached very low levels, perhaps as low as 200 metric tons, but the trend since that time has been upward, to a high of about 2,630 metric tons in 1974. The history of commercial landings does not present a true picture of changing catches or abundance because bluefish is a popular sport fish. The recreational catch in 1970 from the Gulf of Maine to Cape Hatteras was about 45,315 metric tons, the largest sport catch of any species by weight. This is by far the most abundant estuarine foodfish in the Bight at present. The causes of the wide fluctuations in abundance of bluefish are not known.

SUMMARY AND CONCLUSIONS

The importance of estuarine-dependent species in marine commercial and recreational fisheries in the United States varies widely by region. By weight, almost all the catch in the Gulf of Mexico is made up of estuarine species. In California and New England marine species dominate. In the Middle Atlantic region, including Chesapeake Bay, the fisheries depend heavily upon estuarine species, and because the human population is large and heavily concentrated in the coastal area man-made stresses on the living resources are intense. As might be expected, the importance of estuarine environments to the fisheries is greatest in regions of major estuarine development. It follows that these are the regions in which human environmental effects, other than fishing itself, are most likely to damage the fisheries.

Although there is ample evidence to support the conclusion that man's activities in the coastal zone are detrimental to estuarine fishery resources, actual or potential benefits should not be ignored. The problem is that the benefits, like the adverse effects, are essentially unplanned. More attention needs to be paid to making maximum use of potentials for benefit, in designing and locating engineering projects, sewage treatment plants and outfalls, power plants, and offshore structures.

Human activities have not been favorable to shellfish resources or the industries they support. Shellfisheries provide the best examples of human damage to the environment, with attendant economic, public health, and social-political repercussions. Estuarine shellfish resources and the economic health of the shellfish industry can be useful indices of the condition of estuarine waters, to measure success or failure of estuarine management.

Evidence of human damage to finfish resources, other than the effects of fishing, is more difficult to find, although undeniable examples of some adverse effects can be cited. The evidence comes mainly from laboratory experiments (21) and observation of obvious effects in the natural environment (11, 15) including catastrophic events. The migratory habits of fishes contribute to the problem, so that it is extremely difficult to evaluate the effects of man-made environmental conditions, natural environmental fluctuations, fishing, and other factors. This is not to say that it is easy to sort out the effects of these things upon mollusks. The difference is that most estuarine mollusks do not move about, which makes periodic observation easier.

The obvious conclusion is that man's diverse activities have damaged estuarine fisheries and that man is a constant threat to the living resources. Some bold moves have been made recently: for example, the Federal Water Pollution Control Act Amendments of 1972 (PL92-500) and the Coastal Zone Management Act of 1972 (PL92-583). PL 92-500, among other things, proposes to eliminate all harmful discharges into territorial waters by 1985. This is a worthy objective, but it is unlikely that it can be achieved in an absolute sense. Nevertheless, if it succeeds in improving water quality, or even prevents further deterioration, it will be a step ahead. Some effects have already been noted in important waterways like the Hudson River, if increased abundance of species like American shad (16) and sturgeon is a valid index. Improved water quality is only one objective among many, however. Greatly improved controls over fishing must be developed if all other environmental controls are to produce noticeable fishery benefits.

A recent editorial in *Science* eloquently stated the problems of communication between scientists and legislators (2). Decision makers are extremely unwilling to accept the fact that science often cannot provide unequivocal answers. Fishery legislation usually is urged or enacted on the basis of oversimplified concepts of cause and effect. Many forces operate to change the catch of a particular resource. Some of these, such as overfishing and natural and man-made environmental changes, affect abundance directly. Some, such as changes in ocean current systems, and water pollution which forces closure of shellfish beds, affect availability of the resource to fishermen. Others, such as changes in abundance, cause changes in fishing strategy for economic reasons. Fishery scientists very seldom are able to identify precisely the reasons for changes in fish stocks or to allocate among all possible causes the relative contribution of each. Few fishery situations can be interpreted with the elegant simplicity and finality of predicting the consequences of placing an impassable dam downstream of the spawning grounds of an anadromous species. The probability is fairly high that estuarine fisheries will continue to be managed poorly unless decision makers are willing to act prudently and forcefully without waiting for absolute proof of damage, and unless the people are willing to cooperate.

REFERENCES

1. Clark, J. 1967. Fish and man. Conflict in the Atlantic Estuaries. Am. Littoral Soc., Spec. Pub. 5: vi + 78 p.
2. David, E. E., Jr. 1975. One-armed scientists? Science 189(4204):679.
3. Deuel, D. G. 1973. 1970 Salt-Water Angling Survey. U.S. Dept. Commerce, Nat. Marine Fish. Serv., Current Fish. Stat. 6200:iii + 54 p.
4. Dewling, R.T., K.H. Walker, and F. T. Brezenski. 1972. Effects of pollution: Loss of an $18 million/year shellfishery, p. 553-559. *In* Marine Pollution and Sea Life. M. Ruivo (ed.). Fishing News (Books) Ltd., London.
5. Dovel, W. L., and J. R. Edmunds. 1971. Recent changes in striped bass spawning sites and commercial fishing areas in upper Chesapeake Bay: Possible influencing factors. Chesapeake Sci. 12(1): 33-39.
6. Grosslein, M.D., E. G. Heyerdahl, and H. Stern, Jr. 1973. Status of the international fisheries off the Middle Atlantic coast. A technical reference document prepared for the bilateral negotiations of U.S.A. with U.S.S.R. and Poland. U.S. Dept. Commerce, Nat. Marine Fish. Serv., NE Fish. Center, Lab. Ref. No. 73-4: 117 p.
7. Gunter, G. 1967. Some relationships of estuaries to the fisheries of the Gulf of Mexico, p. 621-638. *In* Estuaries. George H. Lauff (ed.). Am. Assn. Adv. Sci., Pub. No. 83.

8. Johnson, R. K., and T. S. Y. Koo. 1975. Production and distribution of striped bass eggs in the Chesapeake and Delaware Canal. Chesapeake Sci. 16(1): 39-55.
9. Klaassen, Barry J. Personal communication.
10. Knapp, W. E. 1975. Marine commercial fisheries of New York State: An analysis by gear. N.Y. State Sea Grant Inst. (in press).
11. Mahoney, J. B., F. H. Midlige, and D. G. Deuel. 1973. A fin rot disease of marine and euryhaline fishes in the New York Bight. Trans. Am. Fish. Soc. 102(3):596-605.
12. Mansueti, R. J. 1961. Effects of civilization on striped bass and other estuarine biota in Chesapeake Bay and Tributaries. Proc. Gulf Caribb. Fish. Inst., 14th Ann. Sess: 110-136.
13. McHugh, J. L. 1966. Management of Estuarine Fisheries, p. 133-154. In A Symposium on Estuarine Fisheries. Am. Fish. Soc., Spec. Pub. 3.
14. _____. 1972. Marine fisheries of New York State. U.S. Dept. Commerce, Nat. Marine Fish. Serv., Fish. Bull. 70(3):585-610.
15. Mearns, A. J., and Marjorie Sherwood. 1974. Environmental aspects of fin erosion and tumors in Southern California Dover sole. Trans. Am. Fish. Soc. 103(4):799-810.
16. Medeiros, W. H. 1975. The Hudson River Shad Fishery: Background, management problems, and recommendations. N.Y. State Sea Grant Inst., NYSSGP-RS-75-011, Albany, N.Y.:iv + 54 p.
17. Neville, W. C., and G. B. Talbot. 1964. The fishery for scup with special reference to fluctuations in yield and their causes. U.S. Dept. Interior, Fish and Wildl. Serv., Spec. Sci. Rept. Fish. 459: 61 p.
18. Odum, E. P. 1971. Fundamentals of Ecology. W. B. Saunders Co., Philadelphia: xiv + 574 p.
19. Silverman, M. J. 1972. Tragedy at Northport. Underwater Naturalist 7 (2): 15-18.
20. Turner, J. L., and H. K. Chadwick. 1972. Distribution and abundance of young-of-the-year striped bass, Morone saxatilis, in relation to river flow in the Sacramento-San Joaquin estuary. Trans. Am. Fish. Soc. 101(3):442-452.
21. Waldichuk, M. 1974. Coastal marine pollution and fish. Elsevier Sci. Publ. Co., Amsterdam. Ocean Mgmt. 2:1-60.
22. Wallace, D. H. 1971. The biological effects of estuaries on shellfish of the Middle Atlantic, p. 76-85. In Symposium on the Biological Significance of Estuaries. Sport Fish. Inst., Washington, D.C.

TIDAL LOCKS ACROSS THE EAST RIVER:

AN ENGINEERING SOLUTION TO THE REHABILITATION

OF WESTERN LONG ISLAND SOUND[1]

Malcolm J. Bowman
Marine Sciences Research Center
State University of New York
Stony Brook, New York 11794

ABSTRACT: Water quality in western Long Island Sound and New York Harbor is seriously degraded. A major source of pollutants is sewage released into the East River, a cooscillating tidal strait connecting the Sound to the Harbor. Very rapid and significant improvements in water quality could be attained by constructing ship locks across the Upper East River to increase the circulation of the sea through the Harbor and Sound. During ebb tide these locks would be opened, allowing an unhindered flow of Sound water into the Harbor. After six hours the locks would be closed at slack water, blocking the following flood tide from re-entering the Sound. The net result would be a strong pulsating unidirectional flow ($\sim$2500 $m^3 sec^{-1}$) of relatively clean central Long Island Sound water, pumped by the semi diurnal tides through New York Harbor, and the Lower Bay, out into the New York Bight.

Simple models indicate that the concentration of conservative contaminants in the western Sound and the Harbor would drop by $\sim$88% and $\sim$45%, respectively, from present levels, within a month of operation. The accompanying decreases in inorganic nutrient concentrations are calculated and tabulated for both winter and summer conditions.

The major physical effects of blocking Hudson River water from entering the Sound through the East River would be to change the essential estuarine characteristics of western Long Island Sound to those of a coastal embayment, and increasing the salinity of the western Sound and New York Harbor both by $\sim$4°/$\circ\circ$.

[1] Contribution 147 of the Marine Sciences Research Center (MSRC) of the State University of New York at Stony Brook.

28

INTRODUCTION

A vast sum of money (~4.6×10^9) is being spent over the next few years on expanding and upgrading the waste water treatment facilities in the New York, New Jersey and Connecticut areas (12). In the New York City metropolitan region alone, over 90 m^3 sec^{-1} (2,100 MGD) of treated and untreated effluent, from an estimated 15×10^6 persons, are released into the waters around the City (27).

The problem of combined storm and municipal waste sewer systems in New York City remains critical; ". . . The money spent for this construction in large part will be wasted if means of mitigating the effects of combined sewers are not found" (12).

Of the above total, some 15 m^3 sec^{-1} (400 MGD) are transported into western Long Island Sound by the East River (Fig. 1) (1) a co-oscillating tidal strait

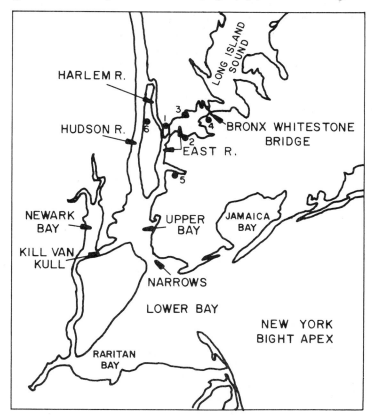

Figure 1. Map of greater New York Harbor. There are 76 sewage treatment plants located within the area shown (12). The six plants referred to in the text are: 1-Wards Island, 2-Bowery Bay, 3-Hunts Point, 4-Tallman Island, 5-Newtown Creek, and 6-North River.

connecting the sound to New York Harbor (2). These sewage effluents have seriously degraded water quality in western Long Island Sound, as judged by shellfishing closures, violations of State and Federal water quality standards, and inorganic nutrient and dissolved oxygen levels. Pollution has markedly decreased the recreational and shellfishing utility of the western sound, where the population density and recreational demand are greatest. Phytoplankton blooms, partly attributable to the high nutrient levels found in the western sound, sometimes occur in the summer, followed by increased degradation of water quality (10, 9, 4).

Presently all of New York Harbor, including the Upper and Lower bays, Jamaica and Raritan bays, and out to the three mile limit, the East River and western Long Island Sound (regions 1-3 of Fig. 2) are proclaimed off-limits for shellfishing (24). Although reliable statistics are unavailable to judge the effects of pollution on sound finfisheries (15), commercial shellfishing losses over the last fifty years have exceeded 5×10^8, with an estimated loss of 1.7×10^7 in 1970 alone (31). Public concern over the deterioration of Long Island Sound led to the funding of a 3.5×10^6, 3 years study which has recently documented over 600 recommendations for improving the environmental quality of the region (16).

Deterioration in water quality in western Long Island Sound is attributable to several factors. First, and most obvious, is the proximity of the western sound to New York City. Four of the largest sewage works in the city discharge directly

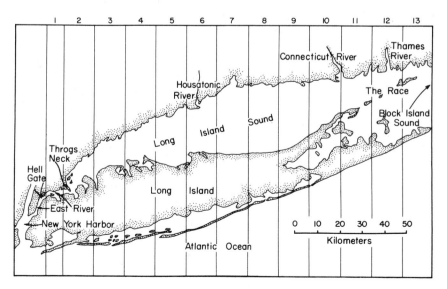

Figure 2. Location map of Long Island Sound illustrating the 13 model sections (from (1)).

into the upper East River (Wards Island, Bowery Bay, Hunts Point, and Tallmans Island; Fig. 3), and collectively represent major contributors of effluent to the western sound (1).

Other factors are related to the physical properties of the sound and the East River. Tidal circulation in the western sound is weak (1), and coupled with the narrowing crossection, provides only poor dispersion of high contaminant concentrations in the western region of the sound. In addition, the tidal excursion of the East River is only ~70% of its own length (2); correspondingly the river also possesses poor dispersion characteristics. The combined effects of these influences are reflected in the high levels of inorganic micronutrients found in the western basin (Figs. 4 and 5; (1)).

Despite huge expenditures on new sewage treatment works, projected improvements in water quality are very modest. For example, maximum improvements of only 13% and 9.5% in dissolved oxygen levels relative to saturation values in the East and Hudson rivers (Fig. 1) have been predicted (17) for the Newtown Creek and North River treatment plants presently under construction at projected costs of 4.6×10^8 and 9.9×10^8, respectively (12).

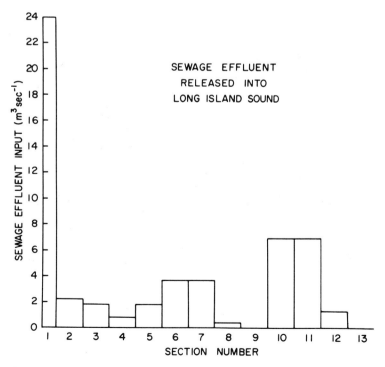

Figure 3. Distribution of sewage effluents released into Long Island Sound (from (1)).

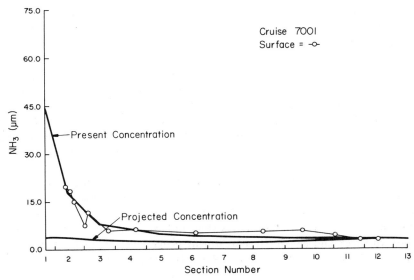

Figure 4. Present and projected ammonia concentrations in Long Island Sound in winter.

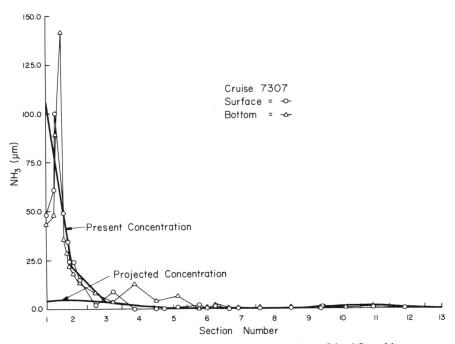

Figure 5. Present and projected ammonia concentrations in Long Island Sound in summer.

TIDAL LOCKS ACROSS THE EAST RIVER

The East River is a 25 km. hydraulic channel whose cooscillating flow arises as a consequence of a 3 hour phase difference in the tides across its ends (2). The tides in New York Harbor (Upper Bay) are transmitted from the New York Bight via the Narrows (Fig. 1), and propagate as progressive waves up the Hudson River (13). On the other hand, the tides in Long Island Sound have a standing wave character (14). The main connection of the sound to the ocean is at the eastern mouth, where the tides are transmitted from the Atlantic through Block Island Sound (Fig. 2). Tidal exchange is intense ($\sim$4 $\times$ 10^5 m^3 sec^{-1}) (34) near the Race, but drops monotonically to $\sim$8,000 m^3 sec^{-1} near the confluence of the Sound and the East River (2).

The concept of diverting rivers to disperse accumulated pollutants can be traced back to Greek mythology. Hercules in one day cleaned out the filthey stables of Augeas, which housed 3,000 oxen and remained unwashed for 30 years, by diverting two rivers through it (32). More recently, in 1922 Reeve (22) proposed a novel method of diluting contaminants in New York Harbor by inducing a large net inflow of clean Long Island Sound water through the East River by installing gates to rectify that river's co-oscillating tidal flow (i.e., to convert a reversing into a uni-directional current).

Briefly, the scheme centered around constructing ship locks across the East River. During ebb tide (i.e., westerly flow), the gates would be fully open, allowing an unhindered flow of Long Island Sound water into New York Harbor. Six hours later the gates would be closed at slack water, completely blocking any return of New York Harbor water back into the Sound. The current would flow to the harbor for approximately six lunar hours (6.2 hours) and then be zero for the next six lunar hours.

Twice each day a pulse of sound water ($\sim$$10^8$ m^3) would displace $\sim$30% of the water in the harbor during the tidal cycle (volume of Upper Bay $\sim$3.7 $\times$ 10^8 m^3; 13). For comparison, the mean discharge of the Hudson River ($\sim$560 m^3 sec^{-1}; 7) displaces $\sim$7%, and the tidal prism ($\sim$0.74 $\times$ 10^8 m^3; 13) represents 20% of the volume of the Upper Bay below mean low water.

Fifty years later, while investigating alternate methods of sewage waste management, I independently derived precisely the same idea, but in my case I was principally interested in reducing pollutant levels in the Long Island Sound estuary. Tidal energy would be harnessed to drastically increase the circulation of the sea through the western sound and the harbor at the same time the locks would block all sewage effluents originating in New York City from entering the western basin.

EXPECTED IMPROVEMENTS IN WATER QUALITY

The expected improvement in water quality in Long Island Sound resulting from such an engineering project was evaluated by estimating the reduction in concentration of dissolved inorganic nutrients (NH_3, NO_2, NO_3), using a mathematical model of the sound (1). Nutrients were chosen as convenient tracers of effluent as they are abundant in sewage, are relatively easy to determine experimentally, and have direct influences on the eutrophication of receiving waters.

The model is constructed around one dimensional, steady state, water, salt, and nutrient balance equations for each of the 13 regions of the sound (Fig. 2). Computed values of nutrient concentrations and transport reflect the combined effects of tidal dispersion, non-tidal advection, local inputs from sewage treatment facilities and agricultural runoff, and first order biochemical decay. Boundary conditions at the western end of the model are estimates of the net fluxes of water, salt and nutrients through the East River. A sufficient condition at the eastern boundary is an oceanic source (or sink) of constant concentration. The model was adapted for the present investigation by modifying the western boundary conditions to reflect increased net fluxes through the East River. The westward transport of salt and nutrients out of the western sound were considered as purely advective fluxes contained in the modified East River flow (equivalent to a steady transport ~2,500 m^3 sec^{-1}).

Tidal transport through the Race is presently some 55 times that through the East River (33). One assumption of the calculations is that rectifying the East River current would not have a major effect on the tidal circulation throughout most of the sound, and hence tidal dispersion coefficients would be expected to remain valid under the changes in western boundary conditions (as it turns out, contaminant concentrations are predicted to become almost uniform throughout the sound and hence the results are relatively insensitive to the exact values of the exchange coefficients). Another assumption is that the computed biochemical decay coefficient k for each nutrient investigated in the original model would remain unaltered. This coefficient k, representing the balance between biological utilization in the surface layers and regeneration of nutrients in bottom sediments, was varied over a significant range to test the sensitivity of the predictions to the value of k.

Experimental data were obtained from two cruises made in Long Island Sound by the Marine Sciences Research Center to gather water quality data under both winter (cruise 7001; January, 1970) and summer (cruise 7307; August, 1973) conditions. Surface and bottom nutrients samples (surface only on cruise 7001) were gathered and concentrations plotted along the longitudinal axis of the sound, together with two theoretical curves; one for existing conditions (1) and the second for the modified boundary fluxes. Results obtained with the model are illustrated in Figs. 4 and 5.

The expected reductions in ammonia concentrations are quite impressive for the western basin (regions 1-5). Concentrations at the western end of the sound are predicted to drop from ~45 μM down to ~3 μM in winter, and remain almost uniform along the length of the sound, with only slight increases at both ends reflecting the effects of local sewage inputs from western Long Island and the Connecticut River basin.

Table 1. Ratio of projected to present maximum nutrient concentrations in Long Island Sound. The uncertainties arise from variation in sewage effluent source concentrations and biochemical decay rates (Bowman, 1975). Summer and winter data based on results of MSRC cruises 7307 and 7001, respectively. Regions expected to be affected are based on those sections where a significant difference was found between present and projected concentrations.
    *The drop in concentration of a conservative contaminant can be estimated by setting k=0. For example, setting C_{13} =0 (Block Island Sound) leads to a projected/present mean concentration ratio in region 1 of 12%.

Nutrient	Projected/present mean concentrations (region 1;%)		k^{-1} (days*)		Regions affected by a reduction in concentration	
	summer	winter	summer	winter	summer	winter
NH_3	3.2±0.4	7.9±1.1	2.9±0.8	22±3	1-4	1-9
NO_2	2.9±0.5	12 ±1.6	2.3±0.5	95±23	1-4	1-13
NO_3	5.6±0.6	36 ±2.6	9.6±2.6	460±92	1-6	1-13

The expected reduction in ammonia concentrations in summer (based on cruise 7307 data) is even greater with levels dropping from ~110 μM to ~3 μM near the western boundary. Concentrations east of region 4 remain essentially unaltered, as most of the ammonia originating in the western sound is utilized by high rates of local phytoplankton uptake.

The expected improvements in water quality as indicated by the changes in concentration of other nitrogeneous nutrients (NO_2, NO_3) are similar and are listed in Table 1, together with the variability introduced by uncertainties in biochemical decay rates and present boundary fluxes.

The essential result of the project on nutrient levels in the western sound would be to reduce concentrations down to values found in the central and eastern basins, which are only slightly higher than adjacent coastal waters (23).

INCREASE IN LONG ISLAND SOUND SALINITY

Westerly advection induced in Long Island Sound by using tidal gates in the East River would increase the salinity, especially at the western end, due both to the elimination of water of Hudson River origin entering the sound, and also to the increased inflow of higher-salinity water from the east.

The increase in mean salinity predicted by the model is shown in Fig. 6, assuming that the salinity in Block Island Sound (region 13) remains constant at 31°/oo. The salinity increases by ~4.5°/oo just east of the locks. The resulting reduction in horizontal salinity gradient would reduce the intensity of the estuarine circulation, especially in the western sound. Gravitational convection is presently a dominant feature of the non-tidal circulation throughout the estuary (34). Conditions in the western basin would tend to approximate those of a coastal embayment with a reduced horizontal density gradient in the central and eastern regions maintained principally by the outflows of the Housatonic and Connecticut rivers.

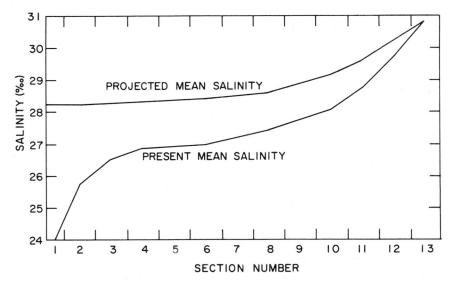

Figure 6. Present and projected mean salinity distribution in Long Island Sound.

Although the improvement in water quality may enable the reopening of significant numbers of shellfishing beds in the Sound, it should be pointed out that the small increase in salinity would also increase the predation of oysters which depend on the low salinity of sound water for their survival.

RATES OF IMPROVEMENT IN WATER QUALITY IN LONG ISLAND SOUND

An estimate of the rate at which water quality in the western sound would be expected to improve was made by calculating the residence time of sections 1-4 by simply dividing the volume of each section by the modified flow through the East River ($\sim$2,500 m³ sec^{-1}). The residence times would be 11 hours, 5 days, 10 days, and 16 days for regions 1-4 respectively, or $\sim$32 days for the entire region. Thus, in about a month a volume of water equal to that in the western sound (regions 1-4) would be pumped through the East River and New York Harbor, and the time scale indicates the period necessary to reduce contaminant concentrations in the water column. Of course, any reduction in the accumulation of organic and toxic substances in the sediment would take much longer.

THE EFFECT ON THE SALINITY AND CONTAMINANT CONCENTRATIONS IN NEW YORK HARBOR

Increasing the flux of water, salt and contaminants into New York Harbor would obviously alter the salinity, effluent concentrations and the circulation. Determination of the changes in circulation patterns and sediment transport are beyond the scope of this paper, but could effectively be evaluated by running tests on the New York Harbor hydraulic model operated by the U.S. Army Corps of Engineers in Vicksburg, Mississippi (25).

The expected changes in salinity and effluent concentrations were estimated to a first approximation by simple mass balance calculations. The harbor (Upper Bay) was assumed well mixed with a mean salinity S_{UB} of $21^{\circ}/_{\circ\circ}$ (20), with a mean Hudson River inflow R_H of 560 m^3 sec^{-1} (7), and an additional 56 m^3 sec^{-1} influx from Newark Bay R_{NB} through the Kill van Kull (19). The mean salinity of continental shelf water S_0 is taken as $31^{\circ}/_{\circ\circ}$ (3).

In steady state and with reference to Fig. 7, the appropriate salt balance equation is

$$(R_H + R_{NB} + R_E)\, \rho\, S_{UB} = M_S + D\,(S_0 - S_{UB}) \tag{1}$$

where $R_E \sim 350$ m^3 sec^{-1} and $M_S \sim 1.2 \times 10^4$ kg sec^{-1} are the present mean non-tidal fluxes of water and salt through the East River (1), ρ is the density of sea water $\sim 10^3$ kg m^{-3}, and D is a tidal dispersion coefficient appropriate to exchange between the harbor and the New York Bight apex. The left side of the equation represents the advection of salt out from the harbor to the ocean, the first term on the right-hand side is the influx of salt from the East River, and the

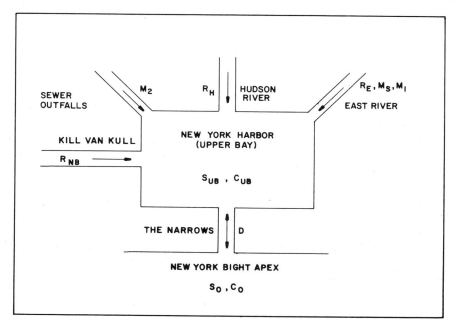

Figure 7. Schematic diagram of New York Harbor for mass balance calculations (see text).

second represents the dispersion of salt into the harbor from the New York Bight.

After rectification of East River flow, the advective flux into the Upper Bay becomes

$$R_H + R_{NB} + R_E' \sim 3{,}100 \text{ m}^3 \text{ sec}^{-1} \ (R_E' \sim 2{,}500 \text{ m}^3 \text{ sec}^{-1}).$$

This should be compared to the present mean tidal flux past the Narrows, the entrance to the Upper Bay, which is about 1.4×10^4 m^3 sec^{-1} (19).

Rewriting equation (1) for the modified fluxes, we get

$$(R_H + R_{NB} + R_E') \rho \, S_{UB}' = M_S' + D \, (S_0' - S_{UB}'). \tag{2}$$

S_0' was computed by setting $S_{UB}/S_0 = S_{UB}'/S_0' = 0.68$ (this is a better approximation than assuming $S_0 = S_0'$; D.W. Pritchard, private communication).

The modified salt flux $M_S' \sim 7.1 \times 10^4$ kg sec^{-1} was determined from the Long Island Sound model, and S_{UB}' and S_0' represent the new values of mean salinity in the harbor and the New York Bight apex.

Solving equation (2) yields $S_{UB}' \sim 26.0°/_{oo}$ indicating that the injection of sound water of salinity $\sim 28°/_{oo}$ pumped through the East River will increase the salinity of the Upper Bay by $\sim 5.0°/_{oo}$. An investigation of each term of equation (2) reveals that the second term on the right hand side is 14% of the first term, and hence the effects of expected changes in the tidal dispersion coefficient D arising from the modified circulation can be neglected to a first approximation.

Elementary effluent balance equations were used to predict expected changes in the concentration of a conservative contaminant. Two components of the total concentration were studied, that arising from sources in the East River and Long Island Sound, and that arising from other sources around the harbor.

Again the harbor was assumed to be well mixed, and the ocean effluent concentration C_0 was taken as a fixed proportion of C_{UB}, i.e., $C_0 = \alpha C_{UB}, 0 < \alpha < 1$. ($\alpha = 0$ implies that the ocean is an infinite sink ($C_0 \equiv 0$), and $\alpha = 1$ is equivalent to stating that diffusion is negligible ($D \equiv 0$). The actual value of α is unknown in the absence of a value for C_0, and is left as a parameter. The results are compared for a range of α.

Let the present effluent influx from the Upper East River into the harbor be M_1 kg sec^{-1} and the influx from other sources around the harbor be M_2 kg sec^{-1}. Further, let the present effluent concentration in the harbor be C_{UB}. Then the effluent balance equation for the harbor is

$$M_1 + M_2 = D(C_{UB} - C_0) + \rho \, (R_E + R_H + R_{NB}) C_{UB}$$

or

$$C_{UB} = (M_1 + M_2)/(D(1 - \alpha) + \rho \, [R_E + R_H + R_{NB}])$$

After rectification, M_1 and R_E increase to M_1' and R_E' respectively, while the other parameters remain unaltered.

Hence, C_{UB} changes to C_{UB}' where

$$C_{UB}' = (M_1' + M_2)/(D(1-\alpha) + \rho \, [R_E' + R_H + R_{NB}]).$$

Taking

$$
\begin{array}{lll}
M_1 & = 0.74 \times 10^4 \text{ kg sec}^{-1} & (12, 1) \\
M_1' & = 3.1 \times 10^4 \text{ kg sec}^{-1} & (12) \\
M_2 & = 4.4 \times 10^4 \text{ kg sec}^{-1} & (12), \text{ and} \\
D & = 8.3 \times 10^5 \text{ kg sec}^{-1} & (\text{eqn. } 1),
\end{array}
$$

leads to $C_{UB} \sim 29°/oo$, $C_{UB}' \sim 19°/oo$, $C_{UB}'/C_{UB} \sim .66$ ($\alpha = 0$) or $C_{UB} \sim 53°/oo$, $C_{UB}' \sim 24°/oo$, $C_{UB}'/C_{UB} \sim .45$ ($\alpha = 1$). The expected improvement lies between these extremes.

Two opposing factors control the drop in effluent concentration: first is the dilution attributable to the influx of cleaner Long Island Sound water, and second is the increase in effluent loading through the addition of contaminants which previously were transported into or were released directly into western Long Island Sound.

The above elementary calculations indicate, fortunately, that the former effect predominates, and effluent concentrations drop in the Upper Bay. Of course it is clear that sewage concentrations in Raritan Bay, the Lower Bay, and the inner New York Bight and along the New Jersey shoreline will also decrease as a simple consequence of the increased dilution of sewage effluent. The increased flow of water out of the harbor simply disperses the same mass of effluent over a larger volume of ocean.

NOTES ON TRAFFIC CONGESTION AND SAFETY

The East River, the second entrance to the Port of New York, has always been a highly congested and dangerous shipping route. A survey conducted by the U.S. Coast Guard (30) showed some 1,065 vessels plying the Hell Gate region (Fig. 2) during one 69 hour period in January, 1972, equivalent to one vessel every four minutes. Another Coast Guard survey during one week in the summer of 1974 measured an average of eight vessels per hour passing the Hell Gate region (28).

Long Island Sound ports handled 2×10^7 tons of cargo in 1964, largely consisting of shallow draft barges (29). This figure should be compared to 1.4×10^7 tons handled annually by the Port of Amsterdam, a port entirely isolated by locks from the sea (5).

During the fiscal years 1969-1972, 46 collisions, rammings and groundings involving 80 vessels occurred. Fifteen of the 46 incidents were considered to have been preventable by some form of shore-based vessel traffic management or assistance (11). The construction of ship locks across the East River obviously would necessitate new traffic management policies. It is clear, however, that traffic congestion is now such that these will probably be introduced in the near future anyway.

An added advantage of constructing locks would be in reduced navigational hazards. Although ships would have to negotiate the locks when closed, currents in the river would be virtually eliminated half of the time, and accordingly the dangers of Hell Gate, where currents up to 3 m sec^{-1} are common, reduced.

ENGINEERING PROBLEMS AND OTHER ASPECTS

A detailed discussion of the engineering design of tidal locks is beyond the scope of this paper. A few factors are worth pointing out, however. The site of a dam would need to be in the upper East River, preferably east of the large sewage treatment plants, certainly east of Hell Gate where strong tidal currents would make navigation on ebb tide through an open lock hazardous. The gates would have to provide minimal resistance to flow when open. As suggested by Reeve (22) this might necessitate the use of self opening and closing louvers across those sections of the dam not served by locks.

The width of the East River is ~900 m near the Bronx-Whitestone Bridge (Fig. 1), with an average depth ~7 m below mean low water (13). The mean tidal range is ~2.0 m; this also represents the maximum height difference that would exist across the locks. Four locks would probably be needed. Two large locks ~60 X 20 m near the center of the channel would handle major shipping in two directions, and two smaller locks ~25 X 8 m near the banks would take smaller commercial traffic and pleasure craft. Each major locking operation would release ~2500 m^3 of East River water into the sound, equivalent to the flow during 1 second while the locks are open, and is entirely negligible.

Unconsolidated and semiconsolidated bottom sediments in western Long Island Sound of Pleistocene and Cretaceous age rest on a Precambrian and Paleozoic crystalline bedrock surface that slopes to the southeast (8); these should provide no major obstacle to the construction of a dam in the area.

In proposing a novel scheme to convert Long Island Sound into the nation's largest fresh water reservoir, Gerrard (5,6) estimated, based on some published figures (18), that two bridge-dams constructed across both the western and eastern Sound (a much greater span) could together cost ~$1-2 X 10^9 in 1966 dollars. Even if the East River project were to cost ~$10^9 today, it still would compare favorably with funds presently being spent on New York City sewage treatment plants ($4.5 X 10^9; 12).

It is interesting to note the existence of two other major engineering projects presently under construction or in planning stage for water control around large cities (26). Floodgates are being constructed across the Thames River at a cost of $3.1 X 10^8 to block storm surges from inundating the City of London. The gates normally will lie flat on the river bed, but will be raised mechanically to block abnormally high incoming tides.

Several famous Venetian monuments are threatened by the long term rise in sea level and by storm surges originating in the Adriatic Sea. Plans have been laid for the construction of hollow water filled caissons which would lie on the floor of the channels linking the Venetian lagoon with the sea. Air would be pumped in during storm weather, making the caissons buoyant and causing their tops to swing upwards, closing off the lagoon. Construction costs have been estimated at 15×10^8.

CONCLUSION

Obviously, the implementation of a radical scheme such as described above would have many political, socio-economic, ecological, sedimentary, navigational, engineering and hydrographic aspects, further discussions of which are far beyond the scope of this paper. However, the potential significant benefits accruing from an engineering solution to the rapid rehabilitation of western Long Island Sound would indicate that the proposal should be given serious consideration.

ACKNOWLEDGEMENTS

I wish to thank all those persons at the Marine Sciences Research Center who have contributed in various ways to the ideas presented in this paper. I also offer my gratitude to D.W. Pritchard and H.H. Carter for their careful readings of and suggestions for improvement of the manuscript.

This work is a result of research sponsored by the New York State Sea Grant Institute, NOAA Office of Sea Grant, Department of Commerce, under Grant No. 2-35281. The U.S. Government is authorized to produce and distribute reprints for governmental purposes notwithstanding any copyright notation that may appear hereon.

REFERENCES

1. Bowman, M.J. 1976. Pollution prediction model of Long Island Sound. *In* Proceedings of Civil Engineering in the Oceans/III, Newark, Delaware, June 9-13, 1975. Am. Soc. Civil Eng. In press.
2. _____. 1976. The tides of the East River, New York. J. Geophysical Res. 81: 1609-1616.
3. _____, and L.D. Wunderlich, 1975. Hydrographic Properties. Marine Ecosystems Analysis Program New York Bight Atlas monograph #1. In press. New York State Sea Grant Institute, Albany, New York.
4. Bromberg, A.W. 1971. Statement. *In* Proceeding of Conference in the matter of pollution of the interstate waters of Long Island Sound and its tributaries-Connecticut-New York. pp 27-44, 56-228. Environmental Protection Agency, Wash., D.C.
5. Gerard, R.D. 1966. Potential freshwater reservoir in the New York area. Science 153: 870-871.

6. _____. 1967. A Long Island Sound reservoir. J. American Water Works Assn. 59 (11): 1351-1356.
7. Giese, G.L., and J.W. Barr. 1967. The Hudson River Estuary. Bulletin 61, State of New York Conservation Department, Water Resources Commission.
8. Grim, M.S., C.L. Drake, J.R. Heirtzlet, et al. 1970. Sub-bottom study of Long Island Sound. Geol. Soc. of Am. Bull. 81: 649-666.
9. Hardy, C.D. 1972. Movement and quality of Long Island Sound waters, 1971. Marine Sciences Research Center Technical Report #17. State University of New York, Stony Brook, N.Y.
10. _____, and P.K. Weyl. 1971. Distribution of dissolved oxygen in the waters of western Long Island Sound. Marine Sciences Research Center Technical Report #11. State University of New York, Stony Brook, N.Y.
11. Hickey, E.J. 1975. U.S. Coast Guard, vessel traffic system branch, Wash., D.C. Private communications.
12. Interstate Sanitation Commission. 1972. 1972 Annual Report, New York, N.Y.
13. Jay, D.A., and M.J. Bowman, 1975. The physical oceanography and water quality of New York Harbor and western Long Island Sound. Marine Sciences Research Center Technical Report # 23. State University of New York, Stony Brook, New York. 72 pp.
14. LeLacheur, E.A., and J.C. Sammons. 1932. Tides and currents in Long Island and Block Island sounds. U.S.C. & G.S. Special Publication #174. Wash., D.C. 187 p.
15. McHugh, J.L., and H.M. Austin. 1973. Prospects for managing the fisheries of Long Island Sound. Marine Sciences Research Center, Stony Brook, N.Y. and New York Ocean Science Lab., Montauk, New York. 3 p. Unpublished manuscript.
16. New England River Basins Commission. 1975. People and the Sound; A plan for Long Island Sound. Vol. I: Summary. Vol. II: Supplement. National Technical Information Service, Springfield, Va. 22151.
17. O'Connor, D.J. 1970. Water quality analysis for the New York Harbor complex, p. 121-144. In A.A. Johnston, (ed.), Water Pollution in the Greater New York Harbor Area. Mayor's Committee on Oceanography. Gordon and Breach, New York, N.Y. 211 p.
18. Pansini, A.J. 1966. The Long Island-Northeast Plan. Eastern Industrial World, 4:28.
19. Parsons, H. deB. 1913. Tidal phenomena in the Harbor of New York. Trans. Am. Soc. Civil Eng., 76: 1979-2106.
20. Pritchard, D.W., A. Okubo, and E. Mehr. 1962. A study of the movement and diffusion of an introduced contaminant in New York Harbor waters. Chesapeake Bay Institute, The Johns Hopkins University Technical Report #31. 89 pp.
21. Rattray, J.E. 1973. The perils of the Port of New York. Dodd, Mead, New York. 302 p.
22. Reeve, S.A. 1922. Cleansing New York Harbor. Geog. Rev. 12: 420-423.
23. Riley, G.A., and S.A.M. Conover. 1956. Oceanography of Long Island Sound, 1952-1954. III. Chemical Oceanography. Bull. Bingham Oceanogr. Coll. 15: 47-61.
24. Rosellini, L. 1973. Clammers lose more of LI Sound. In Newsday, Aug. 6, p. 7.

25. Simmons, H.B., and W.H. Bobb. 1963. Pollution studies for the Interstate Sanitation Commission, New York Harbor model. U.S. Army Engineers Waterways Experiment Station Misc. Paper 2-588, Vicksburg, Miss.
_____, and _____. 1965. Hudson River Channel, N.Y. and N.J. plans to reduce shoaling in Hudson River channels and adjacent pier slips. U.S. Army Engineers Waterways Experiment Station Technical Report #2-694. Vicksburg, Mississippi. 100 p., 18 Tables, 8 photos, 108 plates.
26. Sullivan, W. 1975. Giant floodgates may save sinking cities. *In* The New York Times, Jan. 31.
27. Suszkowski, D.J. 1973. Sewage pollution in New York Harbor: A historical perspective. M.S. research paper, Marine Sciences Research Center, State University of New York, Stony Brook, N.Y. 68 p.
28. Sutherland, S. 1975. U.S. Coast Guard vessel traffic system study. U.S. Coast Guard, Governor's Island, New York, N.Y. Private communication.
29. U.S. Army Corps of Engineers. 1965. Waterborne commerce of the United States, calendar year 1964. Part I, Waterways and harbors Atlantic coast.
30. U.S. Coast Guard. 1972. New York Harbor maritime hazards study. Captain of the Port of New York, U.S. Coast Guard Governor's Island, New York, N.Y. 3 Vols.
31. Ward, G.H. 1971. *In* Estuarine modelling: an assessment. Document 16070 DZV, Environmental Protection Agency, Water Pollution control research series. Wash., D.C.
32. Webster. 1968. Webster's new 20th century dictionary, unabridged second edition. World Publishing Company, N.Y.
33. Weyl, P.K. 1976. The water of Long Island Sound, p. 220. *In* L. Koppelman, P.K. Weyl, M.G. Gross and D. Davies (eds), The Urban Sea: Long Island Sound. Praeger, N.Y.
34. Wilson, R.E. 1976. Gravitational convection in Long Island Sound. Estuarine and Coastal Marine Science, 4: in press.

THE SANTEE-COOPER: A STUDY OF ESTUARINE MANIPULATIONS[1]

Björn Kjerfve
Belle W. Baruch Institute for Marine Biology and Coastal Research
University of South Carolina
Columbia, SC 29208

ABSTRACT: Prior to 1941, the Santee had the fourth greatest discharge of any U.S. east coast river. However, with increasing demand for hydroelectric power, most of Santee's flow was diverted into the tidally dominated Cooper River. Subsequently, Charleston Harbor, South Carolina's largest port at the mouth of the Cooper, began shoaling, primarily due to deposition of riverborne silts and clays. Annual dredging expenditures now amount to approximately $5 million. To alleviate the shoaling problem, most of the Cooper River flow will be re-diverted into the Santee River within the next several years. However, rather than allowing the Santee to follow its former course, the plans call for the construction of an 18.5 km canal between the two river systems at a projected cost of $91 million. Undoubtedly, the shoaling problems in Charleston Harbor will be reduced but at the expense of deteriorating water quality due to less effective harbor flushing. The lower Santee, on the other hand, is likely to experience a return to pre-1941 conditions. Rediversion will halt the severe delta erosion but will at the same time destroy the lucrative hard clam industry and extensive seed oyster beds in the lower river. Man has over the past three and a half decades manipulated the Santee River system extensively—but was it necessary?

DIVERSION

The Santee River originates in the Appalachian Mountains of North and South Carolina. Via a number of tributaries, the river traverses the piedmont, meanders across the coastal plain, and empties into the Atlantic Ocean 75 km northeast of Charleston, S.C. The river basin (Fig. 1) measures 41,000 km² (36)

---

[1] Contribution #127 from the Belle W. Baruch Institute for Marine Biology and Coastal Research.

and is inhabited by 58% of South Carolina's population (26). Until 1941 the Santee, with an annual mean discharge of 525 m³ s⁻¹, was the fourth largest river on the U.S. east coast. However, with the completion of the Santee-Cooper Project at that time, most of the Santee discharge was diverted into the Cooper River in response to increased demands for hydroelectric power.

Unlike the Santee, the Cooper River was a tidally dominated stream, entirely confined to the coastal plain, with a net seaward discharge of only 2 m³ s⁻¹ (32). Charleston Harbor, South Carolina's largest port, is located at the confluence of the Cooper, Ashley, and Wando rivers (Fig. 2). The Ashley and Wando are still tidal streams with insignificant annual mean fresh water discharges of 2 and 7 m³ s⁻¹, respectively (32).

The Santee-Cooper Project, first envisioned in 1915, was finally completed in 1941 by the South Carolina Public Service Authority (PSA) under license (#199) from the Federal Power Commission (33). Dams were built on both the Santee and Cooper rivers. The Wilson Dam, constructed 140 km upstream on the Santee, created Lake Marion, South Carolina's largest lake with a 450 km² surface area. Lake Moultrie (245 km²) resulted from the damming of the Cooper River at Pinopolis, northwest of Charleston (Fig. 2). The diversion scheme came about as a means of effectively harnessing hydroelectric power from the Santee. The water elevation difference is only 6 m at the Wilson Dam between Lake Marion and the Santee, whereas the difference between Lake Moultrie and the Cooper at the Pinopolis Dam measures approximately 23 m. Therefore, lakes Marion and Moultrie were connected through a 12 km diversion canal through which, on the average, 88% of the Santee flow is channeled into the Cooper River. Thus, maximum advantage was made of the steep fall at Pinopolis. With

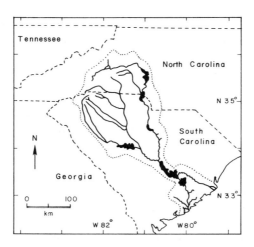

Figure 1. Santee River basin.

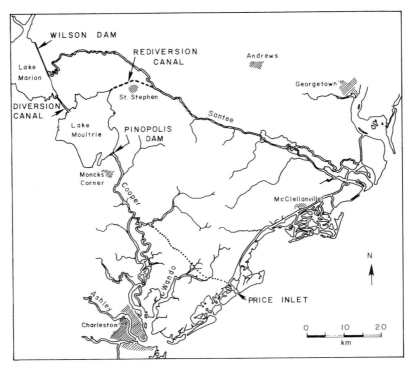

Figure 2. Features of the diversion and rediversion on the Santee and Cooper Rivers. The dotted line to Price Inlet indicates an early, now discarded rediversion route.

diversion, the lower Santee annual mean discharge decreased from 525 to 62 m^3 s^{-1}, whereas the Cooper discharge increased from 2 to 442 m^3 s^{-1} (32, 37).

The Jefferies hydroelectric plant was constructed at Pinopolis along with a fossil steam power plant having an even greater maximum power generating capacity. The Jefferies hydro and fossil steam plants make up 1.51% and 4.85%, respectively, of the generating capacity of all South Carolina power plants (27). Although a hydroelectric plant was installed at the Wilson Dam spillway, its power generating capacity is insignificant and accounts for only 0.02% of the state's total generating capacity (27). The generating capacities of some South Carolina power plants are listed in Table 1.

The 1938-42 Santee-Cooper Project, completed at a cost of $48 million (through 1944), was not the first attempt to connect the two rivers. The Santee Canal, a 35 km long navigation canal with 10 locks was completed in 1800 in the vicinity of the present diversion canal and Lake Moultrie (19, 33). It was, however, abandoned with the coming of the railroad. Although ease of navigation was not a major argument in favor of the Santee-Cooper Project, a 3 m deep

Table 1.   Installed generating capacity of South Carolina power plants in 10^3 kW as of June 30, 1974 (26).

Power Plants	Fossil Steam	Internal Combustion Turbine	Nuclear Steam	Hydro	Total	Rank
Oconee I & II (Duke Power)	−	−	1,732	−	1,732	1
Wateree (SCE&G)	770	−	−	−	770	2
H.B. Robinson (CP&LC)	−	−	769	−	769	3
Williams (SCE&G)	611	60	−	−	671	4
Darlington County (CP&LC)	−	630	−	−	630	5
JEFFERIES[a] (PSA)	412[b]	−	−	128[c]	540	6
Canadys (SCE&G)	473	16	−	−	489	7
WILSON LANDING[d] (PSA)	−	−	−	2	2	40
Total for South Carolina	3,781	1,318	2,501	888	8,488	−

[a]at the Pinopolis Dam

[b]fourth largest fossil steam plant in South Carolina

[c]third largest hydroelectric plant in South Carolina

[d]at the Wilson Dam

navigable channel is now maintained for 196 km upstream from Charleston (29). This navigable channel may in the future be extended another 60 km to Columbia (29).

IMPACT OF DIVERSION

As would be expected, the impact of diversion on the estuaries of the Cooper and Santee rivers was immense. With changes in the amount of fresh water and sediments supplied to the two regions, the salinity regimes, sediment deposition and erosion patterns, flooding characteristics, and floral and faunal communities changed drastically.

The salinity has been determined regularly in Charleston Harbor since at least 1923. Measurements at the Custom House Wharf on the lower Cooper River 8 km from the open ocean (Fig. 3) indicate an annual mean salinity drop from 31 ppt to 16 ppt in response to diversion (38).

The lower Santee, on the other hand, which is less readily accessed because of extensive backswamps, has been studied only to a limited extent. Accordingly, there exist no long-term salinity records from this estuary. Based on information on levee vegetation prior to diversion, it seems likely that the salinity at the mouth of the North and South Santee distributaries (Fig. 4) was approximately 1 ppt or less most of the year. After completion of the project, the salinity in the Santee distributaries increased sharply due to the lowered discharge. Although large temporal variations are common, salinities above 1 ppt can typically be measured 13 and 18 km upriver on the North and South Santee, respectively (5, 6, 31). However, during flood conditions, with discharge in excess of

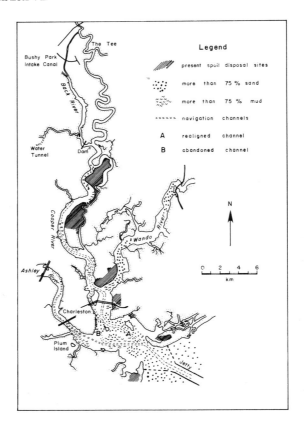

Figure 3. Charleston Harbor.

450 m³ s⁻¹, the salinity is less than 1 ppt downstream to the distributary mouths (9). Also during flood conditions, the fresh-salt water interface has been found to move approximately 3 km along the river channel over one half tidal cycle (9).

The indirect effect of the new estuarine salinity regime in Charleston Harbor turned out to be costly. With increasing discharge, the gravitational circulation became significant (23), and the Cooper River estuary changed from being vertically homogeneous to partially mixed (20), i.e. characterized by seaward directed net² flow in the surface layer, net upriver flow in the bottom layer, and a net flow toward the surface (18). Large quantities of fine-grained suspended sediments were transported into the Cooper River from the Santee basin. In the

---

² "net" implies a time-average over one or more complete tidal cycles.

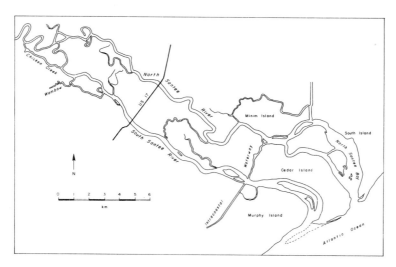

Figure 4. Distributaries of the lower Santee River. The dashed spit on Cedar Island represents the 1942 shoreline.

lower Cooper River and throughout Charleston Habor where the greatest salinity gradients occur, a portion of the suspended sediment load apparently settles out as agglomerates (21).

The mechanisms of sedimentation in estuaries are at best poorly understood (21), and field measurements generally fail to establish reliable suspended sediment budgets (15, 17). Fluvial silts and clays are deposited in Charleston Harbor but so are large amounts of marine sands transported landward with the net bottom current through the inlet (12, 13). The distribution of muds and sands in Charleston Harbor (3) is indicated in Fig. 3. However, independent of mechanism and source, the net result was extensive shoaling of Charleston Harbor beginning immediately following diversion.

Prior to diversion, the Corps of Engineers maintained a 9 m deep harbor navigation channel in Charleston Harbor, which only required removal of 61.2 × 10³ m³ of sediments at a cost of $11,600 annually (22). The Cooper Marl, which underlies Charleston Harbor, is now typically covered by 6-9 m and in places 18 m of unconsolidated sediments (3). To maintain the present 10.7 m navigation channel, the Corps must dredge 7,600 × 10³ m³ of sediments (11) at a cost of $5 million annually. The dredging costs can be expected to increase significantly with the imminent deepening of the navigation channel to 12.2 m and the construction of a new port facility. Further, the question of what to do with the spoil materials is presently a major, unsettled issue, which quickly requires a permanent solution.

To study the shoaling problems at Charleston Harbor, a hydraulic model was built at the Waterways Experiment Station in Vicksburg, MI, in 1947 and operated until 1953. Test runs indicated that realignment of the navigation channel to the north (Fig. 3) would decrease the need for maintenance dredging (25). Realignment took place in 1957, and this portion of the navigational channel has since required no dredging (20, 25). Although the realigned channel bypasses the worst shoaling area at the mouth of the Ashley River, present dredging costs still amount to $5 million and are rising.

The increased Cooper River flow is effective in flushing pollutants from Charleston Harbor. The Cooper water quality is reasonably good above the confluence with the Back River, 34 km upstream from the ocean (16). Below this point, pollution is moderate with high fecal coliform counts (16) and heavy metal concentrations (cadmium, chromium, mercury, lead, and iron) well in excess of U.S. Public Health Service standards for drinking water (16, 28). The recent installation of the Plum Island Sewage treatment plant (Fig. 3) has enhanced the Harbor water quality, primarily with respect to increased BOD and suspended solids removal (26).

Although most of Charleston's drinking water comes from the Edisto River, the Bushy Park intake canal, 55 km upstream from the Custom House Wharf, is also a source of drinking water. Due to the large Cooper River discharge it is only rarely that salt water intrudes into this water supply.

With large quantities of sediments channeled into the Cooper River, the flux of suspended sediments through the lower Santee has decreased drastically subsequent to diversion. This is clearly indicated by severe erosion of the spit on Cedar Island (1, 10, 14) (Fig. 4), a disappearance of land which by far exceeds the expected loss due to relative sinking of the South Carolina coastline (14). Still, the distributary mouths of the Santee remain shallow, usually no more than 3 m, and are not dredged. The only regular dredging occurs where the Intracoastal Waterway intersects the distributaries 8 km from the ocean (Fig. 4).

Because of the shallow water depths, it is likely that the Santee distributaries were vertically homogeneous with respect to salinity and density prior to diversion. In fact, the distributaries were probably totally fresh with the estuarine zone extending as a plume into the nearshore waters similar to the Amazon River (7). At present, the Santee estuary is partially mixed with a weak vertical salt stratification (5), implying a poorly developed gravitational circulation. However, during flood conditions, field measurements show that the Santee becomes vertically homogeneous (9) in contrast to previous speculations of salt-wedge conditions during high discharge (30, 31).

Stands of trunks of remnant trees flank the Santee distributaries, pointing to the increase in river salinity subsequent to diversion. In many cases, these tree trunks are now surrounded by smooth cord grass, *Spartina alterniflora,* which successfully competed for space along the distributaries with upriver salt intrusion. Dead tree trunks and *S. alterniflora* stands may be found upstream of the

U.S. 17 bridge on the South Santee but no closer than 2.5 km seaward of the North Santee bridge (Fig. 4). This is indicative of greater marine influences on the South Santee, in agreement with flow measurements (5), which show that typically only 15% of the Santee discharge finds its way to the Atlantic via the southern distributary.

The change in delta vegetation due to salinity intrusion was a major concern to Santee duck hunters and financial settlements were arranged with the land-owners (2). However, these losses are now easily outweighed by the lucrative Santee shellfish industry, which after diversion was developed under ideal conditions in the first few kilometers of the distributaries. Hard clams, *Mercenaria mercenaria,* have only been harvested for the past two seasons from 3 km^2 of clam producing beds in the Santee (Table 2). During the 1974-75 season, a record 30,805 bushels (1,086 m^3) or 73% of South Carolina's total harvest came from the Santee (8). At a typical price of $9 per bushel, the Santee clam industry is presently an annual $277,000 business and growing. Extensive oyster beds now also exist in the lower distributaries which increase the value of the Santee shellfish industry even further. However, as these beds are primarily harvested for seed oysters to be planted elsewhere along the coast, it is more difficult to evaluate the worth of the Santee oyster industry.

Table 2. Clams and oysters harvested in South Carolina in 10^3 bushels (8).

Season	Clams (*Mercenaria mercenaria*)		Oysters (*Crassostrea virginica*)
	Santee	State Total	State Total
1971-72	—	4.1	247.4
1972-73	—	8.6	301.0
1973-74	2.6	7.3	256.2
1974-75	30.8	42.2	318.5

REDIVERSION

The dredging of Charleston Harbor is a major concern. It is costly, and costs will probably increase further in response to pressures from the South Carolina State Ports Authority and the U.S. Navy to deepen the harbor in the future. The problem of disposition of dredge spoil also exists. Accordingly, the Charleston District of the U.S. Army Corps of Engineers has recently completed plans to redivert most of the Cooper River flow into the Santee in hopes that harbor shoaling will decrease. Test runs have been undertaken on a second hydraulic model at Vicksburg, MI; the final environmental impact statement (33) has been filed and approved; and the project allocation bill is presently pending action in U.S. Congress.

Major features of the rediversion project include the excavation of an 18.5 km long, 114 m wide (bottom width) canal from Lake Moultrie to the Santee River and the building of a new hydro power plant on the canal at St. Stephens (Fig. 2). The canal will consist of a 4.5 km entrance channel in Lake Moultrie, a 6.5 km intake canal to the powerhouse, and an 8 km tailrace canal to the Santee River. It will require construction of one railroad and three highway bridges, building of canal levees, relocation of power lines, design of an automatic fish lift, and the purchase of 9 km² of land (35). Rediversion is scheduled to be completed in four years depending on the annual funding level. The total project cost is $91 million (Table 3) according to a 1 October 1975 official estimate (34). However, the construction expenses can be expected to be much higher, as interest costs have not been included, and the cost estimate presently increases by 12.2% annually (34).

The need for rediversion indeed seems justified. However, the logical solution would be to channel most of the Santee River basin flow into the original alluvial channel at the Wilson Dam (Fig. 2), allowing a reasonable discharge into Lake Moultrie and the Cooper River. This alternative was quickly discarded (33) because of "reduced head for power generation at Wilsom Dam which makes it uneconomical." Secondly, the Corps judged it (33) too expensive, and perhaps not possible, to purchase energy to replace that lost from the Jefferies hydro plant, should an average of 357 m^3 s^{-1} of water be spilled into the Santee River at the Wilson Dam.

Instead, the approved plan calls for the Santee flow to be channeled into Lake Moultrie. From here, on the average 357 m^3 s^{-1} will flow through the rediversion canal to the Santee. The head at St. Stephens will be 15 m, which makes possible the construction of a hydro power plant with a generating capacity of 84,000 kW (35) or 1.0% of the state's total generating capacity. A minimum weekly average of 85 m^3 s^{-1} will in addition be discharged into the Cooper River. This discharge is low enough not to cause a significant gravita-

Table 3.   Cooper River Rediversion Project cost estimate projected to 1 October 1975 (33).

Item	Cost Estimate
Land & Damages	2,192,000
Relocations	6,583,000
Fish & Wildlife Facilities	2,618,000
Power Plant	43,416,000
Roads, Railroads & Bridges	516,000
Channel & Canals	25,231,000
Buildings, Grounds & Utilities	287,000
Permanent Operating Equipment	159,000
Engineering & Design	5,825,000
Supervision & Administration	4,173,000
Total Budgeted Rediversion Costs	$91,000,000

tional circulation in the Charleston River estuary but at the same time sufficiently high to provide cooling of the existing Jefferies steam plant without raising the temperature of the discharge waters above South Carolina legal limits (above 32.2°C or a temperature rise in excess of 2.8°C as compared to the ambient waters (33).

The impact of rediversion on the estuaries of the Cooper and Santee rivers is purely speculation. On the positive side, it seems likely that in the absence of large amounts of fluvial sediments, the shoaling of Charleston Harbor will slow down significantly. The transgression of marine sands into the harbor can be expected to decrease, as well, due to the lack of a developed gravitational circulation. The combined effect should result in less dredging, thus alleviating the need for new spoil disposal sites. Likely negative effects of rediversion on the Cooper include worsening of harbor and river water quality with less efficient estuarine flushing and longer residence times. This problem can and should be solved with adequate industrial waste and sewage treatment facilities; however, construction costs may be staggering. Additionally, the increased salinity of the Cooper River estuary could at times cause contamination of the fresh water at the Bushy Park intake canal. Even without rediversion, salt water has recently intruded within 4 km of Bushy Park, with bottom salinity measuring more than 18 ppt (24). The higher salinity is also likely to adversely affect the seed oyster beds in the Wando River by increasing predation pressure.

With rediversion the Santee should have an average discharge of 419 m^3 s^{-1} or 80% of the river's former flow. The distributary environment should in time revert to pre-diversion conditions. The long-term changes would include a renewed spit-development on South and Cedar islands and the possible halt of beach erosion in the Cape Romain area to the south of the distributary mouths. The extensive *Spartina alterniflora* marshes in the delta will eventually be succeeded, initially by freshwater grasses such as the reed grass *Spartina cynosuroides*.

There are also indications that the South Santee may eventually become the major distributary (9). Measuring from where Chicken Creek (Fig. 4) and the North Santee separate, the distances to the ocean are 28 km and 35 km along the South and North distributaries, respectively. With the steeper gradient along the South Santee and flood discharge measurements of 27% of the total flow through the South Santee (9), it is possible that rediversion could change the relative importance of the two distributaries.

The most adverse effect of rediversion, however, is the potential destruction of the Santee shellfish industry. With increased freshwater input into the Santee distributaries during most of the year, the lucrative hard clam production appears to be doomed in response to the projected low salinity (1 ppt) all the way to the river mouths. The lowered salinity along with the increased suspended sediment load would probably also destroy the Santee seed oyster industry and the nursery grounds of many fish species of commercial and recreational importance.

AN OPINION

According to Cronin (4), "Man's past effects on estuaries have been poorly and incompletely planned, unimaginative and frequently destructive. In view of the many important uses served by these waters. . .it is imperative that a new major human force be utilized in the future—the force of intelligent management." The many manipulations of the Santee River and resulting impacts on the Santee and Cooper estuaries, raise doubts about employment of *intelligent* management in this case.

After completion, at least $139 million will have been spent on diversion and rediversion, not counting the diversion-induced dredging expenses in Charleston Harbor. The end product will not be very different from conditions existing prior to diversion. However, is it necessary to spend $91 million on the rediversion canal project? The same results could be accomplished by allowing the Santee to resume its former course below Lake Marion, with or without a hydro power facility at the Wilson Dam.

According to an unpublished Corps design memorandum, the Cooper Rediversion Project benefit-to-cost ratio is 2.5 (33). The annual project benefits are set at $5.8 million for 50 years, while the annual amortization, interest, and other costs are computed to be $2.3 million (based on a 1974 construction estimate of $78.4 million). However, this annual cost figure assumes an unreasonable 3.25% interest rate. Using a more realistic interest rate of 10%, the annual project cost would rather become $7.9 million (based on $78.4 million) and the benefit-to-cost ratio 0.7. The economic arguments in favor of the planned rediversion are obviously questionable.

Intelligent management of the two estuaries does indeed point to the need for rediversion, in spite of the potential destruction of the Santee shellfish industry. Financial and practical considerations in Charleston Harbor justify the project. Estuarine impacts will be the same whether the Santee River flows through the rediversion canal or across the Wilson Dam. The latter solution seems preferable esthetically and economically. The potential loss in hydro power revenue would be balanced by savings due to decreased need for dredging in Charleston Harbor.

Large scale manipulations of river systems such as the lower Santee, is not an uncommon example of coastal management. A river system was altered; with time, problems developed; solutions were sought; and schemes have been devised to undo the error. *Intelligent* coastal management, on the other hand, would have been to leave the Santee alone in the first place. This case study illustrates the need for thorough understanding of long as well as short-term consequences before man-made modifications of natural systems are undertaken. Let us learn from our mistakes.

REFERENCES

1. Aburawi, R.M. 1972. Sedimentary facies of the Holocene Santee River delta. M.S. thesis. Department of Geology. University of South Carolina. 96 p.

2. Bohlen, A. W. (March 28) 1946. Interoffice communication on "Salinity-Lower Santee River" directed to J. H. Moore, chief engineer. South Carolina Public Service Authority. 20 p.

3. Colquhoun, D. J. 1972. Physical and chemical identification of sediments deposited in Charleston Harbor. Report of Charleston Harbor, South Carolina estuarine values study, contract DACW60-71-C-0007, submitted to the United States Army Corps of Engineers, Charleston District. 53 p.

4. Cronin, L. E. 1967. The role of man in estuarine processes, p. 667-689. *In* Estuaries. G. H. Lauff (ed.). Amer. Assoc. Adv. Sci.

5. Cummings, T. R. 1970. A reconnaissance of the Santee River estuary, South Carolina. Report prepared by United States Geological Survey, Water Resources Division, Columbia, SC. 96 p.

6. Environmental Protection Agency. 1973. Santee River basin, a review and summary of available information on physical, chemical, and biological characteristics and resources. Surveillance and Analysis Division, Athens, GA. 127 p.

7. Gibbs, R. J. 1970. Circulation of the Amazon River estuary and adjacent Atlantic Ocean. J. Mar. Res. 28: 113-123.

8. Gracy, R. C. 1975. Survey of South Carolina's clam resource. PL88-309. National Marine Fisheries Service. In press.

9. Kjerfve, B., and the Marine Science 581x class. 1975. Hydrography of the Santee River distributaries during high discharge. Manuscript in preparation.

10. Lee, C. W. 1973. The recent lithofacies and biofacies of North Santee Bay and the lower reaches of North Santee River, South Carolina, U.S.A. M. S. thesis, Department of Geology, University of South Carolina. 105 p.

11. Little, A.D., Inc. 1974. Analysis of environmental impacts in Charleston Harbor, S.C. 3 volumes. A report submitted to accompany an application of the South Carolina Ports Authority to the Charleston District, United States Army Corps of engineers, for a dredge and fill permit relative to development of a port on the Wando River.

12. Meade, R. H. 1969. Landward transport of bottom sediments in estuaries of the Atlantic coastal plain. J. Sedimentary Petrology 39: 222-234.

13. _____. 1974. Net transport of sediment through the mouths of estuaries: seaward or landward? Mémoires de l'Institut de Géologie du Bassin d'Aquitaine 7: 207-213.

14. Mullin, P. R. 1973. Facies of North Santee inlet and contiguous areas. M.S. thesis. Department of Geology, University of South Carolina. 52 p.

15. Neiheisal, J., and C. E. Weaver. 1967. Transport and deposition of clay minerals, southeastern United States. J. Sedimentary Petrology, 37: 1084-1116.

16. Nelson, F. P. (editor). 1974. Cooper River environmental study. Report No. 117. South Carolina Water Resources Commission, Columbia, SC. 164 p.

17. Pierce, J. W., D. J. Colquhoun, and D. D. Nelson. 1974. Suspended sediment flux, Charleston Harbor estuary to shelf, southeastern United States. Mémoires de l'Institute de Géologie du Bassin d'Aquaitaine 7: 95-102.

18. Pritchard, D. W. 1967. Observations of circulation in coastal plain estuaries. *In* Estuaries, p. 37-44. G. H. Lauff (ed.). Amer. Assoc. Adv. Sci.

19. Savage, H. 1956. River of the Carolinas, the Santee. Rinehart Co., New York, NY. 435 p.

20. Schubel, J. R. 1971. Charleston Harbor: an example of how man altered an estuary's shoaling characteristics by increasing its freshwater input, p. IX 1

- IX 8. *In* The estuarine environment, estuaries and estuarine sedimentation. J. R. Schubel, Convenor. American Geological Institute.

21. _____. 1971. A few notes on the agglomeration of suspended sediment in estuaries. Ibid, p. X 1 - X 29.

22. _____, and D. W. Pritchard. 1972. The estuarine environment. Part 1. J. Geol. Ed. 20: 60-68.

23. Schultz, E. A., and H. B. Simmons. 1957. Freshwater-saltwater density currents, a major cause of siltation in estuaries. Technical Bulletin 2. Committee on Tidal Hydraulics, United States Army Corps of Engineers, Vicksburg, MI.

24. Shealy, M. H., Jr., J. V. Miglarese, and E. B. Joseph. 1974. Bottom fishes of South Carolina estuaries—relative abundance, seasonal distribution, and length-frequency relationships. Technical Report 6. South Carolina Marine Resources Center, Charleston, SC. 189 p.

25. Simmons, H. B. 1966. Tidal and salinity model practice, p. 711-731. *In* Estuary and coastline hydrodynamics. A. T. Ippen (ed.). McGraw-Hill.

26. South Carolina Department of Health and Environmental Control. 1975. Santee-Cooper river basin water quality management plan. 469 p.

27. South Carolina State Budget and Control Board. 1974. Ninety-sixth annual report of the Public Service Commission of South Carolina 1973-1974. 70 p.

28. South Carolina Water Resources Commission. 1973. Wando River environmental studies. An interim report. Columbia, SC. 115 p.

29. Steele, W. J., D. H. Higgins, and W. C. Moser. 1973. The Cooper River, Santee-Cooper Lakes navigation system. Second Interim Report. Report No. 116. State of South Carolina Water Resources Commission. 66 p.

30. Stephens, D. G. 1973. Sedimentary and faunal analyses of the Santee River estuaries. Ph.D. dissertation. Department of Geology, University of South Carolina. 76 p.

31. _____, D. S. VanNieuwenhuise, W. H. Kanes, and T. T. Davies. 1975. Environmental analysis of the Santee River estuaries: thirty years after diversion. Southeastern Geology 16: 131-144.

32. United States Army Corps of Engineers, Charleston District. 1966. Survey report on Cooper River, SC. (Shoaling in Charleston Harbor). Appendix D. Model Studies. Prepared by Bobb, W.H., and H.B. Simmons.

33. _____. 1975. Cooper River Rediversion Project, Charleston Harbor, South Carolina. Final Environmental Statement. 201 p.

34. _____. 1975. Project cost estimate (PB-3), general construction, Cooper River, Charleston Harbor, South Carolina. Approved and projected for 1 October 1975. 3 p.

35. _____. Undated. Pertinent data: Cooper River Rediversion Project, Lake Moultrie and Santee River, South Carolina. 3 p.

36. United States Department of Agriculture. 1973. Santee River Basin, water and land resources, North Carolina-South Carolina. Economic Research Service, Forest Service, and Soil Conservation Service. 167 p.

37. United States Geological Survey. 1975. Water resources data for South Carolina 1974. U.S. Department of the Interior, Water Resources Division, Columbia, SC. 190 p.

38. Zetler, B. D. 1953. Some effects of the diversion of the Santee River on the waters of Charleston Harbor. Trans. Amer. Geophys. Union 34: 729-732.

ZONING –

A RATIONAL APPROACH

TO ESTUARINE REHABILITATION AND MANAGEMENT[1]

J. R. Schubel
Marine Sciences Research Center
State University of New York
Stony Brook, New York 11794

ABSTRACT: The great value of the estuarine zone lies in the multiplicity of uses it serves, but herein also lies its vulnerability. Estuaries can however, serve their many masters, and still remain aesthetically pleasing and biologically productive environments if they are zoned. The ability of an estuary to tolerate each "environmental insult" before suffering significant and persistent ecological or aesthetic damage varies not only from estuary to estuary, but also in different parts of a given estuary. And, within any segment of an estuary it varies temporally. While uniform, stringent, standards for the disposal of wastes may protect the environment, they are environmentally naive and may be wasteful of valuable natural resources; resources that should be used. Estuaries should be zoned. The segments allocated to the various uses are not all mutually exclusive, and the spatial boundaries of the various zones should be defined as a function of time.

The need for a particular type of zoning is imminent; zoning for placement of dredged spoil.

INTRODUCTION

This is an essay—an interpretative literary composition dealing with its subject from a personal point of view. It is an essay on a new approach to estuarine management by an oceanographer who has almost no management experience, but who has spent his professional career studying estuarine processes and problems, who has repeatedly attempted to translate the findings of his research into

[1] Contribution 149 of the Marine Sciences Research Center of the State University of New York, Stony Brook, New York 11794

a form usable by managers and planners, and who has been chronically frustrated by the fact that little of his work, or that of his colleagues, has ever been incorporated into environmental management.

USES AND ABUSES OF THE ESTUARINE ENVIRONMENT

It is in the estuary where man has his most intimate contact with the marine environment and his greatest impact on it. The population density of estuarine areas is approximately twice that of the remainder of the country (7), and these areas serve as the sites of heavy concentrations of industry. Approximately 40% of all manufacturing plants in the country are in *coastal* counties, a large percentage of which are in estuarine areas. Man uses estuaries as transient receivers for his industrial, municipal, and human wastes. He also utilizes them for their extractable resources, both mineral and biological; for shipping and transportation; as a source of industrial-process water; as a source of cooling water for factories and power plants; and for military activities. And he uses estuaries for recreation—for *re*-creation. The great value of estuaries lies in this multiplicity of uses, but herein also lies their vulnerability.

All of these uses are probably "legitimate." Few, if any, of them are inherently prohibitive, and most, perhaps all, need not ever be seriously restrictive below some threshold level of activity. But the demands that the various activities make on the estuarine zone are sometimes in conflict. The conflict arises mainly between those activities, primarily fisheries and recreation, that require the maintenance of certain water quality "standards," and other activities for which water "quality," as we generally define it, is relatively unimportant; activities which in fact frequently lead to a degradation of existing water "quality." There is also a conflict between military and civilian uses that results from the setting aside of large areas of some estuaries for ordnance testing and training. Examples are the use of the upper Potomac by the Dahlgren Proving Ground (U. S. Navy) and portions of upper Chesapeake Bay including part of the Bush River estuary and Romney Creek by the Aberdeen Proving Ground (U. S. Army). Schubel (8) estimated that civilian activities are restricted by military regulations in 17% of the Chesapeake Bay estuarine system. This is nearly twice the cumulative area of shellfish bars closed because of "pollution."

Estuaries *do* have a capacity to assimilate some wastes whether heat, sewage, or dredged spoil, without suffering persistent ecological damage. They *can* support certain levels of shipping and transportation without a significant loss of commercial and recreational fish landings. Some minerals *can* be extracted from the estuarine zone without smothering shellfish beds. The biological resources of estuaries *can* be harvested to certain levels without affecting future yields. Estuaries can serve all of these uses and still remain aesthetically pleasing environments for man's recreation. But an estuary's capacity to support these varied activities is finite. The ability of an estuary to tolerate each "environmental insult" before suffering significant and persistent ecological damage or aesthetic degradation varies not only from estuary to estuary, but also from segment to segment within a given estuary as well. Within any segment of an estuary it varies temporally.

It is apparent that in many estuaries, or at least segments of them, we have exceeded this capacity. It is also apparent, as Gross (5) points out in his paper in this session, that despite the very large expenditures for pollution abatement over the past several decades "the few available long-term data on water quality show little evidence of significant improvement in estuarine water quality in the United States." The ineffectiveness of our efforts is attributable to a variety of factors. Among the more important are our failures to develop regional plans for the management of estuarine systems, and to implement management strategies that have a sound scientific basis. It is not surprising to scientists that the millions of dollars spent on waste treatment in the District of Columbia have had little effect on the water quality of the upper Potomac.

Decisive environmental action does not await a more detailed understanding of estuaries as Holden (6) suggested was the case for Chesapeake Bay. There are certainly many unanswered scientific questions; there always will be. But many of the important features of the prevailing biological, chemical, geological, and physical processes that characterize important estuarine systems are known and understood. Scientific predictions can be made. In many respects scientific information has developed at a faster rate than management's ability to utilize it. Managers and planners rarely have the scientific expertise required for the formulation of plans for effective environmental management, and scientists have been derelict in translating the results of their investigations into a form readily usable by managers and planners. As a result planners have been disillusioned with academicians and have turned to consultants for guidance. The typical planning documents that have resulted are of little value. They form a seemingly endless series of reports outlining the studies that need to be done, but they are of little consequence in effecting solutions.

For a significant improvement in the effectiveness of the management of estuaries and in their condition, a new approach to estuarine management and pollution abatement is required. At a time of fiscal exigencies when we are all being asked to assess the effectiveness of our programs and our personnel, Federal, state, and county environmental protection agencies should do the same. Continuation of present policies, or even acceleration of these policies, will result for many estuaries in little improvement of water quality as generally measured, and will place undue restrictions on estuarine usage. Uniform, invariant regulations and standards whether they are for temperature, bacteria, nutrients, dredged spoil, or turbidity are environmentally naive. The only justification for their enactment is that it simplifies enforcement. A uniform speed limit of 40 km/hr is as irrational as one of 175 km/hr is irresponsible. Uniform estuarine regulations have proven to be ineffective, and are wasteful of natural resources— resources that should be used and used responsibly. The philosophy of those "environmental crusaders," bureaucrats, and politicians who espouse cessation as the solution to all of man's environmental problems is not viable. People live. They eat, they defecate, they procreate, and yes, they also need to re-create. They engage in these activities even during election years. This is not to imply that we should not insist on reasonable levels of waste treatment, on carefully supervised methods of dredging and spoil disposal, on controlled mining, on properly managed fisheries, and on reasonable thermal standards. We should. We should insist on more.

ZONING–A RATIONAL ALTERNATIVE

Estuarine systems should be zoned; zoned into a number of segments in which different water "quality" standards and criteria are applied. These standards and criteria should be consistent with the natural prevailing processes and with the uses that are perceived to be most important for each segment. To date, formal zonation of estuaries has been restricted largely to that associated with military activities, and major shipping channels. Man zones his terrestrial environment into residential and industrial areas, and sets aside portions of it as parks and forests for recreation. He identifies other segments of it for the disposal of his waste products. He does not make it an official policy to spread his garbage and trash uniformly over the landscape. He neither demands nor expects all parts of his terrestrial environment to be of equal "quality." Should he expect to be able to swim and harvest seafood in every part of every estuary? I think not. Segments of some estuaries *should* be designated as spoil disposal areas, as receiving waters for municipal and industrial wastes, as sinks for the heated effluents from power plants, as spawning and nursery areas, as military testing areas, and as fishing and recreational areas. Still others should be preserved, or at least conserved, in a "wild" state. These designations would not necessarily be mutually exclusive; there would be considerable overlap. In addition, some zones might even receive seasonal designations. The identification of a finfish spawning area certainly would not preclude its use as a recreational area for man; indeed many of the activities would probably be similar. But one should not build a large power plant with a once-through cooling system that would use a large fraction of the available water in an important spawning or nursery area—and a STUDY is not required to establish this. If we accept that the primary reasons for "managing" estuaries are to protect their biological resources and to conserve their recreational and aesthetic values, then certain activities should be restricted more severely in some areas than in others and also during those periods when organisms are most vulnerable; presumably during the egg and larval stages.

In one sense, the zoning of estuaries will be more difficult than zoning the terrestrial environment because of the reactivity and mobility of the medium. However, once the desired uses of a segment have been selected, implementation should be simpler since, in general, the water and most of the bottom are publicly owned. The objective however, is essentially the same. Zoning is a formal restriction on use and constitutes a police power. The primary purpose of zoning is to manage. But manage for what? . . . for whom? Management must be directed at some goal or goals if it is to be effective. Good managers, like good scientists, must set significant but realistic goals—goals which if attained will produce worthwhile and desired results, and goals which have a reasonably high probability of being attained. Environmental management is an exercise in decision theory, and the stakes are too high to leave to any one, or even several, special-interest groups.

A prerequisite to the establishment of any zoning plan then is the assignment of priorities to the various uses for the specific areas under consideration. This is the most difficult task in any zoning procedure. In terms of gross monetary return the most "important" uses of the estuarine zone are for military activities, for shipping, and for industry (7). However, the monetary values of commercial fisheries and of recreational activities are also very high, although they are much more difficult to assess. And if communication with nature is indeed one of man's ultimate sources of happiness (4), then the recreational and aesthetic value of estuaries cannot be measured in dollars and cents.

The establishment of priorities clearly involves not only scientific inputs, but social and economic inputs as well. The decisions are in large part value judgments, and natural scientists have no peculiar talents for making such decisions. Scientists can neither determine incontestably what uses of an estuary are most "important," nor even which are most desirable. Through science, we can learn to understand estuaries and even to control them in part, but scientists cannot unequivocally and decisively determine the ways in which we should use and control them; neither can politicians, nor "environmentalists," not alone. These decisions must be made by appropriate governmental agencies in response to the needs and desires of the public.

Effective estuarine zoning must not only take into account present and potential uses of a particular segment but must also recognize existing uses of the contiguous coast. For example, one could prohibit point source outfalls in a particular segment by zoning the receiving waters, but such action would have little effect on water "quality" if there were large adjacent non-point sources, from, for example, agricultural runoff or septic field drainage. In many estuaries, these non-point sources have a much greater impact on water "quality" than do point sources.

Formulation and adoption of a comprehensive zoning plan for estuaries would proceed through the same general steps used for zoning of the land. These include:

    I. Determine uses and assign priorities both regionally and for smaller segments.

   II. Formulate goals and objectives.

   III.   Collect and interpret available data, conduct appropriate water surveys:
           Hydrography
           Biota
           Sediments
           Topography
   IV.   Prepare the formal zoning plan.
   V.   Hold public hearings.
   VI.   Prepare and adopt the zoning plan (ordinance) as a legal document.
   VII.   Administer the zoning ordinance.

The zoning ordinance would consist of a series of maps delimiting the various zones and a text. The text would explain the goals and objectives of the zoning system and the rationale behind the designation of the various zones. A detailed discussion of zoning procedures is beyond the scope of this paper; these have been described at some length in a variety of publications (12) and will not be commented upon further.

ZONING FOR DISPOSAL OF DREDGED SPOIL—A CRITICAL NEED

The need for one particular type of zoning is imminent; zoning for disposal (placement) of dredged spoil. The rapid sedimentation rates characteristic of estuaries pose a serious immediate and enduring threat to one important estuarine activity—shipping—and therefore to the "quality" of many people's lives. Most of the nation's major ports are located in estuaries, and in fiscal terms shipping is the second most "important" use of estuaries (7). According to the Baltimore Port Authority (3), approximately one-half of all jobs in Maryland are dependent either directly or indirectly on the Port of Baltimore. While this figure may be inflated, it is clear that disruption of the activities of the Port of Baltimore, or of any other major port, results in serious economic perturbations.

Shipping is an important and legitimate activity. Most shipping channels require periodic dredging even to maintain their project depths. The intensity of the dredging, and the disposal of the dredged materials have created a great deal of concern, discussion, and speculation about the impacts of these activities on the "quality" of the estuarine environment. The magnitude of dredging activity in the United States is staggering. According to Boyd et al. (1), there are currently about 35,000 km of waterways and 1,000 harbors (including the Great Lakes) that must be kept open to support the nation's waterborne commerce. Each year approximately 230,000,000 m^3 of maintenance dredging is carried out. An additional 61,000,000 m^3 of material is dredged in conjunction with new projects, or to increase the capacity of existing systems.

There have been several recent reviews of the effects of dredging and spoil disposal on the estuarine milieu and biota (10, 11). Shipping channels occupy a very small fraction of the total area of most estuaries and even if these channel areas were totally sacrificed—which is very improbable—it is unlikely that the losses would be biologically significant; and in any event the economic benefits of the channels probably far outweigh any potential "environmental" losses. It is clear that the greatest potential environmental impact of dredging is not from the actual removal of the material, but rather from the disposal of it. Any effects of disposal are clearly a function of the mass and character of the material, the method and time of disposal, and the character—physical, chemical, biological, and geological—of the host environment. Assessment of the probable impacts of disposal depends upon a knowledge of all of these factors.

In an attempt to mitigate the impacts of spoil disposal on the aquatic environment, Federal agencies have established chemical criteria for determining the acceptability of dredged materials for placement in open waters. At the time of this writing the criteria are being re-evaluated. The criteria presently in use were intended to be "environmentally conservative," but they appear to be unduly restrictive with respect to certain designated parameters, and completely disregard a number of other potentially important contaminants including PCB's, pesticides, and others. The criteria clearly do not have a sound scientific basis (2, 9).

Criteria governing the disposal of dredged materials should not be based on total *concentrations* of contaminants, but rather they should be based on the total *masses* of contaminants in the dredged material that are available for biological uptake; the masses of the reactive fractions. The elutriate test is an attempt to assess the concentration of the available fraction, but the test appears to be of little value in predicting long-term ecological impact. Even with the formulation of "appropriate" criteria and standards for placement of dredged materials, decisions on dredging and spoil management should be based on the biological, chemical, geological, and physical characteristics of the particular estuary. The uniform application of Federal criteria and standards has little merit other than simplicity of enforcement.

Estuary-wide dredging and spoil disposal programs should be developed to ensure that maintenance channel dredging can be carried out without prolonged delays. The plans should also be flexible enough to provide a mechanism for decision making on requests for other types of dredging permits. Such plans should include the designation of a variety of types of sites (overboard, diked, etc.) for disposal of different types (quantities and "qualities") of dredged materials. Not all dredged material is "spoil," and certain types of dredged materials may have a greater environmental impact if disposed of in aerobic diked areas than if disposed of by more conventional overboard methods within oxygen-deficient areas of an estuary. The loss of valuable fringing wetland areas through filling must also be controlled. There is little doubt that scientifically defensible regional disposal plans could be developed. Whether such plans would be politically acceptable is quite another matter.

If regional dredging and spoil management plans are not developed promptly, the activities of a number of major ports may be significantly affected, resulting in serious economic perturbations. The observation that a number of our major ports are poorly located is to some extent correct but the suggestion that they should be moved is naive at best. Major ports could not be relocated without serious economic upheaval, and the lead time to implement any such proposals would be decades. The growth of some large ports located near the heads of estuaries should, however, probably be controlled. Baltimore may be such a port.

CONCLUSIONS

If a new approach to estuarine management is not adopted, there will probably continue to be little evidence of improved water quality even if standards are made more stringent and expenditures for pollution control are increased. And the true value of the estuarine zone will continue to be diminished not only because of loss in water "quality," but also because of increased restrictions and prohibitions on non-recreational and fisheries uses of estuaries. A regional estuarine management plan should be developed for each estuarine system that is based on the prevailing biological, chemical, geological, and physical processes that characterize that system. The management plan should be based upon a zonation of the estuary which is compatible with the uses of the adjacent coast. The development of an effective estuarine zoning plan depends upon the assignment of priorities to the various uses, and a partitioning of these uses among various segments of the system. Zonation does not eliminate the need for good waste treatment and environmental standards and criteria; rather it would replace the present indiscriminate approach which is not technologically, scientifically, or economically sound with a concept that calls for adjustment of water quality criteria and standards to characteristic processes of the environment and to the uses of the environment that are perceived to be most important.

ACKNOWLEDGMENTS

Preparation of this report was supported by the Marine Sciences Research Center of the State University of New York. I thank H. H. Carter for his helpful comments and his editorial assistance, and D.W. Pritchard for stimulating discussions over the years. This does not imply that these individuals necessarily agree with all of the ideas expressed in this essay.

LITERATURE CITED

1. Boyd, M. B., R. T. Saucier, J. W. Kelley, R. L. Montgomery, R. D. Brown, D.B. Mathis, and C.J. Guice. 1972. Disposal of dredged spoil, problem identification and assessment and research program development. Technical Rept. #72-8, U.S. Army Engineer Waterways Experiment Station, Vicksburg, Mississippi. 121 p. plus plates and appendix.
2. Carpenter, J. H. 1975. Limiting factors that control dredging activities in the estuarine zone. *In* Estuarine Pollution Control: A National Assessment.

Proceedings of Conference. To appear in June 1976. (Prepared for EPA report on the condition of the nation's estuaries).

3. Douglas, H. 1971. Maryland Port Authority, personal communication.
4. Dubos, R. 1968. So Human an Animal. Charles Scribners Sons, New York. 267 p.
5. Gross, M. G. 1976. Estuarine cleanup—can it work? p. 3-14. *In* M. L. Wiley (ed.) Estuarine Processes, Academic Press, N.Y.
6. Holden, C. 1971. Chesapeake Bay: "Queen of Bays" is a rich commercial and recreational resource, but march of progress imperils her health. Science 172:825-827.
7. National Estuarine Pollution Study. 1970. Report of the Secretary of the Interior to the United States Congress. Document no. 91-58. U.S. Government Printing Office, Washington, D.C. 633 p.
8. Schubel, J. R. 1972. The physical and chemical conditions of Chesapeake Bay. Jour. Washington Acad. Sci. 62:56-87.
9. _____. In press. Sediment and the quality of the estuarine environment: some observations. To be published in "Fate of Pollutants," a book in the John Wiley series "Advances in Environmental Science and Technology."
10. _____, and R. H. Meade. (in press). Man's impact on estuarine sedimentation. *In* Estuarine Pollution Control: A National Assessment. Proceedings of Conference. To appear in June 1976. (Prepared for EPA report on the condition of the nation's estuaries).
11. Sherk, J. A., J. M. O'Connor, D. A. Neumann, R. D. Prince, and K. V. Wood. 1974. Effects of suspended and deposited sediments on estuarine organisms, Phase II. Natural Resources Institute, University of Maryland, Ref. no. 74-20. 267 p.
12. U. S. Department of Commerce. 1966. Zoning for small towns and rural counties. A publication of the Economic Development Administration, U.S. Government Printing Office.

NUTRIENT CYCLING IN ESTUARIES

Convened by:
James J. McCarthy
The Biological Laboratories
Harvard University
Cambridge, Massachusetts 02138

In recent years the development of new approaches, both conceptual and methodological, has provided impetus to studies which address the regulation of productivity in coastal and estuarine waters. In particular, the issue of nutrient cycling and its impact on plankton production is currently being addressed on several fronts, and consistent with the theme of this symposium, we wish to present the recent findings of five research groups which are attacking different aspects of this problem.

The first paper is that of Dunstan and Atkinson, and it serves as a fitting introduction by reminding us that cycling processes which are grand in scale, both in time and space, such as hydrodynamic intrusions from offshore, cannot be ignored in comprehensive nutrient budgets for coastal waters. They discuss the significance of the intrustion phenomenon, nitrogen fixation, and river input to the nitrogen budget of the Georgia Bight, and compare their findings with those of others working in both the open ocean and estuarine regions.

Many estuaries receive waste discharges from metropolitan centers, and given the average nitrogen to phosphorus ratio in sewage, one might expect that if a major nutrient were to "limit" primary productivity in these waters, it would be nitrogen. Taft and Taylor demonstrate, however, that phosphorus may be of significance in regulating phytoplankton productivity within the Chesapeake Bay. The cycle of phosphorus within this estuary is particularly interesting in that over an annual cycle the maixmum orthophosphate concentrations within the water column occur in the summer. This paper also addresses the question of phosphomonoesters as a source of phosphorus for estuarine phytoplankton.

Within the last decade some investigators have directed attention to urea as a potentially significant source of nitrogen for estuarine and marine phytoplankton. Webb and Haas have examined in detail some of the more intriguing aspects of this problem such as the size distribution of phytoplankton which are capable of utilizing urea, diel periodicity in their capacity, the fate of urea carbon, and the overall significance of nitrogenous nutrition derived from urea.

In balancing the equations for nitrogen and phosphorus flux within coastal and estuarine planktonic systems, one has to consider exchange processes between the sediment and the overlying water, and in particular the vertical transport of soluble forms of nitrogen and phosphorus which can be utilized by phytoplankton and bacteria. Hartwig has measured the flux across the sediment-water interface on a coastal sand bottom, and he puts his data into perspective by considering the horizontal movement of water through his study area. His results are both timely and provocative.

We conclude this program with a discussion of the extensive field efforts of Furnas, Hitchcock, and Smayda in their study of nutrient (primary nitrogen) availability, utilization, and supply in Narragansett Bay. This contribution shows dramatically the dynamic nature of nutrient cycling, and it demonstrates that even when, with the best of available techniques, one quantifies all of the processes which one thinks to be important in nutrient cycling, the picture remains incomplete.

SOURCES OF NEW NITROGEN FOR THE

SOUTH ATLANTIC BIGHT

William M. Dunstan
Larry P. Atkinson
Skidaway Institute of Oceanography
P.O. Box 13687
Savannah, Georgia 31406

ABSTRACT: The region off the coast of the southeastern U.S. is characterized by a broad shallow shelf bounded on east by the Gulf Stream and on the west by a coastal plain drained by several large rivers. Because of its geography, dynamics of production in this area is different from other intensively studied coastal areas like Southern California, Oregon (Columbia River), Gulf of Maine and Long Island Sound. The Georgia Bight is particularly suited for studies concerned with the sources of new nitrogen for primary production. Using recent field measurements, we will examine in detail the mechanisms whereby new nitrogen enters the system from three sources—runoff, fixation and intrusion.

While river input in certain areas may be negligible compared to the volume of coastal water, in the Georgia Bight runoff can replace as much as 10% of the shelf water in a year. Salt marshes have been considered as nutrient traps; however, our measurements from the Savannah River indicate that large concentrations of nitrogen particularly DON become available for shelf production.

Blue-green algae represent an impressive portion of the plant biomass on the shelf throughout the year, as indicated by filament counts from 20 stations occupied 4 times during the year. The nitrogen fixation by these plants is still small in comparison to the river runoff and intrusions.

Stefansson and Atkinson (1971) drew attention to the potential of the northerly flowing Gulf Stream to "intrude" on the shelf along the southeast U.S. coast bringing plant nutrients from deep water into the euphotic zone. Part of our ongoing research is concerned with predicting the frequency and magnitude of intrusions and associating the intrusion with primary and secondary production. Biological and chemical data from our recent cruises illustrate the potential magnitude of the intrusion and suggest a rapid but complex relationship between production and the changing concentrations of nitrate in the water column.

Using the production and biomass figures from Haines and Dunstan (1975) and the recent data presented here, the relative importance of the three sources of new nitrogen to the total nitrogen budget is calculated.

INTRODUCTION

The availability of nitrogen for phytoplankton growth is considered one of the major factors controlling production in coastal regions (11). While its regeneration may provide the bulk of that required to sustain production the uniqueness of a particular region is determined by the new nitrogen which enters the productive cycle of the area. The concept of characterizing primary production as "new" or "regenerated" was provided by Dugdale and Goering (4).

In this paper we will consider three sources of new nitrogen to the broad shallow shelf off the S.E. United States coast—river input, nitrogen fixation and instrusions. While Dugdale and Goering's primary concern was whether production is based on ammonia or nitrate, we consider that nitrogen introduced in all forms are new additions to shelf production, particularly in the case of that contributed by rivers.

The Georgia Bight is the shallow shelf area off Charleston, Savannah and Jacksonville which runs 100 km out to a depth of 50-60 meters where at the shelf break it is bounded by the north flowing Gulf Stream (Fig. 1). Relative to other oceanic or coastal areas data on the Georgia Bight is sparse and therefore we are presently involved in studies of hydrographic factors and their relationship to biological productivity on this shelf. The data presented here is taken from the results of six offshore and numerous nearshore cruises taken over the past two years. We will discuss these new data and what they indicate about the mechanisms for providing nitrogen. As studies progress there will be refinements in the quantitative aspects of the nitrogen inputs, nevertheless we have an important data base concerning the process involved.

Nitrogen fixation

The abundance and distribution of the blue-green algae *Oscillatoria (Trichodesmium) thiebautii* and *O. erythraeum* over the Georgia shelf is impressive. Filaments are present in all seasons, in the coastal, central and outer shelf and at times reach high levels of living biomass in sub-surface waters. Ignoring the bloom conditions indicated by visible surface scums, inverted microscope counts of 20-30 stations from each of four cruises revealed *Oscillatoria* in 43% of the September stations, 36% of the December stations, 57% of the April stations and 76% of the July stations (Fig. 2).

Using the most recently published rates of N-fixation of *Oscillatoria thiebautii* of .069 pg cell^{-1} h^{-1} at 1 meter (3), and our counts of filament/cell numbers at 1 meter depth averaged over the entire shelf (Table 1), we find that the average addition per year from N-fixation is 1.14 µg at N liter^{-1}. This value may be high since we considered the cell numbers at depths where *Oscillatoria* is

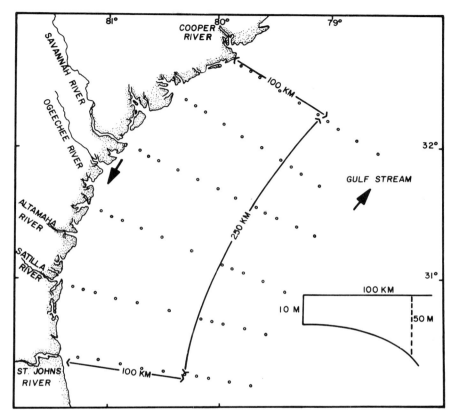

Figure 1. The Georgia Bight study area with stations occupied during four cruises in 1974. The major rivers are indicated as well as a generalized depth profile.

Table 1. Cell concentrations and potential nitrogen fixation based on four cruises in the Georgia Bight.

	Mean cells liter^{-1} $\times$ 10^4 over entire shelf	N fixation μg at^1 liter^{-1} yr^{-1}
September	1.91	.412
December	.52	.112
April	8.20	1.77
July	10.46	2.26
Mean	5.27	1.14

[1] Using Carpenters and McCarthy's mean rate cell^{-1} at 1 meter of .069 pg cell^{-1} h^{-1} and a 12 h day.

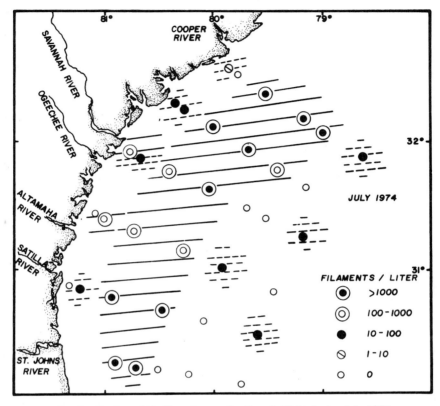

Figure 2. The distribution of *Oscillatoria* over the Georgia Bight during July 1975.

most abundant yet Carpenter and McCarthy (3) mention that higher fixation rates occur in stratified Caribbean water. But even if the fixation rates are doubled the vastly abundant *Oscillatoria* can potentially add only a small amount of nitrogen to the Georgia Bight system. This is not to underestimate its importance to the overall productivity of the Bight since it represents a relatively large biomass and as yet undetermined value as a food source to higher trophic levels.

River Input

The limited data available on the concentrations of nutrients in rivers in most cases only covers the inorganic forms of nitrogen (14). Studies which have considered dissolved organic nitrogen agree that it is often orders of magnitude higher than the inorganic forms; however, this input is often dismissed as being largely unavailable to plant production (c.f. 9, 6). Data from the Savannah River

Table 2.   Concentrations of nitrogen (μg at liter^{-1}) at the mouth of the Savannah River and at a station 8 km off the Georgia coast.

	Nitrate	Ammonia	Dissolved Organic N	Particulate N	Urea	Total N
Mouth of the Savannah River[1]	1.2	2.7	19	6.9	4.5	34.3
5 miles off Wassaw Sound, Georgia[2]	.6	1.1	15	8.2	1.0	25.9

[1] Mean of 2 sets of samples at 3 stations.
[2] Mean of monthly samples between February and July.

and from a station 8 km offshore indicate how significant the concentrations of nitrogen other than nitrate and ammonia can be in the Georgia Bight (Table 2). A portion of the dissolved and particulate organic nitrogen will enter the biological cycle at rates which vary depending on the compound and degree of microbiological activity. We do not know the rate at which nitrogen becomes available for primary production through remineralization in the nearshore sediment and water column but the large concentrations of organic nitrogen in rivers must be considered as important nutrient inputs into the Georgia Bight.

The major rivers flowing into the Georgia Bight are the Cooper, Savannah, Ogeechee, Altamaha, Satilla and St. Johns. The yearly run-off calculated for 1973 was 20.5, 13.7, 3.7, 15.2, 2.7, 6.6 liter^{-1} y^{-1} respectively for a total of 62.4×10^{12} liter^{-1}y^{-1} (14), a figure that can vary as much as 30% from year to year. The shelf area that we are considering has a volume of about 660×10^{12} liter^{-1} so that river input exchanges about 10% of the shelf volume in a year. If we assume that 80% of the total nitrogen (including some regeneration of DON and PN in the euphotic zone) coming in from a river is new nitrogen available for production the rivers could add about 3 μg at N liter^{-1} y^{-1}.

Intrusion

Stefansson et al. (12) and Blanton (2) drew attention to the potential for meander oscillations of the Gulf Stream (15) to intrude on the continental shelf along the southeast U.S. coast. These intrusions bring plant nutrients from subsurface waters into the euphotic zone.

The significance of the intrusion in bringing NO_3 and PO_4 onto the shelf was shown in many of the transects during a series of six cruises in 1973-1975. Fig. 3 shows typical profiles observed during an intrusion. Intrusions result in a stable two layer system with relatively high nutrients in the lower layer. Resulting increases in plant biomass were best shown in a transect in April 1975 over the narrower shelf off St. Augustine, Florida (Fig. 4). Increase in both integrated NO_3 and chlorophyll over the shelf break was again shown in a transect off Jacksonville in February of 1975 (Fig. 5).

Although the data (Figs. 3 thru 5) clearly indicate advection of nitrate and the resultant biomass increase on the shelf they do not reveal the real temporal and geographic extent of the nitrogen input. Our research is concerned with determining the frequency and magnitude of intrusions with a 70 km × 70 km station grid repeatedly sampled over a two week period.

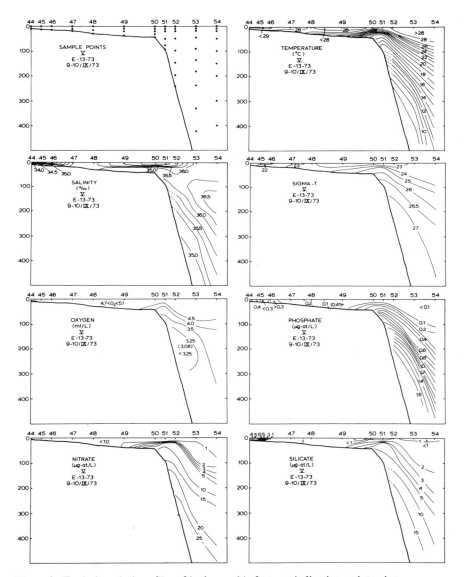

Figure 3. Typical vertical profiles of hydrographic features indicating an intrusion.

There are estimates available from earlier work (2, 13, 1) that indicate the intrusion process occurs on a 1-2 week period and can cover from 1/8 to 1/2 of the outer shelf area. Stefansson et al (12) did not observe intrusions in winter or spring off North Carolina and felt that they were partially wind controlled and characteristic of the summer and early fall.

For the purposes of comparison with the inputs of other new nitrogen to the shelf, we can estimate that the intrusion system can occur every 2 weeks over a

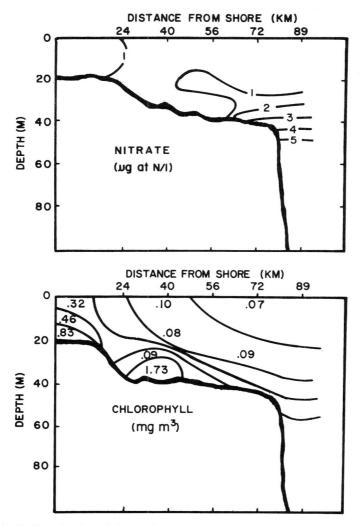

Figure 4. Profiles of chlorophyll *a* and nitrate in a transect off St. Augustine, Florida in April 1975.

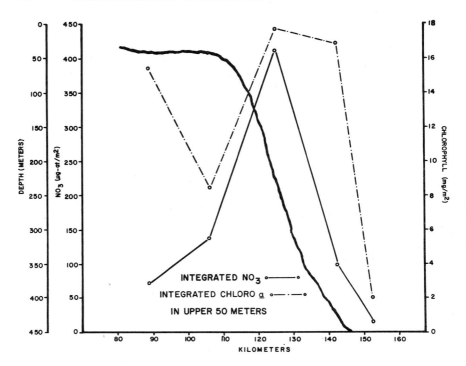

Figure 5. Integrated chlorophyll and nitrate at the shelf break off Jacksonville, Florida during February 1975.

4-5 month period or 10 times per year. The deeper Gulf Stream water, with a concentration of about 10 μg at N liter^{-1} covers about a third of the shelf in a shoreward direction with a north-south horizontal dimension, roughly estimated from the bottom nitrate data (Fig. 6), of 60 km or 25% of the north-south length of our Georgia Bight section. Using these conservative figures the intrusion adds about 13.3 μg at N liter^{-1} to the shelf waters in a year.

CONCLUSIONS

Considering the data available the intrusion process potentially provides most of the new nitrogen for primary production in the Georgia shelf. Using the mean productivity (209 g C m^{-2} y^{-1}; 7) and a carbon to nitrogen ratio of 10 to 1, it appears that intrusions could supply as much as 25% of the nitrogen required for production. This contrasts with other systems such as the Pamlico estuary (10) and the central North Pacific Gyre (5) where regeneration accounts for all but 5 or less percent of the nitrogen required for production. The Georgia shelf may be more similar to the Eastern Bering Sea (8) where 40-50% of the nitrogen is supplied by nitrate imported into the system.

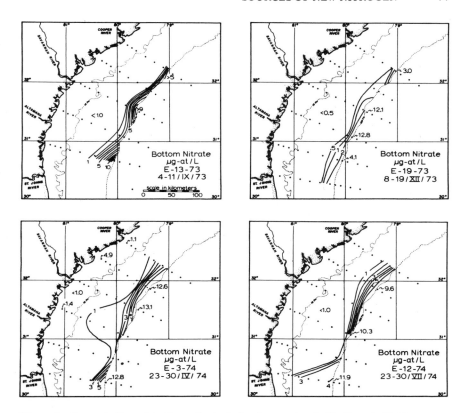

Figure 6. Concentration of bottom nitrate during four cruises in the Georgia Bight area.

It can be hypothesized then, that primary production on the Georgia shelf (excluding a nearshore zone of less than 15 km offshore) is controlled by periodic inputs of nitrate from sub-surface Gulf Stream water. This sort of system would operate like the "batch" type of production and the physiological condition of populations could be expected to substantially change within a period of a few weeks. Also in view of our current understanding of intrusions it is probable that production is very patchy and in contrast to some other areas, has a relatively large dependence (25%) on new nitrogen.

ACKNOWLEDGEMENTS

We thank H.L. Windom and W.S. Gardner for their help and acknowledge the technical assistance of G.L. McIntire, D. Wynne and J.E. Hosford. The research was supported by NSF Grant DES 74-14917 and ERDA 605 and ERDA 905. Shiptime was provided on the R/V *Eastward*, NSF GD-32560.

REFERENCES

1. Atkinson, L.P. 1975. Estimates of volumes, frequency and chemical and biological importance of Gulf Stream intrusions in the South Atlantic Bight. American Society of Limnol. and Oceanogr. 30th Annual Meeting, Halifax, N.S.

2. Blanton, J. 1971. Exchange of Gulf Stream water with North Carolina shelf water in Onslow Bay during stratified conditions. Deep-Sea Res. 18:167-178.

3. Carpenter, E.J., and J.J. McCarthy. 1975. Nitrogen fixation and uptake of combined nitrogenous nutrients by *Oscillatoria (Trichodesmium) thiebautii* in the western Sargasso Sea. Limnol. Oceanogr. 20(3):389-401.

4. Dugdale, R.C., and J.J. Goering. 1967. Uptake of new and regenerated forms of nitorgen in primary productivity. Limnol. Oceanogr. 12:196-206.

5. Eppley, R.W., E.H. Renger, E.L. Venrick and M.M. Mullin. 1973. A study of plankton dynamics and nutrient cycling in the central gyre of the North Pacific Ocean. Limnol. Oceanogr. 18(4):534-551.

6. Haines, E.B. 1974. Processes affecting production in Georgia coastal waters. Ph.D. Thesis, Duke University.

7. Haines, E.B., and W.M. Dunstan. 1975. The distribution and relationship of particulate organic material and primary productivity in the Georgia Bight, 1973-1974. Estuarine and Coastal Marine Science 3:431-441.

8. Hattori, A., and E. Wada. 1975. Assimilation and oxidation reduction of inorganic nitrogen in the North Pacific Ocean, p. 149-162. *In* D.W. Hood and E.J. Kelly (eds.), Oceanography of the Bering Sea. Inst. Mar. Sci., Univ. Alaska, Fairbanks.

9. Hobbie, J.E., B.J. Copeland and W.G. Harrison. 1972. Nutrients in the Pamlico River Estuary, N.C. 1969-1972. Water Resources Research Institute, U. N.C., Contribution 31. Pamlico Marine Laboratory, N.C. State University.

10. _____, and B.J. Copeland. 1973. Sources and fate of nutrients of the Pamlico River Estuary, N.C. Proceedings of 2nd International Estuarine Research Conference, Myrtle Beach, S.C., Academic Press, N.Y.

11. Ryther, J.H., and W.M. Dunstan. 1971. Nitrogen, phosphorus and eutrophication in the coastal marine environments. Science 171:1008-1013.

12. Stefansson, V., L.P. Atkinson and D.F. Bumpus. 1971. Seasonal studies of hydrographic properties and circulation of the North Carolina shelf and slope waters. Deep-Sea Res. 18:383-420.

13. Webster, F. 1961. A description of Gulf Stream meanders off Onslow Bay. Deep-Sea Res. 8:130-143.

14. Windom, H.L., W.M. Dunstan and W.S. Gardner. 1975. River input of inorganic phosphorus and nitrogen to the Southeastern salt marsh estuarine environment. Proceedings of 2nd International Estuarine Research Conference, Myrtle Beach, S.C., Academic Press, N.Y.

15. Von Arx, W.S. 1962. An introduction to physical oceanography. Addison-Wesley, Reading, Mass., 422 p.

PHOSPHORUS DYNAMICS IN SOME

COASTAL PLAIN ESTUARIES

J. L. Taft
W. Rowland Taylor
Chesapeake Bay Institute
Johns Hopkins University
Baltimore, Maryland 21218

ABSTRACT: Data from the literature and from our studies reveal an annual phosphorus cycle in some coastal plain estuaries with maximum concentrations in summer. In the Chesapeake Bay estuarine system phytoplankton biomass seems to be regulated by phosphorus availability in spring but not in summer when inorganic nitrogen availability appears to be a major biomass regulating factor. These conclusions are supported by phytoplankton nitrogen to phosphorus ratios and by alkaline phosphatase enzyme activity which is an indicator of the state of cell inorganic phosphate nutrition. No conclusions can be reached concerning phosphorus regulation of primary productivity during spring from our data because phosphorus monoesters are available in low concentrations to support sustained primary productivity.

INTRODUCTION

Identification of the major factors regulating phytoplankton productivity and regulating the level of phytoplankton biomass are principal objectives of many investigators seeking to understand plankton dynamics in estuaries. Short term nutrient enrichment experiments with phytoplankton assemblages from natural waters reveal the complexity of primary productivity regulation by indicating that inorganic carbon uptake may be unaffected, stimulated, or depressed by nutrient additions (7, 20 for example). However, since the upper limit for phytoplankton biomass is established in the euphotic zone by the total quantity of nutrient elements in the water, identification of major biomass regulatory factors is somewhat simpler. Instantaneous measures of dissolved and particulate nitrogen and phosphorus, and atom ratios of these elements in each pool, provide initial indications of the importance of each element to phytoplankton biomass regulation. In this communication the annual cycle of phosphorus in coastal plain estuaries and its role in regulating phytoplankton biomass are discussed through consideration of the literature and presentation of some recent data concerning the role of phosphorus monoesters in phytoplankton nutrition.

ANNUAL CYCLE

Newcombe and Lang (12) reported an annual cycle of phosphate concentration in Chesapeake Bay in which the highest concentrations were present in summer and the lowest concentrations were found in winter. Smayda (22) made a similar observation for lower Narragansett Bay, R.I., and cited the work of Ferrara who also observed a summer phosphate maximum in upper Narragansett Bay. Smayda termed the cycle "atypical" compared to the annual temperate ocean cycle in which the phosphate maximum occurs in winter. Jeffries (10) observed a summer phosphate maximum in Raritan Bay, N.J. Patten, Mulford and Warriner (13), working in the York River, Va., and lower Chesapeake Bay, observed a late summer and early fall phosphate maximum. Whaley, Carpenter and Baker (29), in an intensive study of upper Chesapeake Bay and its rivers and embayments, frequently found summer phosphate maxima in deep water. Hobbie, Copeland and Harrison (9) and Copeland and Hobbie (4) also observed summer phosphate peaks in the Pamlico River estuary. Soluble Reactive Phosphorus (SRP) data from some of these studies is shown in Fig. 1. The two data points for Delaware Bay (16), included with the Chesapeake Bay data, suggest a summer phosphate maximum occurs in Delaware Bay. A similar summer maximum occurs in some marshes draining into Delaware Bay (18, 21).

The mechanism for the summer phosphate influx into these estuarine waters is poorly understood. It appears the phosphate remineralized by bacterial activity from organic matter in the sediments reacts with iron (III) to form insoluble ferric phosphate which remains in the upper interstitial waters or precipitates on the sediment surface. As summer progresses, oxygen in deep waters is removed faster than it is replaced resulting in hypoxic and anoxic conditions which favor iron III reduction and, subsequently, phosphate release from ferric phosphate into the overlying water. Physical circulation transports the phosphate to the euphotic zone. It has been demonstrated that oxidation of anoxic sediments during sampling significantly decreases the dissolved orthophosphate measured in sediments rich in iron II (1, 27, 28). Furthermore, it has been suggested that inorganic phosphate is adsorbed to or desorbed from suspended sediment particles which thus "buffer" the inorganic phosphate concentration of the water (3, 17). The sites of phosphate binding to suspended sediment have not been identified but, if iron III were involved, phosphate adsorption should predominate in oxygenated waters and desorbtion should predominate in anoxic waters. Other binding sites may be involved if suspended sediment adsorbs and desorbs inorganic phosphate in oxygenated estuarine waters.

Once dissolved in deep water, most phosphate is transported vertically at the same rate as other ions to the euphotic zone where it is taken up by phytoplankton, so its presence is reflected as increased particulate and total phosphorus. We have discussed this observation elsewhere (24).

In fall, the surface waters cool, oxygen input to the deep waters exceeds demand, and phosphate release from the sediments is no longer detectable. Particulate phosphorus in the euphotic zone is flushed out or biologically transported out of the estuary, is remineralized in the water column, or settles onto the bottom. Net phosphate removal from the surface layers of the water column is reflected in decreased total phosphorus concentrations.

Release from the sediments represents the only identifiable inorganic phosphate source for open Chesapeake Bay waters. Most phosphorus entering from major tributaries is organic and must be remineralized to be useful to indiginous phytoplankton assemblages (2, 29). The wetlands draining into Chesapeake Bay are also potential phosphorus sources. At present, the phosphorus input from all wetlands cannot be estimated but the minimum input which could be detected

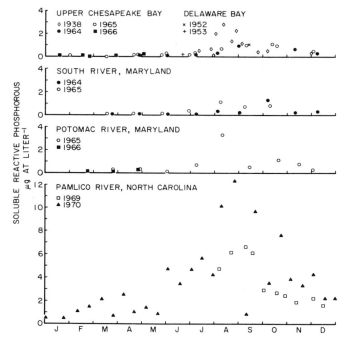

Figure 1. Concentrations of soluble reactive phosphorus during the year in five coastal plain estuaries.

by an instantaneous concentration measurement in open bay waters can be calculated. The Maryland portion of Chesapeake Bay contains about 30×10^{12} liters of water and has about 100 km^2 of tidal marsh draining into it either directly or through tributaries. The marsh water depth at high tide is usually less than 10 cm so a volume of about 1×10^{10} liters flows onto and off the marshes daily. As a yearly average, the total phosphorus concentration of bay waters is about 1 μg-at·liter^{-1} so the instantaneous mass of phosphorus in the water would be about 30×10^{12} μg-at P in the Maryland portion of the bay. A 10% change in total phosphorus, or 0.1 μg-at·liter^{-1}, in the open bay would require a marsh input of 3×10^{12} μg-at P dissolved in 1×10^{10} liters of water, if all marsh water were exchanged on each tidal cycle. The instantaneous concentration in the marsh effluent would then be 300 μg-at·liter^{-1}. However, an annual maximum of total phosphorus at the mouth of a low salinity marsh is about 5 μg-at·liter^{-1} (8). So, if this is typical of other marshes, the marsh input to Chesapeake Bay is small compared to the mass of phosphorus already present. Also, the marsh water may not be completely exchanged on each tidal cycle so some phosphate should be carried back into the marsh (8).

PHYTOPLANKTON BIOMASS REGULATION

The existence of an annual phosphorus cycle in estuaries suggests phosphorus should have maximal influence on phytoplankton biomass regulation in seasons of minimal availability, and minimal influence when it is abundant with respect to phytoplankton requirements. Relative nutrient requirements of the phytoplankton may be estimated by comparing the atom ratios of the elements already in the phytoplankton to ratios for phytoplankton in nutrient rich water. Nitrogen to phosphorus atom ratio for Chesapeake Bay phytoplankton was found to be 14.5:1 in nutrient rich water in winter (25). Ratios were highest in

Table 1. Vertical distribution of combined inorganic nitrogen and soluble reactive phosphorus at Chesapeake Bay station 834G on 28 May 1975. *Limit of detection.

Depth m	NH$_4$	NO$_3$	NO$_2$	SRP	N:P
	μg-at·liter^{-1}				
1	0.4	9.6	0.6	<.03*	387
5	0.9	8.0	0.5	<.03*	313
7	1.9	8.3	0.5	<.03*	357
9	4.6	6.9	0.4	0.06	198
11	5.2	7.6	0.4	0.06	220

Table 2. Vertical distribution of ammonium ion and soluble reactive phosphorus and the nitrogen to phosphorus atomic ratios at Potomac River station P-12 on 19 August 1975.

Depth m	NH₄	SRP	N:P
	μg-at·liter⁻¹		
0	1.2	0.3	4.0
2	0.9	0.3	3.0
4	0.5	0.2	2.5
6	2.0	0.2	10.0
8	19.9	2.0	10.0
10	21.8	2.3	9.5

spring (34:1 to 55:1) and lowest in summer (12:1 to 29:1). A spring nitrate maximum resulting from Susquehanna River input is an annual occurrence in Chesapeake Bay (2). Table 1 shows dissolved inorganic N and SRP concentrations and atom ratios for May 1975. The high N to P ratios in the water *and* in the phytoplankton suggest that phosphorus availability regulates phytoplankton biomass during the spring runoff period.

The lower particulate N to P ratios in summer reflect the decreased nitrate concentrations, as biological activity removes the spring runoff nitrate input, and a shift in the relative availability rates of inorganic nitrogen and phosphorus to the phytoplankton. Summer is the season of maximum phosphate availability in several estuaries and, in Chesapeake Bay, it is also a period of ammonium release from sediments. Table 2 shows ammonium and SRP concentrations for a station in the Potomac River estuary 12 miles from Chesapeake Bay in August 1975 (data from Chesapeake Bay stations were similar for August 1975). Ammonium and SRP concentrations were low in near surface waters and high in bottom waters. Nitrate and nitrite made negligible contribution at each depth and thus were omitted from Table 2. Fig. 2 is a vertical profile of salinity, SRP and oxygen concentration at the same station. In August, deep waters were anoxic and contained high concentrations of both ammonium and SRP. The inorganic N to P ratios (Table 2) were about 10 to 1 in deep water and 3 to 1 in near surface water indicating abundant phosphate, with respect to nitrogen, relative to phytoplankton requirements. It appears, then, that summer phytoplankton biomass is predominantly regulated by inorganic nitrogen availability. The arguments for phytoplankton biomass regulation by phosphorus in spring and nitrogen in summer are supported by measurements of alkaline phosphatase enzyme activity.

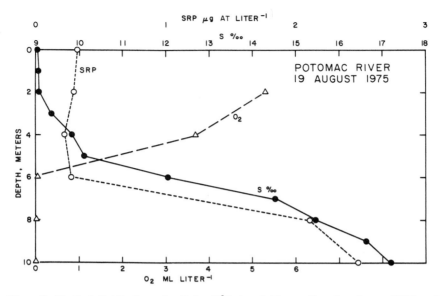

Figure 2. Vertical distribution of salinity (S°/₀₀), soluble reactive phosphorus (SRP) and oxygen at station P-12 in the Potomac River.

ALKALINE PHOSPHATASE ACTIVITY

Many phytoplankton species produce alkaline phosphatase enzymes (11) when intracellular orthophosphate or polyphosphate levels decline below some threshold value. These enzymes, located near the outer cell surface, hydrolyze extracellular organic phosphorus monoesters to release orthophosphate ions for incorporation into the cell. Data in Tables 3 and 4 reveal that phytoplankton in Chesapeake Bay, Potomac River, Delaware Bay (Fourteen Foot Bank) and Pamlico Sound (near Waves, N.C.) produce alkaline phosphatase enzymes when orthophosphate supply is restricted. Alkaline phosphatase activity in natural phytoplankton assemblages is a function of both substrate concentration and pH, and can be approximated by Michaelis-Menten kinetics (26). Table 3 compares maximum uptake velocities (V_m) and Table 4 compares half-saturation constants (K_s) for orthophosphate uptake and alkaline phosphatase activity for natural phytoplankton assemblages in these four coastal regions. Ranges of maximum uptake velocity for phosphorus in both forms are similar for the four regions. The comparable ranges in half-saturation constants for phosphorus utilization in both forms suggests that monoesters may be a significant phosphorus source and may equal orthophosphate in importance when orthophosphate concentrations and resupply rates from outside the euphotic zone are minimal. The concentrations of phosphorus monoester in Chesapeake Bay are

Table 3. Maximum velocity for orthophosphate uptake and hydrolysis of phosphorus monoesters by alkaline phosphatase.

	PO$_4$ Uptake ng-at (μg Chl $a \cdot$hr)$^{-1}$	Alkaline Phosphatase Activity nmoles (μg Chl $a \cdot$hr)$^{-1}$
CHESAPEAKE BAY	3.5 to 156	3.2 to 53.2
DELAWARE BAY	>70	3.6
PAMLICO SOUND		41
POTOMAC RIVER	63	

Table 4. Half-saturation values for orthophosphate uptake and hydrolysis of phosphorus monoesters by alkaline phosphatase. Units are μg-at$\cdot$liter^{-1} of PO$_4^{\equiv}$-P and μmoles$\cdot$liter^{-1} of 3-0-methyl fluorescein phosphate.

	PO$_4$ Uptake μg-at$\cdot$liter^{-1}	Alkaline Phosphatase Activity μmoles$\cdot$liter^{-1}
CHESAPEAKE BAY	0.9 to 1.72	0.14 to 1.0
DELAWARE BAY	0.13	0.50
PAMLICO SOUND		0.06
POTOMAC RIVER	2.5	

usually <0.1 μg-at$\cdot$liter^{-1} indicating that monoesters do not accumulate in the water (26).

Alkaline phosphatase activity in Chesapeake Bay shows some seasonal patterns (26). Hydrolysis rate normalized to phytoplankton biomass is generally higher in spring than in late summer. In August 1975, alkaline phosphatase activity was virtually undectable throughout Chesapeake Bay (Taft, unpublished data). Alkaline phosphatase activity in phytoplankton has been interpreted as a symptom or consequence of phosphorus deficiency (5), and these trends in enzyme activity do reflect the observed differences in the availability of "new" inorganic phosphate with respect to "new" inorganic nitrogen in spring and summer seasons in the bay.

When phosphate is abundant relative to phytoplankton nutritional requirements, cells are able to store it internally as polyphosphate (Poly P). Alkaline phosphatase activity is induced or derepressed when Poly P stores fall below some threshold level (5, 6, 19). Therefore, detectable intracellular polyphosphate stores and the absence of alkaline phosphatase activity associated with the phytoplankton, and a low particulate N to P atom ratio, suggest that phosphorus availability is not a principal factor regulating phytoplankton biomass. These conditions, suggestive of biomass regulation by nitrogen availability, were observed in the Potomac River estuary in August 1975, and the following experiment was performed in an attempt to relate alkaline phosphatase activity to relative nutritional requirements of natural phytoplankton for nitrogen and phosphorus.

Surface water was collected by bucket and screened through 35 μm mesh into a carboy. A one-liter sample was placed in a glass bottle, enriched with 15 μg-at NH_4-N and incubated on deck in natural light attenuated to 73% of the incident value with stainless steel screen. The sample was maintained at the temperature of near surface water flowing through the incubator from the ship's seawater system. Subsamples were taken with time for analyses of ammonium concentration (23), alkaline phosphatase activity (14) and for intracellular polyphosphate. For the Poly P analyses, a 10 ml subsample was pipetted onto a 25 mm Nucleopore® filter with pore size 0.45 μm and then was irradiated with ultraviolet light for 1 hr in a quartz tube containing 10.2 ml distilled water and 0.1 ml 30% hydrogen peroxide. A 5 ml aliquot was analyzed for SRP (23) and the remaining 5 ml was heated with 0.1 ml 40% HC1 in boiling water (23) for 1 hr to convert Poly P to orthophosphate and the SRP produced was measured by the molybdate method. Corrections for sample turbidity were applied to the optical densities. Filter phosphate blanks were significant (0.1 μg-at$\cdot$liter^{-1}), but were constant and small compared to the particulate phosphate retained. Perry (15), however, did not detect phosphate from pre-baked glass fiber filters in a similar procedure. Identical particulate organic phosphate values were obtained in UV irradiated whole water samples and in distilled water containing particles retained in Nucleopore® filters.

The results (Fig. 3) show that when collected, the phytoplankton contained sufficient internal phosphorus stores to suppress alkaline phosphatase induction. As nitrogen was supplied, polyphosphate stores were depleted and, with minimal orthophosphate available in the water ($\leq$0.1 μg-at$\cdot$liter^{-1}), the cells mobilized alkaline phosphatase to hydrolyze extracellular phosphorus monoesters. These results support the conclusion that nitrogen regulates phytoplankton biomass in summer in regions of the estuary where the euphotic zone overlies anoxic bottom waters which contain more "new" inorganic phosphate than nitrogen relative to phytoplankton requirements.

Alkaline phosphatase activity has been suggested as a criterion for identifying "phosphorus limitation" in phytoplankton (5). This usage is correct if "phosphorus limitation" refers to biomass only, in that there is a finite quantity of total phosphorus available. It is necessary to clearly distinguish biomass limitation from productivity limitation because, in nature, primary productivity per unit biomass may be influenced more by the turnover rate of the nutrient in least supply than by the total amount of that element in all forms. In the

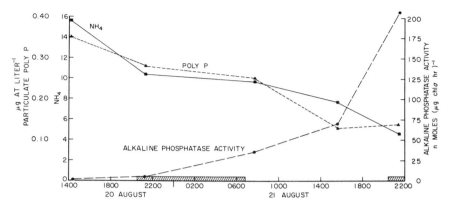

Figure 3. Ammonium concentration change in the water and polyphosphate (Poly P) and alkaline phosphatase activity change in phytoplankton from Potomac River Station P-12.

extreme, alkaline phosphatase activity, which enables phytoplankton to increase the turnover rate of available phosphorus, supports sustained primary productivity during phosphorus regulation of phytoplankton biomass when natural phosphorus monoesters are available for hydrolysis.

ACKNOWLEDGEMENTS

This research was supported by the National Science Foundation, Grant GA-33445, and the U. S. Energy Research and Development Administration contract E(11-1)-3279, Document No. C00-3279-18. We gratefully acknowledge the analytical assistance of R. Loftus and C. McGrath, and the critical comments on the manuscript by Dr. J. J. McCarthy.

LITERATURE CITED

1. Bray, J. T., O. P. Bricker and B. N. Troup. 1973. Phosphate in interstitial waters of anoxic sediments: oxidation effects during sampling procedure. Science 180: 1362-1364.
2. Carpenter, J. H., D.W. Pritchard and R. C. Whaley. 1969. Observation of eutrohpication and nutrient cycles in some coastal plain estuaries, p. 210-221. *In* Eutrophication: Causes, Consequences, Correctives. Nat. Acad. Sci., Washington, D.C.
3. Carritt, D. E., and S. Goodgal. 1954. Sorption reactions and some ecological implications. Deep-Sea Research 1: 224-243.
4. Copeland, B. J., and J. E. Hobbie. 1972. Phosphorus and eutrophication in the Pamlico River estuary, N. C., 1966-1969 — A summary. Water Resources Research Institute Univ. N. C. Rept. No. 65.
5. Fitzgerald, G. P., and T. C. Nelson. 1966. Extractive and enzymatic analyses for limiting or surplus phosphorus in algae. J. Phycol. 2: 32-37.
6. Fuhs, G.W. 1969. Phosphorus content and rate of growth in the diatoms *Cyclotella nana* and *Thalassiosira fluviatilis*. J. Phycol. 5: 312-321.

7. Fournier, R. O. 1966. Some implications of nutrient enrichment on different temporal stages of a phytoplankton community. Chesapeake Sci. 7: 11-19.

8. Heinle, D. R., and D. A. Flemer. 1976. Flows of materials between poorly flooded tidal marshes and an estuary. Mar. Biol. 35: 359-373.

9. Hobbie, J. E., B. J. Copeland and W. G. Harrison. 1972. Nutrients in the Pamlico River Estuary, N.C., 1969-1971. Water Resources Research Institute, Univ. N. C., Rept. No. 76.

10. Jeffries, H. P. 1962. Environmental characteristics of Raritan Bay, a polluted estuary. Limnol. Oceanogr. 7: 21-31.

11. Kuenzler, E. J., and J. P. Perras. 1965. Phosphatases of marine algae. Biol. Bull. 128: 271-284.

12. Newcombe, C. L., and A. G. Lang. 1939. The distribution of phosphates in the Chesapeake Bay. Proc. Am. Phil. Soc. 81:393-420.

13. Patten, B. C., R. A. Mulford and J. E. Warinner. 1963. An annual phytoplankton cycle in the lower Chesapeake Bay. Chesapeake Sci. 4: 1-20.

14. Perry, M. J. 1972. Alkaline phosphatase activity in subtropical central north Pacific waters using a sensitive fluorometric method. Mar. Biol. 15: 113-119.

15. _____. 1976. Phosphate utilization by an oceanic diatom in phosphorus-limited chemostat culture and in the oligotrophic waters of the central North Pacific. Limnol. Oceanogr. 21: 88-107.

16. Pomeroy, L. R., H. H. Haskin and R. A. Ragotzkie. 1956. Observations on dinoflagellate blooms. Limnol. Oceanogr. 1: 54-60.

17. _____, E. E. Smith and C. M. Grant. 1965. The exchange of phosphate between estuarine water and sediments. Limnol. Oceanogr. 10: 167-172.

18. Reimold, R. J. 1965. An evaluation of inorganic phosphate concentrations of Canary Creek Marsh. M.S. Thesis. University of Delaware. 61 p.

19. Rhee, G. 1973. A continuous culture study of phosphate uptake, growth rate and polyphosphate in *Scenedesmus* sp. J. Phycol. 9: 495-506.

20. Schindler, D. W. 1971. Carbon, nitrogen and phosphorus and the eutrophication of freshwater lakes. J. Phycol. 7: 321-329.

21. Shlopak, G. P. 1972. An evaluation of the total phosphorus concentration in the waters of two southern Delaware salt marshes. M.S. Thesis, University of Delaware. 114 p.

22. Smayda, T. J. 1957. Phytoplankton studies in lower Narragansett Bay. Limnol. Oceanogr. 2: 342-359.

23. Strickland, J. D. H., and T. R. Parsons. 1972. A practical handbook of seawater analysis. Fish. Res. Bd. Canada, Bull. 167. 310 p.

24. Taft, J. L., and W. R. Taylor. 1976. Phosphorus distribution in the Chesapeake Bay. Chesapeake Sci., 17: 67-73.

25. _____, _____, and J. J. McCarthy, 1975. Uptake and release of phosphorus by phytoplankton in the Chesapeake Bay estuary. Mar. Biol. 33: 21-32.

26. _____, M. E. Loftus and W. R. Taylor. 1975. Kinetics of phosphate uptake from phosphomonoesters by phytoplankton in the Chesapeake Bay. Manuscript in preparation.

27. Troup, B. N. 1974. The interaction of iron with phosphate, carbonate and sulfide in Chesapeake Bay interstitial waters: A thermodynamic interpretation. Ph.D. Thesis, Johns Hopkins University, 114 p.

28. _____, O. P. Bricker and J. T. Bray. 1974. Oxidation effect on the analysis of iron in interstitial water of recent anoxic sediments. Nature 249: 237-239.

29. Whaley, R. C., J. H. Carpenter and R. L. Baker. 1966. Nutrient data summary 1964, 1965, 1966: Upper Chesapeake Bay (Smith Point to Turkey Point) Potomac, South, Severn, Magothy, Back, Chester and Miles Rivers; and Eastern Bay. Chesapeake Bay Institute Special Report 12, Johns Hopkins University.

THE SIGNIFICANCE OF UREA FOR PHYTOPLANKTON NUTRITION

IN THE YORK RIVER, VIRGINIA[1]

Kenneth L. Webb
Leonard W. Haas
Virginia Institute of Marine Science
Gloucester Point, Virginia 23062

ABSTRACT: The rate of ^{14}C-urea utilization was measured concurrently with in situ $^{14}CO_2$ productivity measurements during four 24 hour stations at the mouth of the York River estuary between July and November, 1974. After two hour incubations, the ^{14}C from urea was recovered from both the CO_2 and particulate phases. Particulate utilization represented 10-55% of the total. Mean utilization rates ranged from 4 ng-at urea N $dm^{-3}h^{-1}$ in November to 175 ng-at $dm^{-3}h^{-1}$ in July. Mean urea-N concentrations ranged from 1.1 to 1.6 μg-at dm^{-3}. Phytoplankton were considered to be principally responsible for utilization of urea because the process was light dependent and subject to inhibition by both DCMU and avidin. Atomic ratios of C:urea-N (productivity/urea-N utilization) ranged from 10 to 170, indicating that at times urea supplied most of the nitrogen required by the phytoplankton.

A rectangular hyperbola adequately describes the relationship between light intensity and both $^{14}CO_2$ and urea utilization. However, urea utilization was saturated at lower light intensities than photosynthesis, resulting in a higher C:N ratio near the surface than deeper in the euphotic zone.

The nanophytoplankton (passing a 15 μm sieve) were responsible for about 80% of both productivity and urea utilization.

INTRODUCTION

Recent studies have emphasized the necessity of including urea as well as the more classical compounds of nitrate, nitrite and ammonia in studies of the nitrogen cycle. Utilization of urea by natural phytoplankton populations has

---

[1] Contribution No. 709 from the Virginia Institute of Marine Science.

been demonstrated (4, 14, 15, 17). Work by Eppley et al., (4) and McCarthy (14) indicates that urea supplies up to about 60% of the nitrogen utilized when the phytoplankton are dependent upon rapidly recycled nitrogen rather than on nitrate from upwelling or other sources.

While participating in field work on the lower York River during 1974, we had the opportunity to carry out in situ productivity and urea utilization measurements at times when availability of other data presumably make the measurements especially meaningful. We wish in this paper to report our results of the urea utilization study; results of the productivity study will be reported in greater detail elsewhere (8).

METHODS

A station near the mouth of the York River Estuary (37°14'42"N, 76° 23'48"W) at the north edge of the channel was occupied on four days for approximately 36 hours starting predawn on 23 July, 21 August, 1 October and 13 November, 1974, and lasting until dusk the following day. The water column was sampled with a submersible pump at 0.5, 1, 2, and 4 meters and thereafter at 2-m intervals to the bottom at two hour intervals for the duration of the station.

Temperature was measured in the pump effluent with a Y.S.I. Thermistor thermometer (Model 43TD). Salinity samples were returned to the laboratory and measured with a Beckman RS-7B Induction Salinometer. Water column light penetration was measured approximately hourly with an underwater photometer (G.M. Mfg. Corp. Model 268WA310). Average incident radiation (langleys min^{-1}) during periods of incubations and total daily light were calculated from a continuous record from an Eppley pyrheliometer located 10 km from the station. Station data are shown in Table 1.

Primary Production Measurements

The water column was sampled with a submersible pump at 0.5, 1, 2 and 4 meter depths every two hours during the daylight hours of the first day of each station. Water samples were fractionated by gravity filtering through a 15 μm mesh Nitex net. Primary productivity was measured by the ^{14}C technique (22). Triplicate 20 cm^3 aliquots of both fractionated and non-fractionated water were placed in 30 cm^3, 25 mm diameter screw top test tubes. These incubators, two light and one dark, were then inoculated with 0.4 to 1.0 μCi ^{14}C sodium carbonate and incubated in situ at the sampling depth for two hours. Total carbonate added as ^{14}C was trivial compared to total inorganic carbon concentrations and not considered in the calculations. Incubations were terminated by addition of neutral formalin (final concentration ca. 2%), and returned to the laboratory for processing.

The plankton were filtered onto prewetted 25 mm diameter, 0.45 μm pore size Celotate® filters, rinsed with filtered seawater and the filter placed in a

Table 1. Summary of environmental conditions and chlorophyll concentrations during sampling periods. Values include mean salinity difference between surface and bottom (Δ ‰) with number of observations in parentheses; mean surface salinity, ‰ and temperature, °C; euphotic zone chlorophyll, μg dm⁻³ for two or three sampling periods encompassing the mid afternoon maximum for total plankton and <15 μm size fraction ± S$\bar{x}$ with number of observations in parentheses; total incident light (langleys/day) including the mean for the previous number of days indicated (mean, days); light attenuation in the water column; mean daily 1% light depth.

	Δ Salinity	$\bar{x}$ Salinity ‰	Temperature °C	$\bar{x}$ Midafternoon chlorophyll a Total	$\bar{x}$ Midafternoon chlorophyll a <15 μm	Incident light (langleys/day)	Light attenuation K (m⁻¹)	Mean daily 1% light depth (m)
22 July 74						803		
23 July	0.18(10)	20.5	25.5	17.8±1.24(12)	12.5± .71(12)	520(627,7)	1.59	2.89
24 July	0.14(9)	20.5	25.0	15.6± .81(8)	—	188		
20 August						558		
21 August	0.07(10)	20.5	25.0	13.2± .73(12)	9.9± .63(12)	285(487,5)	1.22	3.90
22 August	0.03(4)	20.5	25.0			381		
30 September						584		
1 October	3.56(9)	20.0	20.0	20.1±1.54(12)	19.4±1.31(12)	541(641,7)	1.08	4.26
2 October	2.78(9)	20.0	19.5			262		
12 November						98		
13 November	1.06(10)	22.5	15.5	4.81± .119(8)	3.75±2.44(8)	372(292,7)	.76	6.00
14 November	2.19(8)	22.5	15.5			379		

liquid scintillation vial to digest overnight with 0.2 cm^3 NCS® prior to counting by liquid scintillation. The number of samples processed necessitated preservation of samples rather than terminating the incubation by filtration. A small study carried out on the second day of each station indicated that preservation reduced the recovered activity an average of 30% over that of paired, filter terminated samples.

Total carbonate was determined on seawater subsamples which were preserved with $HgCl_2$ and refrigerated until measured with a Beckman 915 infra-red CO_2 analyzer. Primary production was computed following Strickland and Parsons (23). The absolute level of recovery of radioactivity, compared to filter terminated incubations, was not taken into account.

Urea Utilization Measurements

Aliquots of the same fractioned and nonfractioned water used for primary production measurements from the 1 & 4 meter depths of the July station were utilized to determine urea utilization. The 0.5 and 2 meter depths were utilized for the remaining stations. Quadruplicate (three light and one formalin killed control) 15 cm^3 aliquots were placed in the incubators, inoculated with 19.5 nCi ^{14}C-urea and incubated for two hours at in situ depth. Sampling was carried out every two hours during daylight hours. Specific activity of the urea (New England Nuclear, NEC-108H) was 57.5 mCi $(mMol)^{-1}$, thus the added ^{14}C-urea increased the natural urea nitrogen concentration by 45 ng-at N dm^{-3} or about 3 to 4 percent. Incubations were terminated by addition of neutral formalin, and returned to the laboratory for processing.

$^{14}CO_2$ arising from ^{14}C-urea was determined in a fashion similar to the CO_2 collection technique of Hobbie and Crawford (11). The screw caps were replaced by serum vial stoppers containing a plastic center well (Kontes Glass Co. No. K 882320). Inside the well was an accordion-folded filter paper saturated with Hyamine Hydroxide®. Transfer of $^{14}CO_2$ from solution to the filter paper was accomplished by injecting 0.2 cm^3 of 2 N H_2SO_4 through the serum vial stopper and allowing the vials to stand overnight. This method recovered 55% of the radioactivity added as ^{14}C-carbonate to similar samples.

The filter paper was then removed and counted by liquid scintillation. The plankton were recovered for determination of ^{14}C in the identical fashion to that for the primary production measurements. The process of preserving and collecting the $^{14}CO_2$ from the samples prior to collecting the plankton resulted in recovered radioactivity being only 60% of that of paired samples which had the incubation terminated by filtering. Calculation of urea utilization was based on duration of incubation, chlorophyll a concentration, specific activity of ^{14}C-urea in the incubation and the recovered D.P.M. No correction was made for the absolute amount of ^{14}C recovered which is indicated to be 55 and 60% for the CO_2 and the particulate phases respectively.

Liquid Scintillation Counting

A toluene base cocktail containing 4 g PPO and 0.5 g POPOP dm^{-3} was used for both the $^{14}CO_2$ and plankton filters. After addition of 15 cm^3 of cocktail, samples were allowed to rest for several hours prior to counting on a Beckman Model LS150 liquid scintillation counter. This greatly reduced occasional transient chemilluminescense. Samples of the ^{14}C-urea and ^{14}C-carbonate were taken directly into the toluene cocktail with NCS® during each sample inoculation period in the field. Counting efficiency was determined by an external standard which was calibrated with an Amersham/Searle Model 180060 ^{14}C quenched standard set.

Nutrient and Chlorophyll a Measurement

Samples for nutrient analysis and chlorophyll a determination were filtered through glass-fiber filters. The filters were frozen for subsequent chlorophyll a analysis (27). Ammonia was determined immediately by the method of Solórzano (21). Urea samples were preserved with $HgCl_2$ and freezing and analyzed by the di-acetyl-monoxime (DAM) method (18).

RESULTS

A summary of conditions observed during the sampling periods is presented in Table 1. The water column was well mixed during the sampling periods in July and August and stratified during the October and November sampling periods. The water column in this estuary, although previously classified as moderately stratified, is often well mixed shortly after spring tides (7). Chlorophyll a normally exhibits a diel variation in concentration with a maximum near mid-afternoon and a minimum of about 50% of this value from midnight to 0400 hr. Values observed were typical of this estuary ranging from 5 to 20 μg chlorophyll a dm^{-3}. Chlorophyll a values and, consequently, total phytoplankton activity were relatively low during the November sampling.

Ammonia concentrations averaged for the 4 euphotic zone sampling depths showed a strong diel pattern during July, August and October (Fig. 1) with daytime concentrations, ranging from 0.05 to 3.8 μM NH_3, significantly lower than nighttime values ranging from 3 to 6 μM NH_3. This pattern is not really discernable during November when phytoplankton populations and activity were lower. Urea nitrogen concentration means for the two sampling depths used in the urea utilization study ranged from 1.1 to 1.6 μg-at urea N dm^{-3} (Table 2). Urea nitrogen concentrations generally are lower than ammonia concentrations and, except in July, no diel pattern is evident (Fig. 1).

Rates of urea utilization (CO_2 evolved + uptake into the particulate fraction) for the total plankton are presented in Table 2. Nighttime, dark rates were zero. Daytime mean values ranged from a low of 4 ng-at urea N $dm^{-3}h^{-1}$ in November to a high of 175 ng-at urea N $dm^{-3}h^{-1}$ for the surface samples in July. Standard

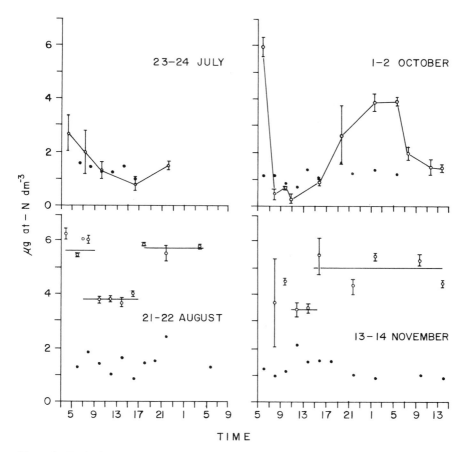

Figure 1. Euphotic zone ammonia (o) and urea nitrogen (•) concentrations at different times during each station. Vertical bars represent $s_{\bar{x}}$ with n = 4. Urea values are means of 2 depths only, each sample measured in duplicate.

errors ($S_{\bar{x}}$) are not presented because rates are light dependent and vary throughout the day in a fashion similar to photosynthesis. The relationship for both photosynthesis and urea utilization versus light intensity are reasonably approximated by a rectangular hyperbola. Values are presented in Table 3 for the light-saturated rate (V) and for the light intensity (K) at which the rate is half maximal for both urea utilization (V_N & K_N) and photosynthesis (V_C & K_C). Values with standard errors ($S_{\bar{x}}$) were calculated by the method of Wilkinson (26), those without $S_{\bar{x}}$ were calculated by the s/v versus s linear transformation. The calculated results are presented which gave the best least squares fit of the experimental points. Values of K_N tend to be lower than those for K_C with the

Table 2. Mean urea-N concentrations and mean daytime in situ urea utilization for total plankton, CO_2 release and particulate phase uptake combined. Surface depth is 0.5 m except for July (1m), sub-surface depth is 2 m (July = 4m). Urea concentration values include ± the standard error and the number of observations in parentheses.

Date	Urea utilization ng-at urea N dm^{-3}h^{-1}		Urea concentration μg-at urea N dm^{-3}
	Surface	Sub-surface	
23 July 74	175.	50.	1.38 ± 0.066(11)
21 August 74	21.7	31.2	1.64 ± 0.24(18)
1 October 74	42.	50.	1.06 ± 0.066(20)
13 November 74	4.27	3.71	1.25 ± 0.109(21)

Table 3. Light saturated rates (V in ng-at C or N (μg chlorophyll a)$^{-1}$ h^{-1} and half saturation constants (K in millilangleys min^{-1}) for the assimilation of CO_2(C) and urea (N) as a function of light level. Values are shown for both <15 μm and total plankton.

	V_N ng-at(μg chl. a)$^{-1}$ h^{-1}	V_C	V_C:V_N	K_N mlangleys min^{-1}	K_C
		Total Plankton			
July	12.5 ± 2.35	322 ± 37.4	25.8	11.8 ± 8.5	155. ± 29.6
August	3.34	187 ± 21.7	56.0	8.8	17.7 ± 4.4
October	2.53	278 ± 27	110.	120.	66. ± 16
November	1.26	184 ± 15.5	146.	19.7	82.5 ± 21.5
		<15 μm Plankton			
July	12.6	262 ± 28.2	20.8	11.7	77.3 ± 17.3
August	2.58	238 ± 15.9	92.	15.0	32.8 ± 4.09
October	2.75 ±.078	236 ± 12.6	86.	10.1 ± 32.	36.8 ± 4.1
November	1.06 ±.24	188 ± 17.2	177.	50.7 ± 46	52. ± 6.5

Table 4. Carbon:nitrogen atomic ratios for in situ ^{14}C productivity and urea N utilization (CO_2 + particulate), mean ± standard error of mean, number of observations in parentheses. Paired value t test indicates no size difference but a significant difference (p = .025) between the surface and sub-surface C:N for both plankton size fractions.

Date	Total plankton		<15 μm plankton	
	Surface	Sub-surface	Surface	Sub-surface
July	15 ± 3.2(6)	4.6 ± 1.62(4)	11.6 ± 1.96(6)	8 ± 3.6(4)
August	48 ± 10.3(6)	26 ± 10.9(5)	70 ± 12.1(5)	27 ± 8.2(5)
October	150 ± 60 (6)	55 ± 10.3(6)	130 ± 41 (6)	80 ± 32 (4)
November	140 ± 23.5 (5)	112 ± 14.9(4)	178 ± 20.8(5)	157 ± 19.5(3)

result that C:N ratios are lower at low light intensities. This trend is also detected in the actual C:N ratios for the individual incubations presented in Table 4 where the subsurface ratios are lower than the surface values.

The fraction of the ^{14}C of urea recovered from the particulate fraction as a percentage of the total recovered is reported in Table 5. The ^{14}C in the particulate fraction ranged from about 15 to 55% of the total for both size fractions of the plankton. The <15 μm size fraction plankton accounted for about 60 to 100% of the CO_2 evolution from urea and 60 to 90% of the ^{14}C from urea recovered in the particulate phase (Table 6).

Both avidin and DCMU statistically significantly reduced (about 30%) the release of CO_2 from urea. A significant decrease for incorporation into the particulate phase was indicated only for DCMU. Measurements of the rate of urea utilization versus urea concentration for these four sampling periods indicated that the uptake process was rate saturated at the measured environmental concentration of urea, i.e. the rate was the same for added concentrations of urea ranging from about 5 to 100% of the observed environmental concentration. Previous to this study, similar experiments with natural plankton samples at this station produced hyperbolic curves relating urea uptake to substrate concentration (Webb, unpublished).

Table 5. Partitioning of ^{14}C from urea between particulate ^{14}C and $^{14}CO_2$ by the total and <15 μm plankton. Units are particulate-^{14}C × 100 (particulate-^{14}C + $^{14}CO_2$)$^{-1}$ ± standard error of the mean. Number of observations in parentheses.

	Total plankton	<15 μm plankton
23 July 74	17.7 ± 1.85(12)	14.1 ± 1.36(12)
21 August 74	50. ± 2.03(11)	44.4 ± 2.79(11)
1 October 74	56.6 ± 1.94(12)	53.5 ± 2.31(12)
13 November 74	33.2 ± 1.59(9)	33.4 ± 1.57(9)

Table 6. Partitioning of both $^{14}CO_2$ and particulate ^{14}C from urea between the total and <15 μm size plankton. Units are <15 μm × 100 (total plankton)$^{-1}$ ± standard error of the mean. Number of observations in parentheses.

	CO_2	Particulate
23 July 74	88. ± 9.5(12)	76. ± 9.9(12)
21 August 74	80. ± 6.8(11)	62. ± 8.8(11)
1 October 74	98.6 ± 9.2(12)	88. ± 9.0(12)
13 November 74	63. ± 5.3(9)	64. ± 5.9(9)

DISCUSSION

Utilization of nutrients by phytoplankton in the light rather than in the dark makes good a priori sense to the plant physiologist because 1) light may directly supply the energy required for the uptake and/or utilization processes or, 2) the supply of raw material is utilized only when other materials supplied by photosynthesis are available. Several papers (5,6,15), both from cultures and from oceanic ecosystems, verify that there is in fact a diel variation in utilization and external concentration of inorganic nutrients. The ecological situation may be further complicated if nitrogen starved cells take up available nutrients whenever available at the expense of stored energy. Hattori (9) reported that the utilization of urea by *Chlorella* in the dark was negatively correlated with cell nitrogen content, with no dark utilization of urea unless cell nitrogen was below about 7.5%. A consideration of diel variations within estuarine ecosystems is probably as important as spatial and seasonal variations.

In the present study, we found that urea utilization followed a diel cycle very similar to that of photosynthesis, i.e. maximal rates during mid-day and zero rates at night. Dark incubations during the daytime without a dark preconditioning period gave measurable rates of urea utilization but at a lower level than the light rates indicating that the process does not turn off as rapidly as photosynthesis. The absence of urea utilization at night is consistent with the interpretation that the utilization is by the phytoplankton rather than bacteria and also that these phytoplankton are not nitrogen starved.

Mitamura and Saijo (17) also report that changes in photosynthesis, incorporation of urea carbon, and liberation of CO_2 from urea all bear a similar relationship to light intensity and indicated that photoinhibition for all three processes took place in Mikawa Bay phytoplankton above about 12,000 lux. In our experiments, we did not observe photoinhibition for either urea utilization or photosynthesis at light levels up to 0.6 langleys/min or about 48,000 lux. We did observe, however, that rates of urea utilization at low light intensities were greater, relative to photosynthesis. This is shown in the data in two fashions: 1. Carbon:nitrogen ratios were significantly lower at our subsurface sampling depth than they were at the surface (Table 4). 2. Our best estimate of a half saturation constant for urea utilization versus light intensity would be about 0.01 langleys/min while that for photosynthesis would be within the range of 0.03 to 0.1 langleys/min (Table 3). The mechanism causing this difference and its implications are not entirely clear. It may be that since the two light driven processes are different they simply have different kinetic characteristics. This suggests that under low light the cells take up carbon and urea nitrogen more in line with the Redfield ratio of 6.6:1 than do cells under high light intensity, which seem to assimilate a disproportionate amount of carbon. Another, although not mutually exclusive, interpretation is more appealing. We (8) as well as others have observed that carbon assimilation ratios (carbon fixed per unit chlorophyll) are not always symmetrically centered around the mid-day light peak but may be greater later in the day (or may also show the "noon-time nap" effect). During October for example, photosynthesis was most efficient in the afternoon while urea utilization was most efficient in the morning. These observations seem to be consistent with a homeostatic mechanism within the cell which channels the energy to the most pressing task, at the expense of other work for which energy could be used, i.e. uptake of raw materials (nitrogen) in the morning and processing them into cell constituents in the afternoon (photosynthesis).

Our results indicate that 10-55% of the total urea carbon utilized was incorporated into the particulate material (Table 5). Mitamura and Saijo (17) report higher values ranging from 38 to 84%. Some of these differences undoubtedly result from different species composition of the plankton. Carpenter, Remsen and Schroeder (2) recovered negligible amounts of ^{14}C in the particulate fraction from a laboratory study of urea utilization by *Skeletonema costatum*. As a result, the field techniques utilized by this group (3, 19, 20) underestimate urea utilization to the extent that carbon incorporated into the particulate phase was not measured.

The mean range of urea concentrations we observed (1.1 to 1.6 μg-at urea N dm^{-3}) appear to be intermediate between those of Chesapeake Bay (<0.05 to 1.0 μg-at N dm^{-3}) (16) and those of Warsaw Sound, Georgia, 2 to 3.8 μg-at N dm^{-3} (19). These differences could in part be a result of the methodologies used. The York River and Georgia estuaries data sets were produced with the DAM method and the Chesapeake Bay data set was produced with the urease method. Walsh has pointed out that the DAM method produced urea values an order of magnitude greater than the urease method when used at the same time but by different investigators in the S.E. Pacific (24). DAM urea values averaged 0.102 μg-at N dm^{-3} higher than the urease values in a direct comparison in Kaneohe Bay, Hawaii, over a concentration range form 0.4 to 1.7 μg-at urea N dm^{-3} (10). Our own attempts (Webb, unpublished) to determine the validity of the DAM urea values utilized an isotope dilution bioassay. Our results were variable indicating that DAM urea values was a good value for some samples and that it was too high for others. In this study, urea concentrations which are slightly too high would result in calculated urea utilization rates which are inflated but not affect the other relationships.

The mean rate of urea-N utilization reported by Remsen's group (19) for Warsaw Sound was about 70 ng-at urea N $dm^{-3}h^{-1}$. This value during the spring is well within the range of values we found during summer and fall (Table 2).

The pathway(s) whereby urea is metabolized and assimilated by plants remains somewhat obscure although a great deal of attention has been paid to the phenomenon. Hydrolysis of urea by urease (EC 3.5.1.5) is the classical pathway. In the past decade, a second pathway, ATP:urea amidolyase (a combination of urea carboxylase, EC 6.3.4.6, and allophanate amido hydrolase, EC 3.5.1.13) has been elucidated (e.g. 13). This latter pathway requires biotin and ATP; the requirement for ATP may be a coupling to the light reactions of photosynthesis in the organisms that utilize this pathway of urea hydrolysis. Incorporation of urea carbon into the particulate fraction would seem to be through the pathway of CO_2 since flushing cultures with 10% CO_2 in air removed particulate uptake of urea carbon (12). On the other hand, both our results and others (17) indicate that if the carbon from urea passes through a

CO_2 pool on its way to the particulate phase, it does not mix extensively with the environmental CO_2 pool. Avidin inhibition also suggests that the pathway requiring biotin is important in the phytoplankton. Inhibition of urea utilization by DCMU presumably is further evidence that light energy is effective through the photosynthetic mechanism although DCMU inhibition of dark uptake of urea has been reported (12).

The assumption that urea carbon is incorporated into the particulate phase via the photosynthetic assimilation of CO_2 which is liberated from urea can be supported by the literature (1, 25). Webster et al. (25), however, note that ^{14}C from urea is incorporated into the free amino acid and protein pools of bean leaves to a greater extent than ^{14}C from ^{14}C-bicarbonate. Our results (Webb, unpublished) with several cultures of marine phytoplankters also indicate preferential labelling of amino acid pools from urea carbon compared to environmental CO_2. It thus appears that CO_2 derived from urea may be delivered to the CO_2 dark fixing pathways rather than to the photosynthetic pathways.

On the basis of the Redfield C:N ratio of 6.6:1, an evaluation of the carbon: urea N atomic ratios (Table 4) indicates that urea could provide between 14 and 45% of the nitrogen required by the phytoplankton during July and August, but only about 5% during October and November. Thus, the July and August estimates are in good agreement with the work of Eppley et al., (4), McCarthy (14) and McCarthy & Eppley (15). Apparently the plankton were not especially active during November when the lowest observed chlorophyll concentrations, lowest temperatures and lowest carbon assimilation ratios were observed. The concentration of ammonia versus time of day for October (Fig. 1) shows that the lowest ammonia concentration occurred at mid-day, suggesting that in fact ammonia was supplying nitrogen. Generally, the diel concentration changes are much greater for ammonia than for urea. Uptake rates can best be estimated from the diel concentration curves for July, where during the morning urea utilization measured by ^{14}C was 110 ng-at urea N $dm^{-3}h^{-1}$ and was 70 ng-at urea N $dm^{-3}h^{-1}$ based on rate of concentration decrease. The difference between these two rates indicates a supply rate of about 40 ng-at $dm^{-3}h^{-1}$. This is about one half of the utilization rate but if continuous for 24 hours should just balance utilization, which takes place only during daylight hours. An ammonia utilization rate calculated from the concentration decrease is 230 ng-at N $dm^{-3}h^{-1}$. Thus, urea utilization is about 23% of the combined urea plus ammonia rates based on concentration changes. This is about one-half the value estimated by using the ^{14}C urea rate, production values and the Redfield ratio of 6.6:1 for carbon:nitrogen. It should be noted that the July plankton seem to be unique in this sampling program. They show by far the greatest absolute urea utilization rates (Table 2) and the highest rates per unit chlorophyll a (Table 3), the lowest C:N ratios (Table 4) as well as the highest fraction of urea-^{14}C liberated as CO_2 (Table 5).

Our results (Table 6) indicate that the small phytoplankton (<15 μm) were responsible for most of the urea utilization and for most of the primary production (9). Thus, there appears to be no special significance, for urea utilization, of large diatoms in this study as suggested by Remsen et al. (19) for the spring phytoplankton flora of Georgia. We are, however, in full agreement that urea must be included in the nitrogen pools that are available for phytoplankton and that utilization of urea appears to be an exception to the general rule that bacteria win the competition with phytoplankton for dissolved organic substrates in the aquatic environment.

ACKNOWLEDGMENT

This research was supported in part by grant NASA - NGL 47-022-005 from the U.S. National Aeronautics and Space Administration and by grant GI-38973 to the Chesapeake Research Consortium, Inc. from the National Science Foundation, Research Applied to National Needs. Special thanks are extended to numerous VIMS personnel who assisted in the field work, including the crews of the R/V Langley and R/V Retriever.

REFERENCES

1. Allison, R. K., H. E. Skipper, M. R. Reid, W. A. Short and G. L. Hogan. 1964. Studies on the photosynthetic reaction. II. Sodium formate and urea feeding experiments with *Nostoc muscorum*. Plant. Physiol. 29: 164-168.
2. Carpenter, E. J., C. C. Remsen and B. W. Schroeder. 1972. Comparison of laboratory and *in situ* measurements of urea decomposition by a marine diatom. J. exp. mar. Biol. Ecol. 8: 259-264.
3. _____, _____, and S. W. Watson. 1972. Utilization of urea by some marine phytoplankters. Limnol. Oceanogr. 17: 265-269.
4. Eppley, R. W., E. H. Renger, E. L. Venrick, and M. M. Mullin. 1973. A study of plankton dynamics and nutrient cycling in the central gyre of the north Pacific Ocean. Limnol. Oceanogr. 18: 534-551.
5. _____, A. F. Carlucci, O. Holm-Hansen, D. Kiefer, J. J. McCarthy, E. Venrick, and P. M. Williams. 1971. Phytoplankton growth and composition in shipboard cultures supplied with nitrate, ammonium, or urea as the nitrogen source. Limnol. Oceanogr. 16: 741-751.
6. Goering, J. J., R. C. Dugdale, and D. W. Menzel. 1964. Cyclic diurnal variations in the uptake of ammonia and nitrate by photosynthetic organisms in the Sargasso Sea. Limnol. Oceanogr. 9: 448-451.
7. Haas, L. W. 1975. The effect of the spring-neap tidal cycle on the vertical salinity structure of the James, York and Rappahannock rivers, Virginia U.S.A. (manuscript).
8. _____, and K. L. Webb. 1975. Phytoplankton dynamics in a temperate estuary. (manuscript)
9. Hattori, A. 1960. Studies on the metabolism of urea and other nitrogenous compounds in *Chlorella ellipsoidea*. III. Assimilation of urea. Plant Cell. Physiol. 1: 107-115.
10. Harvey, W. A. 1974. The utilization of urea, ammonium and nitrate by natural populations of marine phytoplankton in a eutrophic environment. 61 p. M.S. thesis, University of Hawaii.

11. Hobbie, J. E., and C. C. Crawford. 1969. Respiration corrections for bacterial uptake of dissolved organic compounds in natural waters. Limnol. Oceanogr. 14: 528-532.

12. Hodson, R. C., and J. F. Thompson. 1969. Metabolism of urea by *Chlorella vulgaris.* Plant Physiol. 44: 691-696.

13. Leftley, J. W., and P. J. Syrett. 1973. Urease and ATP:Urea amidolyase activity in unicellular algae. J. Gen. Microbiol. 77: 109-115.

14. McCarthy, J. J. 1972. The uptake of urea by natural populations of marine phytoplankton. Limnol. Oceanogr. 17: 738-748.

15. _____, and R. W. Eppley. 1972. A comparison of chemical, isotopic, and enzymatic methods for measuring nitrogen assimilation of marine phytoplankton. Limnol. Oceanogr. 17: 371-382.

16. _____, W. R. Taylor, and J. L. Taft. 1975. The dynamics of nitrogen and phosphorus cycling in the open waters of the Chesapeake Bay, p. 664-681. *In* T. M. Church (ed), Marine Chemistry in the Coastal Environment. ACS Symposium Series, Number 18.

17. Mitamura, O., and Y. Saijo. 1975. Decomposition of urea associated with photosynthesis of phytoplankton in coastal waters. Marine Biology 30: 67-72.

18. Newell, B. S., B. Morgan and J. Cundy. 1967. The determination of urea in seawater. J. mar. Res. 25: 201-202.

19. Remsen, C. C., E. J. Carpenter and B. W. Schroeder. 1972. Competition for urea among estuarine microorganisms. Ecology 53: 921-926.

20. _____, _____, and _____. 1974. The role of urea in marine microbial ecology, p. 286-304. *In* R. R. Colwell and R. Y. Morita (eds), Effect of the ocean environment on microbial activities. Univ. Park Press. Baltimore.

21. Solórzano, L. 1969. Determination of ammonia in natural waters by the phenolhypochlorite method. Limnol. Oceanogr. 14: 799-801.

22. Steeman-Nielsen, E. 1952. The use of radioactive carbon (C^{14}) for measuring organic production in the sea. J. Cons. perm. int. Explor. Mer. 18: 117-140.

23. Strickland, J. D. H., and T. R. Parsons. 1968. A practical handbook of seawater analysis. Bull. Fish. Res. Bd. Canada 167: 1-311.

24. Walsh, J. J. 1975. Utility of systems models: a consideration of some possible feedback loops of the Peruvian upwelling ecosystem, p. 617-633. *In* L. E. Cronin (ed), Estuarine Research, Volume 1. Chemistry, Biology, and the Estuarine System. Academic Press, Inc. New York.

25. Webster, G. C., J. E. Varner and A. N. Gansa. 1955. Conversion of carbon-14-labeled urea into amino acids in leaves. Plant. Physiol. 30: 372-374.

26. Wilkinson, G. N. 1961. Statistical estimations in enzyme kinetics. Biochem. J. 80: 324-332.

27. Yentsch, C. S., and D. W. Menzel. 1963. A method for the determination of phytoplankton chlorophyll and phaeophytin by fluorescence. Deep-Sea Res. 10: 221-231.

THE IMPACT OF NITROGEN AND PHOSPHORUS RELEASE FROM

A SILICEOUS SEDIMENT ON THE OVERLYING WATER

Eric O. Hartwig
Chesapeake Bay Institute
The Johns Hopkins University
Baltimore, Maryland 21218

ABSTRACT: The release (+) or uptake (−) of ammonia (NH_4^+), nitrite (NO_2^-), nitrate (NO_3^+), phosphate (PO_4^{-3}) and dissolved organic phosphorus, nitrogen and carbon (DOP, DON, DOC) from a subtidal (18.3 m depth) siliceous sediment off La Jolla, California was studied over two years. The mean net exchange values (and the total range of rates obtained) were: NH_4^+ $+872\mu$ mol $m^{-2}d^{-1}$ (−47 to +3290); NO_2^- $+34\mu$ mol $m^{-2}d^{-1}$ (−5 to +97); NO_3^- -77μ mol $m^{-2}d^{-1}$ (−720 to +647); PO_4^{-3} $+77\mu$ mol $m^{-2}d^{-1}$ (−438 to +502); DOP $+12\mu$ mol $m^{-2}d^{-1}$ (−28 to +59); DON -75μ mol $m^{-2}d^{-1}$ (−1326 to +1280); and DOC -583μ mol $m^{-2}d^{-1}$ (−30,800 to +23,800). Using published primary production rates and C:N:P ratios it was calculated that 15,000 μ mol N $m^{-2}d^{-1}$ and 935 μ mol P $m^{-2}d^{-1}$ were taken up by the phytoplankton. The benthos released 786 μ mol N $m^{-2}d^{-1}$ and 90 μ mol P $m^{-2}d^{-1}$ or 5% and 10% of the required N and P. It was shown, however, that sediment exchange released only an insignificant fraction of both N (0.4%) and P (0.5%) already contained in the water. By using a conservative mean net advection rate of 1.5 km d^{-1} it must be presumed that the N and P released into the water at this site was superfluous to the needs of the phytoplankton and was, therefore, exported from the area.

INTRODUCTION

Marine benthic communities receive their organic matter from sinking detritus, sinking organisms and benthic autotrophic production. Metabolism, diffusion and resuspension (27, 17, 18, 25, 6, 7) are the major routes of organic matter loss from the benthos, whereas burial is not considered a loss because the organic matter remains in the sediment.

Metabolism results in the assimilation of a portion of the organic matter, the release of another portion and the mineralization of the remainder. Natural bacterial assemblages metabolizing under aerobic conditions mineralize 8-57% of the total organic carbon taken up (10, 29), the remainder being incorporated into new cell material or partially mineralized and released. Many factors determine the rate at which these metabolic products participate in sediment-water nutrient exchange. Diffusion rate, bioturbation, resuspension, metabolic rate, photosynthetic rate, sorption reactions, redox and pH (2, 8, 19, 14, 15, 5, 6, 7, 16) are the more important of these factors.

In shallow coastal waters benthic nutrient exchange can influence the water nutrient chemistry and thereby the productivity of the overlying water (19, 24, 5). The impact of nutrient regeneration on phytoplankton productivity depends on several factors, including the rate of release of the nutrient from the sediment, the depth of the water, the degree of mixing of the overlying water, the *in situ* concentration of the nutrient, and the rate of uptake of the nutrient by the phytoplankton.

In deep waters, such as over the Santa Barbara Basin, California, the importance of benthic nitrogen release to the phytoplankton nitrogen requirement in the upper waters was shown to be negligible, providing only 0.4% of the nitrogen required (23). In Doboy Sound, Georgia, which has a mean depth of 2 m at mean high water and a mean tidal amplitude of 2 m, Pomeroy et al. (19) found that the sediment acted to stabilize the phosphate concentration in the overlying water to 1 μ mol $\mathcal{L}^{-1}$. Hale (5) concludes from direct measurements of nitrogen and phosphorus released from Narragansett Bay sediments (a shallow bay, average depth 8.8 m) that in June, 80% of the nitrogen requirement and more than 200% of the phosphorous requirement of *Skeletonema costatum* (a major component of the phytoplankton community) is released from the bay sediments. In the period of August through November, 780% of the nitrogen requirement was released from bay sediments. Rowe et al. (24) calculated that over the New York Bight benthic nutrient regeneration accounted for more than 200% of the nitrogen required by phytoplankton photosynthesis. The calculations of Rowe et al. (24) were based on many assumptions but show the contribution sediment nutrient regeneration can make even in deeper (80 m) waters.

The purposes of this study were to quantitate the *in situ* rate of nutrient exchange and to analyze the probable impact this exchange would have on the phytoplankton productivity in the overlying waters.

METHODS
Study Site

The study site (Fig. 1) was located in 18.3 m of water in the La Jolla Bight region off the Scripps Institution of Oceanography (SIO). The extended study area was demarcated by its biological and physical similarity to the study site (11, 4). Water movement within this region has been described in detail (21, 1, 12, 9). The size distribution of the sediments was studied extensively by Inman (11). The phi median diameter at this site was 3.50 to 3.24 phi units (88 to 105

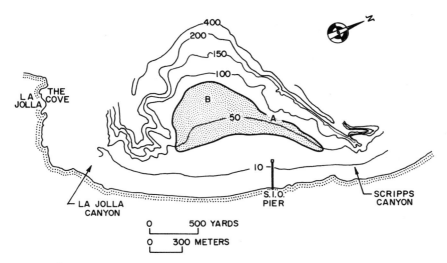

Figure 1. Study site (A) and extended study area (B: stippled area). Contour intervals are in feet.

μm) with a phi deviation measure of 0.3 to 0.5 phi units and a phi skewness measurement of −0.15 to +0.05 meaning the particle sizes were skewed towards smaller phi values (larger particle diameters). Sediment organic carbon content of this siliceous sediment was low (0.06% to 0.21%) (7).

Temperature

Bottom water temperature was taken with a mercury thermometer held approximately three cm above the sediment surface. During portions of the study a Ryan model F-15^R recording thermometer was used. Daily surface and five meter deep temperature measurements were recorded by the SIO Aquarium-Museum.

Sediment Nutrient Exchange

In situ net nutrient exchange was ascertained by analyzing the nutrient concentrations contained in a known water volume overlying a measured sediment area enclosed within plastic boxes over a predetermined time interval. A light and a dark box were used in each experiment to quantitate the exchange rate occurring during the day and night. Daylength was determined from data recorded on a bimetallic actinograph. To determine the spatial variability of nutrient release replicate light or dark boxes were used on four occasions. The time interval between replicate experiments was one to two months in order to estimate temporal averages over a summer and winter. To calculate exchange rates, nutrient concentration changes inside each box were corrected for the volume of water enclosed, the sediment area enclosed, the hours of incubation (generally 6 hours or less) and the hours of daylight and darkness. Each box measured, on its inside dimensions, 30 cm long x 30 cm wide x 15 cm high and when inserted five cm into sediment enclosed 900 cm^2 of sediment and nine liters of overlying sea water.

The ratio of 10 ml of water to 1 cm^2 of sediment is favorable for detecting nutrient changes in short periods of time. This is especially true in sediments of low organic carbon content where biological activity is low. Hale (5) used chambers of a similar ratio in Narragansett Bay, Rhode Island, and incubated them for 2 to 4 hours. In the present study two box types were used, the first box type in the experiments of April 2, 1971, through December 8, 1971, and box type #2 (Fig. 2) in the experiments of January 3, 1972, through February 20, 1973.

To implant the boxes in the sediment several of the silicone stoppers (size #6) were removed and the boxes gently pressed into the sediment to a depth of five cm, taking great care not to disturb the sediment surface. The insertion procedure did not alter nutrient concentrations inside compared to outside the box as long as the sediment surface was not disturbed, if disturbance was observed the box was relocated. The silicone stoppers were left out and a sample bottle was placed onto the sample port tube.

The sample bottle consisted of a one liter polyethylene bottle, the cap was replaced with a silicone stopper (size #6) through which a piece of silicone tubing (16 mm O.D. x 9.5 mm I.D.) protruded by approximately 15 cm. A small hole was made through the bottom of the sample bottle and plugged with a silicone stopper (size #00). The sampling port consisted of a T-connector on the inside of the box connected to a piece of silicone tubing (16 mm O.D. x 9.5 mm I.D.) and glass tubing (8 cm long) on the outside of the box through a silicone stopper. The openings of the T-connector inside the box were oriented parallel to the top of the box and about two cm beneath it. Both the silicone tubing on the sample bottle and sample port were closed by means of a spring clamp.

Water samples from inside the box were obtained by first pushing the silicone tubing of the sample bottle over the glass tubing extension of the sample port. The sample bottle, which immediately before use was flushed with nitrogen gas in the laboratory and at the water surface over the sample site had most of the gas expelled, was almost completely compressed by the nearly two atmospheres of hydrostatic pressure at 18.3 m. Next, the spring clamps were removed from the pieces of tubing and after the sample bottle was one-half full the sample bottle was gently squeezed forcing the water back into the box mixing it completley. Dye experiments in the box showed complete mixing using this sampling procedure.

As the openings of the T-connector were oriented parallel to the top of the box and, therefore, to the sediment surface, gently squeezing water back into the box did not disturb the sediment surface. The sample bottles were then filled until no additional water entered. The small silicone stopper on the bottom of

the bottle was removed and again gently squeezing the bottle, the residual nitrogen was voided. The silicone stopper was replaced and the sample bottle completely filled. To remove the sample bottle the two spring clamps were replaced
and the sample bottle was retracted from the glass tubing.

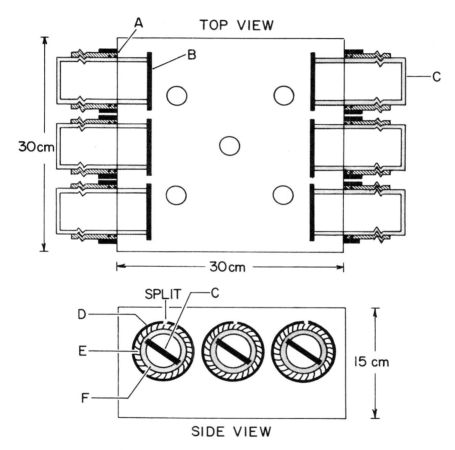

Figure 2. Box type #2 showing "O" ring (A), end stop (B; 7.6 cm diameter clear-acrylic
disc, 3.2 mm thick), plunger handle (C; 6.35 mm clear-acrylic rod), outer plunger
support (D; clear-acrylic tubing 6.95 cm O.D. × 7.3 cm I.D. × 2.5 cm high, split
at a point to allow for O.D. of "O"-ring support), "O"-ring support (E; clear-
acrylic tubing, 7.6 cm O.D. × 6.35 cm I.D. × 9.5 cm high) and plunger (F; 6.35
cm O.D. × 5.1 cm I.D. × 19.0 cm long, is made of clear-acrylic tubing; outside
machined to allow it to pass through "O"-ring support and to be water tight with
use of "O" rings; end stop glued to plunger). Outer support and "O"-ring support
glued to box which is made of 9.5 mm clear-acrylic plastic. Silicone stoppers and
sample port are not shown. [Box type #1 had two polyethylene bags sealed to the
box through one of the top holes instead of plungers.]

When the boxes were first put in position the silicone stoppers in the top were left out allowing the sampled water to be replaced by similar bottom water. After the initial sampling the stoppers were replaced and no new water entered the box. Having the sides of the box sunk five cm assured not only that the boxes were anchored to the bottom but also that outside water would not have time to diffuse into the box or be forced into the box by pressure differentials between the inside and outside of the box created by waves passing overhead. Whereas it is known that waves can cause water movement in submerged sands (28, 22), the degree of movement is dependent on the pressure head (wave height), water depth and the physical properties of the sediments. As these experiments were never performed when wave height was greater than one meter, the amount of "pumping" in these fine sands was not enough to have an effect at a depth of five cm.

The two box types used alternate methods to prevent water entry after the initial sampling. In box type #1, two collapsed polyethylene bags (one inside the other) were placed through one of the box openings, cemented to the box and plugged with a silicone stopper. When taking a subsequent water sample the stopper was removed and the bags filled with water. In box type #2 (Fig. 2), "O"-ring sealed plungers were placed on the sides of the box and during sampling the plungers were pushed into the box to compensate for the water removed.

Controls on the nutrient changes caused solely by the activity of the plankton in the water above the sediment were performed in two-liter glass stoppered bottles. These bottles (a dark and light bottle were used) were filled with bottom water and incubated next to the boxes. For 17 months these controls were routinely used and their activity compared to the sediment activity. It was concluded that the plankton activity did not significantly affect the exchange rates measured in the boxes.

A word of caution on a problem encountered at this exposed coastal station may be useful. Surge activity at times was enough to scour the sand away from the edges of the boxes (sunk five cm) allowing free interchange of inside and outside water. By fashioning deflectors to place around the boxes many experiments were saved, however, approximately 25% of attempted experiments were rendered useless after surge activity had scoured away one or more corners.

Chemical Analysis

Ammonia (NH_4^+), nitrate (NO_3^-), nitrite (NO_2^-), phosphate (PO_4^{-3}), dissolved organic carbon, nitrogen and phosphorus (DOC, DON and DOP) were all analyzed by the methods outlined in Strickland and Parsons (26). All methods were modified to use only 10 ml of sample.

The treatment of the water prior to analysis consisted of filtering the water at a vacuum of 170 mm Hg through combusted (450°C for 2 hr) 2.4 cm ultra-fine glass fiber filters (Reeve Angel 948H) using chromic-acid cleaned and baked

(550°C for 4 hr) glass filtration units. Likewise the 250-ml glass-stoppered bottles used for storing the filtrates at −20°C until analysis were all acid cleaned and baked. These precautions were taken to minimize organic contamination.

RESULTS
Temperature

Temperature data from the SIO pier and the experimental site are plotted in Fig. 3. It is evident that not only was there a large variation of temperature with season but temperature differences between the surface and bottom waters exhibit large variations. Daily temperature recordings also exhibited variations. In winter there was virtually no change in temperature during a day, while in summer it would rise and fall by as much as 7°C in a single 24 hr period. In the spring and fall bottom water temperature oscillations were between these extremes varying only 1 to 4°C daily.

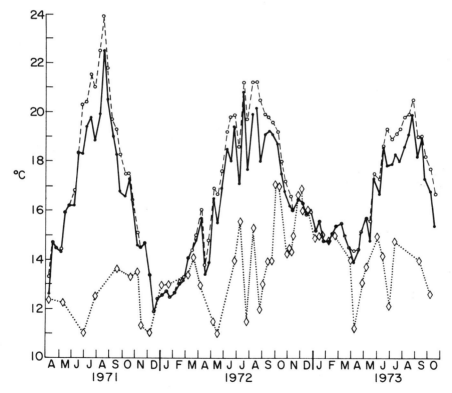

Figure 3. Temperature at experimental site (◊) and at the SIO pier at the surface (○) and 5 meter depth (●).

The large temperature fluctuations observed in the bottom water were due to semidiurnal internal waves and not the heating and cooling of the bottom water mass (13). During winter conditions (mid-November or December to mid-March) the water column was isothermal. In summer, although the water column was stratified, internal waves created, semidiurnally, vertical and horizontal acceleration of the bottom water increasing turbulent mixing. For the purposes of this paper it has been assumed that benthic nutrient exchanges were eventually mixed throughout the overlying water column.

Nutrient Exchange

Figures 4, 5, 6, 7, 8, 9 and 10 show the exchange rates for NH_4^+, NO_3^-, NO_2^-, PO_4^{-3}, DOC, DON and DOP respectively. The mean net exchange value for each nutrient ever the two year period and the respective total ranges of net input (+) or removal (-) of nutrient from the water overlying the sediment are shown in Table 1.

Table 2 tabulates the results of the replicate dark and light box experiments respectively. The percent standard deviation about the mean was greater in the light than in the dark for all inorganic nutrients and DON. This probably results from the patchy distribution of benthic microalgae and their uptake and release of inorganic and organic compounds (6).

DISCUSSION

The nutrient exchanges measured in the course of this investigation allowed a conservative mean net exchange rate to be calculated over yearly time spans. During quiescent periods the rate of exchange was determined satisfactorily by this short term static approach. In periods of surge activity the nutrients contained in the surface layers of interstitial water were mixed into the bottom water. Creating a static situation by placement of a box over the sediment during these periods decreased the actual exchange rate to a degree dependent upon the depth to which the sediment was disturbed, the pre-disturbance concentration of nutrient in the interstitial water, the time taken to re-establish this concentration, and the length of time the box was left in place. Although these experiments were performed under varying intensities of surge they could not be performed during periods of high surge intensities due to the problem of scouring (discussed previously). Also, after a period of no surge activity the onset of surge disturbed the sediment and the contained nutrients were released into the bottom water. The static method utilized did not measure this quantity of nutrient. For the reasons mentioned above the net exchange rates measured were considered conservative.

Eppley et al. (3) give the only known values of primary production for this region over an extended (6 months) period. Using their average phytoplankton production rate of 1.2 gC $m^{-2} d^{-1}$ and a C:N uptake ratio of 106:16 (20) the calculated average nitrogen requirement was 0.21 gN $m^{-2} d^{-1}$. Summing the total

Table 1.   Nutrient exchange between the marine benthos and the overlying water. Values given as release (+) or uptake (−) of the nutrient by the benthos.

Nutrient	Mean ($\bar{x}$)	Range[1]
NH_4^+	$+872\ \mu$ mol m^{-2} d^{-1}	-47 to $+3290\ \mu$ mol m^{-2} d^{-1}
NO_2^-	$+34\ \mu$ mol m^{-2} d^{-1}	-5 to $+97\ \mu$ mol m^{-2} d^{-1}
NO_3^-	$-77\ \mu$ mol m^{-2} d^{-1}	-720 to $+647\ \mu$ mol m^{-2} d^{-1}
PO_4^{-3}	$+77\ \mu$ mol m^{-2} d^{-1}	-438 to $+502\ \mu$ mol m^{-2} d^{-1}
DOP	$+12\ \mu$ mol m^{-2} d^{-1}	-28 to $+59\ \mu$ mol m^{-2} d^{-1}
DON	$-75\ \mu$ mol m^{-2} d^{-1}	-1326 to $+1280\ \mu$ mol m^{-2} d^{-1}
DOC	$-583\ \mu$ mol m^{-2} d^{-1}	-30800 to $+23800\ \mu$ mol m^{-2} d^{-1}

[1] Range—these values represent the highest uptake rates observed and the highest release rates observed.

Table 2.   Results of replicate dark and light box experiments one and two (RD-I, II and RL-I, II). Units are μ mol m^{-2} d^{-1} released (+) or taken up (−). The means ($\bar{x}$) and the standard deviation as ± percent of the mean are given.

	ΔNH_4^+	ΔNO_2^-	ΔNO_3^-	ΔPO_4^{-3}	ΔDON	ΔDOP	ΔDOC
RD-I	+2.78	+0.260	0	+3.00		−0.880	−267
	+1.76	+0.140	0	+5.04	+17.6	+0.660	+ 83
	+3.90	+0.040	0	+3.06	+17.2	−1.10	+ 50
$\bar{x}$	+2.81	+0.147	0	+3.70	+17.4	−0.440	− 45
±	31%	61%	0	26%	0.1%	178%	230%
RD-II	+49.5	+1.46	+3.00	+11.1	+52.3	+1.48	−640
	+22.2	+0.994	+2.21	+ 3.70	+40.5	0	−563
	+47.8	+1.24	+3.60	+10.2	+27.4	+0.356	−328
$\bar{x}$	+39.8	+1.23	+2.94	+ 8.33	+40.1	+0.612	−510
±	31%	15%	19%	40%	25%	103%	32%
RL-I	+15.0	+2.05	−1.09	+2.07	+8.51	−0.63	−146
	+11.4	+1.24	−4.87	+3.84	+4.80	+0.17	−128
	+ 4.34	+1.16	−1.53	+1.85	− 26.6	−0.59	−473
$\bar{x}$	+10.2	+1.48	− 2.50	+2.59	− 4.43	−0.35	−249
±	43%	27%	68%	34%	355%	105%	78%
RL-II	no	+0.638	−1.30	−0.732	−26.3	+0.788	− 78
	data	+0.019	−0.56	−1.29	+17.1	+0.788	+532
		+0.638	+4.5	−0.338	−32.1	+0.300	+423
$\bar{x}$		+0.432	+0.88	−0.787	−13.8	+0.625	+292
±		68%	293%	50%	195%	45%	111%

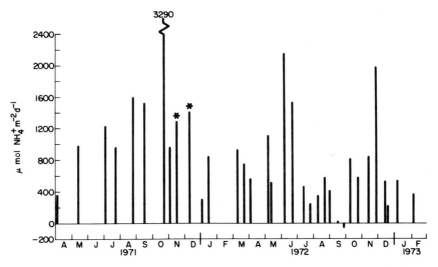

Figure 4. Ammonia release (+) or uptake (−) by the benthos (μ mol NH_4^+ m^{-2} d^{-1}), (*) values used data from only dark boxes, all other values used data from light and dark boxes.

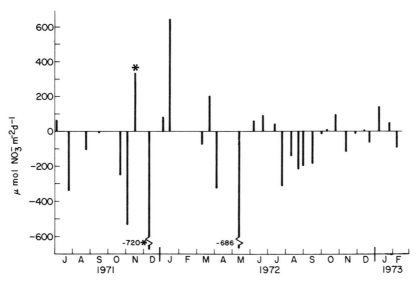

Figure 5. Nitrate release (+) or uptake (−) by the benthos (μ mol NO_3^- m^{-2} d^{-1}), (*) values used data from only dark boxes, all other values used data from light and dark boxes.

benthic nitrogen exchanges and assuming that the DON would be mineralized in the water column to a form utilizable by the phytoplankton yields a net input of 0.011 gN $m^{-2} d^{-1}$ into the water column from the benthos. This amounted to 5% of the daily nitrogen requirement of the phytoplankton. This value is increased to 6% if one does not consider organic nitrogen flux.

Using a C:P uptake ratio of 106:1 (20), the average daily requirement of phosphorus by the phytoplankton was 0.029 gP $m^{-2} d^{-1}$. Summing the phosphorus exchanges and assuming that the DOP compounds present were utilized by

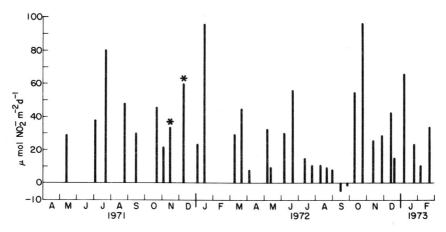

Figure 6. Nitrite release (+) or uptake (−) by the benthos (μ mol $NO_2^- m^{-2} d^{-1}$), (*) values used data from only dark boxes, all other used data from light and dark boxes.

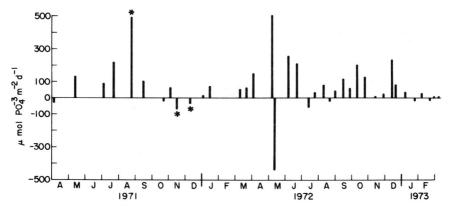

Figure 7. Phosphate release (+) or uptake (−) by the benthos (μ mol $PO_4^{-3} m^{-2} d^{-1}$), (*) values used data from only dark boxes, all other values used data from light and dark boxes.

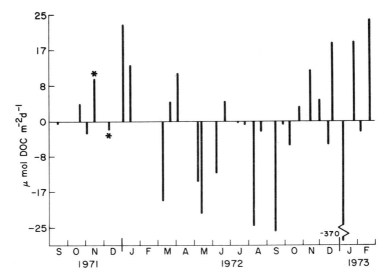

Figure 8. Dissolved organic carbon release (+) or uptake (–) by the benthos (μ mol DOC $m^{-2} d^{-1}$), (*) values used data from only dark boxes, all other values used data from light and dark boxes.

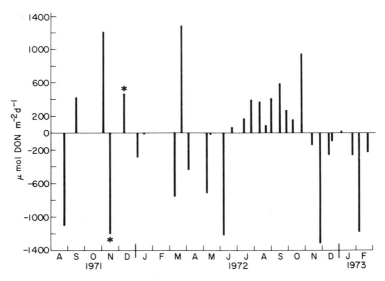

Figure 9. Dissolved organic nitrogen release (+) or uptake (–) by the benthos (μ mol DON $m^{-2} d^{-1}$), (*) values used data from only dark boxes, all other values used data from light and dark boxes.

the phytoplankton yields a net input of phosphorus to the water of 0.0028 gP $m^{-2}d^{-1}$. Therefore, 10% of the daily phytoplankton requirement of phosphorus was released from the benthos. If one considers only inorganic phosphorus exchange this value becomes 8% of the daily phosphorus requirement.

It appears that benthic nutrient exchange of nitrogen and phosphorus provide a significant fraction of the daily phytoplankton requirement. This result would agree with the conclusion of both Hale (5) and Rowe et al. (24). This conclusion, however, may be presumptuous of these data.

If one considers only the water within the La Jolla Bight over the 1.5 km long extended study area and assumes a mean net advection of 1.5 km d^{-1}, which is perhaps a conservative estimate of the actual mean net speed (Dr. Peter W. Hacker, Dept. Earth and Planetary Sciences, Johns Hopkins University), the residence time of water over the study area would be one day. In this region the yearly mean bottom water concentration of $NH_4^+ + NO_3^- + NO_2^- + DON$ was 10.8 μM N or in the entire 18.3 m water column overlying one m^2 of sediment the total nitrogen content equaled 2.0 X 10^5 μ mol NO_3^- and DON each comprised approximately 44% of this, while NH_4^+ and NO_2^- comprised 10% and 2% respectively. The benthic release of 745 μ mol of nitrogen $m^{-2}d^{-1}$, therefore, represented only 0.4% of the nitrogen already in the water. For phosphorus the total water column concentration of PO_4^{-3} plus DOP was 1.8 X 10^4 μ mol (PO_4^{-3} was 82% of the total and DOP 18%) and the benthic release of phosphorus was 89 μ mol $m^{-2}d^{-1}$ or only 0.5% of the phosphorus already in the water.

The interpretation of these percentages in relation to the impact of benthic nutrient exchange on the overlying water nutrient chemistry and phytoplankton

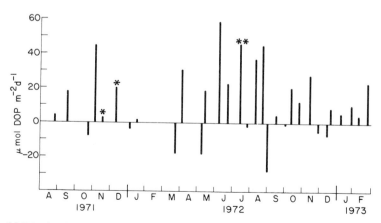

Figure 10. Dissolved organic phosphorus release (+) or uptake (−) by the benthos (μ mol DOP m^{-2} d^{-1}), (*) values used data from only dark boxes, (**) values used data from only light boxes, all other values used data from light and dark boxes.

productivity in this region is that the mean yearly nitrogen and phosphorus benthic exchange resulted in an excess release (as regards phytoplankton requirement) of both nitrogen and phosphorus from the sediments, which, although significant as regards the phytoplankton requirement, was insignificant as regards the total content of nitrogen and phosphorus already in the water. It is recognized that the cumulative effect of benthic nutrient exchange over hundreds of kilometers of coastline can be enormous. Nonetheless, in the extended study area discussed here (Fig. 1) benthic nutrient exchange resulted in a net release of both nitrogen and phosphorus at rates exceeding the uptake ability of the phytoplankton in the time available, resulting in a net export of these nutrients from the region. If this station is not an exception, one can only conclude that in regions such as this there is not a close coupling between phytoplankton production, water nutrient chemistry and benthic nutrient exchange.

ACKNOWLEDGEMENTS

This study was supported by USPHS grant GM-01065, U. S. AEC Contract AT(11-1)GEN 10, P.A. 20, ERDA Contract E(11-1)-3279 and NOAA Sea Grant No. 04-3-158-22, and was part of a thesis for the Ph.D. degree submitted to the University of California, San Diego, Scripps Institution of Oceanography. In particular, I acknowledge the guidance and support of Dr. Angelo F. Carlucci.

REFERENCES

1. Arthur, R.S. 1960. Variations in sea temperature off La Jolla. J. Geophys. Res. 65: 4081-4086.
2. Carritt, D.E., and S. Goodgal. 1954. Sorption reactions and some ecological implications. Deep-Sea Res. 1: 224-243.
3. Eppley, R.W., F.M.H. Reid and J.D.H. Strickland. 1970. In J.D.H. Strickland (ed.), The ecology of the plankton off La Jolla, California, in the period April through September, 1967. Part III. Bull Scripps Inst. Oceanogr. 17: 33-42.
4. Fager, E.W. 1968. A sand-bottom epifaunal community of invertebrates in shallow water. Limnol. Oceanogr. 13: 448-464.
5. Hale, S.S. 1974. The role of benthic communities in the nutrient cycles of Narragansett Bay. M.S. Thesis, Univ. Rhode Island, Kingston, Rhode Island, 123 p.
6. Hartwig, E.O. 1974. Physical, chemical and biological aspects of nutrient exchange between the marine benthos and the overlying water. Ph.D. dissertation, Univ. Calif., San Diego, Scripps Inst. Oceanogr., La Jolla, Calif., 174 p.
7. _____. 1976. Nutrient cycling between the water column and a marine sediment. 1. Organic carbon. Mar. Biol. 34:285-295.
8. Hayes, F.R. 1964. The mud-water interface. Oceanogr. Mar. Biol. Ann. Rev. 2: 121-145.
9. Hirota, J. 1973. Quantitative natural history of Pleurobrachi bachei A. Agassiz in La Jolla Bight. Ph.D. dissertation, Univ. Calif. San Diego, Scripps Inst. Oceanogr., La Jolla, Calif. 192 p.
10. Hobbie, J.E., and C.C. Crawford. 1969. Respiration corrections for bacterial

uptake of dissolved organic compounds in natural waters. Limnol. Oceanogr. 14: 528-533.

11. Inman, D.L. 1953. Areal and seasonal variation in beach and nearshore sediments at La Jolla, California, U.S. Beach Erosion Board, Tech. Memo No. 39: 1-82.

12. Kamykowski, D. L. 1972. Some physical and chemical aspects of the phytoplankton ecology of La Jolla Bay. Ph.D. Dissertation Univ. Calif., San Diego, Scripps Inst. Oceanogr., La Jolla, Calif. 269 p.

13. _____. 1974. Physical and biological characteristics of an upwelling at a station off La Jolla, California during 1971. Est. Coastal Mar. Sci. 2: 425-432.

14. Lee, G.F. 1970. Factors affecting the transfer of materials between water and sediments. Literature review no. 1. Univ. of Wisconsin Water Resources Center Eutrophication Program. Madison, Wis. 50 p.

15. Mortimer, C.H. 1971. Chemical exchange between sediments and water in the Great Lakes: speculations on probable regulatory mechanisms. Limnol. Oceanogr. 16: 387-404.

16. Nichols, F.H. 1974. Sediment turnover by a deposit-feeding polychaete. Limnol. Oceanogr. 19: 945-951.

17. Pamatmat, M.M. 1971. Oxygen consumption by the seabed. VI. Seasonal cycle of chemical oxidation and respiration in Puget Sound. Int. Rev. ges. Hydrobiol. 56: 769-793.

18. _____. 1973. Benthic community metabolism on the continental terrace and in the deep sea in the North Pacific. Int. Rev. ges. Hydrobiol. 58: 345-368.

19. Pomeroy, L. R., E.E. Smith and C.M. Grant. 1965. The exchange of phosphate between estuarine water and sediments. Limnol. Oceanogr. 10: 167-172.

20. Redfield, A.C., B.H. Ketchum and F.A. Richards. 1963. The influence of organisms on the composition of sea-water, p. 26-77. In M.N. Hill (ed.), The Sea, Vol. 2, Interscience, New York.

21. Reid, J.L., Jr., G.I. Roden and J.G. Syllie. 1958. Studies of the Californian current system. Calif. Co-operative Oceanic Fish. Investigations (CAL-COFI) p. 292-324 Progr. Rep. July 1956 - Jan. 1958.

22. Riedl, R.J., N. Huang, and R. Machan. 1972. The subtidal pump: a mechanism of interstitial water exchange by wave action. Mar. Biol. 13: 210-221.

23. Rittenberg, S.C., K.O. Emery, and W.L. Orr. 1955. Regeneration of nutrients in sediments of marine basins. Deep-Sea Res. 3: 23-45.

24. Rowe, G.T., C.H. Clifford and K.L. Smith, Jr. 1975. Benthic nutrient regeneration and its coupling to primary productivity in coastal waters. Nature 255: 215-217.

25. Smith, K.L., Jr. 1973. Respiration of a sublittoral community. Ecology 54: 1065-1075.

26. Stirckland, J.D.H., and T.R. Parsons. 1972. A practical handbook of seawater analysis, 2nd ed. Bull. Fish. Res. Bd. Canada 167: 1-310.

27. Teal, J.M., and J. Kanwisher. 1961. Gas exchange in a Georgia salt marsh. Limnol. Oceanogr. 6: 388-399.

28. Webb, J.E., and J. Theodor. 1968. Irrigation of submerged marine sands through wave action. Nature 220: 682-683.

29. Williams, P.J. LeB. 1970. Heterotrophic utilization of dissolved organic compounds in the sea. I. Size distribution of population and relationship between respiration and incorporation of growth substrates. J. mar. biol. Ass. U. K. 50: 859-870.

NUTRIENT-PHYTOPLANKTON RELATIONSHIPS

IN NARRAGANSETT BAY DURING

THE 1974 SUMMER BLOOM

Miles J. Furnas, Gary L. Hitchcock and
Theodore J. Smayda
Graduate School of Oceanography
University of Rhode Island
Kingston, R.I. 02881

ABSTRACT: Nutrient and ^{14}C uptake by natural phytoplankton populations was measured. Concentrated populations were incubated with uptake saturating concentrations of nitrate, silicate and phosphate, and nutrient uptake was then monitored at 30 min intervals over an eight hour period to yield Vmax, V, ρ_N and ρ_{Si}. Changes in the amount of particulate matter and absolute rates of uptake were related to *in situ* levels of phytoplankton biomass and dissolved nutrients, and replenishment (generation) times (R) estimated. Independent estimates of R were derived from calculations of particulate nitrogen (N_p) and silica production (Si_p) based on ^{14}C uptake and appropriate elemental ratios. Forty-two percent of the annual carbon production of 308 g m^{-2} occurred during July and August when *in situ* nutrient levels were very low. The hourly uptake rates for nitrate (ρ_N) ranged from 0.118 to 0.136; ρ_{Si} was 0.007 to 0.150 μM. The daily N supply for the water column would have to be replenished one to 12 times daily to support N_p and 2.5 to 17 days for Si_p. It would take from 2 to 5 days for the nitrogen excretion rates of the zooplankton (> 153 μm) and benthos to supply the daily phytoplankton nitrogen needs.

INTRODUCTION

The phytoplankton dynamics in Narragansett Bay, Rhode Island have been studied in an ongoing, long-term investigation (19, 20, 23, 24, 13, 14, 3, 9, 10). The winter-spring diatom bloom has been emphasized in studies to date; it begins as early as mid-December and usually terminates in May-June. The summer period is characterized by a series of secondary phytoplankton pulses

118

that exhibit considerable annual variations in time of occurrence, bloom duration, maximum abundance, and dominant species (Smayda, unpublished). The
maximum abundance has ranged from 7,000 cells ml^{-1} (1967) to 59,000 cells
ml^{-1} (1975); initiation of this seasonal bloom has varied from July (1971) to
September (1973); its duration has ranged from a couple of weeks (August
1969) to almost continuously from July through September (1968); annual
differences in dominant species also occur (21).

Some field observations and experimental data suggest that nutrient regulation and grazing influence the erratic summer phytoplankton dynamics in
Narragansett Bay. The present study evaluates the rates of nitrate and silicate
uptake by natural phytoplankton populations during a summer period. These
requirements are then compared to the rates of nitrogen excretion by the
zooplankton and benthos.

METHODS

From May through September 1974 weekly samples were collected by Niskin
bottles in lower Narragansett Bay at station 2 (20) located at 41°34′07″N,
71°23′31″W. Samples were collected near the bottom (ca. 8 m depth), mid-
depth (4 m), and surface at approximately 0900 hr for routine analyses. The
phytoplankton were counted live using a Sedgwick-Rafter Chamber. Dissolved
ambient nutrient concentrations were measured following filtration of the samples through pre-rinsed Whatman GFC glass fiber filters. In the nutrient uptake
experiments, samples were first filtered through pre-rinsed Millipore® (Type
HA) filters. Ammonia was measured by the method of Solórzano (26) and urea
following McCarthy (15). Nitrate + nitrite (30) and silicate (1) were measured on
a Technicon Autoanalyzer II ® and phosphate (6) on a Technicon Autoanalyzer
I®. Zooplankton tows were taken with a 0.3 m diameter #10 (153 μm aperture)
mesh net fitted with a flow meter and towed obliquely.

Primary production (27) was estimated for "pooled" samples obtained by
mixing equal volumes from each sampling depth. Since Narragansett Bay is
usually unstratified year-round (23), the "pooled" sample is believed to give a
representative sample of the "mixed layer" at station 2. The sample was gently
mixed, filtered through a No. 10 net to remove the larger zooplankton, added to
50 ml bottles (in replicate), enriched with 1 μCi ^{14}C, and incubated for 24 hr in
five light compartments of a plexiglass tank (located on the laboratory dock)
through which seawater flowed. The light intensities (created with neutral
density screens) were 100%, 60%, 25%, 10% and 3% of incident radiation. Dark
bottles were prepared as for the light bottles, but wrapped in aluminum foil.
Fixed ^{14}C was measured in a liquid scintillation counter. Chlorophyll was also
measured.

Seven nutrient uptake experiments were carried out between 15 July and 24
September; proximate analyses were begun on 24 June. Phytoplankton was
obtained by concentrating 12 liters of the *surface* sample to 2 to 6 liters by

reverse filtration (2) using Millipore Microweb® filters having a 3 μm mean pore diameter, after first gently prefiltering the sample through a 200 μm mesh net to remove larger zooplankton. The concentrated phytoplankton population ranged from 19,750 to 94,100 cells/ml in the July and August experiments (85 to 99% were diatoms), and 2,875 cells/ml in the September experiment when diatoms represented only 6% of the population. Sample processing began approximately one hour after field collection; about 4 hr were required to concentrate the phytoplankton, which was carried out in a constant temperature room (15 C).

Particulate carbon and nitrogen in the surface concentrate were determined after filtration onto glass-fiber filters, lyophilization, and combustion in a Hewlett-Packard CHN analyzer (22). Particulate silica was determined by Paasche's (17) modification of Werner's (29) method; particulate phosphorus by the method of Menzel and Corwin (16) as modified by Corwin (pers. comm.); and chlorophyll a by Yentsch and Menzel's (31) fluorometric technique using the equations of Lorenzen (11).

Twelve net tows (25 μm mesh) were made during the 1975 diatom bloom (March through May). After straining these samples through a 153 or 102 μm mesh screen to remove zooplankton, the particulate carbon, nitrogen, silicon, phosphorus and chlorophyll contents were determined. The ratios of these constituents were calculated for use as biomass conversion factors in the nutrient uptake experiments. Net phytoplankton was not collected during the summer for this purpose because of ctenophore abundance.

To determine nutrient uptake, 2 to 3 liters of the phytoplankton concentrate were placed into a 3-liter flask fitted with aeration and sampling connections, and incubated at 20 C and 0.14 ly min^{-1}. This temperature approximated ambient conditions. A pyrex flask (soaked in distilled water for one month to remove readily dissolved silica) was used for the first 2 experiments and polycarbonate thereafter (n = 5). Inorganic nutrients were added to give initial experimental concentrations >7.5 μM NO_3, >8 μM SiO_2 and >2.5 μM PO_4. These nutrient concentrations, within the range found in Narragansett Bay (24), gave apparent nutrient saturated uptake rates (Fig. 1). The initial nutrient concentrations were determined immediately following enrichment, and thereafter at intervals of 30 mins or less over the next 5 to 8 hr. Fig. 1 illustrates one of the better data sets obtained. The results of the nitrate and silicate uptake experiments are presented in this paper. For extrapolation to natural populations, the nutrient uptake rates and amounts of C, N and Si in the concentrated material were normalized to ambient bay concentrations.

Based on the NO_3 and Si uptake rates of the phytoplankton concentrate, *in situ* uptakes rates and turnover (i.e., replenishment or depletion) times of plankton biomass and the inorganic nutrients required to support these biomass changes were computed as follows. The maximum average uptake rate (Vmax), calculated (μM hr^{-1}) over the longest time interval where uptake was essentially linear (Fig. 1), was normalized per unit of equivalent phytoplankton nutrient:

$$\frac{Vmax}{PN} \quad \text{and} \quad \frac{Vmax}{PSi} \tag{1}$$

having dimensions of time^{-1}. The concentrations (μM) of living particulate nitrogen (PN) and silicon (PSi) in the surface concentrate were estimated from the chlorophyll content by applying N:Chl and Si:Chl ratios of 6.51:1 and 8.54:1, respectively, derived from the net-tow collections.

In situ uptake rates were calculated using the Michaelis-Menten expression following the procedure of MacIssac and Dugdale (12):

$$V = \frac{Vmax\ S}{K_s + S} \tag{2}$$

where V has the dimension of time^{-1}, and is the velocity of uptake of N or Si per unit time per unit PN or PSi; S is the *in situ* concentration of NO_3-N or Si; K_s is the concentration at which V = Vmax/2. A K_s value of 0.5 μM was used for the NO_3 calculations (Fig. 1) and 2.0 μM for Si, which is about the mid-range of some published values (18, 5). Our experiments did not yield unambiguous values of K_s for silicate uptake.

The total uptake (transport rate), (ρ), of the surface phytoplankton biomass *in situ* was calculated from:

$$\rho_N = V \cdot PN \tag{3}$$

$$\rho_{Si} = V \cdot PSi \tag{4}$$

following MacIssac & Dugdale (12), where ρ is μM of N or Si time^{-1}.

The *in situ* replenishment (R) times of living particulate N and Si, and that of dissolved nutrients, [ ], needed to support this particulate turnover were calculated, respectively, from:

$$R = \frac{PN}{\rho_N} \quad \text{and} \quad R = \frac{[N]}{\rho_N} \tag{5}$$

$$R = \frac{PSi}{\rho_{Si}} \quad \text{and} \quad R = \frac{[Si]}{\rho_{Si}} \tag{6}$$

R does not truly estimate turnover time, but is rather the time that it would take for ρ_N and ρ_{Si} to yield a particulate biomass (PN, PSi) equivalent to that at the start of the experiment, and the time to exhaustion of the initial nutrient concentrations ([N], [Si]) at ρ_N and ρ_{Si}. Given the light dependence of nitrate reductase formation and its diel periodicity (4), ρ_N has been estimated assuming 12 hr uptake for NO_3 per day. Local daylength ranges from 12.5 (September) to

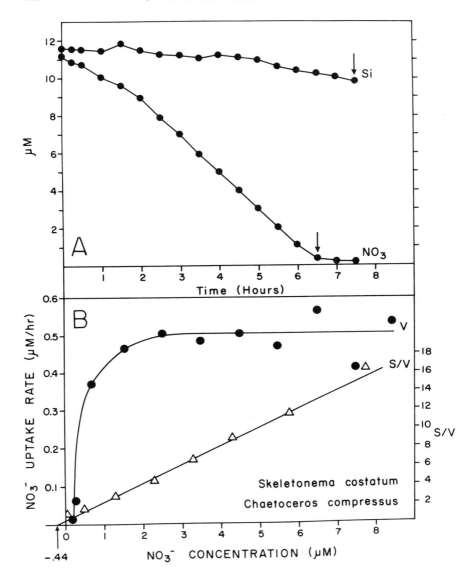

Figure 1. A. Changes in concentrations of Si and NO₃ after enrichment of a phytoplankton population dominated by *Skeletonema costatum* and *Chaetoceros compressus* concentrated from the surface waters of Narragansett Bay. The time span between time 0 and ↓ represents the period over which Vmax was calculated. B. Nitrate uptake rate (V) of this population versus nitrate concentration (S) of the medium. The half-saturation constant (0.44) is given as the negative S-intercept of the linear regression of S/V versus S.

15 hr (June) during the summer. The present, limited data set does not warrant modification of the Michaelis-Menten expression (equation 2) to include the influence of light on nitrate uptake (12). Silicate uptake was calculated using a 24 hr day.

The ^{14}C uptake measurements permit independent calculations of R for particulate and dissolved N and Si *in situ* based on the enhanced nutrient uptake studies. The daily production (μM) of particulate nitrogen (N_p) and particulate silicate (Si_p) is first estimated from carbon production (C_p):

$$N_p = \frac{C_p}{C/N} \quad \text{and} \quad Si_p = \frac{C_p}{C/Si} \tag{7}$$

C/N and C/Si are the atomic ratios in the surface concentrate measured on the day of the ^{14}C experiment. C/N varied from 6.54 to 10.79, and C/Si from 3.98 to 48.57. Replenishment times are then computed as for nutrient uptake:

$$R = \frac{PN}{N_p} \quad \text{and} \quad R = \frac{[N]}{N_p} \tag{8}$$

$$R = \frac{PSi}{Si_p} \quad \text{and} \quad R = \frac{[Si]}{Si_p} \tag{9}$$

Here, the replenishment times for PN and PSi are also measures of the generation times, and the number of replacements of initial [N] and [Si] to yield N_p and N_{Si}, respectively.

Particulate carbon (PC) generation time was calculated after converting the measured chlorophyll of the "pooled" sample into its carbon equivalent from the C/Chl ratio of about 60 found for the net tow analyses:

$$R = \frac{PC}{C_p} \tag{10}$$

RESULTS
Plankton and Nutrient Cycles

The weekly fluctuations in phytoplankton numerical abundance and chlorophyll at 0 m from May through September are shown in Fig. 2A. The population progressively increased by 50-fold from a minimum in mid-May (about 200 cells ml^{-1}) to about 10,000 cells ml^{-1} in mid-June; decreased sharply until early July (600 cells ml^{-1}); then increased precipitously once again to about 20,000 cells ml^{-1}, and generally oscillated between this level and 10,000 cells ml^{-1} until mid-August. The annual maximum (31,200 cells ml^{-1}) for 1974 occurred on 4 September. The population then collapsed and reached its seasonal minimum of 20 to 40 cells ml^{-1} after mid-September.

Skeletonema dominated from mid-May through early June; thereafter through mid-June *Leptocylindrus danicus* dominated and several *Thalassiosira* species in late June. From early July through September the population was again dominated by *Skeletonema costatum*; on 13 August *Chaetoceros compressus* was also abundant. Thus, diatoms dominated the phytoplankton community, exclusive of large populations of the dinoflagellates *Prorocentrum redfieldii* and *Pror. triangulatum*, and of *Olisthodiscus luteus* in mid-June. The chlorophyll cycle generally followed phytoplankton numerical abundance,

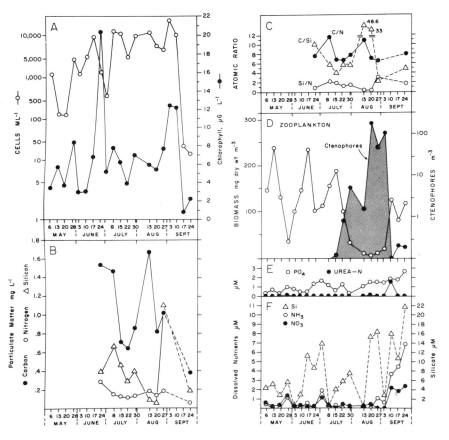

Figure 2. A. The cycles of phytoplankton numerical abundance and chlorophyll at 0 m at Station 2 in lower Narragansett Bay from May to September, 1974. B. The *in situ* concentrations of particulate carbon, nitrogen and silicon based on the phytoplankton concentrate from 0 m and, C, their atomic ratios. D. Mean zooplankton biomass and ctenophore abundance in the 8 m water column. E, F. Nutrient concentrations at 0 m.

except for the annual maximum (20.3 μg liter^{-1}) on 24 June which accompanied a bloom of the microflagellate *Olisthodiscus luteus*.

The fewer measurements of the particulate matter cycles reveal that carbon fluctuated most and nitrogen least (Fig. 2B). Particulate silica progressively decreased from early July through late-August. The oscillations in atomic ratios of the particulate nutrients are shown in Fig. 2C.

The seasonal fluctuations in zooplankton biomass are characterized by major peaks in mid-May, mid-June and mid-July (Fig. 2D). Thereafter, biomass decreased sharply and persisted at its seasonal low throughout August, followed by a September increase. The zooplankton cycle after mid-July was clearly regulated by the ctenophore, *Mnemiopsis leideyii*, which grazes on other zooplankton (Fig. 2D).

The cycles of primary production and assimilation numbers during May - September (Fig. 3) indicate that the phytoplankton population was then extremely viable, notwithstanding the very low dissolved nitrogen (usually <0.5

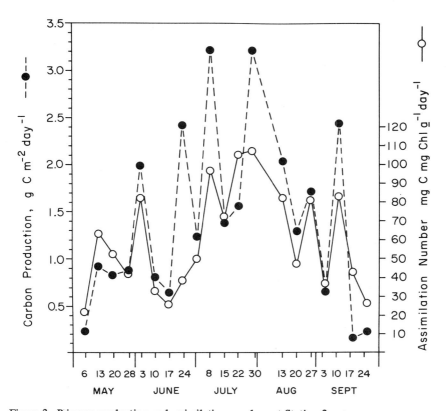

Figure 3. Primary production and assimilation numbers at Station 2.

Table 1.   Primary production (g C m^{-2}) during 1974 at Station 2, based on pooled sample.

	g C m^{-2}	Percent of Annual Total
January - April	51.9	17
May	22.6	7
June	42.2	14
July	65.8	21
August	63.1	21
September	25.7	8
October - December	36.6	12
Annual Total	307.9	

μg-at liter^{-1}) and silicate concentrations (Figs. 2E, F). About 71% of the annual carbon production of 308 g m^{-2} occurred from May through September, with 42% of the annual production occurring in July and August (Table 1). This extraordinary production throughout the summer, notably in July-August, when nutrient levels were persistently low indicates the importance of nutrient turnover to phytoplankton growth during this period.

Uptake and Replenishment of Particulate and Dissolved NO_3 and Si.

The uptake of nitrate and silicate by the concentrated population normalized to the *in situ* surface populations (Fig. 1) and related data are given in Table 2. The hourly, positive uptake (transport) rate for nitrate (ρ_N) ranged from 0.118 to 0.136 μM NO_3-N; ρ_{Si} varied from 0.007 to 0.150 μM.

The replacement times (R) for particulate carbon; the production of particulate nitrogen and silica *in situ*, based on both nutrient and ^{14}C uptake, and the associated replenishment times of dissolved nitrate and silica are presented in Table 3. Growth was very rapid from mid-July through August when carbon doubled in 13 to 31 hr, which corresponds to a daily carbon division rate of 0.78 to 1.79. Particulate nitrogen (based on ^{14}C uptake) doubled every 7 to 32 hr, in good agreement with carbon generation times; but particulate silicon generation times were much more rapid (3 to 5 hr), except for August 13 and 20. Where comparisons are possible, the replenishment (replacement) times of particulate nitrogen (28 to 52 hr) based on nitrate uptake sometimes agreed well with the ^{14}C-based rates and with carbon turnover. Particulate silica turnover, however, was usually slower than its ^{14}C-based rate, and inconsistent with the nitrogen results.

Dissolved nitrate is replenished very rapidly *in situ* to allow the calculated summer production rates of particulate nitrogen. The two independent estimates of this turnover are in good agreement. The *in situ* nitrate concentrations must be replaced every 0.5 to 6 hr to allow the particulate nitrogen produced during ^{14}C uptake, which compares to 2.4 to 6.2 hr based on nitrate uptake. Longer replenishment times (41 to 746 hr) are required to support particulate silica production.

In the last experiment on 24 September the phytoplankton population was at its 5 month low (Fig. 2A); primary production was low (Fig. 3); nutrients were at their maximal levels (Figs. 2E, F), and detectable nitrate uptake did not occur (Table 2). The longer generation times for particulate carbon and nitrogen and replenishment times for dissolved nutrients observed then (Table 3) are consistent with the occurrence of a slow growing population. This suggests that the

Table 2. Standing stocks, nutrient uptake data, and production estimates.

	July			August			Sept.
	15	22	29	13	20	27	24
NO_3 (μM)	n.d.	.39	.16	.24	.68	n.d.	2.40
SiO_2 (μM)	3.91	5.75	7.53	.28	14.23	16.41	21.84
$S°/_{oo}$	31.64	31.64	32.15	31.64	32.15	32.67	31.64
[1] Chl a ($\mu g\ L^{-1}$)	4.72	3.34	4.92	7.53	8.74	10.37	1.95
[2] PN (μM)	2.21	1.57	2.28	3.50	4.07	4.85	.93
[2] PSi (μM)	1.42	1.00	1.50	2.28	2.67	3.17	.61
			Nitrate uptake				
Vmax ($\mu M\ hr^{-1}$)	.209	.270	.530	.417	.008	.302	−.093
Vmax (hr^{-1})	.095	.172	.232	.119	.002	.062	−.100
V (hr^{-1})	?	.075	.056	.039	?	?	−.083
ρ_N ($\mu M\ hr^{-1}$)	?	.118	.128	.136	?	?	−.077
			Silicon uptake				
Vmax ($\mu M\ hr^{-1}$)	.036	−.046	.191	.058	.052	.026	.043
Vmax (hr^{-1})	.025	−.046	.127	.025	.019	.008	.070
V (hr^{-1})	.017	−.034	.100	.003	.017	.007	.064
ρ_{Si} ($\mu M\ hr^{-1}$)	.024	−.034	.150	.007	.045	.022	.039
			Carbon-based production estimates				
[3] PC (μM)	26.00	16.50	36.50	28.50	27.50	37.00	9.80
C_p ($\mu M\ day^{-1}$)	31.20	28.80	65.20	39.10	21.30	49.90	3.60
[4] C/N (atoms)	6.79	6.65	7.70	10.79	7.00	6.54	7.87
[4] C/Si (atoms)	3.98	5.73	5.58	48.57	33.22	2.16	4.83
[5] N_p ($\mu M\ day^{-1}$)	4.60	4.33	8.46	3.64	3.05	7.62	.45
[5] Si_p ($\mu M\ day^{-1}$)	7.84	5.02	11.68	.80	.64	23.09	.73
[6] N_p ($\mu M\ day^{-1}$)	2.92	2.69	6.09	3.65	1.99	4.66	.34
[7] Si_p ($\mu M\ day^{-1}$)	1.90	1.76	3.98	2.38	1.30	3.04	.22

[1] Surface concentrate;
[2] Computed using average N:Chl (6.51) or Si:Chl (8.54) ratios of net-tow material;
[3] In "pooled" sample;
[4] In concentrated sample;
[5] Based on weekly ratios in concentrated sample;
[6] Based on C:N ratio of 10.7 (atoms) for net-tow material;
[7] Based on C:Si ratio of 16.4 (atoms) for net-tow material.

Table 3. Doubling times (hours) of particulate C, N and Si at 0 m based on ^{14}C uptake (a) and nutrient uptake (b) and corresponding *in situ* replenishment times (hours) of dissolved NO_3 and Si; value in ( ) represents equivalent daily C growth rate (doublings day^{-1}).

	Doubling Time					Replenishment time			
	PC	PN		PSi		[NO_3]		[Si]	
	a	a	b	a	b	a	b	a	b
15 July	20 (1.20)	12	–	4	60	1.6	–	20	163
22 July	14 (1.75)	9	28	5	–29	2.2	6.2	28	–170
29 July	13 (1.79)	7	34	3	10	0.5	2.4	16	50
13 August	18 (1.37)	23	52	68	326	1.6	3.4	8	41
20 August	31 (0.78)	32	–	100	59	6.1	–	533	316
27 August	18 (1.35)	15	–	3	144	3.5	–	17	746
24 September	66 (0.36)	49	–24	20	14	128	–2.6	718	559

observed decline in phytoplankton in September from its 1974 maximum (Fig. 2A) not only reflects increased grazing by the growing zooplankton population (Fig. 2D), but is also attributable to a sluggish growth rate caused by unknown factors.

DISCUSSION

In order to support the observed daily primary production rates during the summer of 1974 the *in situ* nitrate concentrations at 0 m would have to be replenished every 0.5 to 6 hr (Table 3). These very rapid replenishment times are confirmed by the nutrient uptake experiments (2.4 to 6.2 hr). However, given the use of surface data and nitrogen dynamics based solely on NO_3 uptake, the local ecological value of these estimations is limited and requires confirmation. Confirmation is also desirable, since particulate nitrogen (N_p) and silica (Si_p) production rates were estimated from carbon production (C_p) and the C:N and C:Si ratios of the surface phytoplankton concentrate, whereas in the nutrient uptake experiments PN and PSi were estimated based on the N:Chl and Si:Chl ratios from the net tow analyses. The average N:Chl ratio of the surface concentrate was about 25:1, compared to 6.51:1 for the net phytoplankton; the corresponding Si:Chl ratios were about 74:1 and 8.54:1, respectively. These differences suggest that ratios derived from the surface phytoplankton concentrate are considerably biased by detritus. Hence, N_p and Si_p were recalculated (equation

7) using a C:N ratio (by atoms) of 10.7:1 and for C:Si 16.4:1; ratios derived from 60:1, 6.51:1 and 8.54:1 to estimate PC, PN and Psi (by weight), respectively, from chlorophyll.

Recalculated N_p is usually less than that based on the C:N ratio of the surface concentrate (Table 2), but the average difference of about 30% has only a modest effect on the required nitrate replenishment times (Table 3). For dissolved silica, the required replenishment times generally increased: 0.1 to 99 days versus 0.4 to 30 days based on the C:Si ratio of the surface concentrate.

Table 4 gives the nutrient replenishment times for the entire water column based on the integrated concentrations of the total dissolved nitrogen ($NO_3 + NH_3 + Urea$), silicon, and primary production. N_p and Si_p were estimated from C_p using the net phytoplankton C:N and C:Si atomic ratios. The very rapid nitrogen replenishment times required to support the observed production rates are confirmed. Exclusive of 24 September, the standing stock of dissolved nitrogen would be depleted in 2 to 21 hr, i.e., the supply would have to be replenished about once to 12-times daily to support N_p. The silicon depletion (replenishment) times were much longer, ranging from 2.5 to 17 days.

The question now emerging is what sources are primarily responsible for replenishing nitrogen several times daily during the 1974 summer. Vargo (28) determined that fed, summer zooplankton populations in Narragansett Bay excreted an average of 19.62 μg NH_3-N mg dry weight^{-1} day^{-1}. Hale (8) studied the nutrient excretion from benthic communities in this estuary. The highest rates of ammonia excretion (as μM m^{-2} hr^{-1}) reported by him, and related to temperature, are expressed by:

$$\Delta\ NH_3\ =\ -48.78\ +\ 12.26T$$

when T is temperature ($^\circ$C). The daily benthic excretion rates were calculated using the bottom temperature and assuming that the hourly excretion rate remains constant throughout the day (i.e., no diel periodicity). These excretion

Table 4. Daily production of carbon (C_p), nitrogen (N_p) and silicon (Si_p); *in situ* concentrations of dissolved nitrogen (NO_3+NH_3+Urea) and silicon, and *in situ* depletion times of N and Si concentrations at the observed production rates.

	Production (mM m^{-2} day^{-1})			Dissolved Nutrients (mM m^{-2})		Depletion Times	
	C_p	N_p	Si_p	ΣN	Si	N (hrs)	Si (days)
15 July	115	10.8	7.0	5.4	53.7	12	8
22 July	130	12.2	7.9	5.9	52.4	12	7
29 July	268	25.1	16.3	1.9	77.7	2	5
13 August	169	15.8	10.3	1.6	24.0	3	2.5
20 August	108	10.1	6.6	8.5	119.0	20	17
27 August	143	13.4	8.7	8.7	129.3	16	14.5
24 September	19	1.8	1.2	73.5	173.9	42	149

Table 5. NH$_3$ excretion rates by the zooplankton and benthos compared to daily phytoplankton needs.

	Np	Zooplankton		Benthic Excretion	Excretion days needed to satisfy Np		
		Biomass	Excretion		ZPL	Benthos	ZPL + Benthos
	(mM m⁻² day⁻¹)	(mg m⁻²)	(mM m⁻² day⁻¹)	(mM m⁻² day⁻¹)			
15 July	10.8	1504	2.11	4.60	5	2.5	1.6
22 July	12.2	800	1.12	4.74	11	2.5	2.1
29 July	25.1	256	0.36	5.01	69	5	4.7
13 August	15.8	80	0.11	5.21	143	3	3.0
20 August	10.1	40	0.06	5.39	168	2	1.9
27 August	13.4	80	0.11	5.57	122	2.5	2.4
24 September	1.8	944	1.32	4.13	1.4	0.5	0.3

rates are compared to the daily phytoplankton nitrogen needs in Table 5. The combined daily excretion of NH_3 by the zooplankton and benthic communities, exclusive of the September experiment, is less than the phytoplankton demands based on production. It would take from 2 to 5 days for their excretion to supply the daily phytoplankton needs, whereas nitrogen replenishment times of several hours are usually needed (Table 4). Nitrate is excreted at much lower rates (28, 8), and cannot make up the difference. Clearly, then, these animal communities cannot replenish through excretion the apparent daily nitrogen demands of the 1974 summer phytoplankton populations. Hale (8) has calculated that on an August day NH_3 supplied into Narragansett Bay through river runoff, sewage and rainfall is 40% of these animal contributions. But inclusion of these additional sources would still not replenish nitrogen at the rates needed for the present observations. The amounts of nitrogen excreted by bacteria, micro-zooplankton, fish and the abundant ctenophore population (Fig. 2D) are unknown. The micro-zooplankton population (13) especially is a potentially important source of excreted nitrogen, and requires future evaluation.

The persistent evidence is that the daily nitrogen supply in lower Narragansett Bay must be replenished frequently during the 1974 summer. This reinforces some observations made the previous summer (3) that the *in situ* inorganic nitrogen + urea concentrations would then have to be replenished every 3 to 4 hr to support the observed primary production. The high and persistently constant phytoplankton abundance during the 1974 summer (Fig. 2A), coupled with very intense primary production (Table 1, Fig. 3) and high assimilation numbers (Fig. 3), reveals that a very viable and rapidly growing phytoplankton population usually occurred then despite the very low nitrogen supplies (Figs. 2E, F). Such behavior is consistent with (indeed, requires) the experimental evidence that nitrogen was then being remineralized very rapidly. We consider the alternative explanation that the phytoplankton are utilizing a reservoir of dissolved organic nitrogen (exclusive of urea) to be unlikely (7).

The most noteworthy feature of the silica results is that the required replenishment times to satisfy daily summer phytoplankton needs are measurable in units of days, whereas for nitrogen it is often in hours (Tables 3, 4).

ACKNOWLEDGMENTS

Ms. Ellen Deason provided the zooplankton data, Mr. James Yoder provided the *in situ* ammonia concentrations, and Ms. Blanche Coyne typed the manuscript and drafted the figures. Dr. Curt Davis has made helpful comments on the experimental procedure. This work was supported by NSF Grant GA 31319X and by a grant from the Office of Sea Grant Programs, NOAA.

REFERENCES

1. Armstrong, F.A.J. 1951. The determination of silicate in seawater. J. Mar. Biol. Assoc. U.K. 30:149-160.

2. Dodson, A.N., and W.H. Thomas. 1964. Concentrating plankton in a gentle fashion. Limnol. Oceanogr. 9:455-456.
3. Durbin, E.G., R.W. Krawiec, and T.J. Smayda. 1975. Seasonal studies on the relative importance of different size fractions of phytoplankton in Narragansett Bay. Mar. Biol. 32:271-287.
4. Eppley, R.W., T.T. Packard, and J.J. MacIssac. 1970. Nitrate reductase in Peru Current phytoplankton. Mar. Biol. 6:195-199.
5. Goering, J.J., D.M. Nelson, and J.A. Carter. 1973. Silicic acid uptake by natural populations of marine phytoplankton. Deep-Sea Res. 20:777-789.
6. Grasshof, K. 1966. Automatic determination of fluoride, phosphate and silicate in seawater, p. 304-307. In L.T. Skiggs, Jr. (ed.), Automation in Analytic Chemistry, 1965 Technicon Symposium. Mediad, Inc., New York.
7. Guillard, R.R.L. 1963. Organic sources of nitrogen for marine centric diatoms, p. 93-104. In C.H. Oppenheimer (ed.), Symposium on Marine Microbiology. C.C. Thomas Publisher, Springfield, Ill.
8. Hale, S.S. 1974. The role of benthic communities in the nutrient cycles of Narragansett Bay. M.S. Thesis, Univ. Rhode Island, 123 p.
9. Hitchcock, G., and T.J. Smayda. 1976. The importance of light in the initiation of the 1972-1973 winter-spring bloom in Narragansett Bay. Limnol. Oceanogr., (in review).
10. _____, and _____. 1976. Bioassay of lower Narragansett Bay waters during the 1972-1973 winter-spring bloom using the diatom *Skeletonema costatum*. Ibid., (in review).
11. Lorenzen, C.J. 1966. A method for the continuous method of in-vivo chlorophyll concentration. Deep-Sea Res. 13:223-227.
12. MacIssac, J.J., and R.C. Dugdale. 1972. Interaction of light and inorganic nitrogen in controlling nitrogen uptake in the sea. Deep-Sea Res. 19:209-232.
13. Martin, J.H. 1965. Phytoplankton-zooplankton relationships in Narragansett Bay. Limnol. Oceanogr. 10:185-191.
14. _____. 1968. Phytoplankton-zooplankton relationships in Narragansett Bay. III. Seasonal changes in zooplankton excretion rates in relation to phytoplankton abundance. Ibid 13:63-71.
15. McCarthy, J. 1970. A urease method for urea in seawater. Limnol. Oceanogr. 15:309-312.
16. Menzel, D.W., and N. Corwin. 1969. The measurement of total phosphorus in seawater based on the liberation of organically bound fractions by persulfate oxidation. Limnol. Oceanogr. 14:280-282.
17. Paasche, E. 1973. Silicon and the ecology of marine plankton diatoms. I. *Thalassiosira pseudonana (Cyclotella nana)* grown in a chemostat with silicate as a limiting nutrient. Mar. Biol. 19:117-126.
18. _____. 1973. Silicon and the ecology of marine plankton diatoms. II. Silicate-uptake kinetics of five diatom species. Mar. Biol. 19:262-269.
19. Pratt, D.M. 1959. The phytoplankton of Narragansett Bay. Limnol. Oceanogr. 4:425-440.
20. _____. 1965. The winter-spring diatom flowering in Narragansett Bay. Ibid 10:173-184.
21. _____. 1966. Competition between *Skeletonema costatum* and *Olisthodiscus luteus* in Narragansett Bay and in culture. Ibid 11:447-455.
22. Sharp, J.H. 1974. Improved analysis for "particulate" organic carbon and nitrogen from seawater. Limnol. Oceanogr. 19:984-989.

23. Smayda, T.J. 1957. Phytoplankton studies in lower Narragansett Bay. Limnol. Oceanogr. 2:342-359.
24. _____. 1973. The growth of *Skeletonema costatum* during a winter-spring bloom in Narragansett Bay, Rhode Island. Norw. J. Bot. 20:219-247.
25. _____. 1973. A survey of phytoplankton dynamics in the coastal waters from Cape Hatteras to Nantucket. *In* Coastal and Offshore Environmental Inventory, Cape Hatteras to Nantucket Shoals, Univ. Rhode Island Marine Publ. Ser. No. 2, 3-1 to 3-100.
26. Solórzano, L. 1969. Determination of ammonia in natural waters by the phenylhypochlorite method. Limnol. Oceanogr. 14:799-801.
27. Steemann Nielsen, E. 1952. The use of radio-active carbon (C^{14}) for measuring organic production in the sea. J. Cons. int. Explor. Mer 18:117-140.
28. Vargo, G. 1976. The influence of grazing and nutrient excretion by zooplankton on the growth and production of the marine diatom, *Skeletonema costatum* (Greville) Cleve, in Narragansett Bay. Ph.D. Dissertation, Univ. Rhode Island, 216 p.
29. Werner, D. 1966. Die Kieselsaure im Stoffwechsel von *Cyclotella cryptica* Reimann, Lewin and Lewin. Arch. Mikrobiol. 55:278-303.
30. Wood, E.D., F.A.J. Armstrong, and F.A. Richards. 1967. Determination of nitrate in seawater by cadmium-copper reduction. J. Mar. Biol. Assoc. U.K., 47:23-31.
31. Yentsch, C.S., and D. Menzel. 1963. A method for the determination of phytoplankton chlorophyll and phaeophytin by fluorescence. Deep-Sea Res., 10:221-231.

POPULATION DYNAMICS

Convened by:
Saul B. Saila
Graduate School of Oceanography
Narragansett Bay Campus
University of Rhode Island
Kingston, Rhode Island 02881
and
Richard C. Swartz
U.S. Environmental Protection Agency
Marine Science Center
Newport, Oregon 97365

Population dynamics, the study of population change, has been visualized as the tension between the tendency of a population to grow and the limits to that growth imposed by the environment. The ways in which various environmental control mechanisms and man's activities are brought to bear on populations are not simple, and they have not been completely elucidated as yet. However a lot of progress has been made in the recent past. The contributions to this session on population dynamics are examples of some of the more recent developments.

It is not coincidence that three of the five contributions to population dynamics relate to the striped bass. This anadromous fish utilizes estuarine feeding and wintering grounds in those areas where it is most abundant. The striped bass spawns in rivers, and concern with the ecological impacts of power plants on rivers and estuaries has resulted in a number of field and modeling studies of this important sport and commercial fish.

The presentation describing the two-dimensional model of the striped bass population in the Hudson estuary combines real-time flow information with a two-layered, age and time-dependent life cycle model to predict the potential impacts of power plants on the population. The analysis of striped bass spawning success in the Potomac Estuary is a very timely and realistic analysis of a very difficult problem. It is known that the fecundity of the striped bass is high and that enormous losses occur during the early life history stages. These losses during early life history undoubtedly influence the relative strength of future

135

year-classes. Finally, with respect to striped bass, it is clear that the concept of a unit stock is fundamental to the study of fisheries population dynamics, and the concept of innate tags is a step in the direction of proper stock identification.

It is regrettable that sufficient vital statistics are not available to model benthic communities with the same degree of sophistication as for some fish species. However, an important contribution to benthic community dynamics in estuaries is made by demonstrating the quantitative variability of these communities in both a seasonal and long term basis.

It is suggested that these contributions to estuarine population dynamics will provide an excellent springboard for advancing the state-of-the-art in this area.

A MODEL STUDY OF STRIPED BASS POPULATION DYNAMICS

IN THE HUDSON RIVER

T.L. Englert, J.P. Lawler,
F.N. Aydin, and G. Vachtsevanos
Lawler, Matusky & Skelly Engineers
415 Route 303
Tappan, New York 10983

ABSTRACT: Present and planned operation of electric power generating stations along the Hudson River may have an impact on the Atlantic striped bass population due to the use of river water in once-through cooling systems at the plants. Withdrawal of river water for cooling purposes can have two prinicpal effects on the young-of-the-year striped bass spawned in the Hudson: (1) Eggs, larvae and early juvenile fish may be entrained in the water which is circulated through the plant's cooling system and returned to the river; (2) later juvenile fish may be impinged on the debris screens at the plant intakes.

Population models of the Hudson River striped bass are useful in making predictions of the impact of entrainment and impingement at the power plants. In the model study presented here, results from a detailed simulation of the young-of-the-year population are input to a model of the adult bass population in order to predict short- and long-range impacts on the population.

The young-of-the-year model traces development of the early life stages of the bass from eggs through the larvae and juvenile stages. Egg production rates calculated from field data are used to initialize the population model. The temporal, spatial and age distributions of the early life stages are simulated by equations which include the effects of hatching period, natural mortality, plant withdrawal rates and the convective and dispersive effects of the Hudson's hydrodynamics. The hydrodynamic simulation is intra-tidal or real-time. The spatial distribution of the organisms is calculated at twenty-nine longitudinal grid points in both the upper and lower layers of the river.

Comparisons of model results and field data provide a measure of the verification of the model.

INTRODUCTION

One of the important uses of models of animal and fish populations is the prediction of the impact of man's activities on such populations. During recent years, a number of models have been developed to attempt to predict the impact of man's activities along the Hudson River on that portion of the Atlantic striped bass population which spawns each spring in the Hudson. This population is of particular interest due to the large Atlantic striped bass sport fishery. As a result of this fishery, the striped bass are of considerable socio-economic importance and have received the attention of citizen groups and regulatory agencies.

Modeling efforts to date have been directed toward predictions of the impact of electric generating stations along the Hudson on the young-of-the-year (y-o-y) striped bass. These generating stations withdraw considerable amounts of water from the Hudson for use in open cycle cooling systems. The withdrawal of river water by power plants has two potential impacts on the y-o-y striped bass: (1) Entrainment of striped bass eggs and larvae through the power plant cooling system, (2) Impingement of later juvenile fish (i.e., fish greater than two inches in length) on the debris screens in front of the cooling water intakes.

Several models have been developed which attempt to provide quantitative predictions of the magnitude of these two potential impacts on the striped bass population. These models range in complexity from sophisticated models developed by the staff of the Nuclear Regulatory Commission (8) and Lawler, Matusky and Skelly Engineers (5, 6) to less complicated approaches like those developed by Clark (1). The model presented here is considerably more complex than any of those presented previously due to the inclusion of real-time hydrodynamics, vertical migration of larvae and complex non-linear mortality functions in the y-o-y and adult stages.

Before explaining the details of the modeling approach used, we present a brief summary of the striped bass life cycle.

DYNAMICS OF STRIPED BASS LIFE CYCLE

Figure 1 presents a schematic of the Hudson River striped bass life cycle, showing the principal life stages and various causes of mortality at each stage.

The life cycle of an individual bass begins with the egg stage. Spawning usually occurs between milepoints 30 and 90, principally above the salt front in the Hudson. The eggs hatch to yolk sac larvae in 1.5 to 3.0 days depending on prevailing temperatures.

The distribution of the eggs and the yolk sac larvae stages are affected by the hydrodynamics in the Hudson estuary. The organisms are moved by the tidal action in the Hudson as well as the freshwater flow and the flow near the plant intakes. However, it appears that the organisms are not transported at the same rate as a chemical or thermal pollutant. This is due in part to their size and, in the larval stage, to some swimming ability.

Larval striped bass exhibit a diurnal vertical migration pattern which causes them to be distributed throughout the water column during the night and concentrated near the bottom during daylight hours. When the post yolk sac larvae reach about half an inch in length they begin migrating to the shoal areas of the Hudson (7).

When the young bass have developed an adult complement of fin rays they are classified as "juveniles." At this time the bass are concentrated in the shoal

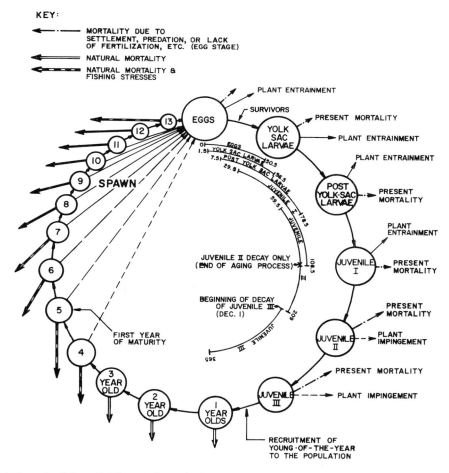

Figure 1. Schematic life cycle for striped bass in the Hudson River.

areas of the river and begin a downstream migration. Their movements are no
longer strongly influenced by the hydrodynamics in the river.

The juvenile bass over-winter in the Hudson and appear to move seaward in
June when the spawning adults return to the Atlantic. The adult stripers reach
sexual maturity at age 5 or 6 and many survive up to 14 years.

YOUNG-OF-THE-YEAR MODEL

The basic modeling approach used in the development of the young-of-the-
year model is described in a recent publication by Englert and Aydin (3). The
approach used is similar to that presented by Rotenberg (8) in that it is deterministic and, like Rotenberg's scheme, it is based on equations similar to those
employed in physics and engineering to model transport processes. The defining
equation for the y-o-y model is:

$$\frac{\partial c^k(x,a,t)}{\partial t} + \frac{\partial c^k(x,a,t)}{\partial a} + \frac{1}{A}\frac{\partial c^k(x,a,t)\cdot Q^k(x,t)}{\partial x} =$$

$$\frac{1}{A}\frac{\partial}{\partial x}\left(EA\frac{\partial c^k(x,a,t)}{x}\right) + M^k(x,a,t) - E^k_R(x,a,t)\cdot c^k(x,a,t)$$

$$- KD(c,a)\cdot c^k(x,a,t) \tag{1}$$

where:

$\frac{\partial c^k}{\partial t}$ = accumulative term

$\frac{\partial c^k}{\partial a}$ = maturation term

$\frac{\partial c^k\cdot Q^k}{\partial x}$ = convection term

$\frac{1}{A}\frac{\partial}{\partial x}\left(EA\frac{\partial c^k}{\partial x}\right)$ = dispersion term

Q = real-time flow

E = dispersion coefficient

A = cross-sectional area of river

$M^k(x,a,t)$ = migration rate

$KD(c,a)$ = mortality rate

$E^k_R(x,a,t)$ = entrainment or impingement parameter

Equation 1 is applied in the egg and larval stages where the mass transport effects of the river's hydrodynamics are operative as well as the juvenile stages during which the behavioral characteristics of the bass determine their longitudinal distribution. The derivation of the migration rate, $M^k(x,a,t)$, and its application to simulation of both vertical and longitudinal migration, is discussed in more detail below.

Convective effects are simulated with sinusoidal functions which account for freshwater flow and tidal action in the upper and lower layers of the river. The specific sinusoidal functions used are given in detail in (3). Figure 2 is a schematic of the simulation of the flows in the upper and lower layers. A dispersion coefficient consistent with the scheme for modeling the convection term is obtained from Elder's equation (2).

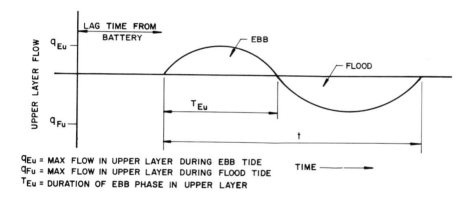

q_{Eu} = MAX FLOW IN UPPER LAYER DURING EBB TIDE
q_{Fu} = MAX FLOW IN UPPER LAYER DURING FLOOD TIDE
T_{Eu} = DURATION OF EBB PHASE IN UPPER LAYER

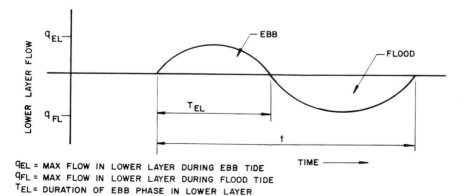

q_{EL} = MAX FLOW IN LOWER LAYER DURING EBB TIDE
q_{FL} = MAX FLOW IN LOWER LAYER DURING FLOOD TIDE
T_{EL} = DURATION OF EBB PHASE IN LOWER LAYER

Figure 2. Schematic of upper and lower layers flow.

Naturally mortality is simulated by a non-linear function having the form:

$$KD = KE + (KE - KO) \left(\frac{C - C_S}{C_S}\right)^3 \qquad (2)$$

where: KD = generalized unit mortality rate, day
 KE = conventional first order or equilibrium rate, day
 KO = minimum unit mortality rate consistent with system biology, day
 C_S = equilibrium population level, fish/unit volume
 C = fish concentration at any point in time and space,

This function is used in order to reflect the expected density dependence of natural mortality.

CALCULATION OF MIGRATION RATE, $M^k(x,a,t)$

The simulation of diurnal vertical migration during the larval stage and the longitudinal migration of juveniles is performed via the same calculation scheme. The basic strategy in the model is to determine a rate of movement between either vertical layers or longitudinal segments which causes the distribution of organisms in the model to approach the distributions calculated from field data.

Table 1 shows vertical migration patterns for yolk-sac and post yolk sac larvae and Table 2 indicates the longitudinal movement of juvenile striped bass. These distribution patterns of the y-o-y striped bass are submitted as data to the model and the following scheme is invoked:

1. Calculate, FD, the fractional distribution of larvae (or juveniles) in each layer (or segment) at each time step in the simulation.

2. At each time step compute the difference between the values of the fractional distributions of organisms in the model and those calculated from field measurements, FM.

Table 1. Vertical migration patterns calculated from 1973 river transect data.

Sampling Stations	Layer	Relative Distribution Yolk sac Larvae		Relative Distribution Post Yolk sac Larvae	
		Day	Night	Day	Night
Cornwall (9)	Upper	0.3070	0.4509	0.4946	0.3295
	Lower	0.6930	0.5491	0.5054	0.7605
Bowline (6)	Upper	0.3082	0.4284	0.3075	0.3518
	Lower	0.6918	0.5716	0.6925	0.6482
Indian Point (6)	Upper	0.2172	0.3688	0.3488	0.3398
	Lower	0.7828	0.6312	0.6512	0.6602
Lovett (6)	Upper	0.3531	0.5166	0.3689	0.4216
	Lower	0.6469	0.4834	0.6311	0.5784
Danskammer/ Roseton (6)	Upper	0.5135	0.5231	0.3633	0.4242
	Lower	0.4865	0.4769	0.6367	0.5758

3. Based on (1) and (2), the migration rate, MR, expressed in terms of numbers of organisms per unit time is:

$$MR = \frac{(FM-FD)\ TN}{DELT} \tag{3}$$

where:

TN   = total number of organisms in segment

DELT = time remaining between present time step and next change in FM

The migration rate, MR, as evaluated above is converted to a concentration flux and distributed on the basis of fractional age distribution to give M^k (x,a,t) in equation (1).

Table 2. Fractional distribution of juvenile fish in the river (1973 Texas Instruments Beach Seine Data, (9))

Segment	Milepoint	Juvenile I[1]	Juvenile II[2]
1	120-130	0.0025	0.0030
2	110-120	0.0028	0.0025
3	100-110	0.0013	0.0003
4	95-100	0.0036	0.0004
5	90- 95	0.0036	0.0004
6	85- 90	0.0034	0.0005
7	80- 85	0.0038	0.0006
8	75- 80	0.0021	0.0002
9	70- 75	0.0021	0.0002
10	68- 70	0.0049	0.0008
11	66- 68	0.0057	0.0009
12	64- 66	0.0053	0.0008
13	62- 64	0.0057	0.0009
14	60- 62	0.0053	0.0008
15	58- 60	0.0159	0.0064
16	56- 58	0.0160	0.0064
17	53- 56	0.0239	0.0096
18	50- 53	0.0239	0.0096
19	47- 50	0.0221	0.0145
20	44- 47	0.0221	0.0145
21	42- 44	0.0148	0.0096
22	40- 42	0.0141	0.0096
23	38- 40	0.0854	0.1412
24	36- 38	0.0854	0.1412
25	34- 36	0.0854	0.1412
26	32- 34	0.0854	0.1412
27	30- 32	0.0854	0.1412
28	20- 30	0.2565	0.1756
29	10- 20	0.1105	0.0260

[1] Entrainable juveniles (25 to 50 mm. in length).
[2] Impingeable juveniles (50 mm. in length).

ADULT MODEL

The young-of-the-year model generates recruits to adult age group 1 (one year olds) from the number of juvenile fish remaining on day 365 after the first spawn. In the calculation of the so-called "equilibrium" or baseline population (i.e., the population level established without plant entrainment or impingement effects), the number of recruits, coupled with adult survival rates is used to generate an adult population for age groups 2 through 14.

Given the number of one-year old fish, the number of fish in successive year classes is obtained as follows:

$$N_i = N_{i-1} \exp(-k_{i-1}\Delta t_{i-1}) \tag{4}$$

in which: N_i = the number of fish in the ith age group at beginning of the ith year.

N_{i-1} = the number of fish in the (i-1) age group at beginning of the (i-1) year.

k_{i-1} = mortality rate to which the (i-1) age is subject.

Δt_{i-1} = one year
$i = 2,3,4 \ldots \ldots 14$

The term $\exp(-k_{i-1}\Delta t_{i-1})$ represents the fraction of the (i-1)th population that makes it through the (i-1)th year. Hence, the number of fish in any given age group is:

$$N_i = N_1 \exp\left(\sum_{j=1}^{i-1} - k_j\Delta t_j\right) \tag{5}$$

in which $j=1, \ldots . i-1$, the number of age groups, from one-year olds to (i-1) year olds.

Using this approach, the total adult fish population at the end of any given year may be expressed as:

$$\sum_{i-1}^{14} N_i = N_1 \left\{ 1 + \sum_{i=2}^{14} e^{-\sum_{j=1}^{i-1} k_j \Delta t_j} \right\} \tag{6}$$

Given the total adult striped bass population, the ratio of females to the total population, and the average fecundity and maturation within each age group, one can predict the total number of eggs that will be produced by the population.

This is written:

$$\text{Total eggs produced} = \sum_{i=1}^{14} N_i \ f_{s_i} \ f_{m_i} \ f_i$$

in which:

f_{s_i} = fraction of females within N_i
f_{m_i} = fraction of sexually mature females
f_i = average number of eggs per female in ith year class.

Data values for these reproduction parameters are given in Table 3. The total number of eggs produced by the adult population is cycled into the young-of-the-year model to generate a new group of one year olds.

Table 3. Selected fertility factors

Age Group	Female Fraction	Female Maturity	Fecundity (Eggs/Fertile Female)
1	0.50	0	0
2	0.52	0	0
3	0.54	0	0
4	0.56	0	0
5	0.58	0	0
6	0.60	0.67	451,000
7	0.62	1.00	780,000
8	0.64	1.00	1,543,000
9	0.66	1.00	1,563,000
10	0.68	1.00	1,841,000
11	0.70	1.00	2,095,000
12	0.70	1.00	2,350,000
13	0.70	1.00	2,269,000
14	0.70	1.00	2,189,000

[1] Based on Texas Instruments, Inc. 1973 data, (9).
[2] Values between age groups 1 and 11 were linearly interpolated from estimates of age groups 1 and 11 female fractions.
[3] Computed by interpolation using data from years 10, 12 and 14.

Previous studies (4) have assumed survivals in the adult stages to include all natural mortality as well as exploitation by sport and commercial fishing. This study, however, separates the total mortality, K_i, into two components—the fishing and natural coefficients. Following Van Winkle (11), P_i, the probability of survival from all causes of mortality for adult striped bass in age class i is calculated as:

$$P_i \; [TOTP(t)] = P_{N_i} \cdot P_{f_i} \; [TOTP(t)] \tag{7}$$

where:

P_{N_i} = probability of survival from causes of natural mortality.

P_{f_i} = probability of survival from fishing stresses.

It is assumed that P_{N_i} and P_{f_i} are independent functions of female biomass available to the fishery at the start of the current year, TOTP(t).

Fishing mortality is assumed to be a function of the size of the fishable population, such that the probability of survival from fishing decreases when the available biomass, TOTP(t), increases. This reflects the hypothesis that high density in striped bass populations is associated with increased vulnerability to the fishery due to schooling and that high levels of abundance result in increasing fishing effort which, in turn, increases mortality due to fishing stresses. When the probability of survival becomes low enough, the size of the population available to the fishery decreases and this causes a decrease in the fishing mortality rate as a result of a decrease in striped bass fishing effort. This allows the population to recover and the available biomass to the fishery increases again.

SOLUTION TECHNIQUE

A finite difference scheme based on the Crank-Nicolson procedure was used to obtain a numerical solution to equation (1). In applying this scheme, the river was divided into 29 longitudinal segments (see Table 2) and two vertical layers of equal depth. During the early life stages, a time step of three hours was used. The details of the finite difference approximation are presented in (3). The solution scheme was programmed in FORTRAN and implemented on a 370/168 IBM computer.

The adult model was implemented in the same manner but it does not require a finite difference approximation. A Newton-Raphson scheme was used, however, to close the cycling loop which connects the adult population and the egg complement.

VERIFICATION OF MODEL PREDICTIONS

In the development of the present model, data collected by Texas Instruments, Inc. (9) were used for verification purposes.

The young-of-the-year model generates egg, larvae, and juvenile distributions in terms of the aging process as a function of river location, and time, for a given layer of the river. When the model predictions of the spatial and temporal distributions of the life stages are compared with field data, it should be noted that the objective is not to match exactly a set of given field data but to predict plant impact.

Perfect agreement between model predictions and field measurements of abundance is not essential to realistic predictions of plant impact on the population since these impacts are presented in terms of percentages. It is important for the model to reproduce the spatial and temporal distribution patterns of the young-of-the-year bass since this will affect the estimates of impact. Clearly, the locations of peak organism abundance relative to power plant sites will affect the impacts. As shown in Figures 3 and 4, the model is successful in reproducing the areas of peak abundance found in the field data.

In fact, Figure 3 shows a good agreement between the measured egg numbers and those predicted by the real-time model. The real-time model produces profiles similar to the field measurements of eggs, with the sampling peaks reproduced very well.

It appears, however, that the transport mechanisms operative in the model may have the effect of smoothing out the peak concentrations by redistributing the eggs longitudinally. This hypothesis is supported by the fact that the areas under the sampling profiles are approximately the same as the areas under the corresponding computed profiles for eggs.

Figure 4 compares post yolk sac larvae observations with model predictions as a function of milepoints for June 18-22, 1973. Again, although the agreement is not exact, the peaks seem to occur in the approximate locations of the sampling peaks.

Model predictions of the fractional distribution of the juvenile fish are in close agreement with the distributions calculated from field data due to the migration term explained above. The migration term assures agreement of the spatial and temporal locations of peak abundances in the model results and the field observations for the juveniles.

At this time, there are very little data available to permit verification of model predictions of the size of the present adult striped bass population. However, since impacts are expressed in terms of percentages rather than numbers of fish, this verification is not essential. One of the important advantages of the model is that it can provide useful estimates of impact with specified assumptions regarding natural mortality and other parameter values (6). The adult model produces a total adult population of about one million Hudson River striped bass under the assumptions of 70% natural mortality per year during years one to three and 54% total mortality per year during years four to fourteen. The latter figure is broken down into 16% due to fishing and 38% natural mortality.

In general, the model prediction of one million adult bass appears to be low. There are several possible reasons for this, including over-estimates of adult mortality and fecundity and an under-estimate of natural mortality during the egg stage. A higher value for egg stage mortality would increase the estimate of total eggs produced and consequently the required number of spawning adults.

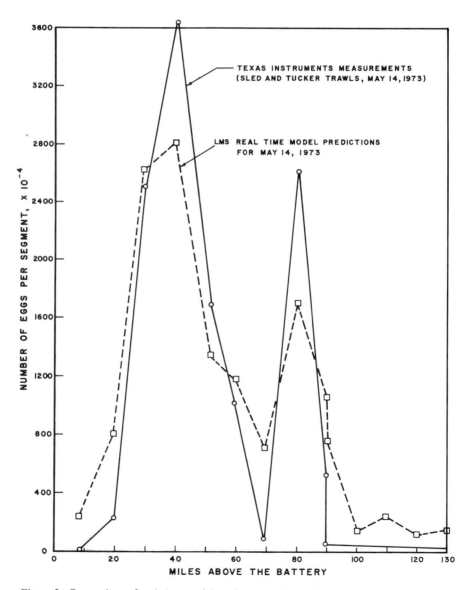

Figure 3. Comparison of real time model predictions with the field measurements.

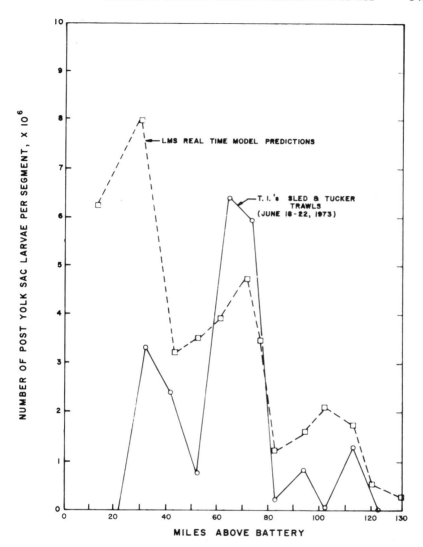

Figure 4. Comparison of the striped bass real time model predictions with the field measurements.

ACKNOWLEDGEMENT

Financial support for this study was provided by Consolidated Edison Company of New York under contract No. 115-049 with Lawler, Matusky and Skelly Engineers.

REFERENCES

1. Clark, John R. (Testimony). 1972. Effects of Indian Point Units 1 & 2 on Hudson River aquatic life. April 5.
2. Elder, J.W. 1959. The dispersion of marked fluid in turbulent sheer flow. J. Fluid Mech. 5:544.
3. Englert, T.L., and F.N. Aydin. 1975. An intra-tidal population model of young-of-the-year striped bass in the Hudson River. Advances in Computer Methods for Partial Differential Equations, Proceedings AICA International Symposium, p. 273.
4. Lawler, John P. (Testimony). 1974. Effect of entrainment and impingement at Cornwall on the Hudson River striped bass population. Project No. 2338.
5. _____. (Testimony). 1972. Effect of entrainment and impingement at Indian Point on the population of the Hudson River striped bass. Docket No. 50-247, October 30.
6. Lawler, Matusky and Skelly Engineers. 1975. Report on development of a real-time, two-dimensional model of the Hudson River striped bass population. LMS Project No. 115-049, prepared for Con-Edison Co. of N.Y., Inc.
7. Nicholson, W. R. and R. M. Lewis. 1973. Briefing paper on the status of striped bass. Atlantic Estuarine Fisheries Center, National Marine Fisheries Service, National Oceanic and Atmospheric Administration.
8. Rotenberg, M. 1972. Theory of population transport. J. Theor. Biology. 37:291.
9. Texas Instruments Incorporated. 1973. Hudson River program fisheries data summary. Reports prepared for Consolidated Edison Company of New York, Inc., Vol. I, II and III.
10. U.S. Nuclear Regulatory Commission. 1975. Final environmental statement related to operation of IP3, Con Ed Company of N.Y., Inc. Docket No. SO-286, Vol I & II.
11. Van Winkle, et al. 1974. A striped bass population model and computer programs. Oak Ridge National Laboratory, ORNL-TM-4578, ESD No. 643.

AN ANALYSIS OF 1974 STRIPED BASS SPAWNING SUCCESS

IN THE POTOMAC ESTUARY[1]

Tibor T. Polgar
Martin Marietta Corporation
Environmental Technology Center
1450 South Rolling Road
Baltimore, Maryland 21227

J. A. Mihursky, R. E. Ulanowicz, R. P. Morgan, II, and J. S. Wilson
Chesapeake Biological Laboratory
Center for Environmental and Estuarine Studies
University of Maryland
Solomons, Maryland 20688

ABSTRACT: The Potomac River Fisheries Program is concerned with the long-term effects of power plant ichthyoplankton entrainment on striped bass *(Morone saxatilis)* recruitment. Since striped bass population fluctuations are determined strongly by environmental conditions during spawning and early development, assessment of power plant-induced ichthyoplankton mortalities must consider the mechanisms controlling spawning success.

Ichthyoplankton distributions for 1974, spawning population abundance and fecundity, and environmental conditions were considered for analysis. Loss of the early part of the spawn (including the peak) accounted for the highest mortalities among ichthyoplankton. This was due to the proximity of these distributions to the salt wedge where transport into regions unfavorable to survival seems to have occurred. The later, successful portion of the spawn occurred further upstream, in fresh tidal portions of the river. The sequence of events leading to an assessment of factors affecting ichthyoplankton survival are evaluated. Due to high early mortalities in ichthyoplankton, 1974 spawning success was low, and a poor yearclass is projected.

[1] Contribution no. 686, Center for Environmental and Estuarine Studies, University of Maryland.

INTRODUCTION

Since the early 1970's, the Maryland Power Plant Siting Program has been sponsoring research to determine the impact of power plant operations on the State's aquatic resources in the Chesapeake Bay and its tributaries. Because, at present, there are two large power plants with once-through cooling systems operating in the upper tidal portion of the Potomac—and another facility proposed in the oligohaline region of the river—a great deal of the power plant-related research has been conducted in the Potomac.

The conservation of commercial and sport fish species is of primary concern. Salinity regimes in the estuary and seasonal habitat use by anadromous and other fish species produce distinct biogeographical zones in which finite populations—or specific life stages of given populations—may be impacted by the operation of even a single power facility. Therefore, a region-wide Potomac River Fisheries Program was initiated in 1974 to investigate the extent of possible fish stock depletions through entrainment of early life stages into power plant cooling systems. Although other species also are being studied, the early life stages of the striped bass *(Morone saxatilis)*, one of the most important fish resources in the State of Maryland, have been singled out for detailed investigation in the program. Historical data and preliminary estimates by Morgan and Wilson (16) indicate that Potomac striped bass contribute approximately 20% of the fishable stock in the Chesapeake Bay region.

Several previous studies in the Potomac have demonstrated that the striped bass population here is an identifiable and indigenous race, continually returning to the river to spawn (12, 15, 17). Here, as in other systems with striped bass populations, the ichthyoplankton abundances of striped bass vary considerably from year to year, as does the catch of adult stock. However, except as an indicator of potential egg production, spawning stock size and ichthyoplankton abundances are not clearly related. Analyses of historical catch records by Koo (9) and Bigelow and Schroeder (4) indicate that adult striped bass populations are dominated by single yearclasses, with irregular periodicity. The success of particular yearclasses is attributable apparently as much to the size and fecundity of the parent stock as to environmental conditions prevailing during spawning and early development. It is these conditions that largely determine the extent of survival of ichthyoplankton stages to juveniles. Juvenile abundance is thought to reflect the size of a particular yearclass. The key to understanding population fluctuations in time, therefore, is to identify environmental factors controlling spawning success and to determine the characteristics of the adult stock.

The information and preliminary results of the program discussed here concentrate on the mechanisms that produced the degree of spawning success observed in 1974. The temporal and spatial properties of ichthyoplankton distributions have been analyzed for this year. Relative and absolute adult abundances also have been estimated, together with age, weight, sex, and maturity properties of the population. Using estimates of abundances and mortalities of various ichthyoplankton stages, along with environmental and detailed flow measurements, we have identified factors that influenced overall survival of ichthyoplankton in 1974.

The program has been continued in 1975, and there are plans to extend it at least through the 1976 spawning season. With three years of detailed information at hand, we intend to close the loop between ichthyoplankton production and adult stock size. Simulation modeling of the 1974 observations, which will further refine and quantify the results presented here, are underway.

REVIEW OF SOME FACTORS INFLUENCING SPAWNING SUCCESS

Although there is extensive literature on the natural history and dynamics of striped bass populations, little has been done to comprehensively assess the role of environmental patterns in determining spawning success. Indications from the literature are that a delicate balance exists between the response of the striped bass spawning stock to environmental cues setting off the spawning process and the subsequent time history of key environmental changes (temperature, salinity, and flow) influencing the survival of eggs and larvae. In the critical early life stages, for example, currents may sweep eggs and larvae into temperature, salinity, or other water quality regimes unfavorable to their survival (13). Albrecht (1) has shown that suspension of the semi-buoyant eggs is required for successful hatching and has estimated that water movements on the order of 30 cm/sec are necessary for suspension in estuaries.

In natural systems, observations indicate low occurrences in salinities greater than a few parts per thousand, implying that egg and larval transport into salt wedges may be an important factor in survival. Although laboratory experiments by Bayless (3) indicate that survival and growth of striped bass larvae apparently are enhanced in salinities ranging from 3.5 to 14.0 ppt, eggs and larvae in the field are found at much lower salinities. Eggs have been found in salinities as high as 6 ppt (2, 7), but most investigations reveal that spawning and survival is optimal in fresh or only slightly saline waters: Carlson and McCann (5) reported eggs in salinities less that 0.1 ppt in the Hudson River, and Farley (8) found striped bass egg concentrations in salinities of 0.18 ppt or less in the Sacramento-San Joaquin River system. In the latter system, Radtke and Turner (21) found the spawning process blocked at 0.35 ppt. Transport processes may also affect other biological requisites for survival. For example, larvae may be transported into regions without suitable or sufficient food supplies.

The onset of spawning in relation to the displacement of the upstream migrating adult stock from the salt wedge, and the generally downstream direction of flow, are both critical in early survival. If, for example, a quick temperature rise triggers spawning downstream, close to the saltwedge, the eggs may be subject to high natural mortality by transport into higher salinities or into regions lacking food organisms for the larvae, even though the higher temperatures favor survival by decreasing hatching and larval development times. The optimal range for peak spawning is generally 14-17°C. Shannon and Smith (22) and Pearson (18) demonstrated that egg incubation time decreases as temperature increases: from 58 hours to 28 hours as temperature is increased from 16°C to 30°C, and from 48 hours to 36 hours as temperature is increased from 18°C to 22°C, respectively. Investigations of the synergistic effects of salinity and temperature on hatching showed that eggs do not survive above 1 ppt (23) at higher temperatures.

The magnitude of freshwater runoff (and therefore transport) influences spawning success by affecting both salinity and temperature structures in the upper estuary, as well as plankton productivity and detrital levels. The relationship between amounts of runoff and flow and the development of thermal and salinity regimes is not clear in the Potomac. In combination with variations in spawning stock size, these factors may induce a large range in spawning success, accounting for the appearance of nonlinear dominant yearclass phenomena.

SAMPLING PROGRAM DESIGN

In order to gain as full an understanding of population dynamics relationships as possible, it was necessary to sample all phases of the population: adults, ichthyoplankton, and juveniles. Because spawning and developmental events are serially related, sampling of successive stages also served an experimental control function in the measurement of events in each stage. Maintaining a comprehensive program design, including information on the movements and redistribution of all stages, was especially important in this study because the approximately 60-river mile segment study area was logistically too large for extensive replication of measurements. Therefore, we were forced to rely on consistency assessments of results for control as the spawning process developed.

The sampling grid consisted of 12 transects, comprising a total of 38 sampling stations, which spanned the Potomac from below Morgantown up to Washington, D. C. (Fig. 1). Transects were spaced to have equal water volume between each transect pair (200×10^6 m^3) and separations greater than a tidal excursion distance.

Details of the entire sampling program are described in (14), (10), (24) and (16). In this paper we shall concentrate on ichthyoplankton distributions in deeper river areas, characteristics and movements of the adult stock, and river hydrography.

Absolute abundance of adult bass during spawning was determined by acoustic surveys, a method developed for the program (24), while relative abundances were obtained from gill net sampling of the spawning population. Gill net stands of four mesh sizes (to ensure that no adult classes of striped bass would be

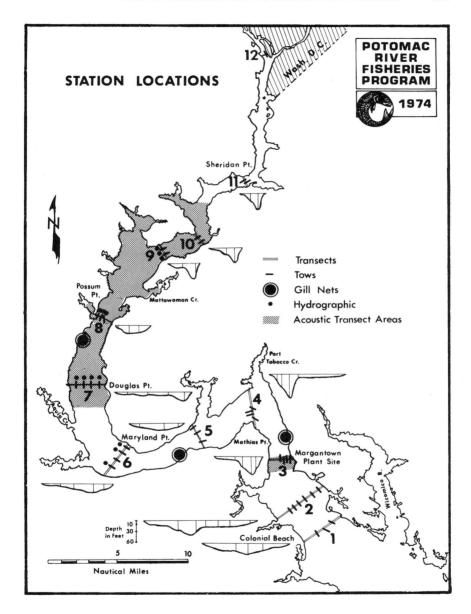

Figure 1. Locations of weekly, oblique, ichthyoplankton tow stations along 12 Potomac
River transects. Current meter stations, gill net locations and acoustic fish survey
areas are also shown.

excluded) were deployed between transects 3 and 4, 5 and 6, and 7 and 8 (Fig. 1). After separation from other species, the striped bass were weighed, their length measured, and the specimens then were dissected to determine sex, age, and sexual maturity—all measures for determining the potential productivity of the population. Acoustic surveys were carried out to count the striped bass adults and trace their movements (24). Detailed work around the gill net stands being sampled provided the necessary calibration of acoustic target strength with size distribution and made extrapolations of striped bass distributions possible in the entire spawning area. The combined gill net sampling and acoustic programs led to estimates of potential egg production and also identified the river areas in which spawning was taking place.

Ichthyoplankton sampling was carried out at all 38 sampling stations weekly throughout the duration of the spawning and developmental periods (from 2 April 1974 to 19 August 1974). Egg and larval distributions across and along the axis of the river and in shore and tributary areas were examined, while hydrographic studies correlated transport processes in the river with the evolution of ichthyoplankton distributions. (A full description of current meter deployment and data recovery and analyses may be found in Refs. 19 and 20.) Starting at the bottom, a 0.5-mm mesh net, 1 m in diameter, was towed through the water column to vertically integrate samples. Water volumes of approximately 50 m^3 were strained at each successive 3-m depth layer. Depending on river depth, total volume strained varied from station to station. Strained volume was monitored and computed from net-mounted flow meters with on-board readouts. At each ichthyoplankton station, temperature, salinity, turbidity, dissolved oxygen, and light intensity data were collected. Ichthyoplankton were identified and enumerated to taxa and species (where possible).

Variations in vertical ichthyoplankton distributions from day to night were investigated with two paired 0.5-m bongo nets that could be lowered to discrete depths, opened for sampling, and then reclosed for retrieval.

Distributions of juvenile bass (30-45 days old, at least 30 mm long) in the river shallows were also determined by shore seine netting at 35 stations.

Data from various program elements were cataloged in a unified data system. Field and laboratory information were recorded on coded forms, integrating physical and biological data in standardized formats.

INTERPRETATION OF ICHTHYOPLANKTON SURVEY RESULTS

Major factors determining the longitudinal patterns of eggs and larvae are the movements of the adult population during spawning and the advective action of net non-tidal river flows in transporting these distributions generally downstream. In fact, the input of eggs by the adult stock is some continuous function of time during a finite period, while adult movement takes place longitudinally and generally upstream. At the same time, net flows are transporting surviving ichthyoplankton uniformly downstream if the process takes place in the fresh-water portion of the estuary. Therefore, a correlation may exist between temporal and spatial variations. In the meantime, large mortalities are experienced by both eggs and larvae.

In order to trace this process, ichthyoplankton samples were separated into the egg stage and three distinct larval phases of development: yolk-sac larvae (up to 12 days old), finfold larvae (up to 24 days old), and post-finfold larvae. Lengths of the larvae were also measured. These properties provided a reasonable capability for tracing the development of distributions. Yolk-sac larvae were the most abundant stage after eggs, and, in some analyses where there were relatively few later developmental stages found, all later larval stages were combined with this category.

Fig. 2 presents composite time and space (longitudinal) patterns of eggs, total larvae, salinity and temperature. The spatial axis denoting transect locations is scaled to true river mile distances between transects. For the purposes of this presentation, density values for stations within each transect are appropriately weighted, taking into account sample volumes that reflect river cross-sectional area and depth. Both temperature and salinity are presented at the 3-m level. The river throughout the sampling period was generally isothermal with depth, and detailed temperature-salinity analyses are presently underway.

Spawning began around 10 April 1974 at transects 4 and 5. It is noteworthy that at this time temperatures ($\sim 10^\circ$C) were below those reported as optimum for spawning and hatching of eggs. We believe that a rapid temperature rise triggered premature spawning. Salinities in these areas were below 1 ppt at this time, a normal range for spawning. The general positive slope of the egg time-space distribution implies that spawning proceeded continually further upstream in time. As the spawning process developed, temperatures increased over the whole spawning region, generally monotonically and uniformly. With the exception of one sampling date (cruise 5-30 April, 1974), all eggs were found above the 1 ppt salinity isopleth.

The spawning peak occurred at transect 7 on cruise 5 (30 April 1974). Approximately 70% of the eggs obtained during the whole sampling period were collected on this date at this transect. Temperature reached 15°C, and the salinity was essentially 0 ppt during the peak spawning activity.

Analysis showed that both the onset and peak of spawning were related to the rate of temperature rise. It seems, therefore, that temperature gradients in time, rather than the value of temperature alone, may also be responsible for discrete spawning events. After peaking, spawning activity proceeded at a continually reduced rate, moving upstream in time. From the egg distribution data, it could not be determined whether the upstream shifting pattern was due to the upstream movement of the spawning stock itself, or to the upstream propagation of spawning activity through the stock already distributed in a stationary pattern.

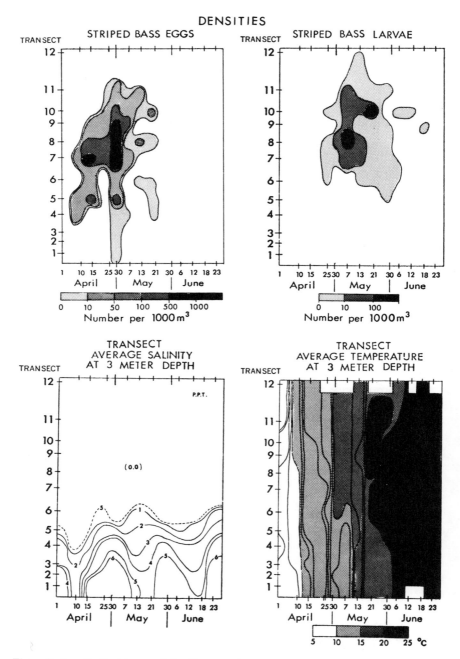

Figure 2. Longitudinal-temporal distributions of eggs, larvae, salinity and termperature.

Larvae were found upstream of the egg distributions in spite of the measured downstream transport in the river. The larval distribution also displayed a tendency towards a negative slope, indicating downstream drift of larvae in time. Since all significant larval densities were upstream of the major egg peak in space, no traces of larvae were detected for over 80% of the eggs spawned, indicating an almost complete lack of hatching success of the peak spawn downstream, i.e., the relatively low egg densities observed above transect 8, produced near the end of the spawning process, were responsible for almost all of the larvae that survived.

The egg distribution in Fig. 2 displays unimodal behavior because of the large spawning peak. In contrast, larval distributions are bimodal, with peaks near transects 8 and 10 over an extended period of time. This implies differential survival rates for eggs and larvae in time, and, because of the apparent upstream movement of adults during the process, also in space. The observed phenomenon is probably related to decreased hatching and developmental times for the later spawn as temperature increased over the whole spawning area in time.

Until 30 May 1974, yolk-sac larvae were associated predominantly with salinities of less than 1 ppt. Because of the disappearance of the egg peak and the sharp tapering off of both larval and egg distributions at the salt wedge (in spite of a tendency to be dispersed in the longitudinal direction by turbulent processes), it seems that unidirectional downstream transport of both eggs and larvae into the salt wedge caused the highest mortalities in all early ichthyoplankton stages. The lack of survival of egg and yolk-sac stages in saline waters may have been due to a lack of food organisms required by developing larvae in the saltwedge region as well as to salinity stress. Successful development upstream may have been both a function of temperature (due to more rapid development at later, higher temperatures) and food availability. No definitive conclusions were drawn from food habit and availability studies (10) to resolve this ambiguity.

Subsequent presentation of abundance data will show that egg and yolk-sac distributions were sharply truncated at the salt wedge (transect 6). Flow investigation revealed that longitudinal dispersion processes were important in spreading distributions, with dispersion coefficients attaining values as high as 1.5×10^6 cm^2/sec. Therefore, the truncation of egg and larval distributions implies fairly rapid mortalities as distributions encountered the salt wedge—on a time span much less than the weekly sampling rate.

The acoustic fish density surveys conducted during the ichthyoplankton sampling periods supported the hypothesis of the upstream movement of spawning adults in time. Two acoustic surveys were made in the entire region shaded in Fig. 1. When the analysis of only large acoustic targets was considered, the results displayed the density distributions of adult fish greater than 40 cm in length. These fish were striped bass only, as corroborated by the concurrent, twice-weekly gill net collections of Morgan and Wilson (14). Fig. 3 is taken from

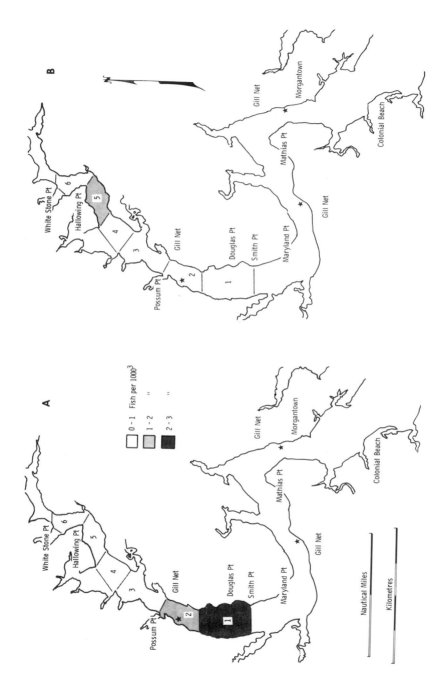

Figure 3. Acoustically determined densities of fish targets greater than 40 cm in length. See text for explanation of parts A and B.

Zankel et al. (25), and indicates the measured densities of large targets in six longitudinal river regions (not corresponding to transect designations). Distribution A (one million adults) was obtained during peak spawning, and the highest concentration of spawning adults corresponds to the time and location of major spawning activity. The second distribution, B (0.5 million adults), was obtained two weeks later, and the highest spawning adult concentrations occurred near transects 9 and 10. The surviving larvae upstream of the location of major spawning can be traced to a reduced adult population spawning upstream. The reason that the bimodal larval peaks were traceable only to adult distribution maxima rather than to eggs is the high mortalities eggs experienced throughout the process, as evidenced by the change in density of egg and larval distributions.

ESTIMATES OF ICHTHYOPLANKTON ABUNDANCES AND MORTALITIES

From weekly sampling of ichthyoplankton distributions, we can calculate densities using tow volumes and counts. These densities may then be extrapolated into river volumes represented by each tow to yield total abundances for the day the sample was taken. However, both the sampling interval of seven days and the hatching and developmental times must be taken into account to interpolate net production between sampling times and obtain abundance estimates over the entire process. Since larval developmental times exceeded the periods between sampling, the calculation between weeks must also take into account only the fraction of the developmental time spent near a particular transect location.

The following general expression was used for abundance estimation of all stages:

$$A_{stage} = 168 \sum_{i=1}^{W} \sum_{j=1}^{N} \sum_{k=1}^{J_k} V_{jk} \times D_{ijk} \, C_{ijk}$$

where

W = number of sampling weeks
N = number of transects
J_k = number of tows at the jth transect
D_{ijk} = density of life stage during ith week at tow location k of the jth transect
168 = the mean time between sampling, in hours
C_{ijk} = the average developmental time of each stage, in hours
C_{ijk} = $-4.69 \times T_{ijk} + 131.6$ for eggs
C_{ijk} = 286 for yolk-sac larvae
C_{ijk} = 264 for finfold larvae
T_{ijk} = temperature during the ith week at section k of the jth transect
V_{jk} = river prism volumes into which tow densities were extrapolated.

The hatching time is a decreasing function of increasing temperature derived from Doroshev (6) and Mansueti (11). There is no temperature-dependent information on other developmental times; however, these are realistic averages from the literature.

The time-integrated absolute abundance estimates centered on each transect location, and the percentage relative abundances are presented in Table 1. The dominance of the singular peak event in spawning on cruise 5 is evident in the abundance of eggs at the transect 7 location over time, as well as in the densities shown previously. However, in the time-integrated abundances of the two larval stages, it is seen that the major egg peak did not contribute to spawning success, or total abundance in those stages. Even though the abundances are integrated over time in the table, both transport downstream and adult movement upstream may be seen. The upstream shift from eggs to yolk-sac larvae is indicative of spawning stock movement. The downstream shift from yolk-sac larvae to finfold larvae is consistent with measured net river flow. The persistence of bimodal distributions from yolk-sac to finfold larvae as the larvae shifted downstream in time (Table 1) indicates that mortalities became uniform over the populations in time, and that survival increased in time.

Potential egg production of the spawning population was estimated (16) from adult age distribution, sex ratio, maturity, and weight of the spawning stock. The absolute stock abundances of Zankel et al. (25) (about 1-3 adult fish/1000 m^3, or one million fish during peak spawning and one-half million 2 weeks later) were applied to this information to arrive at a potential egg production of 75 billion eggs. Stage-to-stage mortalities were directly obtainable from the abundance estimates given in Table 1:

Potential to calculated (apparent mortality)   93.84%
Egg to yolk-sac mortality   98.40%
Yolk-sac to finfold mortality   95.34%

Insufficient numbers of post-finfold larvae were collected to calculate finfold to post-finfold mortality.

Table 1.   Abundance estimates on striped bass eggs and larvae in Potomac Estuary, 1974.

Centered on Transect	Eggs × 10^5	%	Yolk-Sac × 10^5	%	Finfold × 10^5	%
5	1410	3.10	0	0		0
6	401	.90	2.6	.35	11.4	33.5
7	33885	74.80	149.5	20.5	1.0	3.3
8	3692	8.00	264.0	36.2	11.4	33.5
9	4380	9.70	60.6	8.3	5.4	15.9
10	1402	3.10	239.4	32.8	3.8	11.2
11	229	.5	13.8	1.9	.9	2.7
TOTAL	45290	100.1	730	100.0	34	100.1

The first "mortality" figure above could be a consequence of unfertilized and immature egg production effects, perhaps producing rapid disintegration and less than complete extrusion of all eggs, respectively. More importantly, however, overall efficiency of egg sampling was undetermined, which leads to the possibility of a large difference between the potential and accountable figures from sampling. The latter two estimates for mortalities have been obtained, however, by the same sampling method, with roughly equal efficiencies. The abundance distributions of the three stages and the previous interpretation imply that egg and yolk-sac mortalities were considerably higher through the peaking of the spawning process in the lower part of the spawning grounds than subsequently upstream. Further work is necessary to partition time- and space-dependent mortalities in each stage. The timing and locations of biological events, in combination with net longitudinal river flow, have been shown to be primary determinants of high mortalities, especially in the egg stage.

Considering the high mortalities that may occur within the first year of life, and the 3.4 million finfold larvae estimated, it is safe to speculate that 1974 spawning success was poor and will produce a low yearclass. The spawning stock size (25) and preliminary reports of catch in 1974 indicate that the spawning stock had relatively high abundance. Subsequent juvenile collections at 35 stations, repeated biweekly for three months, obtained only 22 striped bass juveniles (10). Although no quantitative historical comparisons have been made with juvenile data from the Maryland Department of Natural Resources, the juvenile abundances this year were very low compared to other years.

CONCLUSIONS

Almost all egg samples were obtained in fresh waters. The peak of spawning occurred sufficiently near the salt wedge for transport of eggs into higher salinities. The largest apparent mortalities during transition from egg to yolk-sac stage were connected with this transport, and were attributable either to low salinity tolerances of eggs and early yolk-sac larvae or to other indirect effects of higher salinities, such as the absence of food organisms necessary for survival and development in this region. Distributions of both eggs and larvae were sharply truncated near the salt wedge, in spite of large longitudinal turbulent dispersion. This indicates rapid mortality as distributions encountered the salinity gradient.

Surviving larvae were found upstream of major egg densities. This distribution shift was due to the movement of the spawning stock upstream, and the rate of this movement exceeded the downstream drift of water. The upstream location of surviving larvae, representing the tail of the spawning process in time, was favorable for survival because it allowed early development in fresh water with possibly higher abundances of food organisms. In addition, increasing river temperatures in time shortened both hatching and developmental periods.

Large ichthyoplankton mortalities occurred during development. The low abundances of post-finfold and juvenile stages indicate a poor yearclass for 1974.

Although no causal mechanisms have been identified for peak spawning close to the salt wedge, this phenomenon—which promoted transport of most early-spawned eggs and yolk-sac larvae into saline regimes—seems to be responsible for the poor spawning success observed. Work is continuing to resolve some of the questions raised and to refine conclusions reached here.

ACKNOWLEDGEMENTS

The Potomac River Fisheries Program is supported by the Department of Natural Resources of Maryland. The authors are indebted to John Cooper, K. Wood, R. Prince, E. Gordon and R. Block of the Chesapeake Biological Laboratory, who were responsible for ichthyoplankton field data collection, sorting and identification. G. Krainak of the Martin Marietta Corporation is acknowledged for organizing the PRFP data base and for programming and statistical analyses involved in these investigations. Capt. William Keefe and Mate Clellen Keefe were also instrumental in enabling us to meet our field objectives.

REFERENCES

1. Albrecht, A. B. 1964. Some observations of factors associated with the survival of striped bass eggs and larvae. Calif. Fish and Game 50:101-113.
2. Bason, W. H. 1971. Ecology and early life history of striped bass, *Morone saxatilis*, in the Delaware Estuary. Ichthyological Associates Bulletin, 4, 122 p.
3. Bayless, J. O. 1972. Artificial propagation and hybridization of striped bass, *Morone saxatilis* (Walbaum). South Carolina Wildlife and Marine Resources Department, 135 p.
4. Bigelow, H. B., and W. C. Schroeder. 1953. Fishes of the Gulf of Maine. Fishery Bulletin of the Fish and Wildlife Service, Vol. 53. U.S. Government Printing Office, Washington, D. C.
5. Carlson, F. T., and J. A. McCann. 1968. Hudson River Fisheries Investigations 1965-68. Report to Consolidated Edison Company, 50 p. *In* G. P. Howells and G. J. Lauer (eds.), Proceedings of the Second Symposium on Hudson River Ecology, Tuxedo, New York, October 1969, 320-372, 473 p. (Hudson River Policy Committee).
6. Doroshev, S. I. 1970. Biological features of eggs, larvae and young of the striped bass (*Roccus saxatilis*, Walbaum) in connection with the problem of its acclimatization in the U.S.S.R. J. Ichthyol. 10(2):235-248, VNIRO, Moscow, U.S.S.R.
7. Dovel, W. L. 1971. Fish eggs and larvae of the upper Chesapeake Bay. NRI Spec. Rept. No. 4, 71 pp. University of Maryland, NRI.
8. Farley, T. C. 1966. Striped bass, *Roccus saxatilis*, spawning in the Sacramento-San Joaquin River system during 1963 and 1964. Calif. Fish and Game Bull. 136 pp.
9. Koo, T.S.Y. 1970. The striped bass fishery in the Atlantic States. Chesapeake Sci. 11:73-93.
10. Loos, J. 1975. Shore and tributary distributions of ichthyoplankton and juvenile fish, with a study of their food habits. Final Report, Academy of Natural Sciences of Philadelphia, Maryland Power Plant Siting Program.

11. Mansueti, R. 1958. Eggs, larvae and young of the striped bass, *Roccus saxatilis*. Maryland Dept. Res. and Education, Contr. No. 112.
12. _____. 1959. Winter and spring tagging of striped bass concluded in Potomac River. Maryland Tidewater News 15:11-12.
13. May, R. C. 1974. Larval mortality in marine fishes and the critical period concept, p. 3-19. *In* J. H. Blaxter (ed.) The Early Life History of Fishes Springer-Verlag, Berlin.
14. Mihursky, J., R. M. Block, K. Wood, R. Prince and W. E. Gordon. 1974. Interim Report on 1974 Potomac Estuary horizontal ichthyoplankton distributions. University of Maryland Center for Environmental and Estuarine Studies, Ref. No. 74-170.
15. Morgan, R. P., II. 1971. Comparative electrophoretic studies on the striped bass *(Morone saxatilis)*, and white perch *(Morone americana)*, and electrophoretic identification of five populations of striped bass in the upper Chesapeake Bay. Ph.D. Thesis, University of Maryland.
16. _____, and J. Wilson. 1974. Interim report on 1974 spawning stock assessment. Potomac River Fisheries Program, Maryland Power Plant Siting Program.
17. Nichols, P. R., and R. V. Miller. 1967. Seasonal movements of striped bass *(Roccus saxatilis)* tagged and released in the Potomac River, Maryland, 1959-61. Chesapeake Sci. 8(2):102-124.
18. Pearson, J. C. 1938. The life history of the striped bass, or rockfish, *Roccus saxatilis* (Walbaum). Bull. U. S. Bureau Fish 49:825-851.
19. Polgar, T. T., and R. E. Ulanowicz and D. A. Pyne. 1975. Preliminary analyses of physical transport and related striped bass ichthyoplankton distribution properties in the Potomac River in 1974. Environmental Research Guidance Comm. Report, Maryland Power Plant Siting Program, Ref. No. PRFP-72-2.
20. _____, _____, _____, and G. M. Krainak. 1975. Investigations of the role of physical transport processes in determining ichthyoplankton distributions in the Potomac River. Power Plant Siting Program, Ref. No. PPRP-11/PPMP-14.
21. Radtke, L. D., and J. L. Turner. 1967. High concentrations of total dissolved solids block spawning migrations of striped bass, *Roccus saxatilis*, in the San Joaquin River, California. Trans. Am. Fish. Soc. 96:405-407.
22. Shannon, E. H., and W. B. Smith. 1967. Preliminary observations on the effect of temperature on striped bass eggs and sac fry. Proceedings of the 21st Annual Conference of S. E. Assoc. Game and Fish Comm., p. 257-260.
23. Turner, J. L., and T. C. Farley. 1971. Effects of temperature, salinity and dissolved oxygen on the survival of striped bass eggs and larvae. Calif. Fish and Game 57(4):268-273.
24. Zankel, K. L., L. H. Bongers, T. T. Polgar, W. A. Richkus and R. E. Thorne. 1975. Size and distribution of the 1974 striped bass spawning stock in the Potomac River. Martin Marietta Corp., Ref. No. PRFP 75-1.
25. _____, R. E. Thorne, L. H. Bongers, T. T. Polgar and W. A. Richkus. 1975. An assessment of the Potomac River striped bass spawning stock by acoustic methods (submitted to J. Fish Res. Bd. Can.).

THE USE OF INNATE TAGS TO SEGREGATE SPAWNING STOCKS

OF STRIPED BASS (*MORONE SAXATILIS*)

Thurman L. Grove[1] and Thomas J. Berggren
Texas Instruments Incorporated
P.O. Box 237
Buchanan, New York 10511
and
Dennis A. Powers
Department of Biology
The Johns Hopkins University
Baltimore, Maryland

ABSTRACT: The striped bass (*Morone saxatilis*) is an anadromous fish that makes extensive migrations along the Atlantic coast and utilizes the Atlantic coastal estuaries as spawning and nursery areas. Industrial usage of the estuaries, commercial fishing pressure and increased recreational use of the fishery have stimulated interest in the relative contribution of striped bass spawning stocks from various estuaries.

Studies have been performed to assess the feasibility of using meristic, morphometric and biochemical characters as innate tags to segregate striped bass from various spawning populations. Representative samples of the spawning populations of the Roanoke River, four tributaries of Chesapeake Bay (Potomac, Rappahannock, Choptank and Elk Rivers), and the Hudson River were collected. Data were collected from each fish on up to forty-two meristic and morphometric characters, forty-five protein characters and twenty-eight enzyme systems involving fifty-two genetic loci. Linear and quadratic discriminant function analyses were employed to evaluate the discriminative power of the meristic and morphometric characters. Biochemical characters were evaluated using univariate techniques.

---

[1] Present address: Beak Consultants, Inc., Cornell Industry Research Park, Brown Road, Ithaca, New York 14850.

A set of three meristic and morphometric characters was derived to provide maximum separation among the various spawning stocks. It was not possible to separate the stocks within the Chesapeake estuary due to overlap in the character sets. The sample values from the Chesapeake tributaries were pooled and entered in a discriminant function with values from the Hudson and Roanoke populations. The character sets allowed separation of the resultant three spawning stocks with an overall correct classification of seventy-two to seventy-four percent.

INTRODUCTION

The striped bass (*Morone saxatilis*) is an anadromous fish that makes extensive migrations along the Atlantic coast and utilizes the Atlantic coastal rivers and estuaries as spawning and nursery areas (14, 20). Significant sport and commercial fisheries exist for striped bass in the estuaries and along the coast from Cape Hatteras north to central Maine (13). Studies dealing with striped bass migratory behavior have generally concluded that the major fraction of the coastal fishery is produced in the Chesapeake Bay system (1, 14, 18, 21, 22). The results of legal proceedings concerning power plant construction on the Hudson River[1,2] have suggested that Hudson River striped bass may also contribute significantly to the Atlantic fishery.

Assessment of the relative contribution of various spawning stocks to the fishery, which provides a useful base of information for management of the fishery, requires that the natal stock of any fish at large in the fishery can be identified. Studies by Raney et al. (19, 20) have demonstrated that meristic characters allowed seventy to eighty percent separation between particular age specific year classes of striped bass originating in the Hudson River and those originating in the tributaries of Chesapeake Bay. Furthermore several studies (5, 7, 11, 15, 16) have shown that biochemical characters allow identification of fish from various spawning stocks. This paper discusses the results of a study designed to evaluate the feasibility of using biochemical, meristic, and morphometric characters as innate tags to identify striped bass spawning stocks among the major spawning areas of the Hudson River, tributaries of Chesapeake Bay, and the Roanoke River.

MATERIALS AND METHODS
Field Collection

The use of innate tags for stock discrimination implies breeding isolation of spawning aggregations and a continuity (either genetic or otherwise) in characteristics of the stocks. Consequently, the following assumptions were made:

---

[1] Clark, J. 1972. Testimony before the USAEC. Docket 50-247. Oct. 30
[2] Goodyear, C.P. 1974. Testimony presented to the Committee on Merchant Marine Fisheries of the U.S. House of Representatives. Feb. 19.

first, striped bass tend to return to their natal streams to spawn; and second, any sexually ripe striped bass collected in the spawning area of a particular river during spawning season originated in that river.

During the spring spawning season of 1974, striped bass were collected from the spawning areas of the Rappahannock (156 fish), Potomac (202 fish), Choptank (91 fish), Elk (250 fish), and Hudson (192 fish) rivers. During 1975 additional striped bass were collected from the Rappahannock (70 fish), Potomac (53 fish), Choptank (52 fish), Elk (57 fish), Hudson (168 fish), and Roanoke (99 fish) rivers. Various types of fishing gear were used including pound nets, haul seines, drift, stake and anchor gill nets. Specimens were collected throughout the entire spawning period and across the size distribution of the fish represented in the catches.

Specimen Processing in Field

A numbered jaw tag was attached to each fish; and blood, liver, and muscle tissue samples were obtained in the field from each fish for isozyme analysis. One to two ml of blood was collected with a syringe from the cardiac region, placed in a test tube, and centrifuged to separate the cellular and serum fractions. The fractions were separated and stored at 0° to 4.4°C until processed. A strip of muscle tissue excised from the region between the first and second dorsal fins across the lateral line, and a lobe of liver tissue, were placed in separate whirl pack bags and stored in liquid nitrogen until processed. Only liver tissue was collected in 1975 for subsequent isozyme analyses.

Scale samples for age and growth analysis were obtained from a key location above the lateral line between the first and second dorsal fins. Sex and state of maturity were recorded for each fish. Specimens in 1974 collections were preserved in 20% formalin and transferred to the laboratory where meristic counts and morphometric measurements were taken after one month storage in formalin. Meristic counts and morphometric measurements were performed in the field on fresh fish in 1975.

Specimen Processing in Laboratory

Thirty meristic counts and morphometric measurements were taken on each fish in 1974 collections. Scales along lateral line, scales above lateral line, scales below lateral line, scales around caudal peduncle, spines on first dorsal fin, and soft rays on the second dorsal and anal fins were enumerated in the manner prescribed by Hubbs and Lagler (10). All rays including rudiments were counted on the left and right pectoral fins. Gill raker counts were made on the first gill arch, excluding and including rudimentary rakers. A gill raker was considered rudimentary if its height was less than the diameter of its base. A gill raker located in the angle of the arch was counted with the lower arm.

Total length, standard length, snout length, length of upper jaw, head length, orbit to angle of preopercle length, length of orbital, interorbital width (least fleshly width), predorsal length, length of caudal peduncle, depth of caudal peduncle, length of base of second dorsal fin, length of first spine of second dorsal fin, length of base of anal fin, and length of first spine of anal fin were also measured as prescribed by Hubbs and Lagler (10). Fork length was measured from the most anteriorly projecting part of the head to the deepest fork of the caudal fin. The internostril width was the least fleshy distance between the excurrent nares. All measurements were taken to the nearest millimeter.

The ten meristic counts and morphometric measurements which showed discriminative potential in 1974 data were taken on each fish in 1975. These were number of lateral line scales, left and right pectoral ray count, second dorsal ray count, anal ray count, upper gill raker count, fork length, snout length, head length, and internostril width.

Three scales (nonregenerated) from each specimen were cleaned and permanently mounted on acetate cards with a heat press. Age determinations and measurements from the focus to the first and second annuli were made on the magnified scale images (43.5X, 45X, 47.5X) which were projected by a calibrated scale projector.

Isozyme Analyses

Standard starch gel electrophoresis was employed to screen all protein characters that might be useful in discriminating among striped bass spawning stocks. Forty-five protein systems, including sixteen serum proteins and hemoglobins, were examined. In addition, twenty-eight enzyme systems involving fifty-two loci were elucidated (Table 1). The discriminative potentials of the variant enzyme systems were analyzed with univariate techniques.

Analytical Techniques

Spawning stocks contain fish of all mature age classes, differentially sampled by various gear. The biases presented by variations in year class strength and gear selectivity were avoided by eliminating from further analysis any character that was significantly correlated with length.

Linear and quadratic discriminant function analyses were used to classify individuals from a mixed sample into their respective spawning stocks. The methods differed in that the linear function required multivariate normal distribution of data and a common variance-covariance matrix. The quadratic function was more general and did not require a common variance-covariance matrix. Thorough treatments of linear and quadratic discriminant analyses are found in Anderson (3) and Kendall and Stuart (12), respectively.

RESULTS

Laboratory processing on specimens collected in 1974 provided data on 42 meristic and morphometric characters. Fifteen of these characters (Table 2) were used in the discriminant analyses of the 1974 data. Twenty-one characters were rejected because of correlation with length. Two additional characters were highly cross correlated with other characters and similarly removed. Spine counts on the first dorsal fins were constant for all but eight fish, scale-to-scale variation was observed when scale measures were not used as ratios, and bias was observed on the upper arm gill raker counts because of problems in the dissection and removal of the gill arch from preserved fish. Consequently, these characters were removed from further analysis. Eight characters (Table 2) were used in the discriminant analyses in 1975.

A low overall probability of correct classification occurred when segregation of 5 (1974) or 6 (1975) spawning stocks was attempted (Table 3). Adequate segregation of spawning stocks within the Chesapeake region was not possible; therefore, data from the Chesapeake rivers were pooled and new discriminant functions were determined for 2 regions (Hudson and Chesapeake) with 1974 data and 3 regions (Hudson, Chesapeake, and Roanoke) with 1975 data.

Table 1.   Enzyme systems investigated.

Enzyme System	Abbreviation	No. of Loci
α-napthyl acetate esterase	α-nap. acetate-EST	1
α-napthyl butyrate esterase	α-nap. but-EST	4
Serum esterase	ser-EST	1
Phosphoglucomutase	PGM	1
Phosphohexoseisomerase	PHI	2
Isocitrate dehydrogenase	IDH	1
Alcohol dehydrogenase	ADH	2
Glucose 6-phosphate dehydrogenase	G6PDH	1
Alkaline phosphatase	ALK PHOS.	1
Acid phosphatase	ACID PHOS.	1
Glucokinase	GK	1
Glutamate dehydrogenase	GDH	2
α-glycerophosphate dehydrogenase	α-GPDH	2
6-phosphogluconate dehydrogenase	6PGDH	2
Lactate dehydrogenase	LDH	2
Superoxide dismutase	SOD	2
Leucine aminopeptidase	LAP	1
Fructose 1,6 diphosphatase	F1,6DiPhos	2
Creatine kinase	CK	2
Adenylate kinase	AK	2
Aspartate aminotransferase	AAT	3
Xanthine dehydrogenase	XDH	1
Sorbitol dehydrogenase	SDH	2
Glyceraldehyde 3 phosphate dehydrogenase	G3-P DH	2
Monoamine oxidase	MO	2
Malate dehydrogenase	MDH	2
Peroxidase	Per	5
β-hydroxybutyrate dehydrogenase	βOHbutDH	2

Total systems-28   Total loci — 52

Table 2.   Initial variables used in character sets.[1]

Scales	MERISTIC Fin Rays	Gill Rakers
*Along lateral Line	*Left Pectoral	**Upper Arm
Above lateral Line	*Right Pectoral	Lower Arm
Below lateral Line	*Second Dorsal	
Around Caudal Peduncle	*Anal	

MORPHOMETRIC
Length Ratio

Snout/Head	Head/Fork
Internostril/Head	Base Anal Fin/Fork
*Snout/Internostril	

GROWTH

*First Annulus to Second Annulus/Focus to First Annulus Measure Ratio

[1] No asterisk designates that variable was used with 1974 data only. Single asterisk designates that variable was used with 1974 and 1975 data. Double asterisk designates that variable was used with 1975 data only.

Table 3.   Correct classification percentages of various spawning stocks by quadratic discriminant analysis.

Year	Hudson	Chesapeake				Roanoke	Overall
		Rapp.	Pot.	Chop.	Elk		
1974	78.7	39.3	38.5	52.3	57.9		54.3
1975	50.6	37.1	26.4	73.1	21.1	68.7	48.7

When either 2 or 3 regions were analyzed the linear and quadratic discriminant techniques showed that three characters were the "best" discriminators between spawning stocks and that additional characters did not significantly improve the overall discrimination. The three characters, in order of importance were snout length/internostril width ratio, first annulus to second annulus measure/focus to first annulus measure ratio, and lateral line scale count. Therefore, subsequent classification matrices refer only to discriminant analyses involving this set of three characters.

An overall correct classification of approximately 76 percent between the Hudson and Chesapeake spawning stocks is shown in Table 4. Based on tests for common variance-covariance matrices, the quadratic function was more valid for the 1974 data and the linear function was more valid for the 1975 data. The close agreement between the 1974 and 1975 data indicated that year-to-year variation in discriminative potential of the character set was low.

Table 4.   Correct classification percentages of Hudson, Chesapeake, and Roanoke spawning stocks.

Year	Type of Analysis	Hudson	Chesapeake	Roanoke	Overall
1974	Linear	78.1	77.3	–	77.5
	Quadratic	83.1	74.6	–	76.4
1975	Linear (2 stocks)	78.6	74.6	–	76.2
	Linear (3 stocks)	78.0	65.5	84.8	73.5
	Quadratic (2 stocks)	76.8	73.7	–	75.0
	Quadratic (3 stocks)	76.2	62.9	87.9	72.3

Of the 52 isozyme loci examined, only two were polymorphic: α-glycerophosphate dehydrogenase (α-GPDH) and isocitrate dedydrogenase (IDH). Serum transferrin also varied, but was too labile to be used. Frequencies for alleles in α-GPDH and IDH are shown in Table 5. Genetic variance was low in all spawning stocks but varied clinally with latitude. IDH was fixed in the Hudson River spawning stock, while the frequency of the less common allele (C) increased in southern spawning stocks. The degree of variation of α-GPDH was greater in the Hudson River than in southern spawning stocks.

Fixation at the IDH locus in the Hudson River striped bass provided a mechanism to uniquely classify a fraction of fish as "non-Hudson" in origin. Chesapeake fish misclassified as Hudson in the discriminant analyses which possessed the C IDH allele were redefined as "non-Hudson" or Chesapeake fish. This increased the overall correct classification to approximately 78 percent.

An overall correct classification based on meristics and morphometrics of approximately 73 percent among the Hudson, Chesapeake and Roanoke spawning stocks is shown in Table 4. Tests for common variance-covariance matrices

Table 5.   Gene frequencies for each allele.

River	Year	Gene Frequency				
		α-GPD		IDH		
		A	B	A	B	C
Hudson	1974	0.117	0.883	0	1.000	0
	5/14-5/23 1975	0.120	0.880	0	1.000	0
	5/24-6/5 1975	0.031	0.969	0.065	0.924	0.011
Elk	1974	0.090	0.910	0	0.945	0.055
	1975	0.018	0.982	0	0.991	0.009
Choptank	1974	0.068	0.932	0	0.977	0.023
	1975	0.038	0.962	0	0.990	0.010
Potomac	1974	0.024	0.976	0	0.966	0.034
	1975	0.011	0.989	0	0.978	0.022
Rappahannock	1974	0.017	0.983	0	0.944	0.056
	1975	0.019	0.981	0	0.948	0.052
Roanoke	1975	0.010	0.990	0.025	0.950	0.025

indicated that the quadratic function was more valid. A reduction of 4 percent in the overall percentages of correct classification occurred when 3 regions were analyzed. This was due to misclassification of Chesapeake fish into the Roanoke River. The Chesapeake correct classification percentage dropped from approximately 74 percent to 63 percent with the addition of the Roanoke River spawning stock, whereas the Hudson correct classification percentage remained stable.

The gene frequencies for each allele of α-GDPH and IDH in 1975 are shown in Table 5. The clinal changes are comparable to those observed in the 1974 data. A new IDH allele (A) was found in Roanoke fish which uniquely separated individuals possessing it from the Chesapeake and Hudson spawning stocks. Redefinition of Hudson and Chesapeake misclassifications possessing C and A alleles respectively increased overall correct classification by one to two percent.

DISCUSSION

Multivariate discriminant analyses when applied to meristic and morphometric characters, first and second year growth rates, and appropriate transformations of those characters, demonstrated the feasibility of using such characters to identify spawning stocks of striped bass.

It was possible to produce a character set for each spawning stock that was independent of size, sex, and time of capture. It was not possible to segregate among striped bass from tributaries of Chesapeake Bay.

Three characters, first annulus to second annulus distance, focus to first annulus distance, snout length/internostril distance, and number of scales along lateral line, provided virtually all the discriminative power. The probability of correctly classifying a fish with three characters was 72-74%.

The use of discriminant function analysis with meristic characters for separating populations of fish has been successful in numerous studies. Hill (9) correctly classified 81% of shad into their respective Hudson and Connecticut river populations based on six meristic characters. Fukuhara et al. (8) correctly classified 77% of sockeye salmon into their respective Asian and North American populations based on seven meristic characters. Amos et al. (2) correctly classified 72% of pink salmon into their respective Asian and North American populations based on only three meristic characters. Most recently, Parsons (17) contrasted autumn and spring herring, correctly classifying from 80.6 to 86.2% for autumn herring spawners and from 79.4 to 90.7% for spring herring spawners based on three meristic characters.

Thre results for striped bass compare favorably with those of previous stock discrimination studies in which meristic and morphometric characters were employed when two stocks were compared. When three stocks were compared, the overall probability of correct classification was slightly less than many of the previous studies. The *a priori* probabilities for correctly classifying an individual are 0.50 for a two-population case and 0.33 for a three-population case. Consequently, the increase beyond *a priori* levels compares favorably.

There are potential limitations in using the results to assess the relative contribution of various estuaries to the Atlantic fishery. Fish at large in the fishery come from spawning areas other than those sampled in the current study. Indeed, striped bass spawn in most of the major rivers from Florida north to the Hudson (18). Results of tagging studies (1, 6, 21, 22, and ongoing American Littoral Society programs) suggest that fish from all spawning sources north of Cape Hatteras utilize the entire coast north of their respective spawning areas to Maine. The relative contribution of all other spawning areas to the Atlantic fishery is unknown and undoubtedly will affect the probability of correctly identifying fish of Hudson, Chesapeake, or Roanoke origin. Further detailed studies are necessary to assess the contribution from other areas.

The biochemical genetic structure of striped bass is one of the most homogeneous ever studied. The fixation of IDH in the Hudson, the presence of the IDH-A allele in the Roanoke, and the clinal nature of both IDH and α-GPDH provide discriminative power beyond the meristic and morphometric characters. IDH fixation allows unique "non-Hudson" or "non-Hudson-non-Chesapeake" classification of a fraction of specimens. The clinal nature of enzyme frequencies, when extrapolated to more southerly stocks, may allow correct identification of a greater number of "non-Hudson-non-Chesapeake" fish originating in the more southerly spawning areas.

The appearance of A and C IDH alleles in the Hudson collections after May 24, 1975 requires explanation. Sampling in 1974 continued in the Hudson through June 2 and no individuals collected possessed either an A or C allele. Similar results were obtained in 1975 through May 23. We believe that the appearance of the A and C IDH alleles was caused by an influx of southern migrants into the lower Hudson. The α-GPDH B allele was present in the 1975 stock with a frequency similar to 1974 values (0.880 and 0.883) through May 23. However, after May 23, it increased to 0.969, a value similar to that in more southerly stocks. The IDH-A allele simultaneously appeared in late May in specimens collected off the northern New Jersey coast (Powers, unpublished data). The northerly spring migration is well documented (14, 18, 20) and Koo (13) demonstrated striped bass first appear in Long Island and New England commerical landings in May.

Even though southern migrants enter the Hudson, the probability of significant levels of genetic mixing is remote. All specimens possessing IDH A and C alleles were collected in lower Haverstraw Bay, the southern fringe of the spawning region. Additionally, the Hudson spawning peak occurred prior to May 23.

ACKNOWLEDGEMENTS

This study was funded by Consolidated Edison Company of New York, Inc. as part of the Hudson River Ecological Survey. Thanks are due to Messrs. Joel Lieberman and John Bennett and the other members of the Texas Instruments

Data Center for assistance in data analysis. The cooperation of commercial fishermen, too numerous for individual credit, allowed collection of the large number of fish necessary for the study. To the laboratory and field technicians, Eddie Baldocchi, Dana Grass, Michael Locke, Edwin Manter, Ronald McGratten, Thomas Orvosh, Martin Otter, Diane Powers, George Roth, and Donald Strout, we owe special thanks for their devoted hours spent gathering the data. We thank Dr. Peter F. Brussard for constructive criticisms of the manuscript.

LITERATURE CITED

1. Alperin, I.M. 1966. Dispersal, migration, and origins of striped bass from Great South Bay, Long Island, New York. N.Y. Fish and Game J. 13(1): 79-112.
2. Amos, M.H., R.E. Anas, and R.E. Pearson. 1963. Use of discriminant function in the morphological separation of Asian and North American races of pink salmon, *Oncorhynchus gorbuscha*, (Walbaum). Bull. 11. Int. North Pac. Fish Comm., p. 73-100.
3. Anderson, T.W. 1958. An introduction to multivariate statistical analysis. John Wiley and Sons, Inc., New York. 374 p.
4. Dixon, W.J. (ed.). 1971. BMD. Biomedical computer programs. Univ. of Calif. Press, Berkeley. 600 p.
5. Drilhon, A., J.M. Fine, P. Amouch, and G.A. Boffa. 1967. Les groupes de transferrines chez *Anguilla anguilla*. Etude de deux populations Mediterraneenes d'origine geographique differante. Compt. Rend. Acad. Sci., Paris. 265:1096-1098.
6. Florence, B. 1974. Tag returns from 1375 large striped bass tagged in two Maryland spawning rivers. Outdoor Message Oct. Organized Sportsmen of Mass.
7. Fujino, K. 1969. Atlantic skipjack tuna genetically distinct from the Pacific specimens. Copeia 1969 (3):626-629.
8. Fukuhara, F.M., S. Murai, J.J. LaLorre, and A. Sribhibdadh. 1962. Continental origin of red salmon as determined from morphological characters. Bull. 8. Int. North. Pac. Fish. Comm., p. 15-109.
9. Hill, D.R. 1959. Some uses of statistical analysis in classifying races of American shad (*Alosa sapidissima*). U.S. Fish. Bull. 147. Vol. 59. p. 269-286.
10. Hubbs, C.L., and K.F. Lagler. 1947. Fishes of the Great Lakes region. Bull. Cranbrook Inst. Sci. 26. 186 p.
11. Jamieson, A. 1967. Two races of cod at Faroe. Heredity 22: 610-612.
12. Kendall, M.G., and A. Stuart. 1968. The advanced theory of statistics. Vol. 3. Hafner Publ. Co., New York. 557 p.
13. Koo, T. S.Y. 1970. The striped bass fishery in the Atlantic States. Chesapeake Sci. 11(2): 73-93.
14. Merriman, D. 1941. Studies on the striped bass, *Roccus saxatilis,* of the Atlantic coast. U.S. Fish and Wildl. Serv. Fish. Bull. 50(35): 1-77.
15. Moller, D. 1966. Genetic differences between cod groups in the Loften area. Nature 212:824.
16. Morgan, R.P., T.S.Y. Koo, and G.E. Krantz. 1973. Electrophoretic determination of populations of the striped bass *Morone saxatilis* in the upper Chesapeake Bay. Trans. Am. Fish. Soc. 102 (1): 21-32.

17. Parsons, L.S. 1972. Use of meristic characters and a discriminant function for classifying spring and autumn spawning Atlantic herring. Int. Conn. N.W. Atl. Fish. Res. Bull. 9. p. 5-9.

18. Porter, J., and S.B. Saila. 1969. Final report for the cooperative striped bass migration study. Contract #14-16-005, Bureau of Sport Fisheries and Wildlife, U.S. Fish and Wildl. Ser. p. 1-33.

19. Raney, E.C., and D.P. deSylva. 1953. Racial investigations of the striped bass, *Roccus saxatilis* (Walbaum). J. Wildl. Mgt. 17 (4): 495-509.

20. _____, W.S. Woolcott, and A. G. Mehring. 1954. Migratory patterns and racial structure of Atlantic coast striped bass. Trans. 19th N.A. Wildl. Conf. p. 376-396.

21. Schaefer, R.H. 1968. Size, age composition and migration of striped bass from the surf waters of Long Island, New York. N.Y. Fish and Game J. 15(1): 1-51.

22. Vladykov, V.D., and D.H. Wallace. 1952. Studies of the striped bass, *Roccus saxatilis* (Walbaum), with special reference to the Chesapeake Bay region during 1936-1938. Bull. Bingham Oceanog. Coll. 14(1):132-177.

THE DYNAMICS OF ESTUARINE BENTHIC COMMUNITIES[1]

Donald F. Boesch, Marvin L. Wass and Robert W. Virnstein
Virginia Institute of Marine Science
Gloucester Point, Virginia 23062

ABSTRACT: Populations of most macrobenthic species in a polyhaline mud-bottom community were quite variable seasonally and over longer time periods. The dynamics of the populations reflect life histories of the species and long-term habitat changes. Few common species were persistent and most were either irruptive, annuals, or euryhaline opportunists responding to habitat changes. Of the two dominant species in an oligohaline community, one had seasonally variable populations, and the other persistent populations.

Estuarine benthic communities include a predominance of r-strategists, but overgeneralization about the relationships of life histories to environmental constancy and predictability can obscure insight, because most communities include species with a wide variety of adaptive life history strategies. Benthic communities in temperate coastal and estuarine environments are generally characterized by wide fluctuations in abundance of many constituent species but more persistent qualitative composition. The dynamic nature of estuarine benthic communities seriously limits the usefulness of short-term baseline and impact studies.

INTRODUCTION

Understanding seasonal dynamics has been a central aim of plankton ecology for many years. Fisheries science has similarly been concerned with assessment of the nature and causes of temporal variations in stocks. However, benthic ecology has lagged in its consideration of the seasonal and long-term dynamics of communities. Although temporal changes have long been recognized (31, 42) and the factors responsible discussed (43) (e.g. climatic changes, variations in recruitment and survival, and biogenic habitat modifications), a general condition of static equilibrium can be inferred from most descriptions of soft bottom benthic communities.

[1] Contribution No. 768 from the Virginia Institute of Marine Science

177

Recent attention to the seasonal dynamics of shallow-water macrobenthic communities (1, 9, 11, 26, 41, 46) has shown the existence of strongly seasonal patterns of recruitment and mortality, which effect markedly changing community structure throughout the year. Although few have had the luxury of studying benthic population variability for more than a year or two, several assessments of long-term variability have shown similarly striking changes occurring over periods ranging from 1 to 45 years (6, 9, 12, 20, 39, 44, 50, 51).

Knowledge of variations in populations and community structure is necessary for the empirical testing of hypotheses relating community stability and persistence to complexity and diversity (2, 18, 29, 32). Knowledge of the nature and causes of natural fluctuations is also critical in assessing the effects of man's activities on estuarine communities, and in the future use of the vast amount of environmental baseline data currently being collected.

This report summarizes the results of long-term investigations of the macrobenthic communities at a polyhaline mud-bottom site and in an oligohaline environment in the Chesapeake Bay system. Observations are made on the factors generally responsible for seasonal and long-term variability in estuarine benthic communities, on the relationship of adaptive life-history strategies to community persistence and stability, and on the implications of the dynamic nature of communities to environmental impact assessments.

DYNAMICS OF A POLYHALINE COMMUNITY

A mud-bottom site in 9 m of water in the lower York River estuary in the southern Chesapeake Bay (Fig. 1) was sampled frequently (biweekly, monthly or bimonthly) from November 1960 to July 1963, twice in 1964, once each in 1965 and 1966, and quarterly from May 1972 to July 1975. The bottom salinity at this site is typically polyhaline, ranging from 18 $^{\circ}/_{\circ\circ}$ in spring to 24 $^{\circ}/_{\circ\circ}$ in fall, but this pattern varies depending on freshwater discharge. The years 1965-1971 were unusually dry and salinity at the site did not drop below 19 $^{\circ}/_{\circ\circ}$, but Tropical Storm Agnes reduced salinity there in the summer of 1972 to 11 $^{\circ}/_{\circ\circ}$. The sediments at the study site are predominantly silts and clays with 20-30% very fine sand.

The details of sampling and a thorough discussion of the patterns of abundance of all common species will be presented in another paper (Boesch et al., in prep.). In this report we will generally describe the patterns found based on collections from a 1.0 mm sieve. Finer sieves were used during portions of the study.

Fig. 2 depicts population trends of species which, although seasonally fluctuating, were more or less consistently abundant from year to year. The opisthobranch *Acteocina canaliculata* was more abundant during winter and spring when small individuals recruited to the community during summer and fall grew large enough to be taken on the 1.0 mm screen. Successful recruitment occurred in late 1972 following the disruptions caused by Agnes (4), but recruitment in

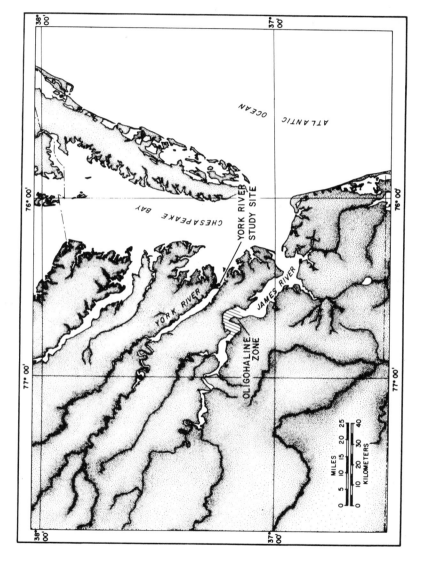

Figure 1. Lower Chesapeake Bay showing the location of the polyhaline study site in the lower York River and the oligohaline study zone in the James River.

1973 and 1974 was poor. The caridean shrimp *Ogyrides limicola* was more abundant during the fall and winter following annual recruitment in the late summer and fall. Extremely poor recruitment occurred during 1972 following Agnes but the usual recruitment pattern has since resumed. Although this shrimp is infaunal, movement of swimming adults may be responsible for some of the observed fluctuations. The amphiurid brittle star *Micropholis atra* demonstrated the most persistent population levels of any species. Small individuals were rare, suggesting that the populations consisted of several year classes with low mor-

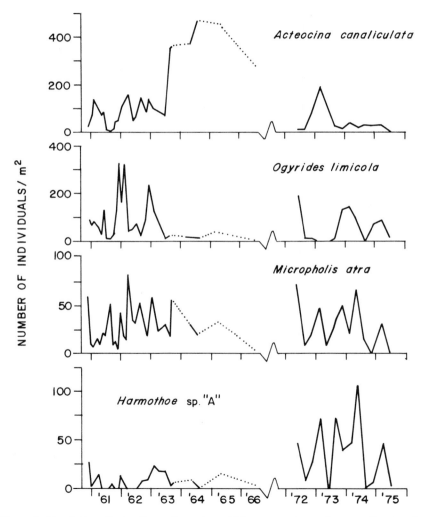

Figure 2. Population fluctuations of species which were more or less equally abundant during 1960-66 and 1972-75 at the polyhaline York River site.

tality and recruitment rates. The scale worm, *Harmothoe* sp. "A" appears to be a commensal of *Micropholis*. Its abundance is strongly correlated with that of *Micropholis* and thus shows no regular seasonality.

Fig. 3 depicts population trneds of species which have become much less abundant since 1972. The polychaete *Nephtys incisa* was a regular dominant

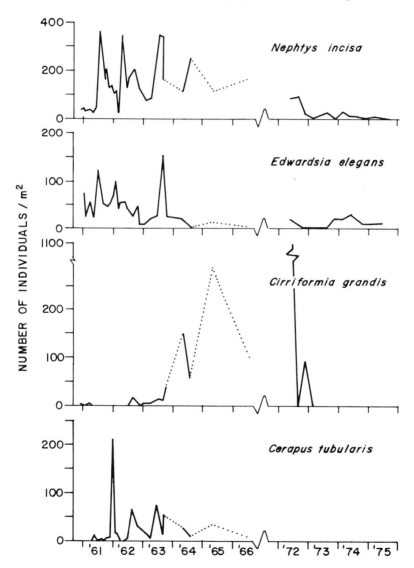

Figure 3. Population fluctuations of species which were much reduced in abundance since 1972 at the polyhaline York River site.

during 1960-66 when it exhibited strong seasonal pulses with highest abundance during spring. The population densities of *N. incisa* declined markedly throughout the lower York River following Agnes (4) and recruitment during 1973, 1974, and 1975 has been very poor. Populations of the small actinian *Edwardsia elegans* showed no regular seasonal variability. *E. elegans* was not collected from November 1972 to August 1973, and although it has since recovered somewhat, densities remained much lower than those found during 1961-62. The large cirratulid polychaete *Cirriformia grandis* was only abundant during the infrequently sampled period from late 1963 to mid-1972. *C. grandis* was usually only abundant at deeper, higher salinity channel locations during the early 1960's. Its increase at this sampling site was coincident with the increase in salinity during the dry years from mid-1964 to mid-1972, and its decline was coincident with the events following Agnes. The amphipod *Cerapus tubularis*, although seldom abundant, occurred with great frequency during 1961-66. No specimens have been collected at this site since the sampling was resumed in 1972.

Fig. 4 presents population trends for three species with very irruptive populations. The polychaete *Pectinaria gouldii* experienced large increases in population during the spring and early summer of both 1962 and 1963. However, these population levels quickly declined during middle and late summer. Blue crabs, *Callinectes sapidus*, can decimate *P. gouldii* populations in shallow sand habitats when the crabs become active in the summer (45), but it is not known if the crabs are responsible for the drastic mortalities witnessed in this mud bottom. The phoronid worm *Phoronis muelleri* was common but only very sporadically abundant. Only during June 1966 were dense concentrations (over 1400 m^{-2}) found.

The bivalve *Mulinia lateralis* exhibited ephemeral irruptions for which it is now well known (1, 22, 23, 34, 37, 40). These irruptions took place in the winter and spring during some years (1961, 1963, 1973) but not others. The largest irruption (to over 2000 m^{-2}) occurred in 1961 and was due to recruitment both during November and December and April and May. The dense population was virtually eliminated sometime in July and data on seston concentrations at precisely the same location (30) during 1961 suggest exclusion of the filter feeding *Mulinia* by high near-bottom turbidity. Such conditions may develop as a result of increased activities of deposit feeders during the summer and may be responsible for similar mortalities of *Mulinia* in Long Island Sound (23, 34). In the Chesapeake Bay, *Mulinia* populations probably average 2-3 generations per year. *Mulinia* is very fecund, grows rapidly, matures quickly and as such is adapted for the opportunistic exploitation of resources.

Fig. 5 presents density estimates for five species which have become much more abundant since Agnes. Although all of these species are widely distributed in Chesapeake Bay, they are typically most abundant in mesohaline habitats (3, 4), i.e. upestuary of the present sampling site. In the polyhaline zone these species are often abundant in polluted or disturbed habitats and this led Boesch

(3) to refer to the polychaetes *Paraprionospio pinnata*, *Pseudeurythoe* sp., *Heteromastus filiformis*, and *Glycinde solitaria* as "euryhaline opportunists."

Paraprionospio pinnata has been a numerically dominant species in the community since the fall of 1973, when a large increase in population density was observed. A similar pulse was observed in the fall of 1974. This same seasonal pattern of abundance has been observed in Hampton Roads (1) and in the mesohaline York River 12-34 km upestuary of the present site (Boesch, unpublished data). The amphinomid *Pseudeurythoe* sp. has been the numerical co-dominant of *P. pinnata* since the summer of 1973. Although densities found

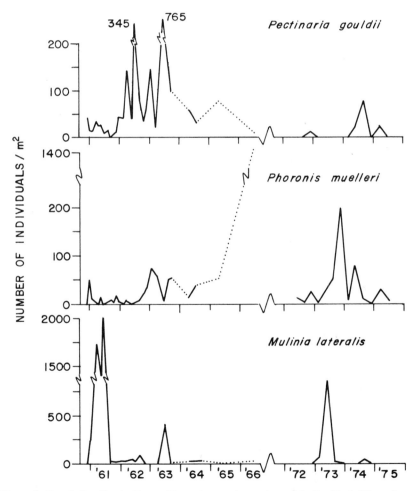

Figure 4. Population fluctuations of irruptive species at the polyhaline York River site.

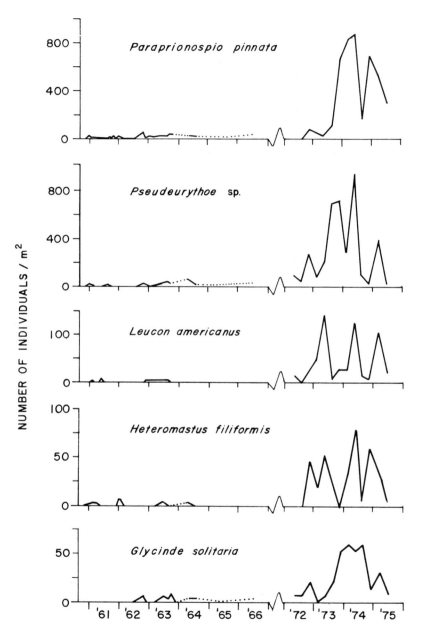

Figure 5. Population fluctuations of species which have become much more abundant since 1972 at the polyhaline York River site.

during 1960-66 were always less than 70 m^{-2}, samples taken in 1970 at a nearby site yielded up to 680 m^{-2} (Boesch, unpublished data). Consistent seasonal patterns of abundance were not apparent during 1973-75.

The cumacean *Leucon americanus* increased in abundance beginning in the fall following Agnes. Spring abundance peaks and summer lows occurred during each subsequent year. In mesohaline mud bottom habitats, densities of *L. americanus* are usually much higher than reported here and peak abundance of the 1.0 mm sieve populations usually occurs in winter (Boesch, unpublished data). The spring peaks at this site may be a result of downestuary movements of populations in response to the low spring salinities. The rarity of the capitellid polychaete *Heteromastus filiformis* during 1960-66 is more surprising than its increase in abundance since 1972. *H. filiformis* is one of the most ubiquitous macrobenthic animals in the Chesapeake Bay, and densities which occurred since 1972 are not unusual in other polyhaline habitats (1). No regular seasonal pattern of abundance was discovered. The small goniadid *Glycinde solitaria* is similarly ubiquitous but it is usually only abundant in shallow sandy habitats or, in mud, only under mesohaline conditions.

The Agnes floods of the summer of 1972 affected the community at this site through reduction of both salinity (to 11 $^{\circ}/_{oo}$) and dissolved oxygen (<1 mg l^{-1}). Although salinity and dissolved oxygen have more or less returned to 1960-63 conditions, the composition of the community since 1972 has remained different than in the early part of the study period. The new composition resembles that typical of mud bottoms several km upestuary during 1969-70 (3), suggesting a downestuary shift in faunal zonation patterns in response to salinity reduction. The minimum (spring) salinities experienced during the years 1973-75 were similar to those in 1960-62, but the winter salinity has been lower (16-18 vs. 20 $^{\circ}/_{oo}$) and the winter temperature warmer (minimum 3-5°C vs. 1-2°C) in 1973-75 compared with 1960-62.

Alternately, the change in composition may be a result of poor recruitment of a species (e.g. *Nephtys incisa*) which previously excluded the insurgent species (e.g. *Paraprionospio pinnata, Pseudeurythoe* sp., etc.). The species which have declined in abundance since 1972 are relatively stenohaline species; several are near their southern distributional limits in the Chesapeake Bay. The species which have increased in abundance are more euryhaline and opportunistic and several are near their northern limit. Thus, it is impossible to conclude which has been most responsible for the changes in the community: the change in salinity, the change in temperature, or adjustments in the biotic "balance."

The persistence of community composition may be considered separately from the persistence of populations of component species (2, 32). One approach is to examine total assemblage similarity, or homogeneity, over time (32). A similarity analysis was performed on the collections from the York River site using log-transformed abundance as an importance value and the Bray-Curtis

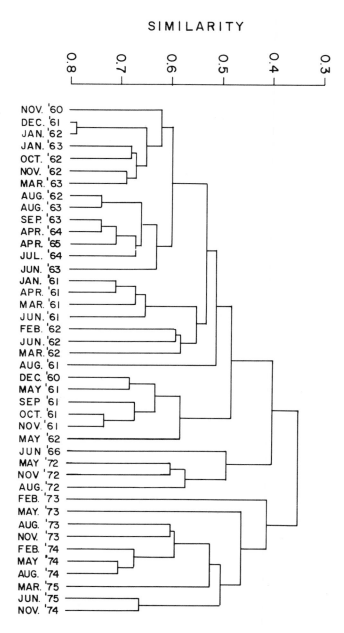

Figure 6. Dendrogram showing patterns of quantitative similarity among the collections at the polyhaline York River site.

similarity index followed by agglomerative clustering of collections using group-average sorting (7). The resulting dendrogram (Fig. 6) clearly shows the major change in community composition which has occurred since 1972. The patterns of similarity among samples during the period 1960-1965 generally reflect the presence of epifaunal species or the overwhelming dominance by *Mulinia lateralis*. The qualitative composition of the community has remained much more persistent than the quantitative composition. Most species collected during the early phases of sampling have been found since 1972.

Species diversity is another property which is undefined for single species populations but is an attribute of the community. At the York River site species diversity as measured by Shannon's information formula (33) varied widely over the period of study, ranging from 1.2 to 4.4 bits/individual. Variations in species evenness rather than in species richness were primarily responsible for the fluctuations in species diversity. Thus, low diversity was found when the collections were numerically dominated by one or two species, e.g. *Mulinia lateralis, Pectinaria gouldii, Phoronis muelleri,* or *Paraprionospio pinnata.*

The trophic structure of communities may be more persistent than their qualitative (species) composition (16, 32). Except during short-lived irruptions of the suspension-feeding *Mulinia lateralis* and *Phoronis muelleri*, the York River community was dominated by deposit feeders. Although the dominant *Nephtys incisa* has declined in abundance, other deposit feeders, e.g. *Paraprionospio pinnata* and *Pseudeurythoe* sp., have increased in its place. However, the apparent trophic persistence is more a consequence of the available food resources than of some internal homeostatic mechanism within the community.

DYNAMICS OF AN OLIGOHALINE COMMUNITY

For comparison with the polyhaline York River community, we have assembled some data on population fluctuations in an oligohaline soft-bottom community in the nearby James River estuary (Fig. 1). The data come from periodic sampling of 16 stations over 7 years by the Virginia Institute of Marine Science. Salinity in the area sampled ranged from 0 to 10 $^\circ$/$_{\circ\circ}$ and is usually lower than 5 $^\circ$/$_{\circ\circ}$. The benthic community is of very low diversity, with less than 15 common species and only two species were consistently abundant in the samples taken: the mactrid bivalve *Rangia cuneata* and the spionid polychaete *Scolecolepides viridis.*

The mean densities of both species over all 16 sites for each sampling period are presented in Fig. 7. The two species exhibit differing levels of persistence, reflecting differences in life histories. *Scolecolepides viridis* is basically an annual species with spring recruitment pulses reflected in the denser spring populations which decline throughout the year. Recruitment regularly occurs but is of variable success. *Rangia cuneata*, on the other hand, is a longevous species with individuals at least 8 years old present in the James River populations. The

Rangia population was composed of several year classes and was not heavily dominated by juveniles. There are no obvious recruitment pulses.

Rangia and *Mulinia*, like many other mactrids, e.g. *Spisula* in the North Atlantic (43, 50) and *Notospisula* in eastern Australia (15), exhibit rapid growth and maturity, high fecundity and population irruptions. However, *Rangia* has

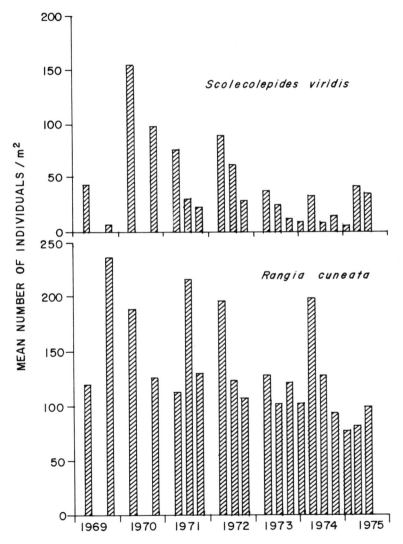

Figure 7. Population fluctuations of the two dominant species in the oligohaline James River.

found a refuge from the high adult mortalitites inflicted on *Mulinia* populations because of its ability to inhabit extremely low salinity or poikilohaline environments. *Rangia* must maintain high reproductive potential because of the waste of pelagic larval dispersal for a species which is successful only in restricted habitats. However, it has greater adult survival than *Mulinia* and consequently a longer period of maturation. Thus, in terms of its life history and persistence *Rangia* seems less of an r-strategist than *Mulinia*.

It is difficult to compare the persistence of the James River oligohaline community with that of the York River polyhaline community. Comparisons based on qualitative or quantitative community composition are difficult to adjust for the great differences in species richness and dominance between the two communities. The composition of the oligohaline community was certainly not obviously less persistent. Comparisons of the persistence of populations of constituent species are likewise difficult because both communities contain species which were persistent as well as those which were not. The relative persistence of populations of the two mactrid bivalves was the reverse of what would be predicted by widely held hypotheses relating greater persistence with increased environmental constancy and higher species diversity (2, 32, 38). The maintenance of an equivalent level of persistence even though confronted with greater environmental variability and the relatively trivial effects produced in oligohaline communities due to the perturbations of the Agnes flood (4) is evidence of the greater stability (resistance and resilience as opposed to persistence) of the low-diversity oligohaline communities (2).

FACTORS CAUSING VARIATIONS IN BENTHIC POPULATIONS

Most coastal soft-bottom benthic communities appear to exhibit substantial seasonal and long-term variability. Especially in temperate latitudes, seasonal patterns of reproduction cause pulses in populations due to the incorporation and subsequent mortality of recruits. Prominent seasonal pulses have been described for nearshore and continental shelf communities (11, 12, 26, 49, 50, 51) as well as for estuarine communities (1, 13, 40, 45). These pulses may be regular and predictable, varying in intensity from year to year, or they may be aperiodic as in the case of *Mulinia lateralis* in this study.

Long-term variability in benthic communities has not been as extensively studied as seasonal variations. Ursin (44) reported large changes in the composition of benthos of the Dogger Bank (North Sea) which had been sampled 29 years earlier. Ziegelmeier (50, 51) presented results of semiannual sampling over 17 years in the German Bight which showed population irruptions (e.g. the polychaete *Spiophanes bombyx*), temporary extinctions [e.g. the bivalve *Tellina (=Angulus) fabula*], and longer term declines (e.g. the bivalve *Nucula nitida*) of common species. Klimova (20) found significant changes had occurred in benthic assemblages of Peter the Great Bay (Sea of Japan) during the 37 years since they were earlier studied. He attributed these to intensified silting which favored deposit feeders over the previously dominant suspension feeders. Frankenberg (11, 12) reported on three year-long studies of macrobenthos on the inner Georgia continental shelf. The communities experienced extreme seasonal variations, but several species had characteristic seasonal pulses of varying intensity during each of the years studied. Other species had less predictable pulses and still others were abundant throughout only one or two of the years. Buchanan et al. (8) sampled a mud-bottom assemblage in the North Sea over a 4 year period and found "conservative" species which had rather persistent populations, "volatile" species which were subject to great fluctuations in abundance from year to year, and "opportunistic" species which could rapidly increase their numbers, probably in response to the elimination of volatile species.

Few seasonal or long-term studies have not uncovered marked fluctuations in benthic populations. Sanders (38) found little seasonal change in the *Nucula-Nephtys* community in Buzzards Bay. The dominant species in this community are long-lived and are represented by several size classes. Fager (10) found surprising constancy of populations in a nearshore sand community; however, he worked only with large, long-lived epifauna. Lie and Evans (24) found high persistence in species composition in infaunal communities in Puget Sound, although relative dominance among numerically important species varied. They present no absolute abundance data for comparison with our results, but Nichols' (28) studies at the same locations show considerable fluctuations in the abundance of at least one important species.

Many complex factors are responsible for short and long term variations in estuarine benthic populations, and space does not allow a detailed discussion of each of them (see also 31). Physico-chemical variables are direct factors in so far as they affect the physiological processes of any life stage of an organism. Salinity is, of course, a particularly important factor in estuaries, and tidal, seasonal, long-term, or aperiodic changes in salinity greatly affect the benthos (3, 47). Temperature is especially important to seasonal programming of reproduction as well to growth and survival. Changes in the nature of the substrate can profoundly affect the composition of benthic communities (9, 19, 20). Dissolved oxygen may be particularly critical in summer because of the high temperatures, organic richness, and density stratification which may exist in some estuaries (21, 46). Finally, man's activities may affect any of the above factors or may introduce toxins which otherwise affect benthic populations (36).

No less important, but often less obvious, are the many biological factors which affect reproduction, recruitment and survival. Seasonally programmed reproduction is adaptive for enhancing survival of offspring, and the timing of reproductive dispersal is probably important in determining survival of the dispersal stages and recruits. Little is known about the effect of mortality of planktonic larvae on setting success. In the Chesapeake Bay, grazing on zooplankton during the summer, particularly by the ctenophore *Mnemiopsis leidyi*, is sufficiently intense to depress the standing stock of the dominant zooplankter,

Acartia tonsa, even though its intrinsic rate of population increase is greatest during the summer (17). Thus, planktivorous grazing may greatly affect the survival of dispersal stages of benthic organisms and may account for the fact that although many benthic animals spawn through the summer in the Chesapeake Bay, recruitment is heaviest in the spring and fall.

Thorson (43) pointed out that although the larvae of some benthic organisms can delay metamorphosis until they sense a favorable habitat, in reality most species set in a much broader variety of habitats than those in which they can survive. Early mortality of new recruits must be very important in shaping the composition of the adult community. Benthic suspension feeders and deposit feeders may act as indiscriminate croppers of juvenile populations and the intensity of this predation may affect population sizes. For example, Ziegelmeier (51) found that when a severe winter in the German Bight eliminated the bivalve *Tellina* (= *Angulus*) *fabula*, which feeds on surface deposits and near-bottom seston, unusually large populations of the polychaete *Spiophanes bombyx* developed. He hypothesized that the intense surface feeding activities of *Tellina* regulate the size of the *Spiophanes* population. Other examples are given by Thorson (43).

Of course, selective predation may also be important in regulating populations. Virnstein (45) found that a shallow sand bottom in the lower York River protected from blue crabs and bottom-feeding fishes developed an assemblage qualitatively and quantitatively different from that on natural unprotected bottoms. Sergerstråle (39) attributed the infrequent recruitment success of *Macoma balthica* on deep mud bottoms in the Baltic to active predation on spat by the amphipod *Pontoporeia affinis*. There are a host of additional examples of predator control in communities of rocky shores (8).

Competition for space or other resources may also effect population fluctuations, but there have been few well documented reports of control by such competitive interaction. Woodin (48) experimentally showed that the presence of tube-building polychaetes reduced the densities of a burrowing polychaete *Armandia brevis*, suggesting competition for space. Presumably, competition is important in the exclusion of euryhaline opportunists and other species characteristic of low salinity habitats from undisturbed polyhaline habitats (3).

Another type of interaction which is of underestimated influence in soft-bottom communities is interference competition or amensalism. Rhoads and Young (35) have shown the existence of a "trophic group amensalism" in muddy bottoms in which the intense reworking and resuspension of sediments by deposit feeders exclude suspension feeders. A similar phenomenon may be responsible for the rapid demise of *Mulinia* irruptions in the York River. Eagle (9) has shown how elimination of dense populations of non-selective deposit feeders by severe turbulence allowed the temporary establishment of more diverse macrofauna. Mills (27) provides another example in which the grazing activities of the gastropod *Nassarius obsoletus* on an intertidal sand flat caused sediment instability and low infaunal densities. The migration of *Nassarius* offshore in winter allowed the dense colonization of the flat the next spring by the tubicolus amphipod *Ampelisca abdita*. The dense mats of *Ampelisca* stabilized the sediments and allowed development of an abundant infauna. The *Ampelisca* mats persisted and excluded *Nassarius* for 2 years until the tubes were abandoned by breeding *Ampelisca*. Examples are also available of an organism's modification of sediments to the point at which it can no longer survive or reinforce its populations in that habitat. For example, Keary and Keegan (19) describe a succession from an *Abra-Venus* community to an *Amphiura* community in which the build-up of a thick horizon of shell fragments rendered the sediment impenetrable for the burrowing bivalves. The surface sediments became finer and were then colonized by the *Amphiura* community.

LIFE HISTORIES AND COMMUNITY PERSISTENCE

Fluctuations in abundance of a species are obviously related to its life history. Species with great reproductive potential, rapid growth and maturation, and high mortality tend to fluctuate widely in abundance. Such species are adapted for short-term exploitation of an unpredictable habitat, and are often referred to as opportunists or r-strategists (13, 25). At the other extreme, equilibrium species or K-strategists have persistent population levels, low reproductive potential, slow growth and maturation and low mortality rates. Of course, in reality most species occupy the middle ground in this multidimensional spectrum of life histories.

Theoretical and empirical evidence accumulated indicates that constant environments favor species with low reproductive and mortality rates and long generation times, and that less predictable environments favor more opportunistic species (14). The benthos of the Chesapeake Bay, an environment which is inconstant and often unpredictable, conforms to this pattern in that it is dominated by species which are short-lived and of widely fluctuating abundance. However, there is a danger in oversimplifying the relationship of environmental constancy and community persistence because, in fact, communities are generally composed of an assortment of both equilibrium and opportunistic species (6). In the estuarine communities we have studied, the ophiuroid *Micropholis atra* and the bivalve *Rangia cuneata* showed much greater persistence than other members of their communities. In the deep sea certain ophiuroids have a much higher reproductive potential, are more heavily represented by juveniles, and probably fluctuate in abundance more than crustaceans and bivalves (14).

Even within a taxocene there may be great differences in life history strategies. For example, Buchanan (6) found that in a sympatric congeneric pair of ophiuroids, one (*Amphiura filiformis*) is short-lived and fast-growing, dies after spawning, and successfully recruits each year, while the other (*A. chiajei*) is long-lived and slow-growing, can breed more than once, and is only sporadically recruited. The dominant infaunal bivalves in Long Island Sound (57) have disparate life history strategies as evidenced by their patterns of survivorship: *Mulinia lateralis* experiences heavy juvenile mortality and lives less than 1 year; *Yolida limatula* experiences heavy mortality of young but survivors may live 5 years; *Pitar morrhuana* has relatively low juvenile mortality but high mortality of 3-4 year old individuals; and *Nucula annulata* experiences extremely low mortalities for the first 7 years but heavy subsequent mortality (D. C. Rhoads, pers. commun.).

Presumably, community persistence follows the same trend as persistence of constituent populations and increases with environmental constancy and predictability. But there seems to be no consensus as to what constitutes community persistence: constancy of constituent populations, constancy in species composition, or constancy of similar trophic or functional "structure" (2). The distinction is an important one, because coastal and estuarine benthic communities generally seem to have highly variable constituent populations, but more constant species composition, and perhaps even more constant trophic structure.

Persistence must not be confused with community stability, i.e. a community's ability to maintain or return to its initial state after external perturbation. Boesch (2) argued that estuarine communities, although perhaps lacking persistence, possess high stability in their resistance to, and resilience from, disturbances.

IMPLICATIONS FOR ENVIRONMENTAL IMPACT ASSESSMENTS

Many studies are presently being conducted to provide baseline information on natural communities prior to some development or to assess the impacts of man's activities through field sampling. The great population variability exhibited in the communities we studied points out the extreme limitations of baseline and impact studies of short duration. Without a detailed knowledge of community dynamics, natural variations may be mistaken for the effects of a pollutant, or, worse, vice versa. Knowledge of life histories is essential in the interpretation of the results of impact surveys. More attention should be placed on effects on the equilibrium species in a community rather than on opportunistic "pollution indicators" which can also sporadically exploit pristine habitats.

Research support should be redirected to regionally representative studies of long-term variations in natural and altered communities in preference to the temporally inadequate, facility-specific studies currently proliferating. Only through such long-term studies can we gain the necessary insight into dynamic processes necessary to interpret site-specific surveys.

ACKNOWLEDGMENTS

Portions of this work were supported by the Virginia Electric and Power Company and the U.S. Environmental Protection Agency (Grant No. R803599-01-0). Michael E. Bender kindly allowed the use of the James River data. We are grateful for the assistance provided by James A. Kerwin, John C. McCain, Robert J. Diaz, and Kenneth A. Dierks in collection and processing of samples and data analysis.

We thank J.F. Grassle, D. Frankenberg, D. Maurer, C.H. Peterson, S.D. Pratt, and R. Rosenberg for reviewing the manuscript and providing helpful suggestions for its improvement.

REFERENCES

1. Boesch, D.F. 1973. Classification and community structure of macrobenthos in the Hampton Roads area, Virginia. Mar. Biol. 21:226-244.

2. _____. 1974. Diversity, stability and response to human disturbance in estuarine ecosystems. Proc. First Intern. Congr. Ecol. (Pudoc, Wageningen, The Netherlands):109-114.

3. _____. in press. A new look at the zonation of benthos along the estuarine gradient. *In* B.C. Coull (ed.), Ecology of Marine Benthos. Univ. South Carolina Press, Columbia.

4. _____, R.J. Diaz, and R.W. Virnstein. in press. Effects of Tropical Storm Agnes on soft-bottom macrobenthic communities of the James and York estuaries and the lower Chesapeake Bay. Chesapeake Sci.

5. Buchanan, J.B. 1967. Dispersion and demography of some infaunal echinoderm populations. Symp. Zool. Soc. Lond. 20:1-11.

6. _____, P. F. Kingston, and M. Sheader. 1974. Long-term population trends of the benthic macrofauna in the offshore mud of the Northumberland coast. J. Mar. Biol. Ass. U.K. 54: 785-795.

7. Clifford, H.T., and W. Stephenson. 1975. An Introduction to Numerical Classification. Academic Press, New York and London. 229 p.

8. Connell, J.H. 1972. Community interactions on marine rocky intertidal shores. Ann. Rev. Ecol. System. 3:169-192.

9. Eagle, R.A. 1975. Natural fluctuations in a soft bottom benthic community. J. Mar. Biol. Ass. U.K. 55:865-878.

10. Fager, E.W. 1968. A sand-bottom epifaunal community of invertebrates in shallow water. Limnol. Oceanogr. 13:448-464.

11. Frankenberg, D. 1971. The dynamics of benthic communities off Georgia, U.S.A. Thalassia Jugosl. 7:49-55.

12. _____, and A. S. Leiper. in press. Seasonal cycles in benthic communities of the Georgia continental shelf. *In* B.C. Coull (ed.), Ecology of Marine Benthos. Univ. South Carolina Press, Columbia.

13. Grassle, J.F., and J.P. Grassle. 1974. Opportunistic life histories and genetic systems in marine benthic polychaetes. J. Mar. Res. 32:253-284.

14. _____, and H.L. Sanders. 1973. Life histories and the role of disturbance. Deep-Sea Res. 20:643-659.

15. Green, R.H. 1968. Mortality and stability in a low diversity subtropical intertidal community. Ecology 49:848-854.

16. Heatwole, H., and R. Levins. 1972. Trophic structure, stability and faunal change during recolonization. Ecology 53:531-534.

17. Heinle, D.R. 1974. An alternate grazing hypothesis for the Patuxent Estuary. Chesapeake Sci. 15:146-150.

18. Holling, C.S. 1973. Resilience and stability of ecological systems. Ann. Rev. Ecol. Systematics 4:1-23.

19. Keary, R., and B. F. Keegan. 1975. Stratification by in-fauna debris: A structure, a mechanism and a comment. J. Sed. Petrol. 45:128-131.
20. Klimova, V.L. 1974. Year-to-year changes in the bottom fauna on the shelf in the center of Peter the Great Bay (Sea of Japan). Oceanology. Acad. Sci. U.S.S.R. 14:137-139.
21. Leppäkoski, E. 1975. Macrobenthic fauna as indicator of oceanization in the southern Baltic. Merentutkimuslait. Julk. Havsforskningsinst. 239:280-288.
22. Levinton, J.S. 1970. The paleoecological significance of opportunistic species. Lethaia 3:69-78.
23. _____, and R.K. Bambach. 1970. Some ecological aspects of bivalve mortality patterns. Amer. J. Sci. 268:97-112.
24. Lie, U., and R.A. Evans. 1973. Long-term variability in the structure of subtidal benthic communities in Puget Sound, Washington, U.S.A. Mar. Biol. 21:122-126.
25. MacArthur, R.H. 1972. Geographical Ecology; Patterns in the Distribution of Species. Harper & Row, New York. 269 p.
26. Massé, H. 1972. Quantitative investigations of sand-bottom macrofauna along the Mediterranean north-west coast. Mar. Biol. 15:209-220.
27. Mills, E.L. 1967. The biology of an ampeliscid amphipod crustacean sibling species pair. J. Fish. Res. Bd. Canada 24:305-355.
28. Nichols, F. H. 1975. Dynamics and energetics of three deposit-feeding benthic invertebrate populations in Puget Sound, Washington. Ecol. Monogr. 45:57-83.
29. Orians, G.H. 1975. Diversity, stability and maturity in natural ecosystems, p. 139-150. In W.H. van Dobben and R.H. Lowe-McConnell (eds), Unifying concepts in ecology. Junk, The Hague.
30. Patten, B.C., D.K. Young, and M.H. Roberts, Jr. 1966. Vertical distribution and sinking characteristics of seston in the lower York River, Virginia. Chesapeake Sci. 7:20-29.
31. Pérès, J.M. 1971. Considerations sur la dynamique des communautes benthiques. Thalassia Jugosl. 7:247-277.
32. Peterson, C.H. 1975. Stability of species and of community for the benthos of two lagoons. Ecology 56:958-965.
33. Pielou, E.C. 1975. Ecological Diversity. Wiley-Interscience, New York. 165 p.
34. Rhoads, D.C. 1974. Organism-sediment relations on the muddy sea floor. Oceanogr. Mar. Biol. Ann. Rev. 12:263-300.
35. _____, and D.K. Young. 1970. The influence of deposit-feeding organisms on sediment stability and community trophic structure. J. Mar. Res. 28:150-178.
36. Rosenberg, R. 1972. Benthic faunal recovery in a Swedish fjord following the closure of a sulphite pulp mill. Oikos 23:92-108.
37. Sanders, H.L. 1956. Oceanography of Long Island Sound 1952-1954. The biology of marine bottom communities. Bull. Bing. Oceanogr. Coll. 15:345-414.
38. _____. 1960. Benthic studies in Buzzards Bay. III. The structure of the soft bottom community. Limnol. Oceanogr. 5:138-153.
39. Segerstråle, S.G. 1960. Fluctuations in the abundance of benthic animals in the Baltic area. Soc. Scient. Fennica. Comm. Biol. 23(9):2-19.
40. Stickney, R.R., and D. Perlmutter. 1975. Impact of intracoastal waterway

maintenance dredging on a mud bottom benthos community. Biol. Conserv. 7:211-226.

41. Stripp, K. 1969. Jahreszeitliche Fluktuationen von Makrofauna und Meiofauna in der Helgoländer Bucht. Veroff. Inst. Meerestorsch. Bremerh. 12:65-94.

42. Thorson, G. 1957. Bottom communities (Sublittoral and shallow shelf). Geol. Soc. Amer. Mem. 67(1):461-534.

43. _____. 1966. Some factors influencing the recruitment and establishment of marine benthic communities. Neth. J. Sea Res. 3:267-293.

44. Ursin, E. 1952. Change in the composition of the bottom fauna of the Dogger Bank area. Nature 170:324.

45. Virnstein, R.W. 1976. The effects of predation by epibenthic crabs and fishes on benthic infauna in Chesapeake Bay. Ph.D. Thesis. College of William and Mary, Williamsburg, Virginia. 87 p.

46. Watling, L. 1975. Analysis of structural variations in a shallow estuarine deposit-feeding community. J. Exp. Mar. Biol. Ecol. 19:275-313.

47. Wolff, W.J. 1971. Changes in intertidal benthos after an increase in salinity. Thalassia Jugosl. 7(1):429-434.

48. Woodin, S.A. 1974. Polychaete abundance patterns in a soft-sediment environment: The importance of biological interactions. Ecol. Monogr. 44:171-187.

49. Zavodnik, D. 1971. Contribution to the dynamics of benthic communities in the region of Rovinj (Northern Adriatic). Thalassia Jugosl. 7:447-514.

50. Ziegelmeier, E. 1963. Das Makrobenthos im Ostteil der Deutschen Bucht nach qualitativen und quantitativen Bodengreiferuntersuchungen in der Zeit von 1949-1960. Veroff Inst. Meeresforsch. Bremerh. Sonderbd, 1 (3. Meeresbiol. Symposion):101-114.

51. _____. 1970. Uber Massenvorkommen verschiedener makrobenthaler Wirbelloser während der Wiederbesiedlungsphase nach Schadigungen durch "katastrophale" Umwelteinflusse. Helgoländer wiss. Meeresunters. 21:9-20.

POPULATION DYNAMICS AND ECOLOGICAL ENERGETICS

OF A PULSED ZOOPLANKTON PREDATOR, THE

CTENOPHORE *MNEMIOPSIS LEIDYI*

Patricia Kremer[1]
Graduate School of Oceanography
University of Rhode Island
Kingston, R.I. 02881

ABSTRACT: The seasonal cycle of abundance of the ctenophore *Mnemiopsis leidyi* in Narraganaett Bay, Rhode Island, is characterized by a dramatic biomass increase of several orders of magnitude reaching a peak of 15-60 g wet weight/m^3 in the late summer, followed by a rapid autumn decline. To investigate the important controlling mechanisms of this annual pattern, results of studies of fecundity, feeding, and metabolism of *M. leidyi* were synthesized in formulating computer simulation models of individual and population biomass. Model predictions were compared to observations of individual growth and estimates of the field population biomass. In the early summer, when the smaller zooplankton are abundant, the ctenophore growth and reproductive rates seem to be sufficient to account for the tremendous observed biomass increase. In the late summer the zooplankton prey becomes depleted, individual ctenophore growth and fecundity depressed, and the ctenophore population ceases to grow. These features are reflected well in the simulation models, but without including predation on the ctenophores, the autumn decline is not adequately represented. In nature, metabolic demands combined with predation and flushing, probably decrease the standing stock of ctenophores in the fall, and cold winter temperatures along with low food availability postpone growth of the population until the following summer.

[1] Present address: Allan Hancock Foundation, University of Southern California, Los Angeles, Calif. 90007.

INTRODUCTION

In many locations ctenophores have been observed to occur in strongly seasonal pulses (21, 2, 7, 1 19, 10). The annual biomass extremes often span several orders of magnitude and consequently the population dynamics are characterized by periods of extremely rapid increase and decline. In spite of high densities and rapid changes, only a few studies have attempted to quantify biomass and to investigate the processes hypothesized to be the most important in producing the marked seasonal pattern in ctenophore abundance. Working with different species of the genus *Pleurobrachia*, Greve (8) and Hirota (9) were both able to maintain laboratory cultures of the animals. Both studies involved observations of growth, feeding, and reproduction, and Hirota also looked at respiration. Baker (1) has investigated growth, fecundity, and respiration of *Mnemiopsis mccradyi*. Miller (19) studied respiration, feeding, and growth of *M. leidyi* and, based on a calculated energy requirement, concluded that the animals must be receiving a supplemental food source in addition to zooplankton, perhaps detritus or phytoplankton. All these investigations were directed at a quantitative definition of basic ecological processes, and Hirota (9) and Miller (19) used their results to formulate static carbon and energy budgets for the organisms.

Mnemiopsis leidyi is the most common ctenophore of Narragansett Bay, Rhode Island. This study has quantitatively investigated not only the distribution and abundance of the population, but also the feeding, metabolism, fecundity, and growth of these ctenophores. This paper attempts to synthesize the results of these investigations in computer simulation models of individual growth and of the whole population. These simulations of *M. leidyi* in Narragansett Bay respond to changes in temperature and food availability, and explore the controlling mechanisms of the dramatic seasonal biomass cycle.

The formulation of energy and material budgets is a classical tool in autecology. The value of this approach comes in seeing if the measurements of the parts make sense when put together. In general, past studies have derived steady-state snapshots of the energetics for a given set of environmental conditions, or long term averages. It is a logical extension to look at the growth and decline of individuals and populations through time as conditions change. The present study of ctenophore energetics and population dynamics attempts to do this by using a deterministic approach in formulating both individual and population models based on empirical data, and comparing the results with independently derived measurements of growth and field biomass. Such modeling may be a rigorous test of the data since measured errors in rates may become large when integrated over long periods of time.

Observed Seasonal Pattern

Narragansett Bay is near the northern limit of the range of *Mnemiopsis leidyi*. The seasonal pattern of abundance is characterized by a pronounced peak in late

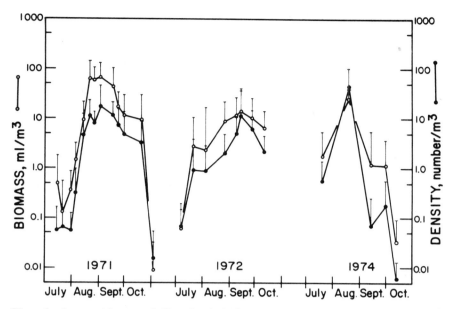

Figure 1. Average biomass of *Mnemiopsis leidyi* over three sampling seasons, in ml/m³ (open circles) and number/m³ (closed circles). Bars represent positive standard deviations of the data from approximately twenty sampling stations.

summer when the biomass may reach a baywide average of more than 50 ml/m³ (Fig. 1). Although the three years of sampling show roughly the same general pattern, there were marked differences in both the timing and magnitude of the maxima. For example, in 1974 the population dropped dramatically in September, a period of high biomass in the other two sampling years. This decline coincided with an increase in abundance of another ctenophore, *Beroë ovata*, known to be a predator on other ctenophores (12, 2, 25). This is the first recorded observation of this species in Narragansett Bay.

All three years showed similar patterns of early rapid population increase during July and August. Analysis of the relative spatial distribution of ctenophores during this period indicated that the animals became numerous first in the upper bay (15). Intensive sampling confirmed the existence of a remnant population (1-2 animals per 10^4 m³) throughout the winter which might represent the necessary seed stock for the summer population explosion.

Fecundity and the Rapid Biomass Increase

For a small overwintering population to be responsible for the large observed summer biomass, fecundity must be quite high during the early summer. Measurements of egg release by the animals during this critical period confirmed a large reproductive potential (Fig. 2). To evaluate quantitatively whether these reproductive rates were sufficient to account for the observed biomass increase, a mathematical model was formulated to represent this early summer period. The details of this model are discussed elsewhere (14). In summary, the model satisfactorily predicted the biomass increase given the observed overwintering biomass as an initial condition (Fig. 3). The results of this model indicated that even with juvenile survival rate of only 0.1 to 1.0 percent, and adult mortality of 10% per day, the growth and reproductive potential of *Mnemiopsis* during the early summer was sufficient to account for the observed biomass increase. This preliminary model was valuable in assessing the credibility of the overwintering hypothesis, as well as the validity of growth and reproductive rate measurements and juvenile mortality estimates. However, it was preferable to analyze more than just early summer increase, and to model explicitly both growth and reproduction rather than using them as constant inputs.

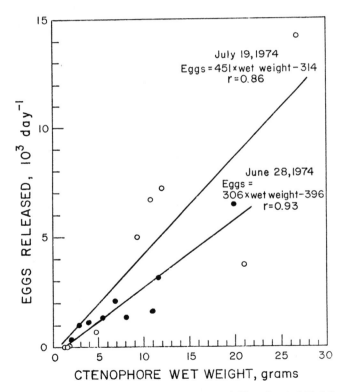

Figure 2. The trend of egg release with wet weight of *Mnemiopsis leidyi* for two dates during the period of rapidly increasing ctenophore biomass. Equations and correlation coefficients for the least squares regression lines are given for each date. (From 14).

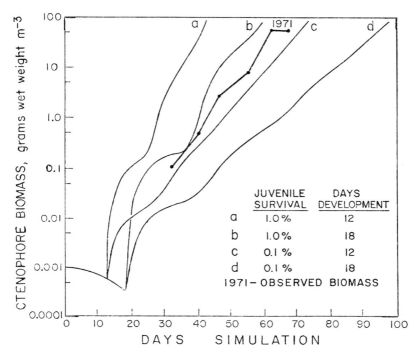

Figure 3. Comparison of simulation results for the early summer increase with observed data. Daily adult mortality equaled 10% for all runs, and juvenile survival and development time to sexual maturity were varied as indicated. (From 14).

SIMULATION MODEL FOR INDIVIDUAL GROWTH

Before attempting to create a population model for *M. leidyi* in Narragansett Bay, growth and reproduction of single individuals were simulated. Results of studies on feeding, respiration, excretion, and fecundity for various sizes of *M. leidyi* over a range of temperatures (10-25 C) provided the basis for the mathematical representation of the important processes expressed in the following equation:

Growth of Individual   = (Ingestion x Assimilation Efficiency) – Metabolism
                          – Reproduction

Ingestion

Consistent with the results of others (19, 27), my determinations of feeding rates for *M. leidyi* from Narragansett Bay have shown that, unlike many plankton-feeding animals, the clearing rate of *Mnemiopsis* is independent of prey

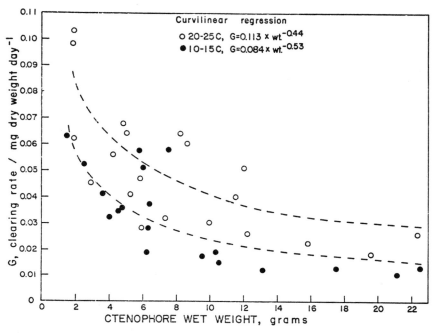

Figure 4. Results of feeding rate determinations for various sizes of *Mnemiopsis leidyi* over a range of temperatures. Lines were fit by curvilinear least squares regression. In general, the observations at lower temperatures (solid circles) yielded lower feeding rates although the two groups were not significantly different by an analysis of covariance on the log-transformed data. (From 14).

concentration (14). However, the clearing rate does appear to be a function of both temperature and organism size (Fig. 4). Curvilinear least-squares regression on the data for the two temperature groups (Fig. 4) determined coefficients for the allometric equation:

$$G = a\,W^b \qquad [1]$$

where   W = grams wet weight of the ctenophore
        G = liters cleared/mg dry weight per day.

In this equation the negative exponential coefficient (b) shows the trend for decreasing weight-specific clearing rate with size, while a comparison of the values of the other coefficient (a) demonstrates the increased feeding rate at higher temperatures. Since physiological processes generally follow an exponential relationship with temperature, the following equation was chosen to represent this change in a

$$a = a_o\,e^{K\,T} \qquad [2]$$

where a_0 and K are constants, and T is temperature, C. The value of K = 0.05 C^1 used in the model is representative of medium sized ctenophores and is equivalent to a feeding rate Q_{10} of 1.7. The choice of a_0 equal to 0.04 l/mg day results in values of a comparable to the empirical regression values.

For given combinations of temperature and organism size, the weight-specific clearing rate, (G), is calculated from an expression of the general form of eqn. 1 with a defined by eqn. 2. For an individual ctenophore, daily carbon ingestion is given by:

$$I = G \times F \times DW \times dt \qquad [3]$$

where   I   =   ingestion, mg C/day
       G   =   volume cleared, liters/mg dry weight· day
       F   =   food concentration, mg C/liter
   DW   =   dry weight of ctenophore, mg
    dt   =   time step of one day.

Dividing this ingestion by the carbon content of the individual ctenophore (1.7% of dry weight) determines the fraction of body carbon ingested per day.

Assimilation

An assimilation efficiency of 75% has been determined for *Mnemiopsis mccradyi* at ecologically reasonable food levels (27), and in the model is assumed for *M. leidyi*. These results are consistent with the generally high assimilations found by many other investigators for a variety of marine zooplankton (16, 4, 5, 3, 26).

Metabolic Demands

Measurements of respiration and excretion were made for over 100 individual ctenophores. Several chemical properties were monitored, including dissolved oxygen, and inorganic and organic forms of dissolved phosphorus, nitrogen and carbon. The details of the methods and results have been presented elsewhere (14) and this paper will discuss only the portions relevant to the carbon-based *Mnemiopsis* growth model. Carbon dixoide release was found to be a linear function of size for animals tested at a variety of temperatures (Fig. 5a). This finding justifies combining the weight-specific carbon dioxide data for all temperatures in a single exponential relationship (Fig. 5b).

In many cases the increase in dissolved organic carbon was as large or larger than the CO_2 increase but did not show any clearcut pattern. The mean ratio (by atoms) of carbon dioxide release to total carbon (CO_2 + DOC) for 37 animals was 0.62 (s.e. = 0.12). Thus, on the average, the total increase in dissolved carbon (CO_2 + DOC) was 1.67 times the CO_2 release. To convert the

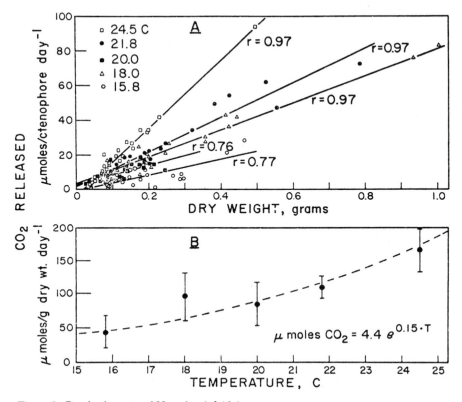

Figure 5. Respiration rate of *Mnemiopsis leidyi*.

    A.   The daily release of carbon dioxide by individuals at a variety of tempera-
tures. The generally high correlation coefficients of the least squares regressions indicate a fairly constant weight-specific respiration rate (μM CO_2
released/gram dry weight per day) for all sizes of individuals.

    B.   Means and standard deviations for the weight-specific respiration rates at a
variety of temperatures presented in A. The equation was obtained by least
squares linear regression on the log transform of these data. The exponential
coefficient of 0.15 represents a Q_{10} of 4.5, emphasizing the sensitivity of
the metabolism of these animals to temperature.

units of the equation in Fig. 5b (μM CO_2/g dry weight) to the fraction of body
carbon respired daily, it is necessary to multiply by

$$\left(\frac{12\,\mu g\,C}{\mu M\,CO_2}\right) \times \left(\frac{1\,mg}{1000\,\mu g}\right) \times \left(\frac{\text{Ctenophore tissue:}}{\dfrac{\text{g dry weight}}{17\,mg\,C}}\right) = 7.06 \times 10^{-4}\ \text{g dry wt}/\mu M\,C$$

Multiplying this result by 1.67 produces the total daily fraction of body carbon
catabolized (respiration plus excretion). Table 1 shows the results of the calculations for various temperatures.

Table 1. Predicted fraction of *Mnemiopsis* body carbon required for respiration and total metabolism (including excretion) at several temperatures. These data are based on equation in Fig. 5B, determinations of organic carbon excretion (av. 38% total dissolved carbon loss), and the carbon content of the body tissue (av. 1.7% of dry weight).

Temperature, C	Daily Respiratory Demand	Total Carbon Release
0	0.003	0.005
10	0.014	0.023
15	0.029	0.048
20	0.060	0.100
25	0.126	0.210

Reproduction

In the simulated growth of an individual ctenophore, reproduction occurs only when assimilation (0.75 x ingestion) is greater than metabolism. This formulation was based on fecundity observations in Narragansett Bay which showed that egg release was greatly depressed in areas of low food concentration (14). Other studies of ctenophores (1, 6) have also suggested a direct coupling.

If the value of assimilated carbon in the model exceeds the computed metabolic demands, the remainder is divided between growth and reproduction. Since the question of growth versus fecundity as a function of animal size and food concentration was not examined empirically in this study, it was assumed that the fraction of the unmetabolized assimilated carbon which goes to reproduction varies with the size of the ctenophore. For smaller animals, relatively more carbon goes to growth, while for larger animals, more goes to reproduction. Hirota (9) and Greve (6) observed this shift in other species of ctenophores, *Pleurobrachia bachei* and *Beroë cucumis*, and results with *Daphnia* further support this pattern (24).

Based on field observations, a minimum reproductive size of 1.2 grams wet weight is designated in the model. At this threshold only 1% of any unmetabolized assimilation goes to reproduction. As the animal grows, this fraction increases exponentially, reaching 100% at 23 mg carbon or 40 grams wet weight, the maximum size attainable in the growth model. The following equation expresses this trend:

$$E = 0.01 \, e^{0.115 \, w} \tag{4}$$

where E = fraction of unmetabolized assimilated carbon put into reproduction

   w = grams wet weight of individual ctenophore.

In Narragansett Bay, fecundity was observed to be highest in the early summer, when ctenophore egg production was equivalent to about 2% per day of the body weight of the animals. For appropriate food levels and temperatures, the growth model which uses the above formulation (Eqn. 4) predicts similar values of 1-2% per day for comparably sized animals.

Growth

In the model the carbon which is finally allocated to growth results from the calculation of all the processes discussed above. When assimilation exceeds the metabolic demand, growth is positive and the animal increases in size as well as producing eggs. However, if the animals do not ingest enough to meet their metabolic requirement, they shrink an appropriate amount. Such shrinkage is a common occurrence for ctenophores which are underfed in laboratory conditions (1; pers. observ.), and field observations of the size distribution of the ctenophore population in Narragansett Bay have also suggested a size diminution in response to food limitation.

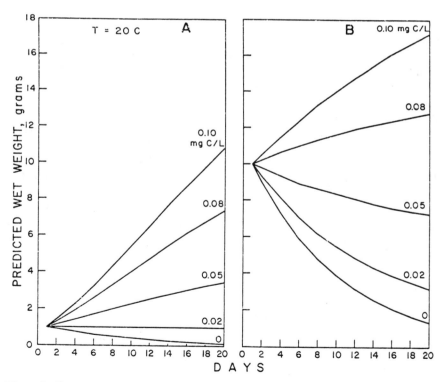

Figure 6. Simulation of growth of individual *M. leidyi* for 20 days at 20 C demonstrating the effect of initial size and food concentration. The smaller ctenophore (A, initially 1 gram), is able to sustain itself on a much lower level of food than the larger animal (B, initially 10 grams).

Simulation Results

Supplied with input data for ambient temperature and food concentration, the individual growth model predicts the change in size of a single ctenophore through time (Fig. 6). Although the growth model is based on carbon, the output can be readily converted to wet weight or length using the following:

carbon, mg = 0.574 X grams wet weight (based on 117 measurements of wet-weight-dry weight and 4 carbon: dry weight analyses)

length, mm = 12.3 X (g wet weight)$^{0.574}$ (based on a regression for 347 individuals).

Depressed growth rates for low food concentrations and larger sizes are obvious features of the model. Because of the higher weight-specific ingestion rates for smaller animals and the constant weight-specific metabolic rate, the smaller animal (Fig. 6a) is able to maintain itself on about one third the food concentration necessary to avoid shrinkage in the larger animal (6b).

In order to test the validity of the growth model, the results were compared with laboratory growth observations of selected *M. leidyi* held at 20 C and a variety of food concentrations (Fig. 7). After this verification of the individual

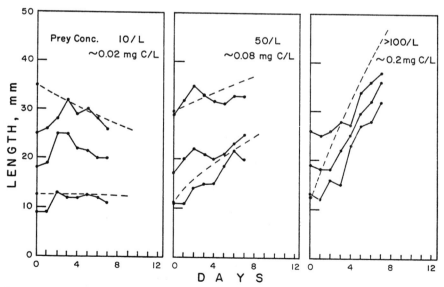

Figure 7. Observed and predicted ctenophore growth at 20 C for three food concentrations of mixed zooplankton. Output of the growth model simulation (smooth line) was converted to units of length to be directly comparable to the laboratory measurements (data points). The carbon levels of the zooplankton were not measured directly but calculated from counts using the results of Martin (17) for mixed Narragansett Bay zooplankton.

growth model with observed rates under static conditions, the model was used to predict growth of individuals in changing temperature and food regimes representative of Narragansett Bay (Fig. 8). The values used for temperature and zooplankton were based on the results of several studies (17, 11, 22) and attempt to portray "typical" summer conditions. A series of model runs were made which examined growth of individuals throughout the summer and early

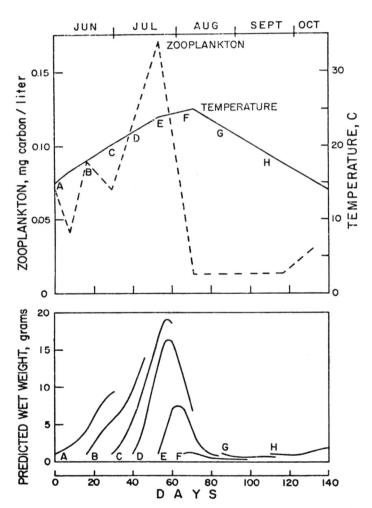

Figure 8. The temperature and zooplankton concentration for a "typical" summer in Narragansett Bay (top) used as forcing functions in simulating individual ctenophore growth (bottom). At each of eight times during the simulated season (A-H) the model predicts growth of an individual from an initial weight of 1.0 g wet wt. over the next 30 days.

fall (Fig. 8 a-h). In all simulations, the animals were initially one gram wet weight, and their weight was predicted for a period of 30 days. These simulations dramatically confirmed that late June and July were the periods of most intense ctenophore growth, due to the tight dependence on food availability.

THE POPULATION MODEL

The model of individual growth is the basis for a population model which predicts the total biomass of *Mnemiopsis leidyi* in Narragansett Bay throughout an entire season. All the relationships and equations developed for the growth model were directly incorporated into the population model, with a few modifications and additions.

Ingestion

The equations for calculating the weight-specific ingestion rate were the same as in the individual growth model, but because an entire population was being represented, a different method of solution was required. Although this model did not attempt to represent realistic dynamic feedback of predation on the zooplankton, increased competition within the ctenophore population for available resources was approximated by the use of the exact integral of the differential equation (Eqn. 3). The explicit evaluation of the instantaneous grazing rate (G) accounts for this competition effect:

$$I = \text{Food} \left(1 - e^{-G \times \text{Adults} \times dt}\right) \qquad [5]$$

where   Food, G, and dt are the same as equation 3

  Adults = the population biomass of mature ctenophores, mg dry wt./liter

  I = ingestion by the population, mg C/liter.

Reproduction

Reproduction in the individual growth model represents a drain of carbon from the animal which reduces the amount of growth which can occur. In the population model, however, reproduction becomes the major means by which the population increases. All the eggs released by the population on a given day are pooled in a single compartment. These individuals remain in a group as they hatch, grow and are subjected to mortality and tidal flushing. In making the simplifying assumption of homogeneously distributed food and uniform temperature, the model assures that all organisms which hatch on the same day necessarily have identical life histories. This approach ignores patchiness which certainly exists and is undoubtedly important in individual life histories. However, the assumption of homogeneous distribution is intended only to yield valid averages for the bay as a whole.

Hatching and Juvenile Development Time

At summer temperatures eggs hatch about twenty-four hours after release (18; personal observation). The development time from egg to reproductive maturity proved to be an important parameter in the model, although it was not investigated directly in my laboratory studies. Working with *M. mccradyi*, Baker (1) found that animals started to reproduce a minimum of twelve days after hatching. Since it is the only data on this subject for *Mnemiopsis* spp., this same value of 12 days has been used as the minimum development time in the population model.

Development time is also modeled as a function of food, since juveniles as well adults may be food-limited, presumably retarding their growth. The effect of food concentration on development time was formulated using a rectangular hyperbola of the familiar Michaelis-Menten form:

$$F = [Food] / \left(K_s + [Food]\right) \qquad [6]$$

Development time, D, can then be expressed:

$$D = D_{min}/F$$

Where D_{min} is 12 days, the minimum development time. The rate at which the fraction, F, approaches 1.0 as the food concentration increases, depends on the choice of K_s, the food concentration at which F = 0.5. In the model the value for K_s was chosen to be 0.02 mg C/l, which is equal to the maintenance food ration for a one gram animal.

The food concentration used to determine the development time for newly released eggs is calculated as the simple average of the predicted development time based on the food concentration at hatching, and a similar prediction made for the later time. The average of these two times should more realistically reflect the actual field development time than either single value. The eggs are placed in a compartment which enters the adult pool of the model the specified number of days later with each animal weighing 1.2 grams wet weight.

Mortality

Although the model formulation included potentially variable adult predation, in the base line run mortality for the adult ctenophores was a constant 2% per day, to approximate only tidal losses (13). Predation by both the ctenophore *Beroë ovata* and butterfish *Peprilus triacanthus* have been observed to occur, but in the absence of any quantitative measurements, these factors were not included in the basic population model.

Juvenile mortality is undoubtedly much higher than for the adults. A 99% attrition from egg to reproductive maturity was used in the population model and seemed reasonable based on the results of the more simplified model of the early summer increase (Fig. 3).

Summary of Formulations

The separate processes used in the growth model were combined with these formulations of ingestion, reproduction, development time, and mortality, and synthesized into the population model. A diagrammatic representation shows the flows and interactions of the model and the corresponding equations (Fig. 9). Zooplankton concentration and temperature are the two forcing functions of the model. All processes, with the exception of tidal flushing and juvenile mortality, are modeled with equations based primarily on quantitative empirical observations.

Simulation Results

In the population model, the forced values for temperature and zooplankton are the same as those used for the seasonal simulation of individual growth (Fig. 8). Simulations were initiated on June 1 with a biomass of 0.001 mg carbon per cubic meter, which was approximately the ctenophore density observed during winter-spring sampling (15). In the standard run of the population model, the initial biomass remained constant, but the assumed size of the organisms was varied from 1 to 10 grams wet weight. This difference in the initial size altered the results producing a "simulation envelope." In general the simulation agreed

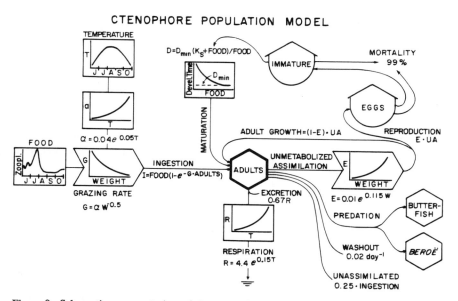

Figure 9. Schematic representation of the ctenophore population for Narragansett Bay. All major flows and functional relationships considered in the model are included with the appropriate equations and coefficients. Although this formulation includes predation by butterfish and *Beroë*, mathematical expressions have not been specified for this predation and it was not included in the baseline run.

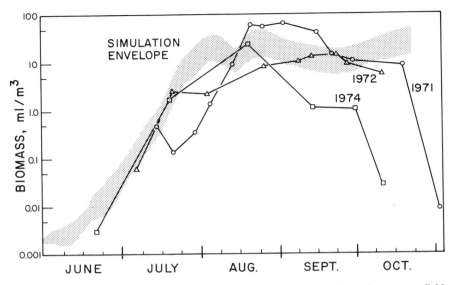

Figure 10. Biomass of *M. leidyi* predicted by the population model vs. the average field
biomass estimates for three years. Simulation envelope results from assuming
different size distributions for the initial biomass.

well with the range of average abundance observed in the field sampling (Fig.
10). The model predicted the biomass peak a little early, but the magnitude was
very similar.

After the second week of August, when food availability became low, the
model predicted a cutoff of reproduction, and shrinkage of the larger organisms.
The result in the model was an overwhelming dominance of small animals (less
than 1 gram wet weight) during late August through October. Field observations,
particularly in the upper part of the bay, confirmed this trend towards lower
fecundity and smaller animals later in the season, although the field results were
much less dramatic than the model. Presumably this inconsistency is at least in
part due to a patchy food distribution in nature which was not included in the
model.

In the fall, the model failed to predict the rapid decline which was observed
in the field sampling. It is likely that this discrepancy was primarily due to the
lack of predation on adults in the model. In Narragansett Bay, the butterfish
Peprilus triacanthus is seasonaly abundant during the late summer and fall, and
has been observed to feed voraciously on *Mnemiopsis* (23). It is likely that
during the early fall butterfish may exert a substantial predation pressure on the
ctenophores, especially in the lower bay where the fish are most abundant. The
predatory ctenophore *Beroë ovata* seems to occur only irregularly in Narragansett Bay, thus minimizing its importance as a *Mnemiopsis* predator. In three

years of quantitative field sampling, *B. ovata* was present only one year and their peak biomass was about an order of magnitude less than that of *M. leidyi.* When *Beroë* was abundant, however, the *Mnemiopsis* population was severely depleted by mid-September (1974) while in other years the collapse did not occur until mid-October (Fig. 1). While butterfish and *B. ovata* are the only well documented predators on *M. leidyi* for Narragansett Bay, it seems likely that there may be others, since the question has not been studied directly.

Starvation and parasitism are additional factors which may contribute to the population decline. During the late summer when food is scarce, the ctenophores may become physiologically weakened due to starvation. In addition to this stress, a large proportion of the population has been observed to be heavily infested with the coelenterate parasite *Edwardsia leidyi* (= *Fagesia lineata*). The energy drain to these parasites has not been quantified but a single ctenophore frequently may contain more than ten large individual parasites and their potential influence is considerable.

Perturbation and Sensitivity Analysis

Several runs of the population model were made to test its sensitivity to changes in key parameters. Changes in the initial ctenophore biomass used, resulted in directly comparable changes in the peak biomass, leaving the timing unchanged. However, manipulating the development time from egg to reproductive maturity plus and minus a few days resulted in large changes in the biomass peak, illustrating the critical importance of the precise value of the time lag in basically exponential growth systems.

Likewise the exact choice of input food concentration had a tremendous influence on the resulting ctenophore biomass. Because the zooplankton concentration is one of the forcing functions and is not explicitly modeled, there was no daily feedback from the ctenophores biomass to the zooplankton standing stock, an increasingly important feature as ctenophore biomass grew. In spite of this limitation, the individual and population models presented here are still valuable because they dynamically examine the internal consistency of several critical processes within the context of an appropriate, defined environment. The sensitivity of the model to arbitrary changes in the input parameters primarily illustrates the tight coupling which unifies the natural ecosystem.

CONCLUSIONS

Other studies of ctenophore population dynamics (2, 8, 10) have concluded that predation is the primary mechanism of population regulation. By contrast, the results of the present study and population model indicate that the population density is primarily a function of the food supply and the role of predation is mainly in bringing about the fall biomass decline, not in limiting the biomass maximum. With abundant food, the enormous growth and reproductive potential of *M. leidyi* could overcome all but the most intense predation pressure. In

Narragansett Bay *Mnemiopsis leidyi* seems to exploit its environment fully and through its ability to increase its population size so rapidly, it well adapted to the large seasonal fluctuations in food and temperature of the northern estuary.

The simulation of individual growth and population dynamics of *Mnemiopsis leidyi* in Naragansett Bay has provided a synthesis and rigorous evaluation of the field and laboratory observations of major ecological processes for this seasonally important carnivore. It should be emphasized that the computer runs of the model used parameter values and initial conditions previously determined from laboratory and field observations and specified prior to the simulations. Within these rather severe constraints, the model's prediction of the population increase and peak biomass level agreed quite well with field observations. Thus it seems that mechanistic simulation models can be effective tools in the synthesis and analysis of information and hypotheses about component parts of ecosystems.

ACKNOWLEDGMENTS

The material presented in this paper represents a synthesis of a variety of laboratory and field studies carried out over the past four years. Dr. Scott Nixon, as well as Dr. C. A. Oviatt, Dr. H. P. Jeffries, and fellow graduate students, particularly my husband James Kremer, were influential in shaping the direction of this research and in guiding its evolution. Dr. Nixon was particularly helpful in his encouragement of a quantitative synthesis of results into the simulation models presented here.

The project was supported in part by a NDEA Title IV fellowship for graduate study and in part by the Office of Sea Grant, U. S. Department of Commerce.

LITERATURE CITED

1. Baker, L.D. 1973. The ecology of the ctenophore *Mnemiopsis mccradyi* Mayer, in Biscayne Bay, Florida. Ms. Thesis, University of Miami. 131 p.
2. Burrell, V.G., Jr. 1968. The ecological significance of a ctenophore, *Mnemiopsis leidyi* (A. Agassiz) in a fish nursery ground. Ms. Thesis. College of William and Mary. 61 p.
3. Clutter, R.I., and G.H. Theilacker, 1971. Ecological efficiency of a pelagic mysid shrimp. estimates from growth, energy budget, and mortality studies. Fish. Bull. U.S. 69:93-115.
4. Conover, R.J. 1966. Assimilation of organic matter by zooplankton. Limnol. Oceanogr. 11:338-345.
5. Corner, E.D.S., and C.B. Cowey. 1968. Biochemical studies on the production of marine zooplankton. Biol. Rev. 43:393-426.
6. Greve, W. 1970. Cultivation experiments on North Sea ctenophores. Helgoländer Wiss. Meeresunters. 20:304-317.
7. _____. 1971. Ecological investigations on *Pleurobrachia pileus*. I. Field Studies. (in German, English abstr.) Helgoländer Wiss. Meeresunters. 22:303-325.

8. _____. 1972. Ecological investigations on *Pleurobrachia pileus.* II. Laboratory Investigations. (in German, English abstr.) Helgoländer Wiss. Meeresunters. 23:141-164.
9. Hirota, J. 1972. Laboratory culture and metabolism of the planktonic ctenophore, *Pleurobrachia bachei* A. Agassiz. *In* Biological Oceanography of the Northern North Pacific Ocean. Motoda Commemorative volume, p. 465-484. ed. by A.Y. Takenouti. Tokyo, Japan, Idemitsu Shoten.
10. _____. 1974. Quantitative natural history of *Pleurobrachia bachei* A. Agassiz in La Jolla Bight. Fish. Bull. 72:295-335.
11. Hulsizer, E.E. 1976. Zooplankton of Lower Narragansett Bay 1972-1973. Chesapeake Sci. 17: in press.
12. Kamshilov, M.M. 1960. Feeding of ctenophore, *Beroë cucumis* (Fabr.). Dokl. Akad. Nauk. 130:1138-1140.
13. Kremer, J.N. 1975. Analysis of a plankton-based temperate ecosystem: An ecological simulation model of Narragansett Bay. PhD Thesis. Univ. Rhode Island. 369 p.
14. Kremer, P. 1975. The ecology of the ctenphore *Mnemiopsis leidyi* in Narragansett Bay. PhD Thesis. Univ. Rhode Island. 311 p.
15. _____, and S.W. Nixon. In press. Distribution and abundance of the ctenophore *Mnemiopsis leidyi* in Narragansett Bay. Estuarine and Coastal Marine Science.
16. Lasker, R. 1966. Feeding, growth, respiration, and carbon utilization of a euphausiid crustacean. J. Fish. Res. Bd. Canada 23:1291-1317.
17. Martin, J.H. 1968. Phytoplankton-zooplankton relationships in Narragansett Bay. III. Seasonal changes in zooplankton excretion rates in relation to phytoplankton abundance. Limnol. Oceanogr. 13:63-71.
18. Mayer, A.G. 1912. Ctenophores of the Atlantic Coast of North America. Pub. 162 Carnegie Inst. of Washington. 58 p.
19. Miller, R.J. 1970. Distribution and energetics of an estuarine population of the ctenophore, *Mnemiopsis leidyi.* PhD Thesis. No. Carolina St. Univ. 78 p.
20. _____. 1974. Distribution and biomass of an estuarine ctenophore population, *Mnemiopsis leidyi* (A. Agassiz). Chesapeake Sci. 15:1-8.
21. Nelson, T.C. 1925. On the occurrence and food habits of ctenophores in New Jersey inland coastal waters. Biol. Bull. 48:92-111.
22. Nixon, S.W., J.N. Kremer, and C.A. Oviatt. In preparation. Narragansett Bay systems ecology program data report.
23. Oviatt, C.A., and P. Kremer. In press. Predation on the ctenophore *Mnemiopsis leidyi* by butterfish *Peprilus triacanthus* in Narragansett Bay, Rhode Island. Chesapeake Sci.
24. Richman, S. 1958. The transformation of energy by *Daphnia pulex.* Ecol. Monogr. 28:273-291.
25. Swanberg, N. 1974. The feeding of *Beroë ovata.* Mar. Biol. 24:69-76.
26. Taguchi, S., and H. Ishii. 1972. Shipboard experiments on respiration, excretion, and grazing of *Calanus cristatus* and *C. plumchrus* (Copepoda) in the Northern North Pacific. *In* Biological Oceanography of the Northern North Pacific. p. 419-431. ed. by A.Y. Takenouti, Tokyo, Japan. Idemitsu Shoten.
27. Walter, M.A. 1975 The ecological significance of the feeding behavior of the ctenophore *Mnemiopsis mccradyi.* Ms. Thesis. University of Miami.

WETLANDS USES

Convened by:

Convened by:
Armando A. de la Cruz
Department of Zoology
Mississippi State University
P.O. Drawer Z
Mississippi State, Mississippi 39762

The ecologic and economic value of estuarine wetlands is reflected in the diversified functions derived from this ecosystem. These functions can be categorized at two levels. First, at an ecosystem level, that is, the use of wetlands as a whole including the intertidal zone and the immediate dryland and neighboring littoral areas for various system functions. For examples, the use of wetlands as pasture and grazing lands, as wildlife sanctuaries, as fishery nursery grounds, as recreational areas; uses in coastal aquaculture, in tertiary sewage treatment, in highway, waterway, and runway construction, and petroleum exploration. Second, at a component level, that is, the use of the major biotic components, mainly the vascular plants, for various purposes. Marsh plants growing abundantly in the wetlands are the source of organic detritus that provides the source and/or agents of nutrition for marine heterotrophs, of materials used in various craft industries, of pulp and other cellulose derivatives, of chemical compounds with potential pharmacological value, and of commodities directly consumed by man.

The materials presently harvested from marshlands and the great potential they have for other forms of human uses, all point to the ultimate importance of estuarine wetlands. Whether coastal wetlands are used by man in their natural state or managed for various other uses, the vital issue is the preservation of estuarine wetlands from permanent alterations that may lead to the deterioration of its natural functions.

GRAZING ON WETLAND MEADOWS

Robert J. Reimold
Marine Resources Extension Center
The University of Georgia
P.O. Box 517
Brunswick, Georgia 31520

ABSTRACT: Marshlands owners have been interested in "profitable" use of their property for centuries. Through ditching, diking and draining, these marshes were visualized as potentially productive pastures. One shortcoming in this planning was the oversight of "cat-clay" problems associated with large quantities of organic matter and anaerobic conditions. The resultant complex polysulfides when oxidized resulted in extremely acidic soil conditions, detrimental to all vegetative growth.

The extent of compaction of the substrate and the species of vegetation in the wetland meadows and prairies determine the feasibility of grazing the area by ungulates. In grazed *Spartina* marshes, net primary production was 50% of the production occurring on ungrazed marshes. Grazing also had a significant negative impact on the macro-invertebrate populations and resulted in differences in plant successional trends. Grazing also significantly reduced detritus production.

Marshes appear to recover from grazing once it is terminated. Since even low density grazing does produce a significant impact on the wetlands ecosystem, additional research is needed to document the optimum invertebrate population, and the minimum detritus production necessary to sustain secondary production in adjoining systems.

In many parts of the world, wetlands have been one of the last remaining frontiers to be "conquered" by man. These wetlands represent a relatively rare habitat, contrasted with forests, grasslands, etc., and are becoming the focal point of quantitative and qualitative ecological assessment. In most places, the philosophy that only the present counts has resulted in little interference of individual or corporate resource exploitation of the wetlands.

219

In the eighteenth and nineteenth century, the coastal wetlands of southeast United States were the site of expanding rice culture. During the period 1800 to 1860, these states produced ninety percent of the United States rice crop. Due to hurricanes, abolition of slavery, and lack of a labor force, the rice industry fell.

During the beginning of the second third of the twentieth century, agricultural interests viewed the wetlands as a potential pasture for cattle production. By diking, draining, and ditching, wetland owners envisioned easy establishment of "instant pastures." In one area, 600 hectares of wetlands were drained and diked. Neely (19) reported that the rather high soil salinity was leached adequately to permit planting of pasture grasses by the end of the first year. By the third year, the area won an award in the "Green Pastures" contest. By the fifth year, the area was devoid of any living vegetation due to the formation of "cat-clays."

The earlier work of Edelman and von Staveren (10) documented the nature of the "cat-clay" problem. When soils with high organic matter content are exposed to sea water under anaerobic conditions, the sulfates become reduced to sulfides which complex with various elements in the soil to become polysulfides. As long as the soils remain wet, these polysulfides do not affect the suitability of the soil for vegetation. When the soils are drained and become dry, the anaerobic zones become aerobic. The resultant oxidation of the sulfides releases sulfuric acid as one of the products of the reaction. In many of these "cat-clay" soils, the pH often reaches 2.5 or below (11). The preceding centuries of rice culture were possible without resultant "cat-clay" problems because the fields were kept flooded or the soils wet, thus preventing sulfide oxidation.

Many wetlands along the Gulf Coast of the United States supported cattle herds of early settlers. During the dry seasons, the cattle grazed on the marsh; during wet periods, the cattle utilized the old dune ridge community (cheniers) for a grazing area (3, 7, 31). Problems associated with these wetland pastures included heat, winds, tides that submerged grazing areas, unstable soils where cattle would bog, lack of shelter, insects, disease and heavy rainfall. Other secondary problems included overgrazing, lack of fences, inadequate potable water for the livestock, lack of reserve food for critical periods and overabundance of inedible plants.

In the fresh water marshes along the U.S. Gulf coast, *Panicum hemitomon, Zizaniopsis miliacea,* and *Phragmites communis* are the most nutritious plants for grazing purposes (31). They remain green and productive throughout the year, but decline when overgrazing takes place. When overgrazing occurs, reedy plants including *Polygonom* sp., *Alternanthera philoxeroides,* and *Daubontonia drummondi* invade the fresh marsh pasture (31). These plants make the fresh marsh pasture less desirable for grazing and consequently the livestock are moved to the salt marsh pastures, if available.

The most desirable pasture plants in the salt marshes include: *Spartina patens, Spartina cynosuroides, Spartina alterniflora, Sporobolus virginicus* and *Distichlis spicata. Paspalum vaginatum* and *Paspalum lividum* are also of range value on some salt marshes, but their presence is usually an early indicator of overgrazing in salt marshes (31). *Juncus roemerianus* and *Iva frutescens* invade the salt marsh pasture and are not conisdered usable for grazing. Occasionally, *Scirpus olneyi* and *Scirpus robustus* are found in the salt marsh pasture. Although they are consumed by livestock, their greatest food value is for muskrat (*Ondatra zibethicus*) and some waterfowl, mainly snow geese (*Chen hyperborea*) (22). In New England, large quantities of *Spartina alterniflora* are consumed by Canadian and snow geese. Ducks feed on the widgeon-grass (*Ruppia maritima*) (4) while the favorite grazing food of the Brent goose is *Aster* sp. (27).

Williams (31) reports that salt marsh pastures will support "one animal unit" per 1.6 to 4.8 hectares depending upon the range quality. The distribution of the old dune ridges has a singificant influence on livestock distribution in the marshes. The ridges are used when the marshes are covered by water. They also provide a rest area for the cattle and calves. The ridges allow partial relief from mosquitos and also serve as a place where livestock can be fed and tended during emergencies.

In order to attain optimum grazing potential, Williams recommended the construction of cattle walkways (30, 32). These earthen levees serve as trails and encourage more uniform grazing. The borrow pits are staggered from one side of the levee to the other at thirty-meter intervals so cattle can move off either side of the levee. Whenever these walkways cross drainage channels, culverts or bridges are installed. Since Williams (32) found that cattle would graze about 0.4 kilometers from the levee when water was on the marsh, optimum spacing of the artificial walkways is at 0.8 km intervals.

These artificial cattle walkways alter the marsh in several ways. Each 1.6 kilometers of walkway consumes about 0.4 hectares of marsh (the area where the fill is placed) and creates 1.1 hectares of open water (borrow pits). These walkways are not, however, a recent design. Petch (23) reported that this concept was in use in grazing marshland pastures in the British Isles for over a century.

Along southeastern United States, Shanholtzer (29) considered grazing in wetlands. Seaside sparrows, sharp-tailed sparrows and red-winged blackbirds as well as the rice rat, deer, marsh rabbits, rodents, cattle, burros, horses and pigs all utilize the protein and energy of *Spartina alterniflora*.

The impact of grazing on these selected southeastern United States marshes has been assessed by Reimold et al. (28). Using representative salt marsh areas grazed by ungulates, areas formerly grazed, ungrazed control areas, and areas simulatively grazed by mechanical means, the effects on primary production, detritus production, mineral content, and macroinvertebrate populations were assessed. In these wetland meadows, *Spartina alterniflora* and *Distichlis spicata* primary production was reduced nearly 70% by grazing as compared to the ungrazed system. The formerly grazed areas (where grazing was prohibited for 1 year by electric fence exclosures) was about half as productive as the control area. The predominant marsh macroinvertebrate, the fiddler crab, *Uca pugnax*, was less abundant in the grazed areas than elsewhere. The ribbed mussel, *Modiolus demissus*, was absent from the grazed areas, presumably due to the effects of trampling by ungulates. Quantification of the high ground area and marsh (with a compactness suitable to support livestock) revealed that the marsh accounted for 44% of the total grazing area. The grazing density was one animal unit per 1.54 hectare. Although the study computed the economic value of the marsh for grazing, the significance was not in terms of dollars per year, but rather the fact that the marsh and the high ground were nearly equal in terms of their value as potential grazing areas for livestock.

The results of this southeastern study (28) document reduced primary production, reduced detritus production, and reduced invertebrate populations due to grazing. The study assumed the amount of material eaten by cattle to be the difference between the grazed and ungrazed marsh. This assumption as well as documentation of the energy loss from decomposition of animal feces and urine are still essential ingredients for future study.

In addition to grazing cattle on these southeastern U.S. salt marshes, small quantities of the plants are harvested to feed cattle. Burkholder (2) and Yarlett (35) documented the high protein content (12 to 13% dry matter) in young *Spartina alterniflora* shoots. They also demonstrated the presence of adequate amounts of the B vitamin complex for other life forms. Burkholder (2) further documents the low concentration of arginine, methionine, histidine and valine which supposedly would make the quality of the *Spartina alterniflora* protein low and of "somewhat limited biological value." This would suggest that the quality of the protein would be substandard for grazing purposes. This awaits further definitive documentation.

In northeastern U.S. Atlantic coastal areas from Nova Scotia to Delaware Bay, *Spartina pectinata* marshes are currently grazed. In these areas, significant quantities of *Spartina patens* were harvested for use as hay (1). Large quantities of hay were harvested from mid-western U.S. *Spartina cynosuroides* marshes during the early twentieth century (17). These midwestern *Spartina cynosuroides* marshes were also formerly used as grazing areas (18). Lynch et al. (16) reported that the effects of burning *Spartina patens* marshes along the United States Gulf Coast improved the pasture for cattle grazing.

Sheep and cattle grazing on salt marshes of other parts of the world have been reviewed in the literature (4, 5, 9, 12, 13, 20, 24, 25, 34). Jakobsen and Jensen (14) described the construction of drainage ditches at a thirty degree angle to the Danish coastline in order to drain the marsh and improve the growth of natural marsh grasses for grazing. In some areas, *Puccinellia maritima* is planted on tidal flats for additional grazing areas. In Denmark, rows of piles or fence posts have also been used to promote silt deposition and growth of vegetation favorable for grazing (14). Jacquet (13) reported the predominant use of *Spartina* marshes was for fattening beef cattle in France. Dairy grazing was not successful since the marsh plants imparted a disagreeable taste and smell to the resultant dairy products. Domestic geese also graze the marshes along the French Channel (4). In Great Britain, Ranwell (24) described marsh grazing areas interspersed with upland ridges and beaches. Although the predominant vegetation grazed was *Spartina townsendii*, *Puccinellia maritima* was also consumed. The grazing actually favored the increase of *Puccinellia maritima*. Absence of grazing on these British marshes in Bridgewater Bay favored the increase of *Atriplex hastata*, *Phragmites communis*, and *Scirpus maritimus*. He estimated that in the upper limits of the *Spartina* marsh, the rate of change from *Spartina* to *Phragmites* (ungrazed) or to *Puccinellia* (grazed) is approximately one decade.

Swedish marshes have been grazed over the past 500 years by goats, sheep and cattle (8). Ranwell (26) reports that the three principal grasses consumed are *Agrostis stolonifera*, *Festuca rubra*, and *Puccinellia maritima*, which in the British Isles can support 2 to 3 sheep per 0.4 hectare. In the absence of adequate numbers of sheep to maintain this early successional stage, mowing is frequently employed (6, 33).

Ranwell (25) reported that *Spartina* is grazed in New Zealand to the extent that in some cases grazing has hindered the spread of *Spartina townsendii*. In many parts of the world where *Spartina townsendii* is found occupying wetland areas (New Zealand, Tasmania, Australia, South Africa, South America, North America, France, Netherlands, Denmark, Germany and the British Isles) a wide variety of herbivorous mammals eat the plant, but few herbivorous birds will eat it (25).

Dorset Horn and Dorset Down sheep consolidate the surface mud of the British Isles marshes by trampling and increase the transformation "to the salting pasture state" by grazing on *Spartina townsendii*. These older breeds of sheep find *S. townsendii* palatable while newer breeds ("Culn type") find it much less so (25). Oliver (21) earlier had demonstrated that *Spartina* was both digestable and palatable for certain breeds of sheep.

Animal hooves break up the turf with each step; this disruption of the substrate leads to increased erosion. In Sweden, grazing also has been responsible for elimination of *Aster tripolium*, *Phragmites*, *Scirpus* and *Triglochin* and the areas eventually being revegetated by *Salicornia*, *Suaeda* and *Spergularia* (5). Chapman (4) found that the Swedish marshes, if not overstocked with grazers, are most successfully used as grazing areas when vegetated with *Agrostis* and *Puccinellia*. Other marsh grazers include camels in the marshes adjoining the Red Sea (15) and marsupials and rabbits in New Zealand and Australia (26).

Other than to produce mutton, pork and beef, the wetland plants considered are a basic requisite for secondary production in wetlands. This review of research related to grazing in wetlands has assessed the primary productivity of grazed and ungrazed marshes as well as the successional characteristics following grazing. Few of the studies have considered the area required per unit weight of grazing animal per unit time and none have accurately documented it. One important aspect untouched in research results to date relates to the amount of primary production and detritus production needed. Once the amount of production necessary to sustain life in the estuary at a given level has been documented, then it will be possible to more accurately manage grazing in wetlands. Indeed at this point in our quest for knowledge, limited grazing of the wetlands should be allowed to continue. Marshes appear to recover from grazing once it is terminated. Since even low density grazing does produce a significant impact on the wetlands ecosystem, additional research is needed to document the optimum invertebrate population, and the minimum detritus production necessary to sustain secondary production in adjoining systems.

REFERENCES

1. Bourn, W.S., and C. Cottam. 1959. Some biological effects of ditching tidewater marshes. U.S. Government Research Report 19: 1-30.
2. Burkholder, P.R. 1956. Studies on the nutritive value of *Spartina* grass growing in the marsh areas of coastal Georgia. Bull. Torrey Bot. Club 83 (5):327-334.
3. Chabreck, R.H. 1968. The relation of cattle and cattle grazing to marsh wildlife and plants. Proc. 22nd Annual Conf. of the Southeastern Assoc. of Game and Fish Comm. p. 55-58.
4. Chapman, V.J. 1969. Salt Marshes and Salt Deserts of the World. Leonard Hill Interscience. London 391 p.
5. _____. 1974. Salt Marshes and Salt Deserts of the World. Second, supplemented reprint edition. J. Cramer. Germany. 392 p.
6. Chippendale, H.G., and R.W. Merricks. 1965. Gang mowing and pasture management. J. British Grasslands Soc. 11: 1-9.
7. Corty, F.L. 1972. Agriculture in the coastal zone of Louisiana. La. State Univ. A.E.A. Information Series No. 25. 27 p.
8. Dahlbeck, N. 1945. Strandiviesen am Sudostlichen Oresund. Act. Phytogeogr. Suec. 18.
9. de Vries, D.M. 1956. Report on a visit made by a working party "Grass cover for embankments" to the west coast of Schleswig-Holstein from 13-17 June 1955. Gesten, Meded. 21. Cent. Inst. Landbouwk. Orderz. Wageningen.
10. Edelman, C.H., and J.M. van Staveren. 1958. Marsh soils in the United States and in the Netherlands. J. Soil Water Conserv. 13:5-17.
11. Green, V.E., Jr. 1957. The culture of rice on organic soils—a world survey. Agronomy J. 49:468-472.
12. Haes, E.C.M. 1955. Plants of the salt marshes. J. Sports Turf Res. Inst. 9(31):94.
13. Jacquet, J. 1959. Reserches ecologiques sur le littoral de La Manche. Encycl. Biog. et Ecol. 5. Paris.

14. Jakobsen, B., and M. Jensen. 1956. Unterøgelser vedrørende land vind ingsmetoder i Det Danske Vadehav. Geograf. Tidskr. 55:21-61.
15. Kassas, M. 1957. On the ecology of the Red Sea coastal land. J. Ecol. 45:187-203.
16. Lynch, J.J., T. O'Neill and D.W. Lang. 1947. Management significance of damage by geese and muskrats to Gulf Coast marshes. J. Wildlife Management 2:50-76.
17. Merrill, E.D. 1902. The North American species of *Spartina*. Bull. Bur. Pl. Ind. U.S. Dept. Agriculture. 9:5-15.
18. Mobberley, D.G. 1956. Taxonomy and distribution of the genus *Spartina*. Iowa State College Journal. Science 30:471-574.
19. Neely, W.W. 1967. Planting, diking, mowing and grazing, p. 212-221. *In* J.D. Newsom (ed.), Proceedings of the Marsh and Estuary Management Symposium. Louisiana State University, Baton Rouge, La.
20. Newman, L.F., and G. Walworth. 1919. A preliminary note on the ecology of part of the South Lincolnshire coast. J. Ecology 7:204-210.
21. Oliver, F.W. 1920. *Spartina* problems. Ann. appl. Biol. 7:25-29.
22. O'Neil, Ted. 1949. The muskrat in Louisiana coastal marshes. La. Wildlife and Fisheries Comm., New Orleans. 152 p.
23. Petch, C.P. 1945. Reclaimed lands of West Norfolk. Trans. Norfolk Norwich Nat. Soc. 16:106-109.
24. Ranwell, D.S. 1961. *Spartina* salt marshes in southern England. I. The effects of sheep grazing at the upper limits of *Spartina* marsh in Bridgewater Bay. J. Ecol. 49: 325-240.
25. _____. 1967. World resources of *Spartina townsendii* (sensu lato) and economic uses of *Spartina* marshlands. J. Appl. Ecol. 4: 239-256.
26. _____. 1972. Ecology of Salt Marshes and Sand Dunes. Chapman and Hall. London. 258 p.
27. _____, and B.M. Downing. 1959. Brent goose (*Branta bernicla* (L.)) winter feeding pattern and *Zostera* resources at Scolt Head Island, Norfolk. Animal Behavior 7:42-56.
28. Reimold, R.J., R.A. Linthurst and P.L. Wolf. 1975. Effects of grazing on a salt marsh. Biological Conservation 8:105-126.
29. Shanholtzer, G.F. 1974. Relationship of vertebrates to salt marsh plants. p. 463-474. *In* R.J. Reimold and W.H. Queen (eds.), Ecology of Halophytes. Academic Press Inc., New York, N.Y.
30. Williams, R.E. 1952. Walkways improve grazing distribution. J. Soil Water Conservation. 7(3): 125-127.
31. _____. 1955. Development and improvement of coastal marsh ranges. Yearbook Agriculture. U.S. Dept. Agriculture. p. 444-449.
32. _____. 1959. Cattle walkways. Leaflet U.S. Department of Agriculture 459: 1-8.
33. Wohlenberg, E. 1965. Deichbau und Deichpflege auf biologischer Grundlage. Die Kuste. 13:73-110.
34. Yapp, R.H., D. Johns, and O.T. Jones. 1917. The salt marshes of the Dovey Estuary. J. Ecology 5:65-103.
35. Yarlett, L. 1965. Important native grasses for range conservation in Florida. U.S. Soil Conservation Service Internal Report. p. 53-59.

MANAGEMENT OF WETLANDS FOR WILDLIFE HABITAT IMPROVEMENT

Robert H. Chabreck[1]
School of Forestry and Wildlife Management
Louisiana State University
Baton Rouge, Louisiana 70803

ABSTRACT: Population levels of sporting and commercial wildlife species occupying wetlands are a product of the quantity and quality of habitat available. The rapid loss or modification of wetland habitat by other land-use practices has increased the need for special management to maintain or improve habitat quality. Habitat management practices should be designed to regulate the species composition, density, and distribution of plants. Major factors affecting plant growth in coastal wetlands are water levels and salinity, and the management practices applied should be those best suited to local conditions for maintaining these variables within acceptable limits. Special attention must be given to other needs of the wildlife species involved, because factors other than plants are also important components of favorable habitat. Management practices currently used in coastal wetlands are shallow water impoundments, water control structures in drainage systems, marsh ditches, and artificial potholes. Other treatments frequently applied are marsh burning, herbicidal treatment, and planting.

INTRODUCTION

Population levels of wildlife species occupying wetlands are a product of the quality and quantity of the habitat available. Individual species have specific habitat requirements and greatest populations occur when conditions are within the optimum range. However, the rapid loss or modification of coastal wetlands, as a result of changing land-use practices, has caused habitat conditions to deteriorate and wildlife populations to decline in many areas.

Activities which have had the most damaging impact on wildlife habitat are canal construction associated with oil and gas exploration, pipelines, navigation,

---

[1] Present address: U.S. Fish and Wildlife Service, Coastal Ecosystems Team, National Space Technology Laboratories, Bay St. Louis, Mississippi 39520.

and flood control; permanent drainage for agriculture, industry, and urbanization; modified drainage patterns associated with levee and highway construction and spoil deposits; and dredge and fill operations. Only with careful planning can mitigating measures be implemented to offset the damaging impacts of such projects on wildlife habitat. As more and more demands are placed on coastal wetlands in the future, the implementation of special management practice will become more and more essential, if critical wildlife habitat is to be maintained at a level of high productivity.

The purpose of this paper is to review the marsh management procedures currently in use to improve or maintain favorable conditions for wildlife. The habitat conditions and management practices discussed apply primarily to the Northern Gulf and Lower Atlantic coastal regions.

WILDLIFE SPECIES

The wildlife species discussed in this paper are primarily those classified as game species and commercial species. Game species in the coastal zone include ducks, geese, coots (*Fulica americana*), rails, gallinules, common snipe (*Capella gallinago*), rabbits (*Sylvilagus* spp.), and white-tailed deer (*Odocoileus virginianus*). Past habitat management efforts have been directed primarily at ducks. Marshes along the Northern Gulf and Lower Atlantic coasts are major wintering areas for migratory ducks, and management efforts on private lands have been directed mainly at habitat improvement to attract ducks for the purpose of sport hunting (7). Numerous federal and state wildlife refuges and management areas are situated within the coastal wetlands, and habitat management is practiced not only to improve habitat qualitity, but also to increase the amount of habitat available.

Commercial wildlife species found in coastal wetlands are the fur-bearing animals and the American alligator (*Alligator mississippiensis*). Principal fur-bearers are muskrat (*Ondatra zibethicus),* nutria *(Myocastor coypus*), raccoon (*Procyon lotor*), mink (*Mustela vison*), and river otter (*Lutra canadensis*) (24, 28).

MANAGEMENT PLANNING

Marshes subject to severe tidal action and drastic salinity changes usually provide poor wildlife habitat. Consequently, such areas are of minimal direct value to most waterfowl or fur-bearing animals (20). Only by careful planning and management can the wildlife productivity of marshes be increased to or maintained at a desirable level.

In planning a marsh management project, careful consideration should be given to several factors. First, establish goals which include priorities on the wildlife species to be produced in the marsh or attracted to the marsh. Usually, optimal conditions for one group are not considered optimal for another. Second, obtain detailed information on area environmental conditions such as water quality, water level fluctuation, soil characteristics, and climatic factors. Third, determine the wildlife value and growth requirements of common plants in the area. Fourth, understand completely the wildlife habitat requirements and the factors affecting wildlife abundance locally.

Marsh management is costly and the benefit gained depends upon the amount invested and the skill with which the program is planned. Unfortunately, marsh owners frequently launch development projects without fully understanding the problems which they are trying to correct. Usually, planning by persons familiar with the problem will reduce the cost of the program and increase its effectiveness.

The ecological processes in coastal marshes are very complex and involve the action and interaction of numerous factors. The objective of management should be to manipulate these processes to produce the desired plant and animal communities. Special projects may be required to accomplish the objective, or protective measures to maintain present conditions may be enough.

The ideal coastal marsh management procedure should reduce water level fluctuation, prevent drastic salinity changes, minimize water turbidity and reduce the rate of tidal exchange. But mainly, the technique should produce stands of desired vegetation in the marsh and marsh ponds in order to increase resident wildlife species density and attract migratory species.

In the management of vast areas of marsh, the first consideration should be the regulation of water levels and water salinities. By manipulating water levels and salinities over a period of time, the marsh manager can either encourage desirable plants or discourage undesirable plants until favorable conditions are created. Under more intensive management, special measures are often used to create special habitat conditions.

All plants have a definite range of tolerance for water salinity and water depth, and generally, plants will grow so long as the conditions are within this range (6, 21). Whenever the water salinity and/or water depth go above or below the tolerance range for an extended period of time, plants unable to tolerate the changed conditions will die out. Then, other species of plants, whose requirements have been met as a result of the change, will invade the area.

WILDLIFE HABITAT

Ideal wildlife habitat conditions in coastal wetlands usually have about three-fourths of the area in marsh and the remainder in scattered ponds. Ponds most desirable are those less than 4 ha. in size.

Brackish marsh management should be directed toward the growth of three-cornered grass (*Scirpus olneyi*), water hyssop *(Bacopa monnieri)* and dwarf spikerush (*Eleocharis parvula*) in the marsh and the growth of widgeon grass (*Ruppia maritima*) in the ponds. Most desirable fresh marsh plants are wild millet (*Echinochloa walteri*), sprangletop (*Leptochloa fascicularis*), fall panicum (*Panicum dichotomiflorum*), and various other annual plants which produce an abundance of seeds. Pondweed (*Potamogeton spp.*), naiad (*Najas quadalupensis*), and duckweed (*Lemna minor* and *Spirodela polyrhiza*) are the most desirable plants of freshwater ponds.

Three-cornered grass roots are a choice food of snow geese (*Chen caerulescens*) and muskrats along the Gulf coast (19). The other plants listed as most desirable in both the brackish and freshwater areas are favorite duck food plants (15).

The marsh manager interested in attracting ducks should improve or maintain conditions for migratory species. The mottled duck (*Anas fulvigula*) is the only waterfowl species nesting in the coastal marshes of the Southeastern U.S. in sizeable numbers (10, 13, 26); consequently, efforts to improve nesting conditions are of secondary importance under most situations. Waterfowl habitat management should be directed toward increasing food production and regulation of water depths to make this food available. Techniques which can be applied by the marsh manager include manipulating water depth, controlling salinity, burning, planting, creating artificial openings and control of undesirable plants.

MARSH IMPOUNDMENTS

In order to regulate or manipulate water depths, the marsh must be impounded or enclosed with a continuous levee. Cost of constructing the necessary levees is prohibitive in most situations; however, in some areas levee systems may already be present and require only slight modification (11). Where water level control is possible, draining the water from fresh marshes during the spring or early summer will permit the soil to dry and grasses, sedges, and other seed-producing annual plants to germinate and grow (1, 3). Reflooding the marsh will attract ducks to the area and make the seeds produced available to the birds. The drying process is essential for plant germination; however, reflooding may be done a few weeks after germination, so long as the new plants are not completely covered with water. Ponds managed for crawfish are also handled in this manner, except that reflooding is recommended during the fall. The management schemes for ducks and crawfish coincide sufficiently so that an impounded marsh can be managed for both (22).

Brackish water impoundments are also used to improve habitat conditions for ducks (3, 17). However, duck usage of brackish impoundments is usually less than that of freshwater impoundments (9). Marsh impoundments may also be managed as permanently flooded freshwater systems. In addition to ducks, these impoundments improve the habitat for a wide array of fish and wildlife species (3).

Although marsh impoundments have been widely used to improve habitat conditions for wildlife, this type of management has certain disadvantages which at times make it necessary to seek other types of management. First of all, impoundments are costly, not only to construct but also to maintain. Also, without facilities for pumping water, years that are unusually wet or dry generally result in poor food production. While impoundments greatly improve habitat conditions for ducks, certain other valuable species such as fur-bearing animals, geese and marine organisms are not benefitted. A limiting factor in the use of impoundments is soil characteristics. Impoundments can be built only in areas which will support a continuous levee. Because of the fluid nature of the subsoil in certain areas, such as Southeastern Louisiana, impoundment use is limited.

Other methods have been used to improve coastal marshes for wildlife. These include placing weirs and earthern plugs in the drainage systems of a particular area and excavating ditches and potholes in the marsh (4, 8). All of these methods have been used with varying degrees of success.

WEIRS

Weirs are beneficial in coastal wetlands managed for fish and wildlife, if properly placed and properly constructed. Weirs resemble low dams and may be constructed of steel or wooden sheet piling. The top or crest of the weir is usually set 15 cm below the natural elevation of the marsh and water flows back and forth across the structure. Weirs do not completely block the flow of water, but prevent total drainage of marshes on low tides and reduce the rate and volume of tidal exchange. The ponds behind weirs produce far more aquatic vegetation than do natural ponds (4) and attract more wintering waterfowl (25). The bayous and ponds behind weirs hold permanent water, and thereby improve access to the marsh for trappers and hunters and improve conditions for fur-bearing animals as well as waterfowl. Hundreds of such structures have been constructed by land owners along the Louisiana coast to improve wetlands for fish and wildlife.

POTHOLES AND DITCHES

Artificial potholes and ditches in marshes create permanent water areas and in most instances are beneficial to wildlife (4). Although such water areas may produce very little food, they attract ducks to the area and make access easier for trappers and hunters. Potholes and ditches provide permanent water during droughts and are important to fur-bearing animals, alligators, nesting ducks, frogs, and small fishes. Fish populations sustained in potholes and ditches during droughts help control mosquitoes when the marshes are reflooded. Special problems often occur when ditches connect to tidal streams. This usually results in excessive marsh drainage, changes salinity levels, and ultimately affects wildlife populations in the area by lowering habitat quality (2).

CATTLE GRAZING

Cattle grazing can be used to some advantage in marshes managed for wildlife (5, 18). Carefully regulated grazing will open up dense stands of vegetation and create conditions favorable for certain species. As for fur-bearing animals, it is generally agreed that marshes managed for maximum fur animal production should not be grazed (19). Grazing will reduce available cover, and cattle trample dens and underground tunnels and compete with herbivores for food.

Grazing which can be easily controlled is usually beneficial to ducks. Not only will grazing remove dense stands of vegetation, but it also sets back plant succession and increases food production. Certain low value plants, such as marshhay cordgrass (*Spartina patens*) decrease with grazing; and certain high value plants, such as *Paspalum vaginatum* increase with grazing. To improve grazed marshes for ducks, cattle should be removed from the marsh during July, August, and September to permit annual grasses and sedges to grow and produce seeds. The marsh should be flooded from October through February with water depths from four to six inches to attract ducks and make the seeds available as a duck food.

Snow geese are attracted to marshes where dense stands of mature vegetation have been removed; consequently, moderate grazing usually benefits this species (5). Geese will feed on new sprouts, plus roots and rhizomes of marsh grasses and sedges. Another species which benefits from cattle grazing is the common snipe. This species prefers areas with exposed mineral soil and no overhead cover, and largest concentrations of snipe are usually found on over-grazed marsh range.

FOOD PLANTINGS

One of the first procedures generally considered to improve marshes for wildlife is to plant some type of vegetation in the marsh to produce food. Such plantings are usually made without site preparation, and almost invariably the efforts fail to meet the objective. Usually, the absence of natural food plants in a marsh is a result of unfavorable soil and water conditions or excessive competition from less desirable plants. The same conditions will cause the failure of plantings made in the same marsh area. Plantings in a marsh can never be a substitute for regulating water levels and salinities to produce natural foods. Only with the application of agronomic techniques have favorable results been achieved with artificial plantings (18).

Since three-cornered grass is a choice food plant of muskrat and snow geese in coastal marshes of the Southeastern U.S., there is considerable interest in managing marshes for this species. Procedures for establishing or re-establishing this species through planting have been the subject of intensive investigations (12, 20, 23). The studies have disclosed that this species can be established by transplanting root stock; however, plantings must be made within certain water and salinity levels and protection must be provided against excessive competition from other plants and animal feeding.

MARSH BURNING

Burning has been widely used as a marsh management procedure (14, 19); however, the value of most of this effort is questionable. Under certain situations, burning is important for maintaining stands of vegetation. Three-cornered grass may be replaced by marshhay cordgrass when the two are growing in mixed stands. Burning can be used to give the three-cornered grass an earlier start during the growing season; however, burning alone will not maintain three-cornered grass and should not be substituted for the necessary water levels and salinities in the management of this species of vegetation or any other. Burning is important for removing dense stands of vegetation and is widely used to attract snow geese to a marsh. The geese are attracted to newly burned areas and frequently remain in the areas until the regrowth is well advanced. Also, trappers find that in burned marshes walking is much easier and animal trails are much more noticeable. However, nutria and raccoon often move from a burned marsh because of the lack of adequate cover.

WEED CONTROL

Chemical control of undesirable plant species in marshes has been used on a limited basis in the past (16, 27); however, in the future this method is expected to have greater emphasis as management efforts become more intensive. Also, herbicides are being developed which are more effective and can be applied to marshes at a more economical cost. As with other management techniques, a marsh manager should consult a specialist in this field before attempting wide-spread applications of herbicides to a marsh.

LITERATURE CITED

1. Baldwin, W.P. 1968. Impoundments for waterfowl on South Atlantic and Gulf Coastal marshes, p. 127-133, *In* J.D. Newsom (ed.), Proceedings of the marsh and estuary management symposium. La. State Univ., Baton Rouge.
2. Bourn, W.S., and C. Cottam. 1950. Some biological effects of ditching tide water marshes. U.S. Fish and Wildlife Service, Research Rep. 19. 30 p.
3. Chabreck, R.H. 1960. Coastal marsh impoundments for ducks in Louisiana. Proc. 14th Annu. Conf. of the Southeastern Assoc. of Game and Fish Comm. p. 24-29.
4. _____. 1968. Weirs, plugs and artificial potholes for the management of wildlife in coastal marshes, p. 178-192, *In* J.D. Newsom (ed.), Proceedings of the marsh and estuary management symposium. La. State Univ., Baton Rouge.
5. _____. 1968. The relationship of cattle and cattle grazing to marsh wildlife and plants. Proc. 22nd Annu. Conf. of the Southeastern Assoc. of Game and Fish Comm. p. 55-58.
6. _____. 1972. Vegetation, water and soil characteristics of the Louisiana coastal region. La. Agric. Exp. Sta. Bull. 664. 72 p.
7. _____. 1975. Waterfowl management and productivity—Gulf coast habitat. First Int. Waterfowl Symp., Ducks Unlimited, Chicago. p. 64-72.
8. _____, and C.M. Hoffpauer. 1962. The use of weirs in coastal marsh management in Louisiana. Proc. 16th Annu. Conf. of the Southeastern Assoc. of Game and Fish Comm. p. 103-112.
9. _____, R.K. Yancey and L. McNease. 1974. Duck usage of management units in the Louisiana coastal marsh. Proc. 28th Annu. Conf. of the Southeastern Assoc. of Game and Fish Comm. p. 507-516.

10. Chamberlain, E.G., Jr.. 1960. Florida waterfowl populations, habitats and management. Fla. Game and Fresh Water Fish Comm., Tech. Bull. No. 7. 62 p.
11. Ensminger, A.B. 1963. Construction of levees for impoundments in Louisiana marshes. Proc. 17th Annu. Conf. of the Southeastern Assoc. of Game and Fish Comm. p. 114-119.
12. Hess, T.J. 1975. An evaluation of methods for managing stands of *Scirpus olneyi*. M.S. thesis. La. State Univ., Baton Rouge. 109 pp.
13. Lowery, G.H. 1954. Louisiana birds. La. State Univ. Press, Baton Rouge. 556 p.
14. Lynch, J.J. 1941. Place of burning in management of Gulf Coast wildlife refuges. J. Wildl. Mgmt. 5(4):454-457.
15. Martin, A.C., and F.M. Uhler, 1939. Food of game ducks in the United States and Canada. U.S. Dept. Agric. Tech. Bull. 634. 308 p.
16. McNease, L.L., and L.L. Glasgow. 1970. Experimental treatments for the control of wiregrass and saltmarsh grass in a brackish marsh. Proc. 24th Annu. Conf. of the Southeastern Assoc. of Game and Fish Comm. p. 127-145.
17. Neely, W.W. 1962. Saline soils and brackish waters in management of wildlife, fish, and shrimp. Trans. 25th N. Am. Wildl. Conf. p. 321-335.
18. Neely, W.W. 1968. Planting, disking, mowing, and grazing, p. 212-221. *In* J.D. Newsom (ed.), Proceedings of the marsh and estuary management symposium. La. State Univ., Baton Rouge.
19. O'Neil, T. 1949. The muskrat in the Louisiana coastal marsh. La. Wild Life and Fisheries Comm., New Orleans, 152 p.
20. Palmisano, A.W. 1973. Habitat preferences of waterfowl and fur animals in the northern Gulf coast marshes, p. 163-190. *In* R.H. Chabreck, (ed.). Proceedings second symposium-coastal marsh and estuary management. La. State Univ., Baton Rouge.
21. Penfound, W.T., and E.S. Hathaway. 1939. Plant communities in the marshland of southeastern Louisiana. Ecol. Monogr. 8:1-56.
22. Perry, W.G., Jr., T. Joanen, and L. McNease. 1970. Crawfish-waterfowl, a multiple use concept for impounded marshes. Proc. 24th Annu. Conf. of the Southeastern Assoc. of Game and Fish Comm. p. 506-519.
23. Ross, W.M., and R.H. Chabreck. 1972. Factors affecting the growth and survival of natural and planted stands of *Scripus olneyi*. Proc. 26th Annu. Conf. of the Southeastern Assoc. of Game and Fish Comm. p. 178-188.
24. St. Amant, L.S. 1959. Louisiana wildlife inventory and management plan. La. Wild Life and Fisheries Comm., New Orleans. 329 p.
25. Spiller, S.F. 1975. A comparison of wildlife abundance between areas influenced by weirs and control areas. M.S. thesis. La. State Univ., Baton Rouge. 94 p.
26. Singleton, J.R. 1953. Texas coastal waterfowl survey. FA Report Series No. 11. Texas Parks and Wildlife, Austin. 128 p.
27. Steenis, J.H., E.W. Ball, V.D. Stotts and C.K. Rawls. 1968. Pest plant control with herbicides, p. 140-148. *In* J.D. Newsom (ed.), Proceedings of the marsh and estuary management symposium. La. State Univ., Baton Rouge.
28. Wilson, W.A. 1968. Fur production on southeastern coastal marshes, p. 149-162. *In* J.D. Newsom (ed.)., Proceedings of the marsh and estuary management symposium. La. State Univ., Baton Rouge.

ASSIMILATION OF SEWAGE BY WETLANDS[1]

Ivan Valiela and Susan Vince
Boston University Marine Program
Marine Biological Laboratory
Woods Hole, Massachusetts 02543

John M. Teal
Woods Hole Oceanographic Institution
Woods Hole, Massachusetts 02543

ABSTRACT: Wetlands have attracted attention as potential components of waste treatment systems because of typically large plant production, high decomposer activity, anaerobic condition and large adsorptive areas in the sediments. These properties seem to provide wetlands with the ability to degrade and eliminate contaminants in waste waters.

The effluent-properties of a variety of wetland habitats in very diverse parts of the world (salt and brackish marshes, polders, cypress, papyrus and mangrove swamps, freshwater marshes and bogs) have been studied. Although each habitat type differs, current results from marshes show that nitrogenous and phosphorus nutrients are removed from contaminated or waste waters. Heavy metals from sludges are retained by marsh muds, but the effectiveness varies considerably for different metals. Petroleum hydrocarbons accumulate in marsh muds and may be actively decomposed, particularly at lower levels of contamination. Chlorinated hydrocarbons have high affinities for marsh sediments and can be altered by microorganismal activity. There is preliminary evidence that counts of coliform bacteria may be reduced in tidal water that floods marshes.

The waste-processing ability of coastal wetlands must have an upper threshold. Little is known about these upper limits but studies in chronically contaminated sites show that marshlands function well even in severely polluted areas.

[1] Work supported by grants from the Victoria Foundation and National Science Foundation. S.V. was aided by a fellowship from the Friendship Fund. Contribution No. 3630 from the Woods Hole Oceanographic Institution.

INTRODUCTION

Wetlands have received increased attention as valuable natural sinks for contaminants and as potential components of waste treatment systems (35, 38, 40, 46, 49, 58, 60, 64, 76). Several properties, held more or less in common by various types of wetlands, provide marshes, bogs, mangroves and swamps with an affinity for contaminants.

Anoxic organic sediments bound in place by plant roots are common to nearly all wetlands and are probably the major factor involved in retention of various chemical species. Reducing environments allow the conversion of heavy metals such as iron into relatively insoluble sulfides and promote the elimination of nitrogen through denitrification.

Most wetlands undergo sedimentation, either by accumulation of undecomposed organic matter and sediments as in bogs, or through accumulation of materials as sea level rises, as in salt marshes. Since the majority of contaminants adsorb on particulate matter, sedimentation of particulates leads to scrubbing of the water column and sequestering of contaminants in the sediments, preventing dispersal of the pollutants.

Wetlands usually show high rates of primary production by both macrophytes and algae. Dissolved substances are converted into less mobile forms through plant uptake. These effects are less marked in bogs and cypress swamps which are not as productive as salt or freshwater marshes.

Wetlands are dominated by plant species tolerant to a variety of stesses, such as high and/or variable salinity, turbidity, sedimentation, temperature, water and nutrient supply. These plants rapidly colonize areas bereft of vegetation and also possess other opportunistic adaptations that enable them to survive substantial alterations in their chemical environment. Such an assemblage of plants are able to survive the added stresses of processing some waste effluents up to as yet unspecified limits.

These ecological features allow wetlands to retain a variety of contaminants: nutrients, heavy metals, and chlorinated and petroleum hydrocarbons. The remainder of this review takes up in turn each of these broad categories of contaminants, and illustrates how the properties of wetlands just outlined operate in relation to selected chemical species. A last section discusses the effect of wetlands on bacteria and viruses in contaminated waters.

We emphasize work in salt marshes because of our own interests and because proportionately more studies have been carried out in this kind of wetland. Salt marshes occur mainly in developed countries where the impingement of man has been most prominent. Elsewhere, pollution and destruction are affecting wetland areas such as mangrove swamps. Unfortunately, less developed countries find it difficult to mount research and conservation programs, and awareness of the properties and importance of wetlands may arrive too late for a large proportion of the world's mangrove swamps.

THE IMPACT AND PROCESSING OF PLANT NUTRIENTS

Sewage effluents contain a variety of plant nutrients. Of these, nitrogen and phosphorus have received the most attention because these are the most important macronutrients limiting plant growth in aquatic systems (39).

For salt marshes exposed to tidal flooding, additions of phosphate have no detectable effect on higher plant primary production (Fig. 1) (10, 62, 74). The lack of response by macrophytes to phosphorus is likely due to the large amounts of seawater-borne phosphate adsorbed to the fine sediments present in and sedimenting into most marshes (53). Phosphate may be secondarily limiting

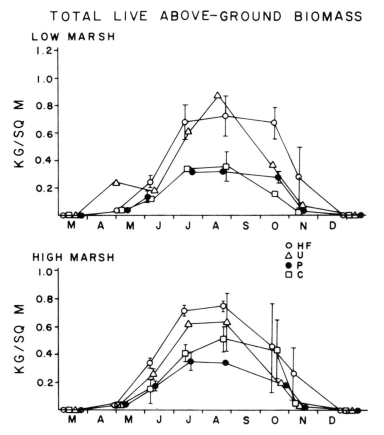

Figure 1. Dry-weight standing crops (mean ± S.E.) of salt marsh vegetation under control (C), phosphate (P), urea (U), and sewage sludge (HF) fertilizations. Low marsh was dominated by *Spartina alterniflora* while high marsh contained mainly *S. patens*.

to marsh plants such as *Spartina alterniflora* grown in coarse sediments since these grain sizes may not hold as much phosphate (10). Tyler (72) established that phosphorus was limiting in his study of Baltic shore meadows, but he was studying the upper reaches of the marsh where tidal flooding was infrequent.

Enrichment with urea yielded increases in growth in a Massachusetts salt marsh (Fig. 1) (74). Similar results with other nitrogen fertilizers have been reported in a variety of geographical sites (10, 52, 61, 62, 72). Tyler (72) found no response to addition of nitrate in Swedish marshes, but oxidized nitrogen species could be quickly lost to the atmosphere through denitrification (71).

Marshall (41) in North Carolina and Nixon and Oviatt (47) in Rhode Island present circumstantial evidence showing that contamination with nitrogen-containing sewage enhanced primary production of grasses in salt marshes. We have obtained two to threefold increases in higher plant production in a Massachusetts salt marsh following experimental enrichment with a fertilizer containing sewage sludge (75, 77). Since urea increases growth while phosphate has no measurable effect, the sewage result is probably due to the nitrogen in the sludge (Fig. 1), and we will emphasize nitrogen in the rest of this review.

We have shown that 80-94% of the nitrogen and 91-94% of the phosphorus added to experimental salt marsh plots in Massachusetts were retained by the plots (76). Similarly, water leaving a *Phragmites* marsh within the Hackensack meadowlands has less nitrogen relative to heavy inputs by Hackensack River water and an upstream sewage treatment plant (42). Phosphate and total nitrogen released from three waste treatment plants were virtually eliminated by passage of the effluent through a *Phragmites* marsh in Hungary (69). Similar results have been reported for nitrate, ammonia and total nitrogen in Fijian mangrove swamps (46), for nitrate and phosphate in East African papyrus marshes (22, 67, 68) and for total nitrogen and phosphorus in bulrush ponds on a Dutch polder (35).

Part of the removal of nitrogen and other nutrients by wetlands is due to plant uptake. In our Massachusetts plots, plant production and standing crops doubled or tripled (Fig. 1), thereby increasing the nutrients stored in the vegetation. Seidel (58) reported marked increases in vegetation in German freshwater ponds treated with sewage effluent.

In the German marsh work and in our experimental plots in Massachusetts, the increase in nutrient delivery led to quantitative changes in plant biomass, but there were no major qualitative changes in the flora of the enriched areas. The species present before the addition of sewage were still present after five years of treatment, although there were changes in plant densities among the salt marsh species involved (77).

In Florida freshwater marshes, Lugo (40) estimated that growth of vegetation in 1500 acres is capable of storing all the nitrogen and 25% of the phosphorus from the sewage of 62,000 people. However, Tyler (72) found that aboveground vegetation contained only 5% of the nitrogen and 2% of the phosphorus added experimentally to Baltic salt marshes. Nixon and Oviatt (47) argue that the growth of *Spartina alterniflora* in Rhode Island salt marshes would only remove a very small fraction of sewage inputs into Narragansett Bay. All these calculations were based on one-harvest sampling of standing crops. However, the amount of nutrients stored in standing crops is an underestimate of the total nutrients processed through the vegetation, since the vegetation of a wetland is continually growing and dying (22, 77, 87). Another problem is that there is seasonal transport of nutrients between below-ground and above-ground plant parts (32, 54) so that harvests of above-ground vegetation may provide misleading values for nutrient retention. Further, there may be losses of nitrogenous materials as well as phosphate (56) through leaching, a common phenomenon in plants (70). To evaluate the extent of fertilizer or contaminant retention it is necessary to use tracer techniques (84) or develop mass balance budgets for nutrients in the sediments and vegetation.

The transport of pore water nutrients into plants, with subsequent losses through leaching, as mentioned above, or through removal of dead plant tissues (48, 65) points out that wetlands are not permanent sinks for nutrients. In our Massachusetts experiments, at least 12% of the nitrogen added could be accounted for in annual above-ground growth. Of this, perhaps half is exported as dead organic matter (65), so about 6% of the added nitrogen may be lost in this fashion. No estimates of leaching losses are available for nitrogen. An additional source of exported nutrients may be the runoff of nutrient-rich interstitial water during receding tides (21). Also, detritus-feeding bivalves in the tidal channels and pools could convert particulate organic matter exported from the marsh surface into dissolved nutrients. Many of the shellfish found in the marsh benthos excrete substantial amounts of dissolved nitrogen (T. Jordan and S. Nixon, pers. commun.). These nutrients, if exported, would be important since nitrogen is the principal nutrient limiting algal production in coastal waters (23, 57, 79).

Current evidence does not show whether wetlands consistently export or import nitrogen. Export of inorganic and dissolved organic nitrogen has been found in salt marsh estuaries in Chesapeake Bay irregularly through the entire year (29). We find that in the more northern latitude of Massachusetts the export of ammonia nitrogen is more seasonal, and follows the onset of senescence of the grass sward, at which time uptake of nitrogen ceases and dead matter accumulates. Exports of dissolved nitrogenous species have also been recorded from salt marshes in Nova Scotia (K. Mann, pers. commun.) and in New York (85). However, others (1, 2) either find no export or an import of nitrogenous species. Further study concentrating on the organic nitrogen fraction, the dominant form of dissolved nitrogen (29, and our own unpublished measurements) is required. The role of meteorologic events such as storms and sudden thaws must also be evaluated. It is likely that major exports of materials out of marshes occur under such circumstances, since we have observed that storms increase nutrient loads of tidal water leaving salt marshes. Seasonal losses of nutrients are also described for freshwater marshes (60). In East African papyrus marshes the rainy season flushes nutrients from sediments into the water column (68).

A second group of mechanisms involves denitrification as the process ultimately responsible for net removal of nitrogen from tidal water. Salt marshes and other wetlands with reducing organic muds and neutral pH's foster high rates of denitrification. Marsh water tends to be eutrophic (76) and usually shows little nitrate relative to ammonia in the water. In rice paddies, an environment very similar to salt marshes, up to 50% of the ammonium added as fertilizer may be lost through nitrification in the aerobic sediment surface, leaching and subsequent denitrification in anoxic layers (71). The addition of sewage-containing fertilizer to our Massachusetts salt marsh plots has enhanced denitrification (Fig. 2), as claimed in other wetlands (9, 46, 68). We have also found that nutrient enrichment in the same plots results in an increase of ammonium in pore water and a consequent reduction of nitrogen fixation (Fig. 3) (78), with the microorganisms using the combined nitrogen provided by the fertilizer instead of going through the energy-requiring process of fixation. Through these alterations of the nitrogen cycle a salt marsh is able to adsorb and dispose of a substantial amount of added nitrogen. Algae and bacteria, through their involvement in nitrogen fixation and denitrification, are the agents responsible for marshes acting as true sinks for water-borne nitrogen. Their importance is in this aspect of marsh ecology, rather than in their modest contribution to nitrogen storage through growth.

RETENTION OF HEAVY METALS BY MARSHES

Heavy metals, particularly lead, zinc, cadmium and manganese, are present in large amounts in sewage sludge (5, 26) and in some sewage effluents (30). One source of sewage contamination for coastal marshes is leachate from sanitary landfill operations. We have examined the concentrations of several metals in marsh surface sediments at increasing distances from a landfill leachate source in Saugus, Massachusetts (Table 1). The concentrations of lead, zinc and cadmium fall rapidly, the lead binding to the sulfide-rich sediments. Zinc and cadmium are taken up by marsh and estuarine sediments via adsorption and ion-exchange processes (19, 30, 53).

The Saugus data and similar results for lead content of marsh sediments in Connecticut (59) suggest that salt marsh sediments are a good sink for heavy metals. However, unlike in deep sea sedimentary deposits where trace elements may be permanently trapped, in salt marshes the physiological adaptations of *Spartina* (66) allow its roots to penetrate anoxic sediments and to pick up metals. We have constructed rough budgets for lead, zinc and cadmium in experimentally fertilized plots in a Massachusetts salt marsh (5) to evaluate the role of

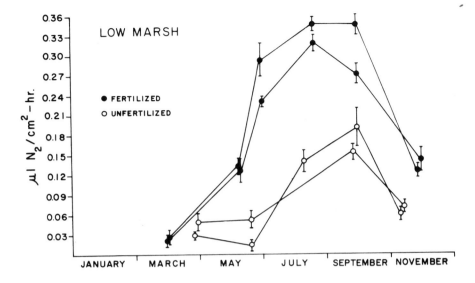

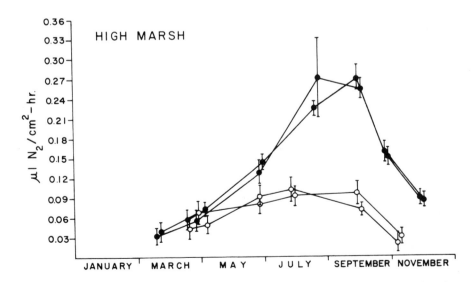

Figure 2. Rate of loss of nitrogen gas from fertilized and unfertilized salt marsh plots for low and high marsh habitats. Data for two plots for each treatment are shown. The fertilizer dosage was equivalent to 7.6 g N m^{-2} wk^{-1}. Lower dosages yielded intermediate rates of denitrification, and are not shown. Data obtained by Warren Kaplan using *in situ* bell jars and gas chromatography.

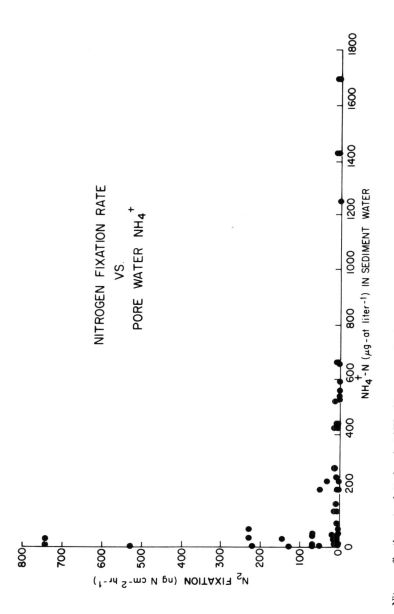

Figure 3. Nitrogen fixation rate plotted against NH$_4$–N concentration in pore water of top 10 cm of sediment in fertilized and control salt marsh plots. Values obtained during 1973. The fixation data were obtained by Charlene Van Raalte using the acetylene reduction method (78).

Table 1. Concentration of selected metals (ppm acid-soluble basis, mean ± std. error) in surface mud of tidal creeks draining leachate away from landfill on Saugus marsh. Values for samples of sediment collected 30 cm below surface are added for comparison to relatively unpolluted marsh sediments. The metal concentrations were obtained by William Kerfoot by atomic absorption spectrophotometry.

| | Approximate distance from source of leachate | | | |
	0 m	61 m	115 m	181 m
Lead	44±1.2	43±2.0	40±1.3	31±2.0
Zinc	398±52	157±12	116±1.2	80±1.7
Cadmium	2.4±0.4	1.3±0.1	1.4±0.03	1.2±0.09

| | Amount of metal at depth of 30 cm | |
	Saugus marsh	Cape Cod marsh
Lead	31.5	30
Zinc	91.0	64
Cadmium	2.7	1.1

the grass in exporting metals from marshes, particularly under conditions of increased nutrient and metal load (Fig. 4).

Nearly all the lead added as sewage sludge fertilizer was retained by the marsh surface sediments. The lead content of *Spartina alterniflora* increased from 2.3 to 4.8 ppm, lower than a value reported for *Spartina* growing near a heavily traveled expressway (4), but similar to that of *Spartina* inhabiting a marsh dredge spoil site in Virginia (18). Metal concentration varies with organic matter content and sediment size (18, 26, 63). The lead concentration of the dredge spoil sediment was about a fourth that of the more organic marsh sediments, suggesting that sediment metal content is not correlated to metal content in vegetation. About 98% of the entering lead was retained by both our control and fertilized plots (Fig. 4). The greater absolute loss of lead from the fertilized plots through tidal export of dead grass (17, 65) was due to the increased lead content of the grass and to the greater biomass of grass produced by the addition of nitrogen in the fertilizer (74, 77).

The sludge-fertilized plots lost only 2-3% of the added zinc through exports of dead grass (Fig. 4), although this value may be an underestimate since dead *Spartina* contains higher levels of zinc than sprouts and mature plants (86). About 16% of the zinc was lost through other routes, most likely exports of dissolved zinc compounds.

Retention of cadmium by the fertilized plots was even less than zinc (Fig. 4), with over 66% of the cadmium leaving by losses other than grass export. Cadmium initially adsorbed onto sediments in fresh water can be partially remobilized with increasing salinity (30, 34).

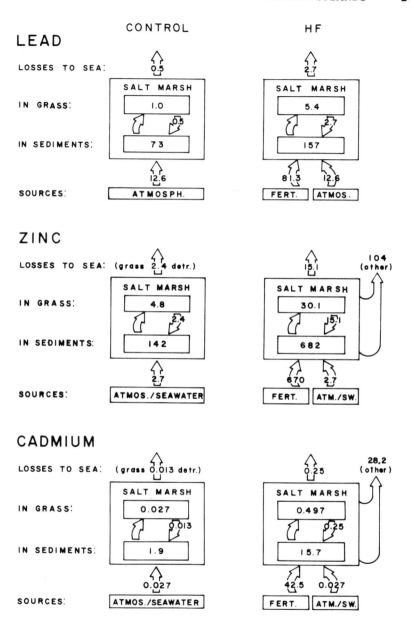

Figure 4. Calculated annual budgets for lead, zinc and cadmium in salt marsh plots under control and sewage sludge fertilization (HF). All values are in mg m⁻². The boxes indicate pools while the arrows show annual fluxes. Adapted from Banus et al. (5).

Shellfish have been suggested to be good indicators of the availability of metals in the water column (31, 51). Clams and oysters placed in the creeks draining the fertilized plots contained significantly higher concentrations of cadmium, but not zinc or lead, than those from control creeks (73). More cadmium than zinc or lead was therefore flushed from the fertilized marsh plots by tidal water.

The detritus feeding crab *Uca pugnax* contained higher concentrations of lead and cadmium in the sludge-fertilized plots (5) than in control areas. Drifmeyer and Odum (18) also reported higher lead levels in detritus feeders from marsh dredge spoil areas than in these animals from natural salt marsh. However, the metal content by dry weight was considerably less for the animals than for the detrital food source, and continued to decrease up the food chain (18.) Baptist and Lewis (6) reported that [65]Zn content also decreased with successive trophic levels. We have not found short or long-term deleterious effects of Pb, Zn or Cd on the animals or plants at the dosages used in our six year Massachusetts studies (37, 73). There is no other information on long-term sublethal effects of heavy metals.

The amount of heavy metals leaving salt marshes and entering food chains is ultimately governed by sediment-water interactions (88). The net flux varies with salinity, temperature, pH, the metal element, and sediment type. Consequently, the ability of marshes to act as sinks for heavy metals will vary with the marsh type and the form of metal-containing sewage it receives.

HYDROCARBON RETENTION AND ITS EFFECTS ON MARSHES

There is little information on the removal and retention of hydrocarbons by marshes. The presence of fine muds held in place by grass roots should provide salt marshes with the ability to remove hydrocarbons from water. Petroleum and modified hydrocarbons such as many pesticides have very low solubilities in water and are readily adsorbed on sediment surfaces. Harvey (27) reported a 10^5-fold concentration of chlorinated hydrocarbons on muds as compared to that dissolved in seawater. Another fate of organic compounds may be uptake and metabolism through marsh plants. *Scirpus*, for example, has been reported to be effective in metabolizing phenols (58).

Our experimental studies on Cape Cod, Massachusetts, yielded circumstantial evidence for the lack of movement of chlorinated hydrocarbons on a salt marsh surface. Fiddler crabs were harmed by chlorinated hydrocarbon pesticides in the sewage applied to marsh plots, but the effect was absent two meters downstream from the experimental sites (37). Use of chlorinated hydrocarbon insecticides is now considerably reduced. However, their replacements, such as organophosphate compounds, are also detrimental to fiddler crabs (81, 82, 83). Since these organophosphorous insecticides are short-lived, there is no information on their retainment by marshes.

Table 2. Concentration (ppb) of polychlorinated biphenyls expressed as Aroclor 1242 in water and mud near a solid waste sanitary landfill area in Saugus, Massachusetts. PCB analysis done by W. G. Steinhauer using the method of Harvey et al. (28) for water and Harvey (27) for sediments. The Pines River provides tidal water for Saugus marsh. Buzzards Bay is an unpolluted body of water. Both these values are included for comparison to the contaminated marsh values.

Inactive landfill area

| | Distance from source of leachate | | | | |
	0m	9m	19m	111m	518m
water	0.34	0.09	0.12	0.03	0.04
mud	1380	725	430	280	290

Active landfill area

| | Distance from source of leachate | | | |
	0m	61m	115m	121m	
water	10.4	0.57	0.42	0.40	*Pines River* 0.40
mud	153	39	78	104	*Buzzards Bay* 2

In a study of a solid waste sanitary landfill in Saugus, Massachusetts, polychlorinated biphenyls in the leachate flowing away from the fill site through a salt marsh were reduced (Table 2). The sediments showed peak concentrations of PCBs near the source of contamination as if much of the PCBs were being adsorbed soon after entering the marsh. The loads of PCBs in the water leaching out of an older inactive landfill area were lower than in an area being actively used for disposal. However, the sediments in the inactive area record much greater contamination than is present so far in the new site.

Hydrocarbons stored in sediments may leach out into overlying waters and can be available for uptake by animals (50). In Wild Harbor, Massachusetts, a salt marsh contaminated by an oil spill, the oil spread to initially uncontaminated sites. This occurred at least partly through release of slightly water-soluble hydrocarbons from the muds (11).

Hydrocarbons may be structurally altered by exposure. For example, a substantial amount of Aldrin added experimentally to marsh plots was microbially epoxidated to Dieldrin (37). Petroleum hydrocarbons in Wild Harbor salt marsh were altered by weathering, leaving residues of complex organic mixtures (7). In both these examples degradation did not lead to detoxification of the contaminants. Although marshes have the potential for effective degradation of hydrocarbons because of high nutrient availability and microbial activity (12), such weathering and degradation also occur in other environments (8) and might not be peculiar to wetlands.

Most oil spills studied have exceeded the tolerance of marsh plants for petroleum contamination and resulted in kills of marsh vegetation, although winter spills were not detrimental to marsh plants (3, 55). However, in most cases recovery of the plants was rapid (15, 16). We observed stands of *Spartina alterniflora* recolonizing oiled areas of Wild Harbor marsh one year after the spill even though the mud was still heavily contaminated with oil. This tolerance of *S. alterniflora* may be very general, since resistance to treatments with an organic arsenic herbicide was marked and rapid regrowth took place (20). *Phragmites* is also resistant to oiling and allows freshwater swamps to retain oil from a refinery effluent in Wales (3a).

FATE OF PATHOGENIC MICROORGANISMS IN WETLAND WATERS

Seidel (58) presents a summary of data showing removal of several species of pathogenic bacteria from water by passage through stands of freshwater macrophytes. Flow though successive stands of three macrophyte species was most effective, although the actual percentage removal or rates of flow of water were not given.

We have measured the number of coliforms in water entering Mill Creek, a small tributary draining *Phragmites communis* marshes on the flood plain of the tidal Hackensack River in New Jersey. Plate counts were obtained using commercially available techniques (Millipore Corp.). The counts of fecal coliform bacteria leaving Mill Creek were substantially lower than the counts in Hackensack River water entering the marsh (Fig. 5). The total coliforms increased again in the ebbing tide, perhaps due to the contribution of the abundant bird life in the reeds. In another set of water samples collected on flooding and ebbing tides in small creeks leading from Mill Creek onto the marsh surface, the fecal and total coliforms in the water were halved by residence over the marsh during tidal inundation (Table 3).

The bactericidal action of seawater has long been known (13, 25, 33, 36, 45, 80). Colberg (14) found that several enteric bacteria *(Escherichia coli, Streptococcus fecalis, Shigella sonnei* and *Salmonella typhimurium)* were reduced in

Table 3.   Mean ± std. error of bacterial counts (cells/100 ml) in flooding and ebbing tides in tidal creeks within Hackensack Meadows, 31 January 1975. Samples collected by C. Mattsen, R. Trattner and W. Kaplan. Numbers in parentheses are the number of creeks involved in each mean.

	Flooding tide	Ebbing tide	t test
Fecal coliforms	667±303 (3)	329±35 (8)	signif. at 0.01
Total coliforms	1820+ ±180 (4)	935±132 (8)	signif. at 0.01

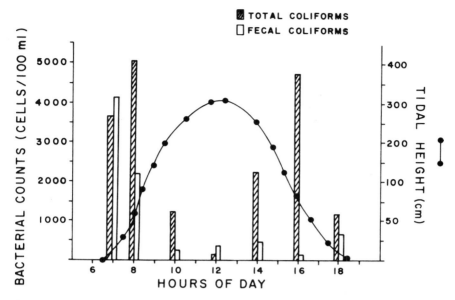

Figure 5. Bacterial counts in water entering and draining a tidal creek in the Hackensack River Marsh.

density by about 4-6 orders of magnitude after about 3-6 days of field incubation in salt marsh waters (salinity about 28-31°/oo). Such a mortality rate could therefore explain the results from Hackensack, without invoking marshes at all. However, the salinity of the Hackensack was only 6-9°/oo, and fresh water has a reduced bactericidal effect compared to sea water (cf. 43). In addition to the elimination of fecal coliform due to exposure to salt water, slow passage over marsh sediments probably adds to the retention of enteric pathogens. This problem requires further study.

We are not aware of studies on the fate of pathogenic viruses transported by contaminated water into marshes.

THE ROLE OF WETLANDS IN ASSIMILATION OF WASTES

Although some freshwater wetlands have been incorporated into waste treatment procedures (35, 58), most studies on the subject are ongoing. The maximum amount of contaminants that can be tolerated by wetlands is unknown. We have added a range of dosages of sewage sludge fertilizer to discover upper tolerances. However, even at our highest dosage (7.6 g N, 4.5 g PO_4, 3 g K_2O m^{-2} wk^{-1}) we have obtained increases in primary production and most secondary producers. The dosages and effects of metals are given in Banus et al. (5) and for chlorinated hydrocarbons in Krebs et al. (37). The amount of chlorinated hydrocarbons in our sludge fertilizer has been reduced recently due to the ban on use of the DDT family of pesticides (44).

Many questions for the practical use of wetlands for waste treatment remain unanswered. It will be necessary to predict what the effect of increased organic loads will be on BOD. Apparently some freshwater marshes can decrease BOD introduced by effluents (24). It will also be useful to know what the effect of adding relatively warm, fresh water will be on seasonally cold or warm, fresh or salt, water wetlands. Further, filling and destruction of salt marshes in the Northeast of the U.S. has reduced the marsh acreage. It is difficult to contemplate devoting much of the remaining area for large-scale waste treatment. South of New Jersey and on the Gulf Coast there are very large areas of marsh remaining, so that plans for using salt marshes in waste-water treatment may be more acceptable and practical.

Wetlands seem to be better processors of wastes than estuaries and coastal waters. It might be feasible to safely dispose of effluents under carefully controlled conditions on marshlands rather than in deeper coastal areas where the elimination of contaminants is not as effective and dispersal of contaminants is more likely. We would like to emphasize, however, that the wetland properties outlined above, and the consequent effects on nutrients, heavy metals, hydrocarbons and pathogens are features of wetlands as they function naturally. They are in fact providing free waste treatment for contaminated waters already. It seems imperative therefore to implement wetland conservation to maintain this subsidy intact.

REFERENCES

1. Aurand, I., and F. C. Daiber. 1973. Nitrate and nitrite in the surface waters of two Delaware salt marshes. Chesapeake Sci. 14:105-111.
2. Axelrad, D. M. 1974. Nutrient flux through the salt marsh ecosystem. Ph.D. Dissertation, College of William and Mary, Williamsburg, Va.
3. Baker, J. M. 1971. Seasonal effects of oil pollution on salt marsh vegetation. Oikos 22:106-110.
3a. _____. 1973. Effects of refinery effluents on the plants of the Crymlyn Bog. Field Studies Council Annual Report. Oil Pollution Research Unit, Orielton Field Centre.
4. Banus, M., I. Valiela and J. M. Teal. 1974. Export of lead from salt marshes. Mar. Pollut. Bull. 5:6-9.
5. _____, _____, and _____. 1975. Lead, zinc and cadmium budgets in experimentally enriched salt marsh ecosystems. Estuarine and Coastal Mar. Sci. 3:421-430.
6. Baptist, J. P., and C. W. Lewis. 1969. Transfer of [65]Zn and ^{51}Cr through an estuarine food chain, p. 420-430. In D. J. Nelson and F. C. Evans (eds.), Proceedings of the Second National Symposium on Radioecology. U.S. AEC Conf. 670503.
7. Blumer, M., and J. Sass. 1972. The West Falmouth oil spill. Data available in November 1971. II. Chemistry. Woods Hole Oceanographic Institution Ref. No. 72-19. Unpublished manuscript.

8. _____, M. Ehrhardt and J. H. Jones. 1973. The environmental fate of stranded crude oil. Deep Sea Res. 20:239-259.
9. Brezonik, P. L., R. G. Bourne, R. L. Nein, Jr. and W. J. Mitsch. 1974. Changes in water quality in the cypress domes, p. 312-375. *In* H. T. Odum, K. C. Ewel, J. W. Ordway, M. K. Johnston and W. J. Mitsch (eds.), Ann. Rept. for 1974, Cypress wetlands for water management, recycling and conservation. Center for Wetlands, Univ. of Florida.
10. Broome, S. W., W. W. Woodhouse and E. D. Seneca. 1975. The relationship of mineral nutrients to growth of *Spartina alterniflora* in North Carolina. II. The effects of N, P and Fe fertilizers. Soil Sci. Soc. of America Proc. 39:301-307.
11. Burns, K. A. 1975. Distribution of hydrocarbons in a salt marsh ecosystem after an oil spill and physiological changes in marsh animals from the polluted environment. Ph.D. Thesis, Massachusetts Inst. of Technology and Woods Hole Oceanographic Institution.
12. _____, and J. M. Teal. 1971. Hydrocarbon incorporation into the salt marsh ecosystem from the West Falmouth oil spill. Woods Hole Oceanographic Institution Ref. No. 71-69. Unpublished manuscript.
13. Carlucci, A. F., and D. Pramer. 1960. An evaluation of factors affecting the survival of *Escherichia coli* in sea water. Appl. Microbiol. 8:243-256.
14. Colberg, P. J. 1975. *In situ* survival studies of enteric bacteria in a salt marsh panne at Great Sippewissett marsh, Massachusetts. Unpublished manuscript, Woods Hole Oceanographic Institution.
15. Cowell, E. B. 1969. The effects of oil pollution on salt marsh communities in Pembrokeshire and Cornwall. J. Appl. Ecol. 6:133-142.
16. _____, and J. M. Baker. 1969. Recovery of a salt marsh in Pembrokeshire, Southwest Wales from pollution by crude oil. Biol. Conser. 1:291-296.
17. Day, T. W., Jr., W. G. Smith, P. R. Wagner and W. C. Stoew. 1973. Community structure and carbon budget of a salt marsh and shallow bay ecosystem in Louisiana. Publ. No. LSU-SG-72-04, Center for Wetlands Resources, Louisiana State University. 80 pp.
18. Drifmeyer, J. E., and W. E. Odum. 1975. Lead, zinc and manganese in dredge-spoil pond ecosystems in Virginia. Environ. Conserv. 2:1-7.
19. Duke, T. W., J. N. Willis and T. J. Price. 1966. Cycling of trace elements in the estuarine environment, I. Movement and distribution of ^{65}Zn and stable Zn in experimental ponds. Chesapeake Sci. 7:1-10.
20. Edwards, A. C., and D. E. Davis. 1975. Effects of an organic arsenical herbicide on a salt marsh ecosystem. J. Env. Qual. 4:215-219.
21. Gardner, L. R. 1975. Runoff from an intertidal marsh during tidal exposure-recession curves and chemical characteristics. Limnol. Oceanogr. 20:81-89.
22. Gaudet, J. J. 1975. Uptake and loss of mineral nutrients by papyrus in tropical swamps. J. Ecology 63:483-491.
23. Goldman, J. C., K. R. Tenore and H. I. Stanley. 1973. Inorganic nitrogen removal from wastewater: effect on phytoplankton growth in coastal marine waters. Science 180:955-956.
24. Grant, R. R., Jr., and R. Patrick. 1970. Tinicum marsh as a water purifier, p. 105-123. *In* Two studies of Tinicum marsh. Conservation Found.
25. Greenberg, A. E. 1956. Survival of enteric organisms in seawater. Publ. Health Rept. 71:77-86.
26. Halcrow, W., D. W. MacKay and I. Thornton. 1973. The distribution of trace metals and fauna in the Firth of Clyde in relation to the disposal of sewage sludge. J. Mar. Biol. Ass. U.K. 53:721-739.

27. Harvey, G. 1975. Biogeochemistry of PCB and DDT compounds in the North Atlantic. Proc. 2nd Int. Symp. Environmental Biogeochemistry. Hamilton, Ontario. In press.

28. _____, W. G. Steinhauer and J. M. Teal. 1972. Polychlorobiphenyls in North Atlantic ocean waters. Science 180:643-644.

29. Heinle, D. R., D. A. Flemer, J. F. Ustach and R. A. Murtagh. 1974. Contributions of tidal wetlands to estuarine food chains. Technical Report No. 29, Water Resources Research Center, Univ. of Maryland. 18 pp.

30. Helz, G. R., R. J. Huggett and J. M. Hill. 1975. Behavior of Mn, Fe, Cu, Zn, Cd and Pb discharged from a wastewater treatment plant into an estuarine environment. Water Res. 9:631-636.

31. Huggett, R. J., M. E. Bender and H. D. Stone. 1973. Utilizing metal concentration relationships in the eastern oyster *(Crassostrea virginica)* to detect heavy metal pollution. Water Res. 7:451-460.

32. Huntjens, J. L. M. 1971. The influence of living plants on mineralization and immobilization of nitrogen. Plant and Soil 35:77-94.

33. Jannasch, H. 1968. Competitive elimination of *Enterobacteriacae* from seawater. Applied Microbiol. 16:1616-1618.

34. Johnson, V., N. Cutshall and C. Osterberg. 1967. Retention of ^{65}Zn by Columbia River sediment. Water Resources Res. 3:99-102.

35. Jong, J. de. 1975. Bulrush and reed ponds: purification of sewage with the aid of ponds with bulrushes or reeds. Paper presented at International Conference on Biological Water Quality Improvement Alternatives. Philadelphia, Pennsylvania.

36. Ketchum, B. H., J. C. Ayers and R. F. Vaccaro. 1952. Processes contributing to the decrease of coliform bacteria in a tidal estuary. Ecology 33:247-258.

37. Krebs, C. T., I. Valiela, G. R. Harvey and J. M. Teal. 1974. Reduction of field populations of fiddler crabs by uptake of chlorinated hydrocarbons. Mar. Pollut. Bull. 5:140-142.

38. Kuenzler, E. J., and A. F. Chestnut. 1971. Structure and functioning of estuarine ecosystems exposed to treated sewage wastes. Ann. Rep. 1970-1971, Univ. of North Carolina. Nat. Tech. Infor. Ser. COM-71-00688, 349 p.

39. Likens, G. E. (ed.). 1972. Nutrients and eutrophication: the limiting nutrient controversy. Amer. Soc. Limnol. Oceanogr. Special Symp. 1:1-328.

40. Lugo, A. 1972. Optimizing the management of the Kissimmee River basin marshes for maximum value to man. *In* The Kissimmee-Okeechobee Basin. A report to the Florida Cabinet. Tallahassee, Florida.

41. Marshall, D. E. 1970. Characteristics of *Spartina* marsh receiving treated municipal sewage wastes. M. S. Thesis, Univ. of North Carolina.

42. Mattson, C. P., R. Trattner, J. M. Teal and I. Valiela. 1975. Nitrogen import and uptake measured on marshes of the Hackensack River in New Jersey. Paper presented at 38th Ann. Meet. Amer. Soc. Limnol. Oceanogr., Halifax, Nova Scotia.

43. McFeters, G. A., and D. Stuart. 1972. Survival of coliform bacteria in natural waters: field and laboratory studies with membrane-filter chambers. Appl. Microbiol. 24:805-811.

44. Meany, R., I. Valiela and J. M. Teal. 1976. Growth, abundance and distribution of larval tabanids in experimentally fertilized plots on a Massachusetts salt marsh. J. Appl. Ecol. In press.

45. Mitchell, R., S. Yankofsky and H. W. Jannasch. 1967. Lysis of *Escherichia coli* by marine microorganisms. Nature 215:891-893.
46. Nedwell, D. B. 1975. Inorganic nitrogen metabolism in a eutrophicated tropical mangrove estuary. Water Res. 9:221-231.
47. Nixon, S. W., and C. A. Oviatt. 1973. Analysis of local variation in the standing crop of *Spartina alterniflora.* Botanica Marina 16:103-109.
48. Odum, E. P., and A. A. de la Cruz. 1967. Particulate organic detritus in a Georgia salt marsh-estuarine ecosystem. p. 383-388. *In* G. Lauff (ed.) Estuaries. Amer. Assoc. Adv. Sci. Publ. No. 83.
49. Odum, H. T., and A. F. Chestnut. 1970. Studies of marine ecosystems developing with treated sewage wastes. Ann. Rep. 1969-1970, Univ. of North Carolina. Nat. Tech. Infor. Ser. PB 199 537. 364 p.
50. Odum, W. E., G. M. Woodwell and C. F. Wurster. 1969. DDT residues absorbed from organic detritus by fiddler crabs. Science 164:576-577.
51. Pentreath, R. J. 1973. The accumulation from water of ^{65}Zn, ^{54}Mn. ^{58}Co and ^{59}Fe by the mussel *Mytilus edulis.* J. Mar. Biol. Ass. U.K. 53:127-143.
52. Pigott, C. D. 1969. Influence of mineral nutrition on the zonation of flowering plants in coastal salt marshes, p. 25-35. *In* I. H. Rorison (ed.), Ecological Aspects of Mineral Nutrition in Plants. 9th Symp. Brit. Ecol. Soc.
53. Pomeroy, L. R., R. E. Johannes, E. P. Odum and B. Roffman. 1969. The phosphorus and zinc cycles and productivity of a salt marsh, p. 412-419. *In* D. J. Nelson and F. C. Evans (eds.), Proceedings of the Second National Symposium on Radioecology. U.S. AEC Conf. 670503.
54. Puckridge, D. W. 1968. Photosynthesis of wheat under field conditions. I. The interaction of photosynthetic organs. Austr. J. Agric. Res., 19:711-719.
55. Ranwell, D. S., and D. Hewett. 1964. Oil pollution in Poole Harbour and its effect on birds. Bird Notes 31:192-197.
56. Reimold, R. J. 1972. The movement of phosphorus through the salt marsh cordgrass, *Spartina alterniflora* Loisel. Limnol. Oceanogr. 17:606-611.
57. Ryther, J. H., and W. M. Dunstan. 1971. Nitrogen, phosphorus and eutrophication in the coastal marine environment. Science 171:1003-1013.
58. Seidel, K. 1971. Macrophytes as functional elements in the environment of man. Hydrobiologia 12:121-130.
59. Siccama, T. G., and E. Porter. 1972. Lead in a Connecticut salt marsh. Bioscience 22:232-234.
60. Spangler, F. L., W. E. Sloey and C. W. Fetter. 1975. Experimental uses of emergent vegetation for the biological treatment of municipal wastewater in Wisconsin. Paper presented at the International Conference on Biological Water Quality Improvement Alternatives, Philadelphia, Penn.
61. Stewart, G. R., J. H. Lee and T. O. Orebamjo. 1972. Nitrogen metabolism in halophytes. I. Nitrate reduction activity in *Suaeda maritima.* New Phytol. 71:263-267.
62. Sullivan, M. J., and F. C. Daiber. 1974. Response of production of cord grass, *Spartina alterniflora*, to inorganic nitrogen and phosphorus fertilizer. Chesapeake Sci. 15:121-123.
63. Swanson, V. E., A. H. Love and I. E. Frost. 1972. Geochemistry and diagenesis of tidal marsh sediment, Northeastern Gulf of Mexico. U.S. Geol. Surv. Bull. 1360.
64. Talling, J. F. 1957. The longitudinal succession of the water characteristics in the White Nile. Hydrobiologia 11:73-89.

65. Teal, J. M. 1962. Energy flow in the salt marsh ecosystem of Georgia. Ecology 43:614-624.
66. _____, and John Kanwisher. 1965. Gas transport in the marsh grass, *Spartina alterniflora*. J. Exp. Bot. 17:355-361.
67. Thompson, K. In press. The ecology of swamps and peatlands in East and Central Africa and their classification for agriculture. Proc. Symp. Classification of Peat and Peatlands, International Peat Soc. Glasgow.
68. _____. 1976. Swamp development in the head-waters of the White Nile. *In* J. Rzoska (ed.), The Nile: Biology Past and Present. Junk Biol. Monogr. 29:177-196.
69. Toth, L. 1972. Reeds control eutrophication of Balaton Lake. Water Res. 6:1533-1539.
70. Tukey, H. B., Jr. 1970. The leaching of substances from plants. Ann. Rev. Plant Physiology 21:305-324.
71. Tusneem, M. E., and W. H. Patrick, Jr. 1971. Nitrogen transformations in waterlogged soils. Louisiana State Univ. Bull. No. 657:1-75.
72. Tyler, G. 1967. On the effects of phosphorus and nitrogen, supplied to Baltic shore meadow vegetation. Botanisk Notiser 120:433-447.
73. Valiela, I., M. D. Banus and J. M. Teal. 1974. Response of salt marsh bivalves to enrichment with metal-containing sewage sludge and retention of lead, zinc and cadmium by marsh sediments. Environ. Pollut. 7:149-157.
74. _____, and J. M. Teal. 1974. Nutrient limitation in salt marsh vegetation, p. 547-563. *In* R. J. Reimold and W. H. Queen (eds.), Ecology of Halophytes. Academic Press, New York and London.
75. _____, _____, and N. Y. Persson. 1976. Production and dynamics of experimentally enriched salt marsh vegetation: below-ground biomass. Limnol. Oceanogr. 21:245-252.
76. _____, _____, and W. Sass. 1973. Nutrient retention in salt marsh plots experimentally fertilized with sewage sludge. Estuarine and Coastal Mar. Sci. 1:261-269.
77. _____, _____, and _____. 1975. Production and dynamics of salt marsh vegetation and the effect of experimental treatment with sewage sludge. I. Biomass, production and species composition. J. Applied Ecology 12:973-981.
78. Van Raalte, C. D., I. Valiela, E. J. Carpenter and J. M. Teal. 1974. Inhibition of nitrogen fixation in salt marshes measured by acetylene reduction. Estuarine and Coastal Mar. Sci. 2:301-305.
79. Vince, S., and I. Valiela. 1973. The effects of ammonium and phosphate enrichments on chlorophyll *a*, pigment ratio and species composition of phytoplankton of Vineyard Sound. Mar. Biol. 19:69-75.
80. Waksman, S. A., and M. Hotchkiss. 1937. Viability of bacteria in seawater. J. Bact. 33:389-400.
81. Ward, D. V. and D. A. Busch. In press. Effect of Temefos, an organophosphorous insecticide, on survival and escape behavior of the marsh fiddler crab, *Uca pugnax*. Oikos.
82. _____, and B. L. Howes. 1974. The effects of Abate, an organophosphorous insecticide, on marsh fiddler crab populations. Bull. Environ. Contamination and Toxicology 12:694-697.
83. _____, _____, and D. F. Ludwig. In press. Interactive effects of predation pressure and insecticide (Temefos) toxicity on populations of the marsh fiddler crab *Uca pugnax*. Mar. Biol.

84. Westerman, R. L., and L. T. Kurtz. 1974. Isotopic and nonisotopic estimations of fertilizer nitrogen uptake by Sudangrass in field experiments. Soil Sci. Soc. Amer. Proc. 38:107-155.

85. Whitney, D. E., G. M. Woodwell and C. A. S. Hall. 1975. The Flax pond ecosystem study: exchanges in inorganic forms of nitrogen between a Long Island salt marsh and Long Island Sound. Paper presented at 38th Ann. Meet. Amer. Soc. Limnol. Oceanogr., Halifax, Nova Scotia.

86. Williams, R. B., and M. B. Murdoch. 1969. The potential importance of *Spartina alterniflora* in conveying zinc, manganese and iron into estuarine food chains, p. 431-439. *In* D. J. Nelson and F. C. Evans (eds.). Proceedings of the Second National Symposium on Radioecology. U.S. AEC Conf. 670503.

87. _____, and _____. 1972. Compartmental analysis of the production of *Juncus roemerianus* in a North Carolina salt marsh. Chesapeake Sci. 13:69-79.

88. Wolfe, D. A. and T. R. Rice. 1972. Cycling of elements in estuaries. Fishery Bull. 70:959-972.

CRAFT INDUSTRIES FROM COASTAL WETLAND VEGETATION

Julia F. Morton, D. Sc., F. L. S.
Director, Morton Collectanea
University of Miami
Coral Gables, Florida 33124

ABSTRACT: Inasmuch as water travel since earliest times has been the easiest mode of movement and transport, useful coastal plants have been prized, some declared sacred, their use regulated, and wanton destruction forbidden. Examples are the portia tree, *Thespesia populnea*, and the related mahoe, *Hibiscus tiliaceus*, both pantropic and both yielding wood and inner bark. Mangroves, particularly *Rhizophora* and *Avicennia* species, provide nearly indestructible wood, flexible branches for fish traps, tannin and dyes. The associated leather fern, *Acrostichum aureum*, furnishes edible fiddleheads and tough fronds for thatching. In the East Indies and Polynesia, *Pandanus tectorius* is an indispensable source of leaf products. Much building material in the form of woven matting is supplied by palms: the nypa (*Nypa fruticans*) over vast saline swamps of Oceania; in Malaysian freshwater swamps, the sago palms, *Metroxylon*; in tidal swamps of tropical America, the timiche, *Manicaria saccifera*. The moriche, *Mauritia flexuosa*, is the main fiber provider in the Orinoco delta. Cattails, *Typha* spp., are utilized on a worldwide scale; *T. domingensis* in saline swamps of the New World and the East Indies. Among sedges and grasses, *Scirpus, Fimbristylis, Spartina* and the universal reed, *Phragmites communis,* have served widely for hut-building, thatching, screens, matting, baskets, brooms and cordage.

INTRODUCTION

Seafarers, coastal dwellers, river voyagers and settlers on borders of lakes and marshes have ready at hand a plenitude of water-associated plants from which to fashion an array of articles to meet their needs. Inasmuch as water travel has been since earliest times the easiest mode of movement and of transport for trade and other purposes, the accessibility of useful coastal plants rendered them of prime importance, some having been considered so vital to survival that they were declared sacred, their use regulated, and destruction forbidden by tribal decrees.

254

Wetland plants are endangered today because waterfront sites are preferred for development and swamps are viewed by developers as wastelands to be dredged or filled. Mangrove swamps are eagerly destroyed yet these are essential to the coastal ecology and contain a vast wealth of plant life. Mangroves and other wetland trees head the following catalog of economic plants of coastal wetlands.

TREES

Avicenniaceae

Avicennia germinans Stearn (*A. nitida* Jacq.), Black Mangrove, inhabits salt marshes and tidal swamps of tropical America and West Tropical Africa. The bark has been used for tanning and dyeing. The wood is hard, heavy, tough, but coarse-grained and tends to split. It is used for fuel, charcoal, fenceposts, piles, boats, flooring, sills, drains and gunstocks (4, 10).

Avicennia marina Vierh. (often erroneously recorded as *A. officinalis*) abounds on the coasts of East Africa, southeast Asia, and from Timor to Australia, New Zealand and Polynesia. Slender branchlets are used as webbing in fish traps (29). The trunks are used for small dugouts, the wood for boat fittings, masts, bedsteads, drums, mallets, rice mortars, chairs, spoons, spade handles; also as fuel for smoking fish and for lime-burning (31). The bark yields a brown dye (9). Ashes of the wood are used in soap-making.

Avicennia officinalis L. occurs on the muddy shores of India, Burma, Ceylon, and from southern China to Formosa and New Guinea. The wood is used for small cabinetwork. Ashes of the wood are mixed with paint to make it adhere well (2).

Combretaceae

Conocarpus erecta (or *erectus*) L. Buttonwood; Buttonwood Mangrove. A common tree of mangrove swamps and rocky shores of tropical America and western Africa from Senegal to the Belgian Congo. It is naturalized in Hawaii. The tree is a source of curiously formed and prized driftwood. "U"-shaped branches are sought for ribs of boats. The wood is hard, heavy, durable in salt water and underground and much used in boat-building and for piles. It has a high calorific value, burns without smoke and has been extensively used as fuel and for making high-grade charcoal (10, 13, 15, 30). The bark is used for tanning hides (17, 26).

Laguncularia racemosa Gaertn. f. White Mangrove. This tree forms thickets in shallow tidal waters or on muddy banks of bays and lagoons from Mexico and southern Florida to Peru and Brazil; also in the Bahamas, the West Indies and along the west coast of Africa. The wood, hard, heavy and durable, is used for fence posts and house-frames. In Cuba, slender branches are soaked in salt water, then twisted into "ropes" on which tobacco leaves are hung to dry (23). The bark is high in tannin and used for curing leather (26).

Lumnitzera littorea Voigt. (*L. coccinea* Wight). Red Teruntum. A familiar tree of tidal swamps and adjacent, occasionally inundated ground on the east coast of Africa and in southeast Asia, throughout Malaysia and the Pacific Islands. The wood, rose-scented when fresh, is fine-grained and heavy, sinks in water. It is in demand for saltwater piling (27)—which has been found intact after 50 years—and bridges, boat-building, flooring, cart axles and tool handles. A red dye is obtained from the bark (7).

Lumnitzera racemosa Willd., White Teruntum, ranging from East Africa to India, Malaysia, northern Australia and Polynesia, is less common than *L. littorea*, smaller, not exceeding 10 m. in height. Because of its small diameter, the wood has few uses apart from serving as house posts and as fuel (7, 9).

Rhizophoraceae

Bruguiera gymnorrhiza Lamarck (*B. conjugata* Merr.). Many-petaled Mangrove. An abundant tree in the depths (18) and on the landward margin of tidal swamps from East Africa and tropical Asia through Malaysia to northern Australia, Japan and Polynesia. The heartwood is dark-red, very hard, heavy, durable and is valued for furniture, posts, piles, rafters and planks (7). The bark is high in tannin and employed in tanning and dyeing (24). The showy red calyces are strung for leis in Hawaii and last 3 weeks if refrigerated (21).

Bruguiera parviflora W. & A. abounds in the interior of swamps on the coasts of India, Malaya and the Philippines and is a pioneer invading overcut areas and holding soil for *Rhizophora* species which follow (7). Its wood is employed for fish stakes and mine timbers.

Bruguiera cylindrica Blume (*B. caryophylloides* Blume) is a less common species of the same regions as *B. parviflora* but favoring higher ground. Its wood has a peculiar odor which repels fish and is not used for fish traps. It is not durable in the ground and its use is limited to the upper parts of structures. The bark has a strong turnip odor and is of no value for tanning (7).

Bruguiera sexangula Poir. (*B. eriopetala* W. & A.) is native to the coasts of India and southern China, is found in most mangrove swamps of the Philippines (5) and is naturalized in Hawaii (13). The wood serves as fuel and for poles.

Ceriops tagal C. B. Rob (*C. candolleana* Arn.). Spurred Mangrove. This mangrove is plentiful in coastal swamps and on muddy shores from East Africa to Queensland. Fishermen make much use of it. Young straight shoots form the webbing of fish traps (29). The wood is a superior fuel, is valued for the knees of boats, for house posts and mine props (11). All parts of the plant are very rich in tannin. The leaves as well as the dark-red bark are much employed in India for curing leather (2). Dye from the bark is used for coloring and preserving fishing nets and is important batik- and mat-making dye in Java (7).

Ceriops roxburghiana Arn. is a less common species in mangrove swamps from India to the Philippines. In stature it is no more than a shrub or small tree. When of sufficient size, it serves the same purposes as *C. tagal.*

Rhizophora mangle L. Red Mangrove; American Mangrove. This much-studied and publicized species fringes shores and estuaries and forms offshore islands over a vast area extending from Baja California and southern Florida t‹) southern Ecuador and northern Brazil, also Bermuda, the Bahamas and West Indies. Introduced into the Pacific in 1902, it has become naturalized from Hawaii to Tonga. Its land-building function is well recognized. Its wood is very hard and heavy, sinks in water, has been much used in South America for boat building and construction; was promoted as material for textile shuttles but found too brittle. In the Bahamas, golf club heads were made from it after long seasoning in salt water. The mangrove stands of Puerto Rico, under the control of the Insular Forest Service, are a major source of high-grade charcoal. The bark has been considerably exploited as tanning material. In tropical America, a bark extract serves as stain for floors and furniture, the root extract as a preservative for fishing nets, and the juice of the hypocotyl is applied to fishing lines. Young shoots and roots give a red dye for fabrics and leather in Central American handicrafts (20).

Rhizophora mucronata Lam (*R. candelaria* W. & A.) is the commonest element of coastal swamps and estuaries from East Africa to Australia. Its features are similar to those of *R. mangle*. It is an important source of wood for house posts and framing and supplies the preferred material for fence-type fish traps, masts for canoes and small dhows (29). Easy to split and of high calorific value, the wood is prized for fuel and for charcoal. Aerial roots are made into bows in Fiji. Split stems are fashioned into baskets for sifting grain (31). The bark is used locally to preserve fishing lines and nets and is commercially harvested for tanning leather. The red sap serves as a hair dye and is an ingredient in paints for pottery (19).

Rhizophora racemosa G. F. Mey. is being applied to the Red Mangrove of west Tropical Africa where it is essential to the way of life of coastal peoples, having all the uses of the two preceding species (10, 28).

Sonneratiaceae

Sonneratia acida L. f. (*S. caseolaris* Engl.). Crabapple Mangrove. This species abounds in the front ranks of mangrove swamps and estuaries from East Africa to the Philippines. The root system sends up dense masses of conical pneumatophores to 1.5 m. tall. These are cork-like in texture and highly useful—for lining insect-cases, for making sandals, shoe soles, innersoles, floats for fishing nets and also corks. The wood of the tree is moderately hard and heavy, useful for piles, posts, house- and ship-building, furniture and musical instruments. Because of its salt content, it must be fastened with copper nails and screws (5).

Sonneratia alba Sm. of East Africa, often forms islands offshore. It is much like *S. acida* but may attain 15 m. and the pneumatophores may be 25 cm. thick at the base. They are used in making floats and the wood in carpentry (9). Branches are often fashioned into boat ribs (29).

Bignoniaceae

Dolichandrone spathacea K. Schum. The Mangrove Trumpet Tree dwells in tidal, swampy land adjoining the mangroves and extends into coastal rice fields from southern India through Malaysia to New Caledonia. The soft, white wood has been made into saddles, wooden shoes and floats for fishing nets. A coarse, dark fiber is extracted from the bark (7).

Enallagma latifolia Small (*E. cucurbitina* Baill.). Black Calabash. This tree occurs in tidal swamps and on borders of estuaries from southern Mexico to Colombia and Venezuela, also on some islands of the West Indies and on the Florida Keys. The wood is fairly hard and heavy and has been used to make tool handles, ox-yokes and plows (23).

Euphorbiaceae

Excoecaria agallocha L. Milky Mangrove; Blind-your-eyes. A common tree of tidal forests and banks of estuaries from India to Australia and Polynesia. It possesses a copious milky sap, exceedingly irritant and toxic, which has been used as a fish poison and for criminal purposes. To avoid the sap, woodcutters remove the bark before felling the trees. The lightweight wood is much employed in carpentry—for making tables, bedsteads (7), packing cases, clogs, toys and matches (6). Fishing floats are fashioned from the roots (24). The wood is also burned for charcoal but the smoke inflames the eyes of the burners (7).

Leguminosae

Mora oleifera Ducke (*M. megistosperma* Britt. & Rose). Nato. Abounding in tidal estuaries of northern Ecuador, the Pacific Coast of Colombia, Panama and Golfo Dulce, Costa Rica, this is a giant tree to 50 m. with buttresses up to 1.5 m. high. The wood, more durable than oak (25), is locally made into furniture. It was formerly exported for shipbuilding. Crossties made in Colombia were used in Peru for the first railways and streetcar tracks (23).

Pterocarpus officinalis Jacq. Bloodwood. Favoring tidal swamps and often forming nearly pure stands just back of the mangroves, this tree ranges from southern Mexico through coastal Central America and northern South America; also in Jamaica and Puerto Rico. The whitish wood, varying from soft and light to fairly hard, is not durable but used for interior construction and articles for indoor use (23).

Lythraceae

Pemphis acidula Forst. Ironwood. This interesting species is found in salt marshes, on flats and coral outcrops washed by high tides, from East Africa to Malaya and the Gilbert Islands. Attaining from 5 to 10 m., it has gray-fuzzy foliage and down-covered branches which are strong, flexible, springy and prized for making eel traps, shark hooks and for fashioning a ring on which to thread coconut shells to splash in the ocean—a method of "calling" sharks (18). The wood is so strong that it is used instead of a piece of shell for a point on a native drill. The trunks are used as posts to support huts and platforms for the dead. A portion of trunk serves as an anvil for splitting coconuts (18). Fence stakes, anchors (7) and pandanus-leaf pounders are made from the wood and it is also esteemed for fuel, making a very hot fire. Tobacco pipes are carved from the roots. The bark yields a red dye for canoe sails (18).

Malvaceae

Hibiscus tiliaceus L., Mahoe, flourishes on muddy or rocky shores behind the mangroves and on the borders of tidal streams throughout the tropics. In Polynesia, it has been revered and protected as a sacred tree because of the utility of its inner bark (bast), highly valued for hula skirts, coarse tapa cloth, waistbands (19), fish traps, strong and heavy rope, cord, fishing and harpoon lines, nets, hammocks, mats, bags (1, 7) and tow for caulking boats (8). The wood is buoyant and useful for floats, small boats, outriggers for canoes; and it also ignites readily from friction (12).

Thespesia populnea Soland. Portia Tree. This tree occurs on the inland edge of mangrove swamps and in sandy scrub around the tropical world. Fiber from the bark is second to that of the mahoe for cord but is tough and much used for rope, bags and caulking. The wood is prized. It is strong, durable under water, fine-grained, often called "Polynesian rosewood", popular for boat-building, furniture, clubs, paddles (17), gunstocks and beautiful bowls (12).

Meliaceae

Xylocarpus moluccensis Roem. The Puzzle-nut Mangrove is found in mangrove swamps, on borders of tidal streams, and on open beaches from Kenya to the Philippines and many islands of the Pacific. The wood is dense, heavy, reddish-brown, often with black stripes; valued for dhow masts (9), furniture, flooring, spokes for wheels (24) and tool handles (27).

Potaliaceae

Fagraea crenulata Maing. The Cabbage Tree inhabits swamps, behind the mangroves, around Sumatra, the Malay Peninsula and Borneo. It reaches over 20 m. in height and owes its name to its large, fleshy leaves. Its wood is hard, heavy, coarse-grained; has been much used for piling (7, 8).

Sterculiaceae

Heritiera littoralis Dry. The Looking-glass Tree frequents the higher levels of mangrove swamps and banks of tidal streams from Kenya to Australia and Guam. Strong, heavy and durable, the wood is valued for masts, canoes, outriggers, boat ribs, piling, house posts, rafters, plows and tool handles (5, 7, 24). Trunks were formerly used in stockades. Twigs are frayed and employed as toothbrushes (7).

SHRUBS AND VINES

Acanthaceae

Acanthus ilicifolius L. The Sea Holly flourishes among the mangroves in brackish swamps and along tidal streams from India and Burma through Malaysia to Polynesia. The chopped and crushed plant is fed to cattle in India (7). The shrub is also burned and lye derived from the ash is used in soapmaking (5).

Leguminosae

Derris trifoliata Lour. (*D. uliginosa* Benth). This climbing shrub or vine abounds in mangrove forests and along tidal creeks and muddy shores from East Africa to India and through Malaysia to Polynesia. The tough stems are used for tying logs to boats. The leaves are used as fodder in India. In Australia, they are pounded and put into shallow water to stupefy fish (19).

Myrsinaceae

Aegiceras corniculatum Blanco (*A. majus* Gaertn.). The Goat's Horn, or River, Mangrove is plentiful on the inshore margin of mangrove swamps and on the banks of tidal rivers in full light, from India and Ceylon to New Guinea and tropical Australia. The wood, varying from white to reddish or brown and from soft to fairly hard, is used to build huts and for fuel (19).

Rubiaceae

Scyphiphora hydrophyllacea Gaertn. f. Chengam. This shrub is very common in mangrove swamps and on muddy coasts from India and Ceylon through Malaysia to Australia and New Caledonia. The wood is dark-brown and very hard. It is prized for making small articles, especially rice ladles (16) and yam-sticks (for digging yams and other tubers and bulbs) (3).

SCREW-PINE

Pandanaceae

Pandanus tectorius Park. (*P. odoratissimus* L. f.). The Textile Screwpine is common in tidal forests of southeast Aisa, Malaysia and Polynesia. The leaves are of the utmost importance for thatching, for making excellent mats, baskets and hats (12) and sacks for coffee, sugar and grain (19). Discarded leaves are used as torches and for caulking (18). In the South Pacific, leaves are split, rolled into lengths of up to 1.2 m., cut and sold as cigarette paper. The people are now being taught to use this material as wallpaper and covering for lampshades.

Fibers from the aerial roots are much used for paint brushes and for binding baskets (19). Trunks and branches serve as house posts and flooring (18).

PALMS

Nypaceae

Nypa (or *Nipa*) *fruticans* Wurmb. The Nypa Palm occurs naturally in great abundance in brackish swamps and along tidal streams from the Ganges of India to Australia. Introduced into Guam, it has become thoroughly naturalized in all brackish waters. The palm is of great economic value; one of the most important palms of the Philippines. The leaflets, stripped from the midrib, are extensively used for thatching and walls, also for making mats, baskets, bags, hats, raincoats (5) and cigarette wrappers (7). At least two leaves are often left to assure survival of the palm. Young leaves serve as food wrapping. The tough petioles are burned for fuel (5) or are chopped and boiled to obtain salt (6).

Palmae

Cocos nucifera L. The familiar Coconut Palm graces sandy beaches and is cultivated far inland all around the tropical world, but it also occurs in low, marshy areas where it receives brief flooding by seawater. The food uses of the nut, unripe and ripe, and as a source of oil, and the tapping of the inflorescence for toddy and sugar, are well-known. In addition, the leaves are of immense value for thatching, walls and screens. Leaflets are woven into baskets, plates, hats, mats and other articles for daily use. Tough petiole bases are burned as fuel. Coconut husks, cut in half lengthwise and supplied with thongs, serve as crude sandals. Fibers (coir) extracted mainly from the retted green husk, are important material for brushes, mats, ropes and stuffing for upholstery and mattresses and for caulking boats (2). Hand-braided coir makes very attractive twine (sennit). Currently, handicrafts are being actively encouraged in the Pacific Islands and the use of coconut wood promoted. Old palms that are no longer bearing are expendable and, in 1969, it was reported that reforestation programs would produce many tons of coconut wood in the Gilbert and Ellice Islands. Properly dried, with its pores filled with epoxy resin, the outer wood of the trunk is durable and usable for bowls, trays, platters and tobacco boxes, with scraps being carved into letter openers and novelty jewelry. Coconut wood mallets are used for pounding pandanus leaves. Coconut shells serve as dishes and are made into all kinds of novelties. Leaf sheaths serve as coconut cloth (7).

Corozo oleifera Bailey (*Elaeis melanocarpa* Gaertn.). The Corozo, or American Oil Palm, is native to coastal-swamp forests and marshy clearings from the lower basin of the Amazon in Brazil to southern Mexico. From the fruits (both pulp and seed) the local people derive oil, tallow and feed for chickens. Fibers from the opened spathe and leaves are used for cordage. The fermented and cleaned inflorescences are used as whisks to drive away mosquitoes (22).

Licuala spinosa Wurmb. Spiny Licuala. This palm inhabits tidal forests, sometimes immediately back of the mangroves, from the Malay Peninsula to the Andaman Islands. It is multi-stemmed, to 3 m. high, with nearly circular, deeply-divided leaves which are used for roofing in Malacca and Singapore (7), and for food wrapping in Cambodia.

Manicaria saccifera Gaertn. Monkey-cap Palm. Large colonies of this palm occur in the tidal swamps from northern Brazil to Venezuela, the lower West Indies and Colombia, and throughout Central America. The durable leaves are commonly employed for roofing and as covering for Indian boats. The brown, fibrous, conical inner spathe (which, after the splitting of the outer spathe, still encloses the inflorescence and the fully formed fruits) is worn as a loincloth or cap, or sold intact as a curiosity. It if often modified to serve as a bag, or opened up to provide "cloth" from which mats and numerous other articles are made for domestic use and sale (14).

Oncosperma filimentosa Blume (*O. tigillaria* Ridl.). The Nibung Palm flourishes in saline or brackish lowlands just back of mangrove stands and in nearby ravines from India and Ceylon to the Philippines. The trunk provides hard, durable wood for bridges, piles, house posts, flooring, roofing and walls, furniture, lances, and frames for fish traps (6, 7). The natives also make from it sharpened spears which are planted at angles pointed toward the garden so as to impale marauding wild pigs when they are chased out of the patches. In Indonesia, the natives when felling leave one stem in each clump to assure regeneration of suckers (6). The spines are used as darts in blow-pipes and as tips on fish spears. Floral spathes serve as kettles for boiling water and as buckets. Baskets are made from the leaves (13).

Raphia taedigera Mart. Jupati; Pine-cone Palm. This palm inhabits marshes and inundated regions of the lower Amazon and northward to Costa Rica. It is a multiple-stemmed palm with short trunks which furnish wood for the walls of native dwellings (25). The cylindrical petioles (to 4 m. long) are used to construct partitions (14). In Panama, the United Fruit Company has used them by the thousands for banana props (1). Fiber from the sheaths at the base of the petioles is used for fishing nets and cordage. Some has been exported to England for tying up hop vines (14).

Raphia vinifera P. Beauv. The Bamboo Palm [not the Wine Palm, now identified as *R. hookeri* Mann & Wendl.] grows in great numbers in tidal bays and creeks of West Tropical Africa. Fiber from the leaf bases has always been used locally to make fishing lines (14), snares for game and cordage (2). Since 1890 it has been harvested and exported (14) as West African Piassava for the manufacture of brooms, and industrial roller brushes. The waste is used as cheap stuffing for upholstery (2). The leaflets are used for thatching and are woven into baskets and hats (14). The tough petioles and midribs serve as "bamboo" for roofing poles, carrying poles and for making canoes, small furniture and arrow shafts.

FERNS

Blechnaceae

Stenochlaena palustris Bedd. Akar Paku; Jagnaya. This fern is plentiful on the inner borders of mangrove swamps and inland around springs from the coasts of southeast Asia to Polynesia. The stems (rhizomes), being durable in salt water, are collected in quantity, dried and marketed in bundles in the Philippines. They are used mainly for tying the framework of fish traps. Occasionally they are fashioned into inferior rope or baskets (5).

Pteridaceae

Acrostichum aureum L. The Coast Leather Fern is abundant in mangrove swamps, along tidal streams and on mud flats along all tropical coasts. In Malaya and the Pacific islands, the fronds are dried and strung on rods to make "ataps" for thatching (7). In West Africa, they are massed upright along fence-like framework placed across streams to catch fish (28).

HERBS

Among plants which have been burned to produce sodium carbonate (barilla) for making soap and glass are:

Aizoaceae

Sesuvium portulacastrum L. Seaside Purslane. This plant is common in mangrove thickets and on muddy flats near the sea throughout the tropics of both hemispheres.

Batidaceae

Batis maritima L. Saltwort occurs in great abundance among mangroves and on estuarine flats of the New World tropics and has been introduced and become naturalized in Hawaii (13).

Chenopodiaceae

Salicornia stricta Dum. (*S. herbacea* auct.), Marsh Samphire, or Glasswort, frequents salt marshes on both sides of the North Atlantic, the Mediterranean, the shores of South Africa and temperate Asia. This and similar glassworts perform essential mud-binding service throughout the coastal regions of the world.

Zosteraceae

Zostera marina L. Eelgrass. This plant grows on the floor of shallow bays, saline pools and brackish streams around the Atlantic and Pacific coasts of North

America, the British Isles, Europe and Asia and the same or a related species in Oceania. Indians use the dried eelgrass for lining baskets, for thatch and for stuffing dolls and other toys. The plant, either harvested or washed ashore, has been widely employed as packing, insulation, filler for upholstery and mattresses and as fertilizer (25).

SEDGES, RUSHES AND REEDS
Cyperaceae

Cyperus malaccensis Lam. Balangot. This is a common plant in tidal mud bordering mangrove and nypa swamps throughout southern Asia, the East Indies and northern Australia. The whole or split stems are used for matting and sometimes for hats. In the Philippines, the making of slippers from this sedge has been a cottage industry since about 1908 (5).

Fimbristylis spadicea Vahl. Esparto. An inhabitant of saline flats and marshes, this sedge is found from Mexico to northern South America and the West Indies. Paper has been made from the stems in southern Mexico (14).

Fimbristylis castanea Vahl. (*F. spadicea* Vahl var. *castanea* Gray) is found from Long Island, N.Y., to Texas; also in Bermuda, the Bahamas and Cuba. It has been harvested with *Spartina* for broom-making and is said to be better than *Spartina* for this purpose.

Juncaceae

Juncus maritimus Lam. The Sea Rush is abundant in salt marshes from the British Isles to South Africa and Australia and in some inland saline regions. It is a very tough plant. The stems are important mat-making material in Spain and Morocco (14).

Typhaceae

While most *Typha* species (Cattails) inhabit freshwater areas, the Southern Cattail, *T. domingensis* Pers. (often erroneously recorded as *T. angustifolia* L.) is more often found in brackish water than in fresh. Its coastal range includes the eastern United States from Maine to Florida, the West Indies, tropical America and Oceania. Its leaves are more leathery than those of other species and the mats, baskets, seats and backs for chairs, and squares for floor-covering, made from these leaves are exceptionally durable. One chair seen on a farm in Venezuela has been in use for 25 years and is still intact.

GRASSES
Gramineae

Spartina pectinata Link, Slough Grass, grows in coastal marshland and on sandy shores, as well as inland in wet prairies from Newfoundland to Texas. *S. spartinae* Merr. flourishes in marshes along the Gulf Coast from Florida to

Mexico. These grasses grow in large tufts to 2 m. high. These and other species of *Spartina* have been commercially harvested in Florida and Louisiana (and *S. pectinata* in Missouri) as substitutes for broomcorn. Often the center (50%) of a broom is *Spartina* and the outer half is broomcorn.

Phragmites communis Trin., the giant Reed Grass, universally distributed in freshwater sites, also occurs in some coastal marshes. It has been extensively used for matting, basketry and thatching and formed the primitive huts of eastern Europe, one of which stands on the shore of the great salt lake, Neusiedler See, in southeastern Austria.

REFERENCES

1. Allen, P. H. 1956. The Rain Forests of Golfo Dulce. Univ. of Fla. Press, Gainesville, Fla. 417 p.
2. Anonymous. 1948; 1950; 1969; 1972. The Wealth of India: Raw Materials. Vol. I, 253 p.; Vol. II, 427 p.; Vol. VIII, 394 p.; Vol. IX, 472 p. Coun. Sci. & Indus. Res., Delhi, India.
3. Bailey, F.M. 1900. The Queensland Flora. Pt. III. The Queensland Gov't, Brisbane, Aust. p. 739-1030.
4. Britton, N.L., and P. Wilson, 1925-30. Botany of Puerto Rico and the Virgin Islands. (Sci. Surv. Puerto Rico & Virgin Isls. 6, Pts. 1-4). New York Acad. Sci., New York. 663 p.
5. Brown, W. H. 1951. Useful Plants of the Philippines. Vol. I. Tech. Bull. 10. Dept. Agr. & Nat. Res., Manila. 590 p.
6. Browne, F. G. 1955. Forest Trees of Sarawak and Brunei. Gov't Ptg. Off., Kuching, Sarawak. 369 p.
7. Burkill, I. H. 1935. Dictionary of the Economic Products of the Malay Peninsula. 2 vols. Crown Agents for the Colonies, London. 2402 p.
8. Corner, E.J.H. 1952. Wayside Trees of Malaya. 2 vols. Gov't Ptg. Off., Singapore. 772 p.
9. Dale, I.R., and P.J. Greenway. 1961. Kenya Trees and Shrubs. Buchanan's Kenya Estates, Nairobi. 654 p.
10. Dalziel, J.M. 1948. Useful Plants of West Tropical Africa. Crown Agents for the Colonies, London. 612 p.
11. Dastur, J.F. 1951. Useful Plants of India and Pakistan. 2nd ed. D.B. Taraporevala Sons, Ltd., Bombay. 260 p.
12. Degener, O. 1945. Plants of Hawaii National Park: Illustrative of Plants and Customs of the South Seas. Author. New York Bot. Gard., Bronx Park, N.Y. 314 p.
13. _____. 1946. Flora Hawaiiensis. (Books 1-4). Author, New York. 1192 p.
14. Dodge, C.R. 1897. Descriptive Catalogue of Useful Fiber Plants of the World. Rpt. #9. U.S. Dept. Agr., Off. Fiber Invest., Washington, D.C. 361 p.
15. Harrar, E.S., and J.G. Harrar. 1946. Guide to Southern Trees. McGraw Hill Book Co., New York. 712 p.
16. Heyne, K. 1950 De Nuttige Planten van Indonesië. Vol. I. W. van Hoeve, 's-Gravenhage/Bandung. 1450 p.
17. Irvine, F.R. 1961. Woody Plants of Ghana. Oxford Univ. Press, London. 868 p.

18. Luomala, K. 1953. Ethnobotany of the Gilbert Islands. Bull. 213. Bernice P. Bishop Mus., Honolulu. 129 p.
19. Maiden, J.H. 1889. Useful Native Plants of Australia. Technological Mus. New So. Wales, Sydney. 696 p.
20. Morton, J.F. 1965. Can the Red Mangrove Provide Food, Feed and Fertilizer? Econ. Bot. 19 (2): 113-123.
21. Neal, M.C. 1965. In Gardens of Hawaii. Spec. Pub. 50. Bernice P. Bishop Mus., Honolulu. 924 p.
22. Perez-Arbelaez, E. 1956. Plantas Utiles de Colombia. Libreria Colombiana, Camacho Roldan (Cia. Ltda.), Bogota. 831 p.
23. Record, S.J., and R.W. Hess. 1948. Timbers of the New World. Yale Univ. Press, New Haven. 640 p.
24. Safford, W.E. 1905. Useful Plants of the Island of Guam. Contrib. U.S. Nat'l Herb., Vol. 9. Smithsonian Inst., Washington, D.C. 416 p.
25. Smith, J. 1882. A Dictionary of Popular Names of the Plants Which Furnish the Natural and Acquired Wants of Man, in all Matters of Domestic and General Economy, their History, Products and Uses. Macmillan & Co., London. 457 p.
26. Standley, P.C., and L.O. Williams. 1962. Flora of Guatemala. Fieldiana. Botany 24, Pt. 7, #2. Chicago Nat. Hist. Mus., Chicago. p. 237-418.
27. Stone, B.C. 1970-71. The Flora of Guam. Micronesica, Vol. 6. (complete). 659 p.
28. Walker, A.R. and R. Sillans. 1961. Les Plantes Utiles du Gabon. (Encyc. Biol. LVI). Editions Paul LeChevalier, Paris. 612 p.
29. Weiss, E.A. 1973. Some Indigenous Trees and Shrubs Used by Local Fishermen on the East African Coast. Econ. Bot. 27 (2): 174-192.
30. West, E., and L.E. Arnold. 1946. The Native Trees of Florida. Univ. of Fla. Press, Gainesville. 212 p.
31. Williams, R.O. 1949. Useful and Ornamental Plants in Zanzibar and Pemba. Zanzibar Protectorate, Zanzibar. 497 p.

PHARMACOLOGICAL POTENTIAL OF MARSH PLANTS

D. Howard Miles, Department of Chemistry
Armando A. de la Cruz, Department of Zoology
Mississippi State University
Mississippi State, MS 39762

ABSTRACT: The presence of benzyl cyanide in the volatile constituents of *Juncus roemerianus* prompted us to initiate analyses of alkaloids in three species of vascular plants that are most abundant in the tidal marshes of Mississippi. An unexpected high amount (2.0 g/kg) of total alkaloidal fraction was found in the adult needlerush *Juncus roemerianus* and as much as 1.5 g/kg in the giant-cordgrass *Spartina cynosuroides,* and the bulrush *Scirpus americanus*. Subsequent analysis of the volatile constituents of *S. cynosuroides* by combined GLC-MS revealed indole among 60 compounds which we could identify. The presence of benzyl cyanide and indole in the volatile constituents of these plants also pointed toward the possible presence of indole alkaloids. Since two of the most potent anti-leukemic agents, VLB and vincristine, are dimeric indole alkaloids, these observations led us to submit crude extracts from these plants to the National Cancer Institute for screening of their possible antitumor activity. Extracts of seventeen species of marsh plants have been screened against P 388 lymphocytic leukemia in BDF_1 mice. These tests have resulted in the detection of significant activity in some twelve species of marsh plants. Further examination of the extract of *Juncus roemerianus* has resulted in the isolation of the new tumor inhibitor, juncusol. Antitumor agents are currently being isolated from the other marsh plants which have demonstrated significant activity.

INTRODUCTION

The importance of plant-derived compounds in modern medicine is often underestimated. Useful compounds such as digitoxin, rutin, papain, morphine, codeine, atropine, quinine, reserpine, ergotamine, and caffeine present a broad and representative range of pharmacological activities. Drug extracts from *Digitalis purpurea* leaf and *Rauwolfia serpentina* root are constantly prescribed by many physicians. A recent survey has pointed out that 47% of some 300 million (4) new prescriptions written by physicians in 1961 contained as one or more active ingredients, a drug of natural origin. Furthermore, between 1950 and 1960, prescriptions containing drugs of natural origin increased by 7.7% (5).

Investigators that deal with plant drugs are natural product chemists. Their problems are complex and differ distinctly from those of the organic chemist who synthesizes or manipulates molecules using structure-activity relationships as his theoretical motivation to design. Natural product investigators must initially select their plants from a total number of about 750,000 available species excluding the bacteria and fungi. The isolation of antineoplastic agents from plants has been the major goal for our research in the Natural Products Laboratory at Mississippi State University for several years. Work that was funded initially by NIH was directed toward examination of the pitcher plant belonging to the genus *Sarracenia*. Our interest in the possible isolation of antineoplastic agents from marsh plants was prompted by our initial extraction of crude alkaloids from (3), and analysis of volatile oils of (9, 11, 12, 13) four species of marsh plants common in the Mississippi Gulf Coast, namely: the needlerush, *Juncus roemerianus* Scheele (Juncaceae), the giant cordgrass *Spartina cynosuroides* (L.) Roth (Poaceae), the bulrush *Scirpus americanus* Persoon (Cyperaceae) and the salt grass *Distichlis spicata* (L.) Greene (Poaceae). To the best of our knowledge, there has been no report in the literature of an extensive study on the organic constituents of these four species, and of marsh plants in general. The only study of an organic compound in marsh plants is that of 2-furaldehyde reported by Brown (2) in *Spartina alterniflora* Loisel.

PRELIMINARY CHEMICAL STUDIES

GLC-MS analyses of the four Mississippi marsh plants mentioned above revealed four volatile constituents common in all the 4 species, namely: 1) *m*-xylene, 2) *p*-xylene, 3) benzaldehyde and 4) 2-furaldehyde. Some 14 compounds are common to at least three of the species. *D. spicata, J. roemerianus* and *S. cynosuroides* have two compounds in common: naphthalene and methylnaphthalene. *n*-Nonane, *n*-decane, 3-ethyl-o-xylene, *n*-tridecane, *n*-tetradecane, *n*-pentadecane, *n*-hexadecane, *n*-heptadecane, phenyl acetaldehyde, *p*-tolualdehyde, phenol, and benzyl alcohol are common to *D. spicata, S. americanus* and *S. cynosuroides*.

These preliminary studies (9, 11, 12, 13) also revealed the presence of benzyl cyanide in the needlerush *Juncus* and indole in the giant cordgrass *Spartina*, and the salt grass *Distichlis*. The presence of benzyl cyanide and indole in these plants prompted us to continue analysis of crude alkaloids in the three species of vascular plants that are abundant in the tidal marshes of Mississippi. An unexpectedly high amount (2.0 g/kg) of total alkaloidal fraction was found in *J. roemerianus* and as much as 1.5 g/kg in *S. cynosuroides* and *S. americanus*. The presence of benzyl cyanide and indole in the volatile constituents of these plants also pointed toward the probability of the presence of indole alkaloids. Since two of the most potent antileukemic agents, VLB and vincristine, are dimeric indole

Table 1. Activity of crude extracts from marsh plants in terms of survival time of experimental animals over control (% T/C).

Extract (Ethanol)	Dosage (mg/kg)	% T/C
Juncus roemerianus	400	115
	200	120
	100	130
Spartina cynosuroides	400	120
	200	100
	100	120
Scirpus americanus	400	120
	200	120
	100	105

alkaloids, these observations led us to submit crude extracts from these plants to the National Cancer Institute for screening. The initial results shown in Table 1 indicated that *J. roemerianus* was active (% T/C above 125) and that *S. cynosuroides* and *S. americanus* had significant activity against P 388 lymphocytic leukemia in BDF$_1$ mice.

EVALUATION OF THE ANTINEOPLASTIC ACTIVITY OF MARSH PLANTS

There are two systems of tumor activity test evaluation (8), namely: 1) survival tumor systems and 2) tumor weight inhibition systems. Survival tumor systems are evaluated on the basis of survival time and are reported in the form of percent T/C. T is the mean or median survival time of the test animal, and C is the median survival time of treated animals to control animals. In general, a minimal increase in survival of treated animals over controls resulting in a T/C $\geq$ 125% is a necessary condition for further experimental work. Tumor weight inhibition systems are evaluated on the basis of tumor inhibition. There are two ways for reporting the activity, percent T/C and ED$_{50}$. Percent T/C is the ratio of the tumor weights of treated animals to control animals expressed as percent. In general a minimal reproducible tumor weight inhibition of test over control animals resulting in a T/C% of 42 or less is necessary for further experimental testing. ED$_{50}$ is the dose that inhibits tumor growth to 50% of the control growth. If a compound were found to inhibit tumor growth to 50% of the control growth at a dose of 4μg/ml or less (ED$_{50}$ $\geq$ 4 μg/ml) it is deemed significantly cytotoxic. A compound is considered to be devoid of activity if the ED$_{50}$ is greater than 100 μg/ml. For crude natural products of demonstrated minimal *in vivo* activity, *in vitro* testing is used, when possible, for biological assay during fractionation leading to isolation of the antitumor material.

The results of our preliminary analyses catalyzed a comprehensive investigation of estuarine marsh plants which grow abundantly on the coastal wetlands of Mississippi. A number of marsh plants (i.e., *Nymthaea odorato, Panicum virgatum, Ambrosia artemisiifolia,* and *Iris virginica*) had been previously submitted to NCI by other investigators and were shown to be active. Other species that were screened, (e.g., *Rumex crispus, Acta alba, Trifolium pratence, Hamamelia virginica, Ilex opaca, Phragmites communis*), proved to be inactive. NCI screening of marsh plants that we collected from Mississippi showed a relative high percentage of active species. Table 2 lists those plants or plant parts which have shown significant activity (% T/C >125) in addition to *Spartina cynosuroides* and *Juncus roemerianus* when fractionated according to Fig. 1. Ten of the seventeen plants that were screened showed activity of % T/C >125 which indicates that 59% of the plants submitted were active. This percentage of screening success is much higher than the 5% rate for all plants which have been screened by NCI. NCI has also requested confirmatory samples for *Scirpus validus* and *Juncus effusus*. The screening of additional estuarine plants will be a continuing aspect of our marsh study.

Table 2.   Marsh plants exhibiting anti-tumor activity (% T/C > 125).

Plant	Sample	Dose (μg/kg)	% T/C
Spartina alterniflora (shoots)	A	400	123
		200	115
		100	93
	B	400	148*
		200	122
		100	130*
Juncus roemerianus (roots)	A	400	147*
		200	130*
		100	97
	B	400	91
		200	138*
		100	132*
Spartina patens (roots)	A	200	118
		100	116
		50	116
		300	141*
	B	400	118
		200	118
		100	121
Sagittaria falcata (roots)	A	400	135*
		200	138*
		100	139*
	B	400	118
		200	129*
		100	135*
Distichlis spicata (shoots)	A	400	110
		200	101
		100	116
	B	400	118
		200	95
		100	135*

Phragmites communis (roots)	A	400	103
		200	103
		100	100
	B	200	121
		100	119
		150	110
		300	133*
Scirpus robustus (shoots)	A	400	116
		200	110
		100	112
	B	400	128*
		200	112
		100	108
Scirpus americanus (shoots)	A	400	103
		200	128*
		100	110
		400	119
		200	111
		100	112
Cladium jamaicense (shoots)	A	400	90
		200	104
		100	104
	B	400	95
		200	131*
		100	113
	B001	300	108
		200	123
		130	123
		90	106

*Fractions with % T/C > 125.

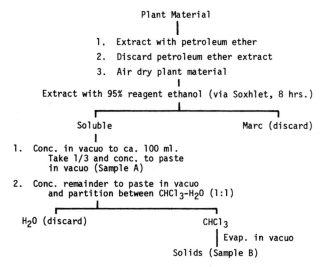

Figure 1. Fractionation procedure for marsh plant materials.

PRESENT EMPHASIS OF INVESTIGATION

On Spartina. As indicated in Table 1, the ethanolic extract of *S. cynosuroides* was only marginally active (% T/C 120) against P-388 leukemia in the initial studies. However, separation of the total alkaloidal fraction as illustrated in Fig. 2 resulted in a fraction (*A*) with a significant confirmed activity as shown below in Table 3 (*A*).

Further fractionation involving extraction of a pH 3 aqueous acid solution with chloroform as shown in Fig. 2 yielded fraction B which has confirmed activity at a high level on two separate samples as shown in Table 3 (B). Fraction C did not show significant activity.

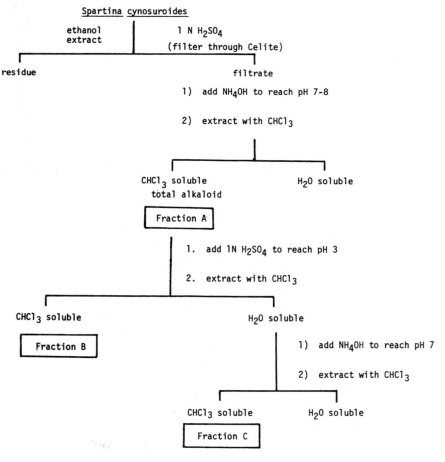

Figure 2. Fractionation procedure for *Spartina cynosuroides.*

Table 3. Anti-tumor activity (% T/C) shown by *Spartina cynosuroides* after further extraction as shown in Fig. 2.

Sample No.	Dosage	% T/C
A	400	133
	200	125
	100	116
	600	133
	400	139
	264	139
	176	–
B	600	164
	400	164
	264	149
	176	141
	400	172
	200	162
	100	135
	50	135

Thus with the activity clearly concentrated in fraction B (% T/C 172), extensive examination of the constituents present in this fraction was initiated. Repetitive chromatography of this fraction has resulted in the isolation of two crystalline compounds. The screening and structure elucidation of these compounds is in progress.

On Juncus. Shoots of *J. Roemerianus* were collected and examined by Mody (1975) during the summer of 1972-73, for extraction with 95% ethanol as outlined in Fig. 3. After removing the solvent *in vacuo*, the residue (ethanol extract) was dissolved in cold 1N sulfuric acid to give basic fraction A and neutral fraction B. Fraction B was dissolved in 50% mixture of benzene and chloroform. Benzene and chloroform soluble part was filtered and evaporated *in vacuo* which yielded fraction C (crude extract). The remaining material was labeled fraction D.

Compound *1* (m.p. 82-84°C) was isolated from the benzene-hexane (1:1) eluate of the Silica Gel G. column chromatography of fraction C followed by crystallization from chloroform-acetone. The IR, NMR, and mass spectrum of this compound were in good agreement with the authentic sample of the saturated long chain alcohol $CH_3(CH_2)_{26}-CH_2-OH$.

Compound *2* (m.p. 137-138°C) (Fig. 4) is the major constituent of 50% benzene-hexane eluate crystallized from hot benzene and methanol (1:1). A green color in the Lieberman-Burchard Reaction (10) indicated a steroid. Compound *3* was shown to be identical with β-sitosterol according to IR, NMR, and mass spectrum comparison with an authentic sample.

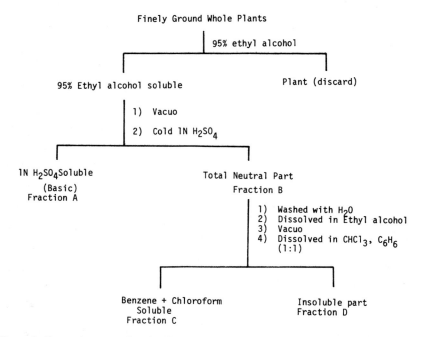

Figure 3. Extraction procedure for *Juncus roemerianus*.

Further chromatography of the 75% benzene-hexane resulted in the isolation of a new phenolic compound, *juncusol* (compound (*3*)) which crystallized from benzene in stout colorless needles, m.p. 176°C. Compound *3* with molecular formula $C_{18}H_{18}O_2$, is sparingly soluble in benzene, moderately soluble in chloroform and highly soluble in acetone and methanol. It gives a deep blue color with Keller's reagent (7) and deep red color with methanolic ferric chloride. Juncusol is also highly soluble in dilute alkali from which it can be regenerated by acidification. That both the oxygen atoms in juncusol are present as hydroxyl functions is confirmed by the formation of a diacetate, compound *5*, $C_{22}H_{22}O_4$ (m/e 350 calculated for $C_{22}H_{22}O_4$), m.p. 110, when juncusol was treated with acetic anhydride in pyridine. Juncusol contains a reducible double bond as indicated by the formation of a dihydro compound *4*, $C_{18}H_{20}O_2$ (M^+ 268), m.p. 167-8°C, when hydrogenated at room temperature over Pd-C. The dihydro compound *4*, on treatment with acetic anhydride in pyridine gives a dihydroacetate, compound *6* $C_{22}H_{24}O_4$ (M^+ 350) m.p. 138°C. The assigned structure was supported by extensive spectral and chemical studies.

The x-ray crystallography structure of juncusol is currently being obtained by Dr. Jerry Atwood at the University of Alabama. Excellent crystals have been obtained and the unit cell data is as follows:

Space group = $14_1/a$
Crystal data - tetragonal
a = b = 16,202(6) Å
c = 21.719(7)$_3$ Å
volume = 5701 Å

Data have, thus far, been collected out to $\Theta = 15°$.

Juncusol (3) has demonstrated confirmed activity in the 9KB test system (Ed_{50} 0.3). Therefore, NCI has requested a substantial quantity (10 g) of juncusol (3) for further testing. Obtaining this quantity of juncusol will require the extraction of some 400 lbs of marsh plants.

9, 10 dihydrophenanthrene is particularly rare as a natural product. So far only a few natural 9, 10 dihydrophenanthrenes have been reported (1, 6). The

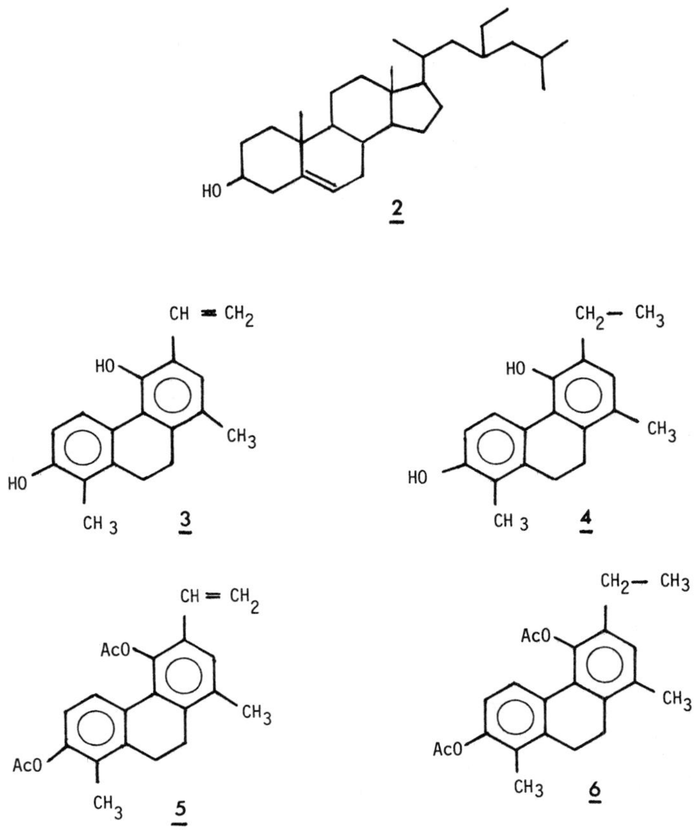

Figure 4. Diagrams of compounds 2, 3, 4, 5, and 6.

7-hydroxy-2, 4-dimethoxy-9,10-dihydrophenanthrene, orchinol *18*, was produced by *Orchis militaris* as a defense substance (14) under the influence of certain morbitic agents.

At least two other neutral compounds which appear to be structurally related to juncusol (*3*) have been detected in *Juncus roemerianus*. The screening of these compounds and their structure elucidation is underway.

In summary, we have shown activity in marsh plants at the rate of 59% against P-388 leukemia. We have isolated and identified one potential drug (juncusol) and have isolated several other compounds from active fractions which are ready for test. The activity of these marsh plants against leukemia may be only the beginning of their possible pharmacological utilization. Arrangements are being made for the screening of their antibiotic, antifungal, cardiovascular, antipsychotic, and anaesthetic activity. Thus the pharmacological potential of marsh plants is extremely promising.

LITERATURE CITED

1. Beroza, M. 1970. Chemical controlling Insect Behavior. Academic Press, N.Y. 170 p.
2. Brown, L.R. 1963. Investigation of salt marsh grass to determine the yield of furfuraldehyde. Ga. Acad. Sci. Bull. 21:20-23.
3. Cruz, A.A. de la, D.H. Miles, and J. Bhattacharyya. 1974. Chemical basis for the detritus food chain in marshes. Assoc. Southeast. Biol. Bull. 21:49.
4. Farnsworth, N.R. 1966. Biological and phytochemical screening of plants. Phar. Sci. 55:225-276.
5. Gosselin, R.A. 1962. The status of natural products in the American pharmaceutical market. Lloydia 25:241-243.
6. Hedin, P.A., A.C. Thompson, R.C. Gueldner, and J.P. Minyard. 1972. Volatile constituents of the boll weevil. J. Insect Physiol. 18:79-86.
7. _____, C. Niemeyer, R.C. Gueldner, and A.C. Thompson. 1972. A gas chromatographic survey of the volatile fractions of twenty species of insects from eight orders. J. Insect. Physiol. 18:555-567.
8. Kupchan, S.M., M.A. Eakin, and A.M. Thomas. 1971. Tumor inhibitors 69: Structure-cytotoxicity relationships among the sesquiterpene lactones. J. ed. Chem. 19:1147-1152.
9. Miles, D.H., N.V. Mody, J.P. Minyard, and P.A. Hedin. 1973. Constituents of marsh grass: Survey of the essential oils of *Juncus roemerianus*. Phytochemistry 12:1399-1404.
10. Mody, N.V. 1975. Investigation of constituents of marsh plants. Ph.D. Dissertation, Mississippi State University, 103 p.
11. _____, J. Bhattacharyya, D.H. Miles, and P.A. Hedin. 1974. Survey of volatile constituents of *Spartina cynosuroides*. Phytochemistry 13:1175-1178.
12. _____, E.J. Parish, J. Bhattacharyya, D.H. Miles, and P.A. Hedin. 1974. The essential oils in *Scirpus americanus*. Phytochemistry 13:2027-2029.
13. _____, A.A. de la Cruz, D.H. Miles, and P.A. Hedin. 1975. The essential oils of *Distichlis spicata*. Phytochemistry 14:599-601.
14. Raney, H.G., R.D. Eimenbary, and N.W. Flora. 1970. Population density of pecan weevil under Stuart Pecan trees. J. Econ. Entomol. 63:697-700.

BEHAVIOR AS A MEASURE OF SUBLETHAL STRESS

Convened by:
Bori L. Olla
National Marine Fisheries Service
Middle Atlantic Coastal Fisheries Center
Sandy Hook Laboratory
Highlands, New Jersey 07732

The search for sensitive and ecologically pertinent measures of pollutant effects on aquatic organisms has stimulated research in a variety of disciplines, including animal behavior. The papers presented at this session deal with the application of behavioral research to current or potential environmental problems. Although few in number, these works reflect the multiplicity of approaches that may be applied to assessing the effects of a variety of man-induced stresses on aquatic organisms.

Although it is obvious that each of these papers differs in scope, there are, nevertheless, common aspects that each investigator has had to consider.

The first was the identification of those behaviors which transcend field and laboratory and are separate and distinct from those induced by the artificiality of the laboratory environment. This is an obvious step in this kind of research and one which entails a careful definition of the "normal" scope of the range of behaviors which are elicited by natural stresses.

A second aspect that each investigator has had to consider, is the role a particular behavior may play in the life habits of an organism. Modification of this behavior could then be directly related to the survival capabilities of the organism or population.

Finally, by establishing norms of behavior prior to any experimentation, the authors have not only determined levels from which departures might indicate a contaminant effect, but also have established criteria for laboratory acclimation. Whatever testing method may be employed, establishment of behavioral criteria of laboratory activities becomes a useful step in evaluating the health and general condition of experimental animals, contributing to the standardization required for experimental replicates.

While significant amounts of time and resources have been expended to define contaminant levels which are deleterious to marine and estuarine ecosystems, it is generally still not possible to predict with any degree of confidence effects other than those due to acute levels causing debilitation or death. If we are to reach a satisfactory state of predictive capability on the subtle effects of contaminants, more effort must be directed towards understanding organisms within the context of their natural environment. The value of using behavior as one method for furthering this understanding should be apparent from these papers.

CRUSTACEAN LARVAL BEHAVIOR AS AN INDICATOR

OF SUBLETHAL EFFECTS OF AN INSECT JUVENILE

HORMONE MIMIC

Richard B. Forward, Jr. and John D. Costlow, Jr.
Duke University Marine Laboratory
Beaufort, North Carolina 28516
and
Zoology Department
Duke University
Durham, North Carolina 27706

ABSTRACT: Both swimming speed and phototaxis by larvae of the crab *Rhithropanopeus harrisii* were monitored upon exposure to the insect juvenile hormone mimic MON-0585 in order to determine their potential as behavioral indicators of sublethal effects. Under the experimental rearing conditions (25°C and $25^{\circ}/_{\circ\circ}$) concentrations above 1.0 ppm caused reduced developmental survivorship. At all zoeal stages (I-IV) neither swimming speeds nor phototaxis for the acetone control treated larvae differed from those for untreated larvae. The lowest concentrations at which a 24-hour exposure to MON-0585 caused swimming speeds to be significantly greater than those for the acetone controls were 0.1 ppm for stage I, 0.5 ppm for stages II and III and 1.0 ppm for stage IV. Continuous exposure to 0.1 ppm and lower concentrations caused significantly faster swimming only at 0.1 ppm in stage II; stages III and IV were normal. The lowest concentrations that caused altered phototaxis after a 24 hr. exposure were 0.5 ppm for stages I and IV and 0.1 ppm for stages II and III. Upon continuous exposure, altered phototaxis occurred only at 0.1 ppm for stage II; stages III and IV were normal. Thus, both swimming speed and phototaxis are altered by MON-0585 at sublethal concentrations. Stage II zoea appear to be most sensitive, and concentrations below 0.1 ppm have no effect upon the measured behaviors.

INTRODUCTION

Recently, insect juvenile hormone mimics or analogues have been proposed as potentially useful compounds for controlling various insect species. The principle of operation for these pesticides is to expose larval insects to substances which imitate the function of the juvenile hormone. Under normal conditions this hormone participates in the endocrine control of differentiation and metamorphosis during post-embryonic development and maturation. The normal developmental pattern of an insect is critically dependent upon the level of juvenile hormone within its body; a high juvenile hormone titer is essential for larval development and low titer or complete absence is a prerequisite for metamorphosis to a juvenile form. By maintaining a high apparent juvenile hormone titer within the larvae, the juvenile hormone mimics are able to prevent successful metamorphosis. Recent literature about juvenile hormone mimics and their effects upon insects is reviewed extensively by Sláma, Romanuk, and Sorm (12).

Phylogenetically, the Crustacea are closely related to insects and most marine forms have planktonic larvae which follow a developmental pattern consisting of a number of discrete stages, with frequent periods of ecdysis or moulting and a final metamorphosis into a juvenile crustacean. Although little is known about endocrine control of crustacean development, within the Brachyura hormones from either the central nervous system or the compound eyes are probably involved (2, 3). Subsequently, crustacean larvae may be affected by juvenile hormone mimics, in a manner similar to insects.

Previous studies with crustacean larvae have concerned the effects of juvenile hormone mimics upon developmental patterns and survivorship. Gomez, Faulkner, and Newman (9) found that the juvenile hormone mimic ZR-512, which is an effective agent against Pea aphids (5), accelerated development and metamorphosis of barnacle larvae. In contrast, Bookhout and Costlow (1) found in a preliminary study that at increasing concentrations of ZR-512 larval development was delayed and mortality increased for the crab *Rhithropanopeus harrisii*. Although these studies can establish concentration ranges at which altered development and survivorship occur, it is possible that at lower concentrations subtle, but potentially detrimental, physiological alterations may also occur and that these could be measured as behavioral changes. Thus, the present study was initiated to determine whether aspects of crustacean larval behavior can be used as quantitative indicators of sublethal juvenile hormone mimic effects.

Larvae from the crab *Rhithropanopeus harrisii* were chosen for study because their developmental pattern and rearing techniques are well established (4). Larvae pass through four zoeal stages which are free swimming in the plankton and molt into a megalopa stage which settles out of the plankton onto the bottom and finally molts into a crab. The behaviors chosen for study are linear swimming speeds and light oriented movement (phototaxis). Previous work has detailed the normal ontogeny of these behaviors through the four zoeal stages when larvae are reared under optimum conditions for developmental success (6, 8). Swimming speeds generally increase with each successive larval stage which perhaps reflects the increase in zoeal size and consequent potential for faster movement. The four zoeal stages are very responsive to light, while the megalopa are totally unresponsive. Physiologically, phototaxis is almost identical for each zoeal stage. The main spectral sensitivity maximum occurs at 500 nm with a smaller peak at 400 nm. The phototactic sign depends upon light intensity, since light-adapted animals are positively phototaxic at high intensities and respond negatively to low intensities (6, 8). Phototactic responses and swimming contribute to horizontal and vertical movements within the water column and probably to predator avoidance (7). Thus any alteration in these behaviors would theoretically reduce the chance for survivorship.

The compound tested in the current experiments was 2,6-di-t-butyl-4-(α,α-dimethylbenzyl)phenol or MON-0585, which has juvenile hormone type activity but is chemically unrelated to insect hormones. This compound is effective against mosquito larvae and was being seriously considered as a potential commercial insect larvicide. Upon exposure all stages of mosquito larvae develop normally, however, metamorphosis is blocked at an early stage of pupation, and the prepupae die in a characteristic compact, stalky, unmelanized form (10, 11). The results of the present study indicate that both swimming speed and phototaxis by *Rhithropanopeus* larvae are altered at concentrations below those causing developmental change. Thus, these behavioral responses can serve as behavioral indicators of sublethal effects.

MATERIALS AND METHODS

Ovigerous female *Rhithropanopeus harrisii* (Gould) were collected intertidally. Using established techniques (4), larvae were cultured at 25°C in 25°/$_\infty$ filtered seawater and on a 12:12 h light-dark cycle. Larvae were only tested on the intermolt days. To avoid complications due to a possible biological rhythm in phototaxis, all experiments were begun four to six hours after the onset of the light phase with larvae light-adapted under room lights plus a 60 W incandescent bulb. Light-oriented movement (phototaxis) and swimming speeds were measured by means of a closed-circuit television system coupled to a dissecting microscope (Olympus), using the techniques previously described by Forward and Costlow (8). The microscope illumination light was interference-filtered to 802 nm, a wavelength which neither alters nor induces phototaxis.

Light stimulation was horizontal to avoid complications due to geotaxis. The stimulus light was a 150 W Xenon arc lamp directed into a monochromator (Farrand model f/3.5) set at 500 nm (full band pass 10 nm). This wavelength was chosen because it is a main spectral sensitivity maximum. The stimulus intensity was regulated by neutral density filters, and the duration, as controlled

by an electromagnetic shutter, was 2 sec in all experiments. The general procedure was to place light-adapted larvae in darkness for 1 minute so they could
assume a random distribution within the test cuvette and then stimulate at 15
sec intervals for a maximum of 6 stimuli. Larvae were then returned to the
culture bowls, and a new sample tested. Positive phototaxis was defined as
swimming towards the stimulus light within the 30° sector centering on the
direction of the light, and negative phototaxis was swimming in the 30° sector
directly away from the stimulus direction.

Sublethal concentrations of the juvenile hormone mimic 2,6 di-t-butyl-4
(α-α-dimethylbenzyl) phenol (MON-0585-obtained from Agricultural Division of
Monsanto Chemical Company, St. Louis, Missouri, USA) were based upon a
developmental study by one of us (J. D. Costlow, Jr., unpublished) in which,
under similar rearing conditions total larval mortality occurred at concentrations
above 1 ppm (i.e., 5 and 10 ppm), with a slight decrease in the percent survival
from hatch to the first crab stage, and an increase in the developmental time
duration at 1 ppm. No effects were observed at lower concentrations. Thus,
sublethal concentrations were considered to be 1.0 ppm and lower.

A stock solution of MON-0585 at 1 ppt was made up in full strength acetone,
and the appropriate test concentrations were mixed daily in seawater. Two
groups of experiments were conducted to test the effects of both acute and
chronic exposure upon all zoeal stages. During the acute tests, larvae were reared
to the appropriate zoeal stage and then exposed to concentrations of 1, 0.5, 0.1
and 0.01 ppm for 24-hours before testing phototaxis and swimming rates. Three
hatches of larvae were separately tested, and the results combined. During the
chronic experiments larvae were continually exposed to concentrations of 0.1,
0.05 and 0.01 ppm. Only stage II, III and IV zoea were tested because exposures
for stage I zoea constituted an acute test. These concentrations were chosen
because the acute exposures for stage I indicated that the lowest concentration
at which behavior was altered was 0.1 ppm. In the chronic exposure two hatches
of larvae were separately tested. For all experiments untreated larvae (seawater
controls) and an acetone control were tested similarly. The acetone control
consisted of exposing larvae to the highest acetone concentration experienced
during the tests with MON-0585, i.e., equivalent to that at 1.0 ppm.

Mean swimming speeds under the various test conditions were statistically
compared to those observed for the acetone control conditions by student's t
test. Phototactic responses were statistically compared by using a 2×2 test for
independence. In both statistical tests significant differences were tested at the 5
percent level.

RESULTS
Swimming Speeds

Fig. 1 shows swimming speeds observed under the microscope illumination
light when the larvae were subjected to the various test conditions. In none of
the experiments did the speed for the acetone-treated control animals differ

significantly from that for the seawater control larvae. Upon acute exposure the lowest concentration of MON-0585 at which swimming speeds were significantly greater than those for the acetone control treatment larvae increased through the

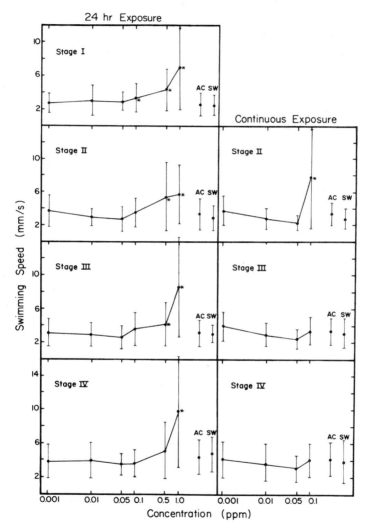

Figure 1. Swimming speed (ordinant) upon exposure to various concentrations of MON-0585 (abscissa). The connected dots show the mean speeds and the vertical lines the standard deviation. AC is the speed for the acetone control and SW that for the untreated larvae in normal seawater. * indicates that the mean speed is significantly faster than the mean speed for the AC larvae. During the 24-hour experiments the average n for each exposure condition was 56 for stage I, 51 for stages II and III and 49 for stage IV. During the continuous experiments the average n was 36 for stage II and 35 for stages III and IV.

zoeal stages, i.e., 0.1 ppm for stage I, 0.5 ppm for stages II and III and 1.0 ppm for stage IV. Chronic exposure to concentrations of 0.1 ppm and lower only cause significantly faster swimming at 0.1 ppm for stage II zoea while stages III and IV were normal. Since stage II is unaffected by this concentration upon acute exposure, the chronic treatment does have an effect. However, this effect does not persist to stages III and IV.

Phototaxis

The normal curve for phototactic responses to different intensities of 500-nm light for light-adapted stage I zoea is shown in Fig. 2. The curves for stages II, III and IV are almost identical (6). Larvae are positively phototactic to high light intensities and become negative to low intensities. Light-adapted acute acetone-treated control larvae show similar responses to different light intensities (Fig. 3).

Alterations in phototaxis by exposure to MON-0585 could involve either a change in the general shape of the response versus intensity curve (e.g., positive or negative phototaxis at different intensities) and/or a change in the response magnitude (percent response) to different intensities. Preliminary experiments indicated that upon exposure the general shape of the response versus intensity curve is not altered as positive phototaxis is greater to high intensities and negative phototaxis to low intensities. Instead the percent response at different intensities changes. Thus in quantitatively evaluating alterations in phototaxis at the various pesticide concentrations, larvae were stimulated with 500-nm light set at various intensities that initiate definite positive and negative phototaxis in

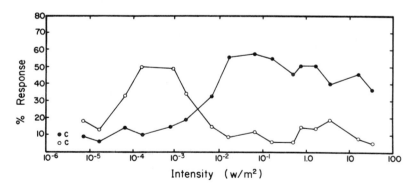

Figure 2. The percent of stage I zoeae (ordinant) showing positive phototaxis (solid dots) and negative phototaxis (open circles) upon stimulation with 500-nm light at various energy levels (abscissa). The control level of positive (C - solid dot) and negative (C - open circles) swimming were determined while the larvae were unstimulated but illuminated with the microscope lamp interference filtered to 744-nm light. The average sample size for each point was 65. Replotted from Forward (6).

untreated larvae. To determine which concentrations actually altered phototaxis, the number of larvae showing positive and negative phototactic responses were statistically compared to the frequency of swimming in these directions when the larvae were moving under the microscope illumination light (i.e., control level for each test condition). For comparative purposes, the percentage of larvae showing positive and negative phototaxis under the various conditions was plotted in Fig. 4, however, the statistics were computed using the original numbers.

Upon acute exposure the lowest concentrations which prevented larvae from displaying either positive or negative phototaxis were 0.5 ppm for stages I, III and IV and 0.1 ppm for stage II (Fig. 4). Although at 0.1 ppm stage III zoeae

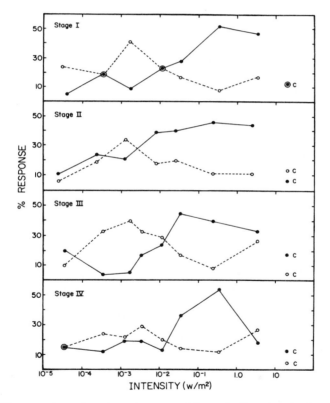

Figure 3. The percentage of acute acetone treated larvae (ordinant) at the four zoeal stages showing positive phototaxis (solid line - dots) and negative phototaxis (dashed line - open circles) upon stimulation with 500-nm light at various energy levels (abscissa). The C - solid dots indicate the control level of positive swimming and the C - open circles that for negative swimming. The average n for each point was 22.

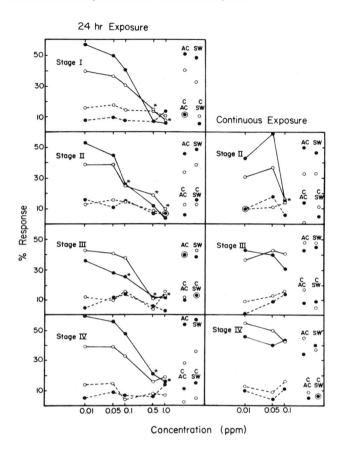

Figure 4. The percentage of light adapted larvae (ordinant) at the four zoeal stages showing positive (solid dots - solid line) and negative (open circles - solid line) phototaxis upon exposure to different MON-0585 concentrations (abscissa). The dashed lines - solid circles indicate the control level of swimming in the positive direction when unstimulated, and the open circles - dashed line that for negative. AC is the acetone control and SW is the untreated larvae in seawater. C - AC and C - SW are the control levels for AC and SW respectively. The average light intensities (W/m²) used to test for positive phototaxis during the 24-hour exposure experiments were stage I - 0.265, stage II - 0.225, stage III - 0.267, stage IV - 0.213 and for negative phototaxis were stage I - 1.33×10^{-3}, stage II - 1.13×10^{-3}, stage III - 1.28×10^{-3}, and stage IV - 1.07×10^{-3}. During the continuous exposure experiments the average intensities for positive phototaxis were for stage II - 0.275, stage III - 0.295, and stage IV - 0.283, and for negative phototaxis were stage II - 1.38×10^{-3}, stage III - 1.48×10^{-3}, and stage IV - 1.43×10^{-3}. The average n for each point in the acute experiments were stage I - 56, stage II - 93, stage III - 71, and stage IV - 61, while those for the chronic experiments were stage II - 36, stage III - 42, and stage IV - 41. The * indicates those responses which are not significantly different from the control.

display normal negative phototaxis, the positive response was not significantly different from random swimming. Upon chronic exposure only stage II at 0.1 ppm displayed no phototaxis while both stages III and IV were normal. Thus a cumulative effect due to continuous exposure is not apparent since stages III and IV are phototactically normal while acute exposure at 0.1 ppm did alter positive phototaxis in stage III.

DISCUSSION

Both phototaxis and swimming speeds can be used as behavioral indicators of sublethal effects of the juvenile hormone mimic MON-0585 upon larvae of the crab *Rhithropanopeus harrisii*. Considering swimming, exposure to higher concentrations results in an increase in speed. This probably reflects a change in the general activity level of the larvae. Since insect respiration is usually altered by juvenile hormone mimics (12), this is not unexpected. In addition, Sacher (11) suggested that the lethal effect of MON-0585 upon mosquito pupae is due to an interference with normal oxygen utilization.

No change in the sign of phototaxis at different light intensities was observed. Instead the magnitude of both positive and negative phototaxic responses was depressed upon exposure to high concentrations of MON-0585. Since phototaxis involves both the reception of light and oriented swimming, altered swimming could result in altered phototaxis. Upon acute exposure, however, the lowest concentrations that produced altered phototaxis and swimming are different (Table 1). Thus something other than a simple change in swimming must be involved in the altered phototaxis. Sacher (10, 11) noted melanization was reduced in prepupae of MON-0585 treated larvae, and that the prepupae appeared whitish. When *Rhithropanopeus* zoeae were exposed to the highest test concentrations (1.0 ppm), they also subjectively appeared to have less body pigmentation than untreated larvae. If pigments within the visual systems were similarly affected, this could contribute to the changes in phototaxis.

The lowest concentration to alter the measured behaviors was 0.1 ppm. The early zoeal stages (I and II) are the most sensitive as both swimming and phototaxis were affected at this concentration in both acute and chronic tests, while behavioral alterations in stages III and IV were only seen at higher exposure

Table 1. Lowest concentrations of MON-0585 that produced alterations in behavior upon acute exposure (In ppm).

	Swimming	Phototaxis
Stage I	0.1	0.5
Stage II	0.5	0.1
Stage III	0.5	0.1 (+only)
Stage IV	1.0	0.5

levels. The lowest concentration that affects larval development is 1.0 ppm. Thus the measured behaviors can be used to indicate sublethal stress.

Since MON-0585 is effective in preventing metamorphosis by mosquito larvae, a consideration of the concentrations necessary for this effect is perhaps instructive. The LD_{50} for MON-0585 upon second, third, and fourth instars of the mosquito *Aedes aegypti* after chronic exposure were 0.065, 0.05, and 0.05 ppm, respectively, while the LD_{90} levels for all three instars was 0.1 ppm (10, 11). These concentration levels are close to those which alter behavior in *Rhithropanopeus* larvae. Assuming that the sensitivity of *Rhithropanopeus* to MON-0585 is representative of crustacean larvae in general, and that the measured behaviors are necessary for larval survival, caution should be exercised in using MON-0585 in areas where crustacean larvae are present.

Thus since both phototaxis and linear swimming speeds by crustacean larvae are altered by a juvenile hormone mimic at concentrations below those which cause developmental change, these behaviors can potentially be used as indicators of the sublethal effects of this type of larvicide.

ACKNOWLEDGEMENTS

This work was supported by Environmental Protection Agency Grant Nos. R-801128-01-1 and R-803838-01-0.

LITERATURE CITED

1. Bookhout, C. B., and J. D. Costlow, Jr. In press. Crab Development and Effects of Pollutants. Thalassia Yugosl.
2. Costlow, J. D., Jr. 1963. The effect of eyestalk extirpation on metamorphosis of megalops of the blue crab *Callinectes sapidus* Rathbun. Gen. and Comp. Endocrinology 3:120-130.
3. _____. 1966. The effect of eyestalk extirpation on larval development of the mud crab *Rhithropanopeus harrisii* (Gould). Gen. and Comp. Endocrinology 7:255-275.
4. _____, C. G. Bookhout, and R. J. Monroe. 1966. Studies on the larval development of the crab *Rhithropanopeus harrisii* (Gould). I. Effect of salinity and temperature on larval development. Physiol. Zool. 39:81-100.
5. Diekman, J. D. 1972. Use of insect hormones in pest control. Presented at National Insect Pest Management Workshop, Purdue University, Product Development Dept., Zoecon Corp., Palo Alto, California.
6. Forward, R. B., Jr. 1974. Negative phototaxis in crustacean larvae: possible functional significance. J. exp. mar. Biol. Ecol. 16: 11-17.
7. _____. 1976. Light and diurnal vertical migration: Photobehavior and Photophysiology of Plankton, p. 157-209. *In* K. Smith (ed.), Photochemical and Photobiological Reviews. Vol. I, Plenum Press, New York.
8. _____, and J. D. Costlow, Jr. 1974. The ontogeny of phototaxis by larvae of the crab *Rhithropanopeus harrisii*. Mar. Biol. 26:27-33.
9. Gomez, E. D., D. J. Faulkner and W. A. Newman. 1973. Juvenile hormone mimics: effects on Cirriped Crustacean metamorphosis. Science 179:813-814.

10. Sacher, R. M. 1971. A mosquito larvicide with favorable environmental properties. Mosquito News 31:513-516.
11. _____. 1971. A hormone-like mosquito larvicide with favorable environmental properties. Proc. 6th Br. Insectic. Fungic. Conf. p. 611-620.
12. Sláma, K., M. Romaňuk, and F. Šorm. 1974. Insect Hormones and Bioanalogues. Springer-Verlag, New York. 477 p.

PCB AND THE ACTIVITIES BUDGET

OF THE CRAB *HEMIGRAPSUS OREGONENSIS*

Walter H. Pearson[1] and Robert L. Holton
School of Oceanography
Oregon State University
Corvallis, Oregon 97331

.BSTRACT: In order to understand the adaptive processes of an estuarine rganism under stress from a pollutant, the behavior of the crab *Hemigrapsus regonensis* was assessed with and without exposure to polychlorinated biphen-'ls (PCB's). Activities budgets were developed from observations of the time :rabs spent in 20 categories of shelter use, posture, and activity while held in habitat models complete with tides. Females spent more time sheltered and feeding but less time displaying than males. For both sexes certain activities predominated at certain tidal stages.

Multivariate discriminant analysis was used to explore differences in activities budgets between ordinary conditions and stressing conditions when PCB-contaminated sand was present. Discriminant functions containing all 20 activities did not clearly distinguish the budgets of stressed and unstressed crabs. Stepwise discriminant analyses, used to choose the most parsimonious functions, indicated where the budgets differed. For both sexes and different trials, feeding consistently appeared in the multivariate discriminant functions and decreased under PCB exposure.

The discriminant analysis suggested that the consequences one would predict for a pollutant-induced stress are functions of the paradigm under which one observes. In a search to predict the consequences of pollution one should be alert not only to the character of the organism's failures, but also to the nature of its successes.

---

[1] Present address: National Marine Fisheries Service, Sandy Hook Marine Laboratory, Highlands, N.J. 07732.

INTRODUCTION

Although predictions of the consequences of various courses of action that will alter estuaries are presently being sought, current biological theory is less than adequate to indicate the fate of populations of estuarine organisms exposed to pollutants and other environmental alterations (13, 14, 17). Up to now the scientist has examined what he could with available techniques and often rested with the demonstration of a pollutant effect. Now the need is for explication of the meaning of these demonstrated consequences in light of developing knowledge of ecosystem structure and function. This study is an attempt to explore not only the kinds and magnitudes of any pollutant-induced change in the behavior of the yellow shore crab, *Hemigrapsus oregonensis*, but also changes of ecological import.

Moving beyond conventional toxicity testing is not simply finding a new response criterion, such as behavioral disruption (10) or reproductive failure (7), but abandoning the implicit view that the living organsim is little more than a reactive machine driven to responses by stimuli (8, 19). We much prefer to see the animal as a goal-directed general open system (8), primarily active rather than reactive (12), and capable of considerable plasticity in reaching its goals (2). Such a broader view of the animal's capabilities leads to methods of observation and experimental designs in which, hopefully, we avoid making robots of our subjects.

Because one productive way to investigate the ecological meaning of behavior is to ask where and how individual crabs spend their time (20, 21), time budgets for the crab's activities were used to assess the behavior of *Hemigrapsus oregonensis* with and without exposure to polychlorinated biphenyls (PCB's). Our initial hypothesis was that an ecologically meaningful behavioral change induced by PCB exposure would become evident as a shift in the activities budget. As we tried to be alert not only to possible disruption of ordinary behavior, but also to innovative behavior that brings adaption to environmental challenge (8, 14), we expected not quantitative predictions of damaging levels of PCB, but the discerning of patterns of events flowing from the crabs' attempt to adapt to the pollutant.

MATERIALS AND METHODS

In hopes of observing ordinary behavior in the laboratory, we constructed two banks of four 40-liter observation and holding chambers, each of which was designed to simulate the crab's mud-rock habitat. A tide generator of composite design (3, 6, 18) maintained a tidal period of 12.4 hours in the habitat replicas. The tide generators and replicas were oriented to give a four-hour period of air exposure and a 20-cm water height at high tide. The timing of the replica tides matched that in Yaquina Bay, Oregon, where the crabs were collected for this study. Daylight-type flourescent lights lit the replicas on a natural photoperiod, and dark-red safe-lights provided illumination for night observations. Fine sand, crushed and whole oyster shell, and barnacled rocks with algal growth provided natural substrate and shelter. Three crabs of each sex were held in each replica and fed commercial shrimp pellets, small blue mussels, and isopods. Before and during the experiments one gram of crushed shrimp pellet was sprinkled over the substrate at low tide every 37 or 50 hours.

A behavioral repertoire (Table 1) developed from several months of observation defined the activities to be timed for the budgets. The amount of time individual marked crabs spent in 56 behavioral units was measured using an electronic metronome (21) that gives an audible tone every five seconds. We watched a crab and, upon hearing a tone, recorded letter codes for the behavioral units seen during the five-second interval. Summing the number of letter codes in a given observational period gives the time spent in the various behaviors.

Because knowledge of an organism's rhythmicity should lead to more appropriate designs for behavioral observations (5), we followed an observation schedule based on the lunar day (24.8 hours). During each 62-minute lunar hour the six crabs in a habitat replica were observed in a random order, and their activities timed for seven minutes. These hourly observation periods began exactly at low tide and continued for two full tidal cycles. The order in which the replicas were watched was also randomized, but the schedule was such that at the end of eight lunar days each crab in every replica had been observed at every hour of the

Table 1.   Behavioral repertoire of *Hemigrapsus oregonensis* with the 20 activities categories into which the behavioral units were grouped.

Behavioral Unit	Activity Category
Shelter Use	
1 Unsheltered	1 Unsheltered
2 Sheltered	2 Sheltered
3 Perched	3 Perched
4 Buried	4 Buried
Posture	
5 Mating	5 Mating
6 Display	6 High
7 High	6 High
8 Low	7 Low
9 Huddled	8 Huddled
Locomotion	
10 Standing	9 Standing
11 Walking	10 Walking
12 Climbing	11 Climbing
13 Running	12 Fast Movement
14 Fleeing	12 Fast Movement
15 Chasing	12 Fast Movement
16 Swimming	*

Social Behavior
17 High Intensity Lateral Merus Displaying   13 Displaying
18 Mid Intensity Lateral Merus Displaying   13 Displaying
19 Low Intensity Lateral Merus Displaying   13 Displaying
20 Claw Shuddering   13 Displaying
21 *Hemigrapsus* Displaying   13 Displaying
22 Walking Leg Contact   14 Walking Leg Contact
23 Piling Up   *
24 Lunging   15 Fighting
25 Striking   15 Fighting
26 Pinching   15 Fighting
27 Pushing   15 Fighting

Prefeeding
28 Claw Probing   16 Prefeeding
29 Dactyl Searching   16 Prefeeding
30 Maxilliped Sweeping Out   16 Prefeeding

Feeding
31 Maxilliped Sweeping Substrate   17 Feeding
32 Claw Feeding   17 Feeding
33 Scraping Feeding   17 Feeding
34 Claw Tearing   17 Feeding
35 Spitting   *

Environmental Manipulation
36 Digging   18 Digging
37 Claw Shovelling   18 Digging
38 Bulldozing   18 Digging
39 Sand Scraping   18 Digging
40 Lifting   18 Digging
41 Ice Tonging   *

Sexual Behavior
42 Grabbing   19 Sex
43 Embracing   19 Sex
44 Struggling   19 Sex
45 Copulating   19 Sex

Body Maintenance
46 Eye Bobbing   20 Cleaning
47 Claw Picking   20 Cleaning
48 Dactyl Picking   20 Cleaning
49 Claw Rubbing   20 Cleaning
50 Leg Rubbing   20 Cleaning
51 Claw Sweeping   20 Cleaning
52 Eye Sweeping   20 Cleaning
53 Antennule Sweeping   20 Cleaning
54 Mouthpart Mashing   20 Cleaning
55 Abdomen Flapping   20 Cleaning
56 Egg Ventilating   20 Cleaning
57 Egg Grooming   20 Cleaning
58 Foam Bathing   20 Cleaning
59 Antennule Flicking   *
60 Scaphognathite Beating   *
61 Molting   *
62 Defecating   *

*These behavioral units were identified for this repertoire but were not used in the construction of the activities budgets.

lunar day. Dividing the number of five-second intervals spent in each behavioral unit by the total number of intervals a crab was observed over the eight lunar day sampling period estimates the proportion of the lunar day the crab spends in each behavior. The observations on the 56 behavioral units were grouped into 20 higher-level categories of shelter use, posture, and activities as indicated in Table 1, and a listing of the time spent by the crabs in each of these 20 categories constitutes the activities budget.

Because *Hemigrapsus oregonensis* readily ingests sand and concentrates PCB from contaminated sand, we thought placing one kilogram of PCB-contaminated sand into the habitat replicas to be a natural mode of exposing the crabs to the pollutant. Eight lunar days' observation gave the time measurements for activities budgets of crabs under control and exposed conditions in two experiments. PCB's (Aroclor® 1260) were sorbed to fine sand by mixing oven-dried sand and an acetone solution of PCB for several days and then evaporating the acetone. In the first experiment the sorption of the PCB to sand varied so that the two highest concentrations could not be distinguished. Consequently the first experiment had two control replicas, two replicas receiving low exposure averaging 30 ppm PCB in the sand, and four replicas receiving moderate exposure averaging 260 ppm. The second experiment, in which the PCB sorbed to sand more consistently, had four control replicas and four PCB-exposure replicas with sand containing an average of 422 ppm. In the second experiment the observer did not know which tanks received the PCB-contaminated sand.

RESULTS AND DISCUSSION

Comparison of the average activities budgets of male and female crabs observed under the eight lunar day schedule (Fig. 1) shows that under ordinary

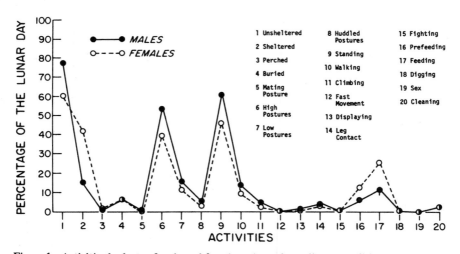

Figure 1. Activities budgets of male and female crabs under ordinary conditions.

conditions with no PCB present in the replicas female crabs spend considerably more time sheltered and feeding than males. The male crabs exhibit slightly more social interaction, displaying, walking leg contacting, and fighting, than the females. The profile for male crabs through the posture and locomotion categories (activities 5 to 12 in Figure 1) is somewhat above that for females and reflects the males being more unsheltered, but both sexes having profiles of the same shape indicates that female crabs budget their unsheltered time among postures and locomotion in the same proportion as males.

The points in Figs. 1 to 3 represent the mean of the replica averages of the percentages of the lunar day spent in the activities. The replica average is the mean of the three male or female crabs and is the basic measurement for the figures and following analysis because the replica is the experimental unit in the statistical sense (3). The variability among replicas is less than the variability among individual crabs within a replica because within a replica a hierarchy based on size develops. One consequence of this hierarchy is that the dominant crab shelters itself less than the others. (See Table 2).

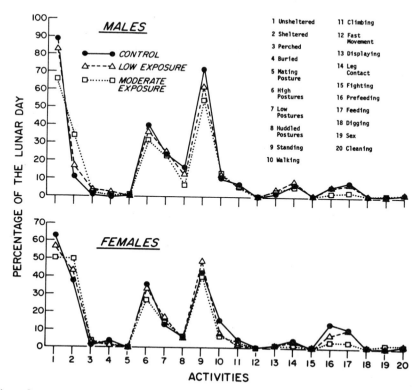

Figure 2. Activities budgets of male and female crabs in the first experiment.

Male crabs not only spend a larger share of time displaying than females, but also give different kinds of displays. Although both sexes commonly give lateral merus displays, in accord with others (15) we observed only the male giving the claw shuddering display. Females typically give the *Hemigrapsus* display described elsewhere (22), but males rarely do.

For both sexes certain activities predominate at certain tide stages. On a rising tide crabs emerge from shelter and feed but are quiet at high tide. General movement, shelter maintenance and displays increase with the falling tide, and the crabs seek shelter during exposure at low tide. At night the crabs are more active, with general movement and feeding occurring from submersion through the high tide and into the falling tide. During the nocturnal low tides crabs often do not shelter themselves but stand quietly.

Similar to fiddler crabs (9, 11), *Hemigrapsus oregonensis* concentrates PCB's from contaminated sediment, and male crabs have the lower whole body burdens. After 31 days at the low exposure in the first experiment, whole body concentrations of PCB averaged 13 ppm for males and 59 ppm for females. At the moderate exposure males concentrated an average of 15 ppm PCB and females 57 ppm. In the second experiment males had a body burden of 36 ppm

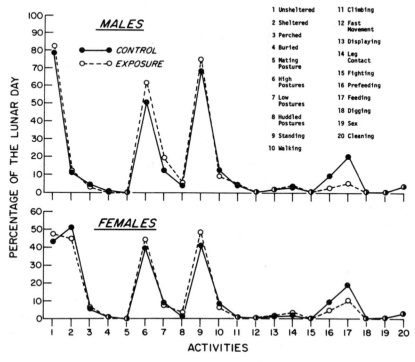

Figure 3. Activities budgets of male and female crabs in the second experiment.

PCB after 23 days while females averaged 79 ppm. A female exhibited the maximum concentration observed in either experiment, 190 ppm PCB.

Examination of Figure 2, the average activities budgets for male and female crabs in the first experiment, reveals that the budgets for control and low exposure conditions differ little except that females spend less time buried and

Table 2. The Proportion of the lunar day male and female crabs spent unsheltered.

Males			Females		
Proportion of time unsheltered		Carapace width (mm)	Proportion of time unsheltered		Carapace width (mm)
Individual	Replica Average		Individual	Replica Average	
.885		28.4	.739		25.5
.870	.869	24.0	.577	.589	21.2
.849		25.1	.451		21.1
1.000		28.7	.654		21.5[3]
.629	.739	24.4[1]	.458	.491	23.9
.589		23.6	.360		19.1
.958		30.5	.662		24.1
.907	.829	24.1	.473	.532	20.5
.623		24.4	.461		20.6
.998		30.0	.479		20.5
.875	.892	24.7	.462	.407	22.0
.803		24.6	.281		21.5
.868		29.5	.476		19.5
.642	.700	24.5	.326	.371	24.5
.591		26.0	.310		21.4
.753		23.6	.912		23.5
.745	.708	25.5	.626	.623	20.5
.626		28.5[2]	.330		21.4[3]
.916		25.5	.796		23.0
.756	.774	26.0	.638	.642	20.1
.649		24.6	.494		22.5
.766		24.2	.540		20.0
.688	.680	26.8	.405	.357	21.5
.606		24.2[2]	.127		21.5
Mean .774	.774	25.9	.502	.502	21.7
STD DEV .136	.081	2.2	.178	.113	1.6

[1] Crab lost both claws.
[2] Crab lost one claw.
[3] Female crab became ovigerous.

prefeeding under PCB exposure. Under moderate exposure both sexes considerably decrease the time spent feeding, and females spend more time sheltered but less buried.

The time consumed in feeding by both sexes is also markedly lessened under exposed conditions in the second experiment (Figure 3). Like the first, the second experiment finds the female crabs buried for a lesser time when exposed to PCB, but, unlike the first, sheltering by females does not increase.

To explore further the differences in activities budgets between ordinary and stressing conditions we performed multivariate discriminant analyses. When all 20 activities categories are used as variables (MANOVA Program of OSU Computer Center), discriminant functions for both sexes in each experiment do not clearly distinguish the budgets of the two conditions.

Next we performed stepwise discriminant analysis (BIOMED Program of OSU Computer Center), which is designed to choose the most parsimonious discriminant function. This discriminant function, in turn, can be used to indicate in which activities the budgets for ordinary and stressing conditions differ as well as to classify newly measured budgets as belonging to ordinary or PCB-exposed conditions. Essentially the BIOMED Program answers the question: If one were to pick n number of variables from the 20 available for a discriminant function, which particular activities would combine to give the discriminant function? The program selects variables stepwise, that is, one at a time, on the basis of the variables with the largest F-value. To enter the discriminant function a variable needed an F-value exceeding that for the 0.01 significance level.

The activities categories selected by the BIOMED Program for stepwise discriminant analysis are listed in Table 3 in the order in which they were selected. The repeated selection of a feeding category as the first variable in the discriminant function for both sexes and both experiments indicates that feeding behavior is consistently influenced by PCB exposure. A posture category also repeatedly appears in the four functions. For the males, locomotion is influenced in both experiments. Female crabs in both experiments show differences between the two budgets in shelter use, digging into the substrate, and cleaning.

At least two perspectives on the discriminant analyses exist. First, the inability to clearly distinguish the budgets when all 20 variables rather than five or six are used suggests that a small change in behavior, in our case in feeding, allows the crab to maintain its general activities. Qualitative observations not reflected in the quantitative data encourage this suggestion. The crabs shifted from feeding on sediment and particles in the sediment under ordinary conditions to feeding by scraping the algal growth from the chamber walls and rocks under exposed conditions. We plan more experiments to refine these observations. Such a shift from feeding on contaminated sand to other apparently less available food sources may serve to lessen the received dose of PCB and has implications for the cycling of this pollutant in the estuarine ecosystem.

Although such pollutant-induced shifts in the kind of feeding behavior found here have not been observed elsewhere, other crustaceans do exhibit depression

Table 3. Discriminant functions resulting from stepwise analyses with the activities categories given with their final coefficients and in the order of their selection.

	Order of Selection					
	1	2	3	4	5	6
First Experiment						
Males	Feeding	Climbing	Mating Posture	Huddled Posture	Fast Movement	
	+3030	+53	-1076	-545	-10232	
Females	Prefeeding	Buried	High Posture	Cleaning	Sheltered	
	-1915	-2391	+214	+848	-113	
Second Experiment						
Males	Feeding	Walking	Cleaning	Huddled Posture	Displaying	Perched
	-254	-2055	-3796	+352	+476	+263
Females	Prefeeding	Low Posture	Digging	Cleaning	Feeding	Unsheltered
	+0	+57	-1856	-79	+56	-0.31

of feeding in the presence of crude oil (1, 16). Exposure of *Pachygrapsus crassipes* to crude oil extracts inhibits the feeding responses (16) which include the similar prefeeding behaviors of *Hemigrapsus oregonensis* found in Table 1. The time which lobsters, *Homarus americanus*, use to find food is increased in the presence of crude oil extracts (1).

The other view of multivariate analyses has implications for the methodology of future pollutant-behavioral studies. If we had initially chosen to observe just the five or six categories selected by the step-wise procedure, we would have observed a considerable shift in behavior. Alternatively, if we had chosen to observe any of the other categories, we might very well have missed observation of a behavioral effect of PCB exposure and our predictions from the same experimental events would be far different. In behavioral studies such as this, one should look initially at the animal in the broadest terms and resist the temptation to focus attention down to a few aspects of behavior.

CONCLUSIONS

Under ordinary conditions male and female shore crabs, *Hemigrapsus oregonensis*, differ in their activities budgets. Under stressing conditions when PCB-contaminated sand is present in the habitat replicas both sexes consistently show a decrease in the time spent feeding but differ in other points of the activities budget where stress-induced shifts occur.

ACKNOWLEDGEMENTS

This work was supported by the U.S. National Science Foundation under Grant No. GX 37350 and by the U.S. Energy Research and Development Administration under Contract No. AT(45-1)-2227, Task Agreement 12 and is assigned number RLO-2227-T12-56.

BIBLIOGRAPHY

1. Atema, J., and L.S. Stein. 1974. Effects of crude oil on the feeding behavior of the lobster *Homarus americanus*. Environ. Pollut. 6:77-86.
2. Beckner, M. 1968. The Biological Way of Thought. University of California Press, Berkeley and Los Angeles. 200 p.
3. Cox, D.R. 1958. Planning of Experiments. Wiley, New York. 308 p.
4. De Blok, J.W. 1964. An apparatus to generate tidal fluctuations and a modification to render constant flow. Neth. J. Sea Res. 2:186-188.
5. Enright, J.T. 1970. Ecological aspects of endogenous rhythmicity. Annu. Rev. Ecol. Syst. 1:221-238.
6. Evans, F. 1964. A tide generating machine for laboratory use. Neth. J. Sea Res. 2:183-185.
7. Grosch, D. S. 1973. Reproduction tests: the toxicity for *Artemia* of derivatives from non-persistent pesticides. Biol. Bull. 145:340-351.
8. Koestler, A. 1967. The Ghost in the Machine. Henry Regnery Co., Chicago. 384 p.
9. Krebs, C.T., I. Valiela, G.R. Harvey, and J.M. Teal. 1974. Reduction of field populations of fiddler crabs by uptake of chlorinated hydrocarbons. Mar. Pollut. Bull. 5:140-142.
10. McLarney, W.O., D.G. Engstrom, and J.H. Todd. 1974. Effects of increasing temperature on social behavior in groups of yellow bullheads (*Ictalurus natalis*). Environ. Pollut. 7:111-120.
11. Nimmo, D.R., P.D. Wilson, R.R. Blackman, and A.J. Wilson, Jr. 1971. Polychlorinated biphenyl absorbed from sediments by fiddler crabs and pink shrimp. Nature (London) 231:50-52.
12. Russell, E.S. 1938. The Behavior of Animals. Edward Arnold, London. 184 p.
13. Slobodkin, L.B. 1968. Toward a predictive theory of evolution, p. 187-205. *In* R.C. Lewontin (ed.), Population Biology and Evolution. Syracuse University Press, Syracuse.
14. _____, and A. Rapoport. 1974. An optimal strategy of evolution. Q. Rev. Biol. 49:181-200.
15. Symons, P.E.K. 1970. Claw-shuddering behavior of the shore crab, *Hemigrapsus oregonensis*. Can. Field-Nat. 84:55-56.
16. Takahashi, F.T., and J.S. Kittredge. 1973. Sublethal effects of the water soluble component of oil: chemical communication in the marine environment, p. 259-264. *In* Center for Wetland Resources, Louisiana State University Publ. No. LSU-SG-73-01.
17. Todd, J.H., D. Engstrom, S. Jacobson, and W.O. McLarney. 1972. An introduction to environmental ethology: a preliminary comparison of sublethal thermal and oil stresses on the social behavior of lobsters and fishes from a freshwater and a marine ecosystem. Progress Report to the U.S. Atomic Energy Commission (AT (11-1)-3567). Unpublished manuscript, Woods Hole Oceanographic Institution. 104 p.

18. Underwood, A.J. 1972. Sinusoidal tide models: design, construction and laboratory performance. J. Exp. Mar. Biol. Ecol. 8:101-111.
19. Van der Steen, W.J. and J.C. Jager. 1971. Biology, causality, and abstraction, with illustrations from a behavioral study of chemoreception. J. Theor. Biol. 33:265-278.
20. Wiens, J.A. 1969. An approach to the study of ecological relationships among grassland birds. Ornithol. Monogr. No. 8. 93 p.
21. _____, S.G. Martin, W.R. Holthaus, and F.A. Iwen. 1970. Metronome timing in behavioral ecology studies. Ecology 51:350-352.
22. Wright, H.O. 1966. Comparative studies of social behavior in grapsoid crabs. Ph.D. Dissert. University of California, Berkeley. 227 p. University Microfilms, Ann Arbor, Mich. (Diss. Abstr. 27B:4184B-4185B).

SUBLETHAL EFFECTS OF PETROLEUM FRACTIONS ON

THE BEHAVIOR OF THE LOBSTER, *HOMARUS AMERICANUS*,

AND THE MUD SNAIL, *NASSARIUS OBSOLETUS*

Jelle Atema
Boston University Marine Program
Marine Biological Laboratory
Woods Hole, Massachusetts 02543

ABSTRACT: Studies by our laboratory on sublethal effects of petroleum fractions on behavior of *H. americanus* and *N. obsoletus* are summarized in an attempt to clarify contradictory results and to gain an understanding of the underlying principles. There appear a surprising number of similarities in the way specific petroleum fractions affect the behavior of a crustacean arthropod and a gastropod mollusc. 1) Acute toxicity, caused perhaps by the branched-cyclic fraction, was evident in short term exposures to whole #2 fuel oil at 50 ppm and kerosene in $\mu l/100$ l quantities. 2) Feeding attraction, caused perhaps by the branched-cyclic fraction in lower concentrations, was seen after exposure to 1 ppm #2 fuel oil, and to kerosene and its branched-cyclic and polar-aromatic fractions in $\mu l/100$ l quantities. 3) Repulsion and feeding inhibition, caused perhaps by the polar-aromatic fraction which contains most of the soluble materials, was observed in response to many concentrations of #2 fuel oil and kerosene, and their soluble fractions down to 1 ppb levels. 4) No effect was apparent in exposures to the lower concentrations of solubles of .1 and .01 ppm #2 fuel oil and to solubles of La Rosa crude at 10 ppb.

The hypothesis is advanced that 1) specific hydrocarbon fractions in specific amounts are responsible for distinct behavioral changes; 2) these fractions are present in varying quantities in different oils; and 3) the changes in behavior are general enough to affect a large number of marine invertebrates in a similar manner.

INTRODUCTION

If we are to determine safe limits for levels of oil pollution in the aquatic environment, we must understand the principal mechanisms by which petroleum hydrocarbons interact with living organisms. Only then can we predict with some measure of accuracy the general and specific problems encountered under the great variety of existing and possible types of oil pollution.

We are still far removed from this goal. Yet, the first steps have been taken, and this paper will attempt to sort out the results from a number of studies done in our laboratory over the last years. We have selected behavioral measurements of the effects of petroleum hydrocarbons because it is largely through behavior that individual organisms make their fitness known at the population and eco-system level: if behavior and fertility are not affected, no change in fitness may result.

Since oil induced changes can take place in any number of behavior patterns, it is important to select such behavior which is highly relevant to survival, is affected by oil, and one which is amenable to laboratory testing. We have generally tried to adhere to the following sequence of experiments: 1) Nature studies or close equivalents to determine the general context and the sensitivity of the animal and of specific types of behavior. 2) Laboratory studies to quantify promising behavioral measures. 3) Detailed, rigorously controlled laboratory tests to collect a data base. 4) Verification of laboratory test results with field studies on the same behavioral effects. This paper will deal with the second and especially the third step of experiments on lobsters and mud snails. These animals were chosen because of an existing body of background knowledge, obvious relevance to the ecosystem, availability, laboratory adaptability, and representation of different taxonomic and ecological groups.

It is the purpose of this paper to bring together the methods and results of these studies in an attempt to see general principles and to formulate hypotheses for experimentation which may eventually lead to the goal of understanding the mechanisms by which petroleum hydrocarbons can interfere with the normal functioning of aquatic animals.

EFFECTS OF VARIOUS PETROLEUM FRACTIONS ON THE BEHAVIOR OF THE MARINE MUD SNAIL, *NASSARIUS OBSOLETUS*

This snail is found in the salt marshes and mudflats of Eastern North America, from New Brunswick in the north to Cape Canaveral in the south, often as the predominant macroscopic species. It experiences a natural temperature range of $2°$ - $40°$ C, with daily fluctuations of as much as $20°$C over a period of minutes. In oil spill areas it has been observed to feed on the corpses of oil victims. In general, *N. obsoletus* is a hardy species which successfully exploits the rough life on the mud flats.

Effects of the Soluble Fraction of Kerosene on Feeding Attraction

This experiment was designed to test the possible interference of a sea water extract of kerosene with the chemically triggered upstream movement of *N. obsoletus*. The chemical attractants used were a chemically purified and stabilized extract of oysters, and a scallop homogenate. Tests were done in a Y-maze in which seawater at 15°C flowed at a rate of 1½ liters per minute in two upstream arms, and out from the third (downstream) arm. One of the upstream arms was randomly chosen to be the stimulus arm for each trial. The soluble fraction, consisting mostly of benzenes and naphthalenes, was prepared as described by Boylan and Tripp (6).

Four treatments were used: control (seawater flow only), kerosene extract at a calculated concentration of 1 ppb or 4 ppb, attractant, and attractant plus kerosene extract. For further details of this experiment see Jacobson and Boylan (7). The results are given in Tables 1 and 2.

Table 1.   Mean ± standard error of number of *N. obsoletus* in stimulus and nonstimulus arms in each treatment at 15 min. * = significantly different at or below the 0.05 protection level using Kramer's modification of Duncan's multiple range test. From (7).

	Attractant (oyster extract)	Attractant + kerosene extract (1 ppb)	Kerosene extract (1 ppb)	Control (seawater)
Stimulus arm x ± s.e.	*4.4±1.0	1.7±0.3	1.0±0.5	0.8±0.8
Nonstimulus arm x ± s.e.	0.6±0.4	1.3±0.3	1.7±0.7	0.3±0.3
Number of trials (10 snails/trial)	5	3	3	4

Table 2.   Mean ± standard error of number of *N. obsoletus* responding positively in stimulus and nonstimulus arms in each treatment. * = comparison of the mean values of attractant and attractant plus kerosene treatment using the Mann-Whitney U test showed that they were significantly different below the 0.05 level. From (7).

	Attractant (scallop homogenate)	Attractant + kerosene extract (4 ppb)	Kerosene extract (4 ppb)	Control (seawater)
Stimulus arm x ± s.e.	*7.0±0	*4.8±0.9	0±0	0.6±0.6
Nonstimulus arm x ± s.e.	0.2±0.2	0±0	0±0	0.3±0.3
Number of trails (10 snails/trial)	5	4	2	3

It appears that a) scallop homogenate is a more powerful attractant than oyster extract, perhaps because attractant compounds were lost in the preparation of oyster extract; b) mud snails do not move much upstream without food stimulus; c) kerosene extract alone does not cause a significant difference in this behavior, although a suggestive "no response" is observed in 4 ppb exposure (Table 2); d) kerosene solubles at 1 ppb effectively block feeding attraction to oyster extract, and interfere significantly at 4 ppb with the more powerful attraction to scallop homogenate. In similar experiments it was found that 0.8 ppb kerosene extract did not interfere with attraction to scallop homogenate (7). Thus, in general, kerosene solubles appear to interfere with feeding attraction in *N. obsoletus* in the range of parts per billion.

A hypothesis for a mechanism of interference is that feeding attraction and kerosene repulsion are balanced by *N. obsoletus*, which would result in the observed partial interference with the more powerful attractant and complete interference with the weaker one. The hunger state of the snail may well change the balance of these inputs: hungry snails may not be inhibited by kerosene repulsion at low exposure levels. More powerful attractants may have the same effects, which could perhaps explain how *N. obsoletus* was observed feeding on oil contaminated corpses. Furthermore, different petroleum fractions, as one can expect due to weathering of different crudes, will most likely have different effects on exposed organisms. The 'input-balance' hypothesis was tested further in the next experiment.

Effects of #2 Fuel Oil and its Soluble Fraction on Feeding and Alarm Response

The purpose of this experiment was to determine the levels of oil exposure necessary to interfere significantly with a) the snail's normal state of visibility, b) its attraction to food, and c) its alarm response. It had been established that *N. obsoletus* often remains buried when unstimulated, that it can be attracted rapidly to crushed mussel (*Modiolus demissus*), and that it will bury in the mud when exposed to an alarm substance which is liberated from a damaged conspecific. These responses are equally powerful in the field (1, 10) and in laboratory assays (5, 11).

For the oil interference experiments the snails were acclimated in the laboratory for 1 week at 20°C in large trays with mud. They were well fed and the water was constantly aerated. Test snails were taken from these trays and given one hour exposure to a certain level of #2 fuel oil [obtained from the reference oils of the American Petroleum Institute (8)] in glass bowls with mud substrate, 1 liter sea water and aeration. Ten snails were tested simultaneously and each test was replicated five times. After one hour exposure, the number of visible snails was counted. Visible snails were defined as those in which the notch of the shell which holds the siphon was visible. Then a half mussel was dropped in the center of the bowl through a glass tube which extended just below the water surface to avoid direct contact of the food with a possible surface film of oil. The number of visible snails was counted after 10 minutes. At that moment a crushed conspecific snail was added through the glass tube and the visible snails were counted again after 10 minutes. The levels of oil exposure were: 0 (control), 50 ppm, 10 ppm, 1 ppm whole #2 fuel oil and the solubles of 50 ppm, 10 ppm, 1 ppm, 0.1 ppm and 0.01 ppm #2 fuel oil. Whole oil was added to the bowls through a microsyringe. Solubles were prepared by slowly stirring an oil slick over seawater for 24 hours in a closed glass jug. The stock solution was made to correspond to the solubles of 50 ppm oil in sea water. Dilutions were made from this.

From the results shown in Table 3, two different and statistically significant effects of oil exposure can be seen: a) super attraction to food after exposure to 1 ppm whole #2 fuel oil, and b) super alarm response after exposure to 10 ppm whole oil and after the solubles of 50 ppm and 10 ppm oil. The second effect is seen also in exposures to 1 ppm whole oil and to the solubles of 1 ppm oil, although the results obtained here are not significantly different from control values. A third effect is not demonstrable with the data representation of Table 3: after exposure to 50 ppm whole oil, many of the snails were observed upside down, half extended from their shells and clearly "sick". Yet some feeding attraction and alarm response still occurred in most of these snails. From their general behavior it appears that these high levels of exposure are most likely toxic and thus beyond the scope of our study of sublethal effects on behavioral performance.

Table 3. *Nassarius obsoletus*: effects of different levels of #2 fuel oil exposure on spontaneous burial, food attraction, and alarm response. Mean and standard error (in parentheses) of number of visible snails. * indicates values significantly different from control (Duncan's multiple range test at p <0.05).

	No stimulus	Food	Alarm
Control (no oil)	3 (.89)	7 (.54)	2.6 (.51)
50 ppm oil	2.4 (.87)	7.2 (.86)	4.4 (.75)
10 ppm oil	2.6 (.60)	7.6 (.60)	*0.0 (.0)
1 ppm oil	2.6 (.75)	*9.4 (.24)	1.4 (.75)
sol. of 50 ppm	3.8 (.73)	8.6 (.68)	*0.2 (.20)
sol. of 10 ppm	1.4 (.24)	7.4 (.60)	*0.4 (.40)
sol. of 1 ppm	2.4 (.68)	6.4 (.40)	0.6 (.40)
sol. of 0.1 ppm	4.4 (.87)	7.6 (.68)	2.6 (.93)
sol. of 0.01 ppm	1.6 (.60)	5.2 (.92)	2.2 (1.02)

In conclusion, in the entire tested concentration range of .01-10 ppm there is no effect of whole #2 fuel oil or its solubles alone on the spontaneous burial state of *N. obsoletus*; there is a toxic effect after exposure to 50 ppm whole #2 fuel oil; there is an effect of increased attraction to food after exposure to 1 ppm whole oil; and there is a clear effect of exaggerated alarm responses at three levels (perhaps at all levels) down to the solubles of 1 ppm oil. The latter interpretation follows from values shown in the "alarm" column in Table 3: at the highest level of exposure the snails are "sick", at the two lowest levels (solubles of .1 and .01 ppm) there is no effect. In the middle range of exposures three values are significantly different from control, while the two others did not differ significantly with our methods.

It should be noted that neither whole #2 fuel oil, nor its solubles caused a reduction of feeding attraction as seen in the previous experiments with kerosene solubles. Exposure levels in the two studies were comparable, if we go roughly by the results of yet another experiment (3) where chemical analyses showed that the solubles of 1 ppm whole oil (in that study La Rosa crude, see below) amount to about 1 ppb. It can be argued that La Rosa crude contains far less soluble fraction than kerosene. However, even exposures to levels of two orders of magnitude higher or lower—solubles of 50 ppm and 0.01 ppm respectively—did not show the feeding reduction effect either. From this, one may conclude that either the food stimulus was much greater in the #2 fuel oil experiment, or that different hydrocarbon fractions in #2 fuel oil and in kerosene cause the difference in results. Since an examination of the methods suggests the former explanation, the input-balance hypothesis is given additional support from this study.

This experiment has led to a second hypothesis that #2 fuel oil contains two different fractions: one which attracts at medium concentrations (1 ppm whole oil) and another which repels at a wide range of medium concentrations (10 ppm whole oil to solubles of 10 ppm, perhaps as low as 1 ppm). These two effects potentiate the naturally occuring attraction to food and repulsion by alarm substance. The hypothesis of the two fractions will be examined in the following experiment with lobsters.

EFFECTS OF VARIOUS PETROLEUM FRACTIONS ON THE BEHAVIOR OF THE LOBSTER, *HOMARUS AMERICANUS*

Homarus americanus inhabits the coastal waters and continental shelf from Newfoundland to Cape Hatteras. This species is of considerable commercial value and the main predator on larger adults is man, whose fishing activities keep their population as small as the fishing laws permit. Oil pollution can not only reduce the size of the lobster population, it can also contaminate specimens marketed for human consumption. Yet it has been the practice in some areas to fish for lobsters with kerosene soaked bricks as bait in the traps. The apparent contrast between attraction to petroleum hydrocarbons as bait and their toxic effects deserved further investigation both from a practical and a theoretical point of view.

Effects of Kerosene Fractions on Feeding and General Behavior

Small groups of 3-5 mature and submature lobsters were kept at 20°C in slowly running seawater in 400 liter aquaria, where a large window permitted clear observation of the behavior of all individuals. They were given shelters and food (pieces of mussel and fish). Testing began when a certain level of social stability had been observed, which took from one to two weeks. Tests were conducted by introducing kerosene fractions on asbestos strips and recording the lobsters' behavior toward these strips. Asbestos was chosen because it was necessary to present a stimulus source within reach of the animals and because asbestos is an inert, absorbent material which can be chemically cleaned to avoid contamination with other materials. Forceps and other anti-contamination precautions were used, since it is known that lobsters can respond with intense feeding behavior to minute quantities of human hand rinse. Blank control strips and strips with 20 μl of kerosene fractions were introduced in a series of trials. The kerosene fractions were: 1) whole kerosene, 2) branched and cyclic fractions, 3) polar aromatics, 4) straight chain aliphatics. Further details can be found in Atema et al. (1973) (2).

The results of these tests are summarized below.

Control: occasional approach, no feeding behavior.
Whole kerosene: searching, feeding and ingestion.
Branched cyclic: searching, feeding and ingestion.
Polar aromatic: searching, repulsion when near, grooming, no feeding.
Straight chain aliphatic: similar to control.

It was concluded that kerosene contains inert (aliphatic), attractive (branched-cyclic, and to some extent polar aromatic), and repellent (polar aromatic) fractions. The feeding response to whole kerosene was less than to the branched-cyclic fraction alone, perhaps due to the presence of the polar aromatics in the former.

In a similar series of tests, lobsters were exposed to kerosene fractions (μl range) soaked on to pieces of brick, which were placed upstream from their shelter (9). Tests were conducted with single mature lobsters at night under low intensity red lights. Some differences with the previous study appeared and the results are listed below for comparison:

Control: lobsters often manipulate foreign objects in their range and control pieces of clean brick were no exception; no abnormal behavior was observed.
Whole kerosene: depressed activity, followed by some attraction and feeding attempts; food ingestion ceased for 3-7 days after kerosene feeding attempts.
Branched-cyclic: ambivalent behavior, aggressive postures toward brick, simultaneous feeding (by the maxillipeds) and rejection (by the pereiopods); followed by alarm and defensive postures, and reduced activity.
Polar aromatic: increased activity, spastic behavior, approaches and fast rejection of brick; food ingestion afterwards faster than normal.
Straight chain aliphatic: depressed activity, prolonged attempts at feeding; food ingestion ceased for 3-4 days after feeding attempts on this fraction.

Effective concentrations cannot be estimated in either of these two tests. The qualitative results, however, confirm the results of *N. obsoletus* tests and the two fraction hypothesis to the extent that both feeding attraction and repulsion are observed in sublethal exposures to various petroleum hydrocarbon mixtures. These lobster studies point in the direction of future research into the mechanisms of interference of specific fractions.

While these two studies were indicative for the different effects observed after exposure to different petroleum fractions, they were qualitative as to levels of exposure and exact behavioral measurements. The following study is an attempt to establish a quantitative methodology for the investigations of oil effects on lobster behavior.

Effects of La Rosa Crude and its Soluble Fraction on Feeding Behavior

In order to quantify both the chemistry and the behavior data, a simple design was adopted. Mature lobsters (about 9 cm carapace length) were kept in individual 100 liter aquaria with standing water and aeration at 22°C. They were acclimated without food for 2 weeks prior to experimentation under dim, ambient light. During the experiment they were fed once a day a half mussel, which was carefully lowered on a string in the opposite corner, so that the lobster responded only to the chemical stimuli of the food. Once a day general behavior was recorded in standardized units (3) for 10 minutes prior to food introduction. Feeding behavior was then timed in seconds from introduction to first alert (Alert), from alert to first movement of the whole animal (Wait), and from movement to first touching the food (Search). Behavior recordings were done for 10 days; the first 5 days without adding oil and the last 5 days under exposure to 1 ml La Rosa crude oil, which was pipetted on top of the water in each tank. Aeration stirred the slick, which practically disappeared over the course of the five days. There were 8 replicates of this treatment. A control group (4 replicates) did not receive oil for the entire 10 day period. A second group of experimental animals (8 replicates) received the soluble fraction of the same amount of crude for the last 5 days. Aliquots of water from the lobster tanks were taken on days 5, 6, 7, 8, and 10 for chemical analysis of petroleum hydrocarbons. Further details—on chemical analyses, etc.—are beyond the scope of this paper, but are given elsewhere (3, 4).

Surprisingly, no effects were found in the feeding and general behavior of lobsters exposed to the soluble fraction. Chemical analysis showed that about 10 ppb (parts per billion) of hydrocarbons were recovered from the seawater in these tests. However, exposure to 10 ppm whole crude had clearly measurable effects. Here, the waiting period of the feeding behavior had doubled; and the movements of some chemosensory related appendages had changed significantly. These changes were interpreted to result in decreased chemical stimulation and increased tactile behavior. The overall result was an increase in the time it took hungry lobsters to find food, which would put it at a selective disadvantage with its competitors in nature.

The exact behavioral measurements of this study were interpreted as support for the input-balance hypothesis: petroleum hydrocarbons (1 ppm) seem to compete negatively with feeding attraction in lobsters as well. This effect was not caused here by the soluble fraction at an effective concentration of about 10 ppb. The hunger state of the animals presents another input, which as yet has not been tested specifically.

GENERAL CONCLUSION OF OIL EFFECTS ON MARINE INVERTEBRATE BEHAVIOR

It is perhaps useful to assess the facts resulting from our work and to state the overall hypothesis derived from these studies. Although only two mature marine invertebrates have been investigated here, there appear a number of similarities in the behavioral responses of a crustacean arthropod and a gastropod mollusk to a number of different hydrocarbon fractions, see Table 4. These effects are:

1) Acute toxicity. Observed in both *H. americanus* and *N. obsoletus*; caused by whole #2 fuel oil at 50 ppm (*N. obsoletus*) and whole kerosene at μ1 per 100 liter quantities (*H. americanus*)[1]. Since this effect was caused also by the branched-cyclic fraction of kerosene, this may be the fraction responsible for the toxic effects. This fraction is similar in kerosene and fuel oils (2, 6).

2) Feeding attraction. Observed in both species; caused by #2 fuel oil at 1 ppm (*N. obsoletus*), and by whole kerosene and its branched-cyclic and polar-aromatic fractions in μ1 per 100 liter quantities *(H. americanus)*. Since the polar-aromatic fraction does not cause ingestion, but instead a strong repulsion at close range, it is likely the toxic branched-cyclic fraction which also causes feeding attraction and even ingestion.

3) Repulsion. Observed in both species; caused by #2 fuel oil at 10 and 1 ppm, by solubles of 1, 10 and 50 ppm #2 fuel oil, by the solubles of kerosene at 1-4 ppb (all in *N. obsoletus*), and by La Rosa crude oil at 10 ppm (*H. americanus*). The solubles are mainly polar aromatics, and it may be essentially this fraction which causes reduced feeding, repulsion, aggression, and the increased alarm responses. Contrary evidence suggested by the lack of effects of La Rosa solubles may be due to differences in concentration of solubles in light (#2 fuel oil, kerosene) and heavy (La Rosa crude) petroleum fractions.

4) No effect. Observed in *H. americanus* feeding behavior with solubles of La Rosa crude at 10 ppb, and in *N. obsoletus* behavior with solubles of .1 and .01 ppm #2 fuel oil. In both cases the input balance may have been in favor of feeding so as to render oil effects undetectable. Perhaps the lower limit of immediate detection by the animals was reached. In long term exposures very different values of upper and lower detectable limits are found.

---

[1] 10 ppm #2 fuel oil is acutely toxic to lobsters within hours.

Table 4.   Effects of different petroleum fractions on marine animal behavior.

Petroleum Fraction	Conc.	Exposure Time	Effects on Behavior	Type** Effect
			Nassarius obsoletus	
#2 fuel oil	50 ppm	1 hr	acutely toxic: interferes w. posture, locomotion	1
"	10 ppm	1 hr	increased alarm response: faster burial	3
"	1 ppm	1 hr	increased feeding attraction; slightly increased alarm response	2+3
solubles of #2	50 ppb?	1 hr	increased alarm response	3
"	10 ppb?	1 hr	increased alarm response	3
"	1 ppb?	1 hr	slightly increased alarm response	3
"	.1 ppb?	1 hr	no effect	4
"	.01 ppb?	1 hr	no effect	4
solubles of kerosene (largely polar aromatics)	1.4 ppb	15 min.	blocks or reduces feeding attraction and upstream movement	3
			Homarus americanus	
Kerosene	10 ppb range	min-hrs	attraction, feeding; depressed activity, stop feeding for days	1+3+2
*br-c of kerosene	"	"	attraction, feeding, aggression, alarm, depressed activity	2(+3+1)
*pol-ar of kerosene	"	"	attraction, repulsion, increased activity	3+2
*str-al of kerosene	"	"	no effect; depressed activity, stop feeding for days after contact	4(+1+2)
La Rosa crude	10 ppm	1-5 days	delayed feeding; change in chemosensory movements	3
solubles of La Rosa crude	10 ppb	"	no effect	4

*   br-c   :   branched cyclic fraction
     pol-ar: polar aromatic fraction
     str-al: straight chain aliphatic fraction

** effects: 1   acutely toxic
               2   attractant, feeding
               3   repellent, noxious
               4   no effect

The resulting hypothesis states that 1) specific hydrocarbon fractions are responsible for these distinct effects; 2) that these fractions are present in different oils in different concentrations; 3) that the effects are general enough in nature as to affect a large number of marine invertebrates in a similar manner; and 4) that biological factors, such as hunger state and stimulus strength, can modify the effects of oil interference (input-balance).

The responsible fractions need to be identified and their specific actions need to be investigated on a number of different organisms under standardized conditions, so that perhaps the mechanisms of interference can be detected. Once the mechanisms are known, the effects of petroleum hydrocarbons on marine life can be predicted with some measure of confidence.

REFERENCES

1. Atema, J., and G.D. Burd. 1975. A field study of chemotactic responses of the marine mud snail, *Nassarius obsoletus*. J. Chem. Ecol. 1:243-251.
2. _____, S.M. Jacobson, J.H. Todd, and D.B. Boylan. 1973. The importance of chemical signals in stimulating behavior of marine organisms: effects of altered environmental chemistry on animal communication, p. 177-197. *In* G. Glass, (ed.) Bioassay techniques in environmental chemistry. Ann Arbor Science Publ., Inc.
3. _____, and L.S. Stein. 1972. Sublethal effects of crude oil on lobsters (*Homarus americanus*) behavior. WHOI Tech. Rept. 72-74, 76 pp.
4. _____, and _____. 1974. Effects of crude oil on the feeding behavior of the lobster, *Homarus americanus*. Environ. Pollut. 6:77-86.
5. _____, and D. Stenzler. (in press) Biological and chemical properties of the alarm substance of the marine mud snail, *Nassarius obsoletus*.
6. Boylan, D.B., and B.W. Tripp. 1971. Determination of hydrocarbons in sea water extracts of crude oil and crude oil fractions. Nature 230:44-47.
7. Jacobson, S.M., and D.B. Boylan. 1973. Effect of sea water soluble fraction of kerosene on chemotaxis in a marine snail, *Nassarius obsoletus*. Nature 241:213-215.
8. Pancirov, R.J. 1974. Compositional data of API reference oils used in biological studies: #2 fuel oil, bunker C, Kuwait crude oil, and South Louisiana crude oil. Report No. AID.IBA.74. American Petroleum Institute.
9. Schmidt, D., and J. Atema. 1970. Effects of kerosene and fractions of kerosene on feeding behavior in the American lobster, *Homarus americanus*. Unpublished report.
10. Snyder, N.F.R. 1967. An alarm reaction of aquatic gastropods to intraspecific extract. Cornell Univ. Agr. Exp. St. Memorandum 403.
11. Stenzler, D., and J. Atema. (in press). Specificity and priority of the alarm response of the marine mud snail, *Nassarius obsoletus*.

AVOIDANCE RESPONSES OF ESTUARINE ORGANISMS

TO STORM WATER RUNOFF AND PULP MILL EFFLUENTS

Robert J. Livingston
Claude R. Cripe
Roger A. Laughlin
Frank G. Lewis, III
Department of Biological Science
Florida State University
Tallahassee, Florida 32306

ABSTRACT: An integrated study was made to determine the relationship of laboratory avoidance reactions of estuarine organisms to specific pollutants and their actual distribution in the field. Experiments tested the reactions of juvenile and adult blue crabs (*Callinectes sapidus*) to storm water runoff and low pH, and the avoidance responses of pinfish (*Lagodon rhomboides*) to bleached kraft mill effluents (BKME). The field reactions of these organisms to point sources of such pollutants were monitored for relatively long periods in the Apalachicola and Apalachee Bay systems (north Florida).

Juvenile and adult blue crabs avoided runoff water from clear-cut fields. They also avoided water with experimentally reduced levels of pH below 6.0. It was found that the low pH of the storm water runoff was a primary factor in the laboratory avoidance reaction of the crabs. However, field studies in areas directly affected by such runoff indicated that although adult blue crabs were usually not found in such water, juvenile crabs actually appeared in higher numbers in portions of the bay characterized by increased runoff and low pH. Although pinfish showed distinct laboratory avoidance responses to low levels of BKME (0.1%, volume/volume), their field distribution did not substantially change after a pollution abatement program was instituted. In this case, incomplete treatment and/or habitat destruction could have accounted for such findings. In both instances, however, it appears that factors other than avoidance response were contributory to the field distribution of the test species.

It was considered that the field response of estuarine organisms to various forms of pollution was the product of complex physical and biological functions such as intraspecific and interspecific competition and predation, habitat alteration, and trophic response. It appears that laboratory studies without associated field information can lead to spurious conclusions.

INTRODUCTION

Avoidance of polluted water by aquatic organisms has been considered to be one of the most important sublethal effects of pollution (31). Various investigations have demonstrated that aquatic species avoid pollutants. This includes reactions to heavy metals (8, 11, 14, 25, 29, 30, 31, 32, 33), pesticides (5, 6, 7, 15), pulp mill effluents (10, 17, 31), and other physico-chemical factors (2, 13, 24, 37). With few exceptions, most of these data are derived from laboratory studies which involve different forms of apparatus, experimental procedures, and test organisms.

The use of such behavioral data as an indicator of environmental contamination should be carefully reviewed. The issue of variable response to different gradients is unresolved (36). Some studies have shown that fishes avoid shallow gradients of sublethal concentrations (14, 35) while others have found such reactions to steep gradients (3, 11, 12, 26, 28, 29, 30, 32, 33, 34). Westlake and Kleerekoper (36) found that the locomotor behavior of goldfish placed in gradients of copper was dependent on the slope of the gradient. Steep gradients elicited avoidance while shallow ones did not. Dose-specific responses have also been described (4, 31) and the possibility exists that avoidance may actually reflect an attraction response for "clean" water rather than avoidance of a pollutant. With increased exposure time, adaptive behavior may lead to reduced avoidance; this reaction can be complicated by reactions to the experimental apparatus and design (1, 9). There are also complications of variable intraspecific behavioral response such as schooling which should be taken into consideration.

Perhaps the most difficult issue to resolve is the extrapolation of laboratory results to field conditions. In this instance, the dichotomous relationship of response to steep and shallow gradients often becomes irrelevant since most laboratory gradients are considerably steeper than those in the field. There have been few quantitative comparisons of field and laboratory avoidance. One such example involved the avoidance of Atlantic salmon to mine wastes (notably zinc and copper) in a Canadian river (26, 29, 32). The level of pollution in the field which caused disturbed movements of the salmon was actually 18 times higher than laboratory thresholds. The variety of external (environmental) factors which influence the behavior of such organisms in their natural surroundings would tend to complicate such comparisons. This is especially true of estuarine organisms which live in a dynamic and extremely changeable environment.

This paper is based, in part, on an integrated field and laboratory program involving two bay systems in north Florida. Laboratory avoidance experiments with storm water runoff and Bleached Kraft Mill Effluent (BKME) have been carried out in conjunction with long-term field surveys in Apalachicola Bay and

Apalachee Bay (19, 20, 23). The main purpose is to compare the laboratory—determined avoidance response with the field distribution of two species of estuarine organisms.

METHODS AND MATERIALS

Avoidance responses of blue crabs (*Callinectes sapidus*) to storm water runoff were analyzed in the Apalachicola Bay System. For the past 2-3 years, extensive portions of upland areas adjacent to East Bay have been clear-cut, ditched, and drained directly into the bay (Fig. 1). During periods of heavy local rainfall (summer-early fall), highly colored water of relatively low pH washed off the cleared fields and into the upper portions of East Bay. Commercial fishermen complained that organisms such as blue crabs became scarce when "black" water

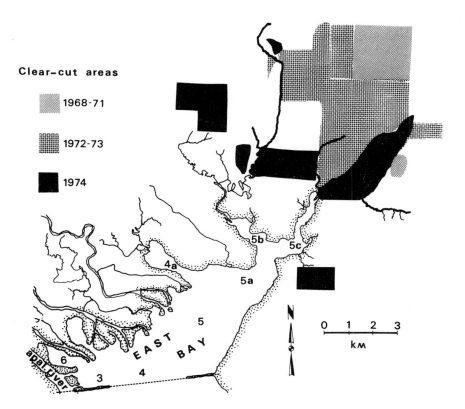

Figure 1. Station locations in East Bay for the blue crab field program. A chronological summary of clearcutting operations in the upland drainage area is also shown (1968 to present). East Bay is part of the Apalachicola Bay System (North Florida), a shallow, highly productive, river-forced estuary.

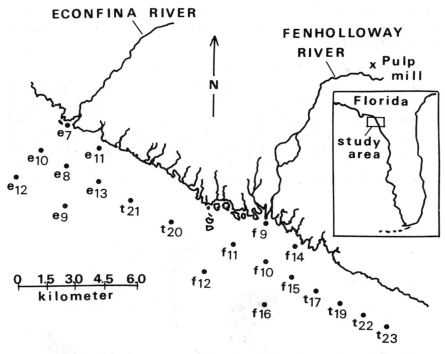

Figure 2. Station locations in Apalachee Bay (North Florida) showing portions of a polluted (Fenholloway) and unpolluted (Econfina) drainage area. Fixed stations in
the two systems are placed in comparable positions with a transect (E10-T23)
running through both areas.

appeared in the bay. A field and laboratory study was initiated (18) to determine the impact of such runoff on the East Bay system. Farther down the coast
(Fig. 2), another long-term field program was underway (19, 22, 39) to determine the impact of BKME on the biota of Apalachee Bay. Results indicated that
runoff from the polluted Fenholloway River caused reductions in the benthic
macrophytes and epibenthic fishes and invertebrates in areas affected by the
BKME (19, 38). Laboratory avoidance experiments were carried out with pinfish
(*Lagodon rhomboides*) to determine possible behavioral reactions of this dominant species to the kraft mill effluents. In both cases, various experimental
approaches were used depending on the size and behavioral repertoire of the
subject organisms.

Avoidance experiments with blue crabs

Control water for avoidance experiments was taken from East Bay (Station
3); the primary source of such water was the Apalachicola River which has a

significant influence on the environmental conditions in the Apalachicola Bay System (18, 20). Water to be tested for comparative effects came from Sandbank Creek located at the head of East Bay approximately 19 Km NE of the Apalachicola River. This creek receives drainage from clearcut areas and flows directly into East Bay (Station 5C). Water was pumped into 3600 liter tanks and transported to the Florida State Marine Laboratory where it was deposited in permanent storage tanks. All water was used within 48 hours of retrieval. The control (Station 3) and experimental (Creek) water was moved to a water treatment system (21) for standardization of key water quality factors (e.g.; temperature, salinity, dissolved oxygen). It then was pumped into a controlled environment room for the behavioral experiments.

All field and laboratory water quality determinations were carried out according to established methodology (23). Temperature was monitored with a stick thermometer while salinity was determined with a temperature-compensated refractometer. Color was measured with a (Hach) APHA platinum-cobalt standard test while turbidity was estimated with a (Hach) model 2100 A turbidimeter. Field water samples were taken with a 1 liter Kemmerer Bottle. Dissolved oxygen was measured with a Y.S.I. oxygen meter and pH was determined with a Leeds and Northrup portable pH meter. Field and laboratory water quality were thus continuously monitored; all experiments were carried out under controlled conditions of temperature, salinity, dissolved oxygen, pH, light intensity and periodicity, etc. (21).

Juvenile (20-60 mm; carapace length) and mature (61-120 mm) blue crabs were used in the avoidance experiments. All experimental organisms were taken with 5 m otter trawls in East Bay. Juveniles were tested in groups of 9-11 individuals in Y-maze experimental troughs constructed of white plexiglass. Dye tests prior to experimentation showed that the apparatus created a steep gradient between the two water sources. Experiments were run initially with control water and highly colored water (acidic) from the upland creek. In addition, a series of tests was run to test for the effect of color where runoff water was buffered to a pH level equivilent to that of control water. Also, the factor of pH was tested by graduated reduction of the pH of control water through metered additions of HCl. Prior to each experiment, subjects were placed in the holding area at the base of the "Y". Water was introduced into each arm at the rate of 4 liters/minute. After a 10 minute holding period, a barrier was remotely lifted and crabs were allowed to enter the Y-maze. Movements were recorded with a Sony AV 3650 videocorder (with camera and video monitor). Thirty minutes later, the water sources were exchanged and the experiment continued for an additional period of 30 minutes under reversed conditions. The adult crabs were tested singly in shortened Y-mazes. After a ten minute holding period, observations were made for 15 minutes under experimental conditions. Water sources were then exchanged and crabs were observed for an additional 15 minute period. Thus, various experiments were run with juvenile and adult blue crabs to determine their reaction to runoff water, experimentally reduced pH, and buffered, highly colored solutions. The experimental range of pH was determined directly from previous field data in the areas of concern.

Avoidance experiments with pinfish

Experiments were carried out with individual fish in a steep gradient trough similar to that used by Sprague (30) and Sprague and Drury (31). Seawater flows of 4 liters/minute entered at either end of the apparatus with drains located at the center. Data were recorded with the closed circuit T. V. as described above. Effluent solutions (BKME), provided daily from a lagoon outfall by the Buckeye Cellulose Corporation (Foley, Florida), were filtered and adjusted to chemical oxygen demand (C.O.D.) of 750 mg/1. Effluent could be introduced at either end of the trough. Controls were run without effluent at the onset of each experiment; the position of the fish was recorded once every 30 seconds for the duration of the experiment (60 minutes). Fish showing a preference for either side of the apparatus were discarded. Those fish displaying no such preference were then exposed to concentrations of 0.01, 0.1, 1.0, and 10.0% (volume/volume) of BKME which was introduced into one end of the trough. Effluent concentrations were tested singly or in a series of increasing concentrations. A more detailed description of the experimental apparatus and procedure is given by Lewis (16) and Lewis and Livingston (17).

Statistical methodology

The avoidance response of juvenile blue crabs was evaluated with the test statistic "Z" which checks for the equality (null hypothesis) of two binomial proportions.

$$Z = \frac{P_1 - P_2}{\sqrt{P_1 (1-P_1) / N_1 + P_2 (1-P_2) / N_2}}$$

where

$$P_1 = \frac{\text{No. organisms in control arm at end of first 30 minute interval}}{\text{No. organisms in both arms at end of first 30 minute interval } (N_1)}$$

$$P_2 = \frac{\text{No. organisms in experimental arm at end of second 30 minute interval}}{\text{No. organisms in both arms at end of second 30 minute interval } (N_2)}$$

$$1 - P_1 = \frac{\text{No. organisms in experimental arm at end of first 30 minute interval}}{\text{No. organisms in both arms at end of first 30 minute interval}}$$

$$1 - P_2 \qquad \frac{\text{No. organisms in control arm at end of}}{\text{No. organisms in both arms at end of}}$$
$$\frac{\text{second 30 minute interval}}{\text{second 30 minute interval}}$$

Animals in the holding area were not included in the count. This statistic becomes more significant with greater values of N_1 and N_2. Avoidance is considered when the number or organisms in the control arm at the end of the first 30 minutes was statistically greater than that found in the same arm at the end of the second 30 minute interval (after the switch of the water sources). This results in an increase in the "Z" value and rejection of the null hypothesis. Significance was based on the following:

$$P < 0.05: \quad Z \geqslant +1.96$$
$$Z \leqslant -1.96$$

$$P < 0.001: \quad Z \geqslant +2.58$$
$$Z \leqslant -2.58$$

Net significant avoidance was calculated by the following:

$$\text{Avoidance Index} = \frac{P_1 - P_2}{P_1} \times (100)$$

The Avoidance Index takes into consideration the proprotions of organisms in an arm without considering the number of animals selecting a particular side of the trough; this means that the "Z" value and the Avoidance Index are not necessarily related.

Statistical analysis of the larger blue crabs was based on the relative time spent in the two areas of the trough. For each 30 minute period, the cumulative frequency distribution of individual visits (time) in the experimental water was compared with that in control areas. Significant difference was tested with the chi^2 approximation of the Kolmogorov-Smirnov, two-sample one-tailed test (27). A cumulative frequency distribution for each sample of observations (e.g., individual visits by each crab) was made using the same intervals for both distributions. The sampling distribution of the chi^2 approximation was used:

$$\chi^2 = \frac{4 D^2 (M_1 M_2)}{(M_1 + M_2)}$$

M_2 = Number of visits in experimental water on a cumulative time-frequency basis

D = maximum deviation between the two cumulative frequency distributions

M_1 = number of visits in control water

Another measure of avoidance involved the amount of time spent in control water as a percentage of the total (less that spent in the holding area). Any value

Table 1. The avoidance response of juvenile blue crabs to runoff water (upper section of table) and to low-pH control water (lower section). Values in parentheses represent the number of organisms that made a choice.

Carapace width ($\bar{X}$ + SD)	Date		Percent of Organisms in		Z	Avoidance index	pH		Color		Turbidity	
			Control arm before switch	Exp. arm after switch			Control	Exp.	Control	Exp.	Control	Exp.
(28.1 ± 4.67)	2/20/75	1.	100 (7)	10 (10)	9.48**	90.0	7.75	5.2	55	245	5.9	6.4
(44.3 ± 7.48)	3/19/75	2.	60 (5)	0 (6)	2.73***	100.0	7.75	5.8	85	300	8.0	25.0
(42.7 ± 7.95)		3.	86 (7)	0 (9)	6.48***	100.0	7.75	5.8	85	300	8.0	25.0
(42.5 ± 5.79)	4/03/75	4.	80 (5)	25 (8)	2.33*	69.0	7.8	5.8	140	140	34.0	34.0
(35.9 ± 2.42)		5.	100 (7)	29 (7)	4.18***	71.0	7.9	5.8	140	140	34.0	34.0
(45.1 ± 3.07)	4/17/75	6.	100 (6)	20 (5)	4.47***	80.0	7.9	6.0	85	83	16.0	16.0
(42.7 ± 3.23)		7.	29 (7)	13 (8)	0.77	55.0	8.0	6.0	85	85	16.0	16.0
(31.5 ± 2.11)	4/24/75	8.	75 (8)	10 (10)	3.69**	86.0	8.1	5.0	135	135	40.0	40.0
(40.4 ± 2.71)		9.	75 (8)	0 (6)	4.90**	100.0	8.2	5.0	135	135	40.0	40.0
(39.3 ± 3.53)	5/01/75	10.	33 (6)	14 (7)	0.82	48.6	8.2	6.45	105	105	18.5	18.5
(48.8 ± 1.75)		11.	100 (5)	100 (7)	0.00	0.0	8.5	8.5	105	105	18.3	18.3

**Significance with p < 0.001
*Significance with p < 0.05

greater than 50% indicated possible avoidance. The switch of the water sources served as a control for possible preference of a given arm of the apparatus. The mean time-response of a given organism in a single test was thus computed, and the average time spent in control and experimental water (e.g., reduced pH or upland drainage water) was also used in the calculations.

The pinfish experiments were also analyzed using a chi^2 significance test (χ^2 > 3.84; P < 0.05). Results of avoidance were expressed as time spent in clean water as a percentage of total test time.

Associated field studies

All field data concerning the distribution of blue crabs in East Bay and pinfish in Apalachee Bay were taken from two long-range programs (18, 19, 20) as described above. All biological collections were based on seven (2-minute) otter trawl samples taken monthly at fixed stations (Figs. 1 and 2). Such collections were preserved in 10% formalin, sorted, identified to species, measured (fishes, standard length; crabs, carapace width), and counted. Data analysis was carried out with an interactive computer program designed for analysis of comprehensive field data (18).

RESULTS
Blue crab studies

Results of the juvenile blue crab experiments are given in Table 1. Organisms avoided the runoff water characterized by low pH and high color; a significant response was obtained in all experiments. There was also significant avoidance to pH conditions lower than 6.0 indicating that reduced pH in the runoff water could be a primary factor in the avoidance reaction. The avoidance index (A. I.) was inversely related to pH. The threshold level (i.e., that stimulus which elicited a positive avoidance by 50% of the experimental population) was estimated at a pH of 6.8 (Fig. 3a). The A. I. of 0 obtained at pH 8.5 was not included in this graph because an insignificant avoidance response was found at pH 6.5 which indicated a leveling of the curve at this point. A comparison of the A. I.'s showed that there was a relatively higher avoidance response to runoff water than the experimentally reduced pH; this indicates a possible reaction to something other than reduced pH. The highest A. I. occurred in colored runoff water at a pH of 5.8.

The results of the experiments with the adult blue crabs are given in Table 2. The runoff water elicited an avoidance response where the crabs spent 92.9% of the time in the control arm. The proportion of organisms showing a significant avoidance reaction to reduced pH (Kolmogorov-Smirnov method) was greater than 70% (Fig. 3b). When analyzed as a ratio of mean time spent in control water to that spent in experimentally adjusted control water (Fig. 3c), there was a similar (pH-directed) avoidance response under both sets of conditions. In each case, the response curve leveled off in a pH range from 6.0 to 7.0. There was an inverse relationship of the mean avoidance and the standard deviation of the response (Fig. 3d). This illustrates the increasingly mixed reaction of the experimental subjects at pH values between 6.0 and 7.0. The test results with buffered runoff (Table 2, Fig. 3b), with a mean % in control water of 56.1, indicates that color *per se* had little influence on the avoidance response. Overall, the adult blue crabs had a similar avoidance reaction to the runoff water as the juveniles; pH seemed to be an important causative agent in such behavior although there was a strong possibility that other factors were involved in this response.

Results of the field studies in East Bay are shown in Figs. 4 and 5. When compared to data taken prior to the major clear-cutting activities (1970-1971),

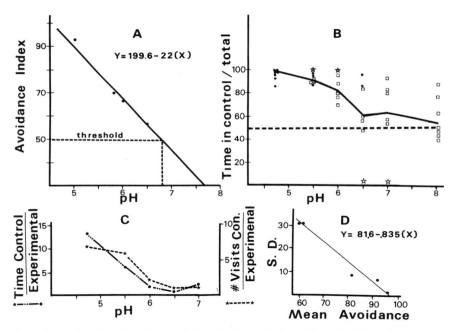

Figure 3. A. Relationship of pH and the Avoidance Index of juvenile blue crabs. Experimental A.I.'s at each pH represent mean figures; data from experiments with upland runoff are excluded.

B. Time-related avoidance response of adult blue crabs to experimentally reduced pH; circles represent significant avoidance; stars represent inapplicable tests of significance; open squares represent no significant response; total time = time spent in both arms of trough.

C. Comparisons of one series of ratios (mean time spent by adult blue crabs in control water to that spent in experimentally reduced pH) with another (mean number of visits by adult blue crabs in control water to that in experimentally reduced pH).

D. Comparison of mean avoidance response (%) to reduced pH by adult blue crabs with the standard deviation of individual reactions. (S.D.)

Table 2. Results of avoidance experiments with adult blue crabs. This includes tests of avoidance with runoff water from the field (*), control water with experimentally adjusted pH (**), and field runoff water with experimentally adjusted pH (***).

pH	Time spent in control water as % of total time spent in both arms of test apparatus	95% confidence interval for response	Mean time (sec.) spent in a visit to: Control	Exp.	Mean # of visits to: Control	Exp.	# of organ. tested	Carapace width ($\bar{X} \pm SD$)	Color Control	Exp.	pH in Control water
4.6	92.9*	(85.5, 100)	–	–	–	–	10	(85.7 ± 6.7)	60	445	8.1
4.6	96.0**	(93.8, 98.1)	174	13	78	11	7	(87 ± 12.5)	40	40	8.1
5.5	92.0**	(85.6, 98.4)	168	28	130	25	7	(94.3 ± 8.2)	50	50	8.1
6.0	81.5***	(74.3, 88.7)	83	33	81	33	7	(87 ± 6.9)	30	30	8.1
6.5	61 **	(33.1, 88.5)	251	246	26	20	7	(82 ± 6.4)	20	20	8.1
7.0	62.7**	(34.8, 90.5)	87	87	49	32	7	(89 ± 10.7)	–	–	8.1
8.2	56.1***	(47.1, 65.1)	144	180.6	23	16	7	(112 ± 13.9)	20	340	8.2

there was a reduction in pH levels in portions of the bay receiving drainage from such areas (5b and 5c) during 1974-75. This was particularly pronounced during March (5c) and the latter portion of the summer (5b, 5c) when local rainfall was heavy. The number of adult blue crabs found in such areas was less than half that taken in the control area (4a). However, juvenile blue crabs did not follow this pattern. At station 5c, there were actually increased numbers of individuals at times when ambient pH levels were low due to upland runoff. At station 5b, the second highest number of blue crabs was taken in water of pH 4. In terms of numbers, both stations compared favorably with control areas (4a) even though there were considerable differences in pH during certain periods of high runoff. The data indicate that although both juvenile and adult blue crabs showed a similar avoidance response to upland water and reduced pH under experimental conditions, field data gave divergent results. Juvenile blue crabs were actually found in increased numbers in areas of the bay characterized by high levels of upland runoff and low pH. Thus, levels of pH which induced a pronounced avoidance under experimental conditions appeared to exert no such effect in the field.

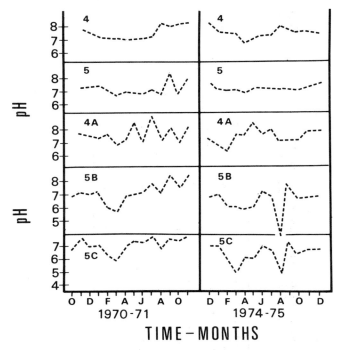

Figure 4. Field levels of pH at selected stations in East Bay prior to (1970-71) and after (1974-75) the clearcutting operations in contiguous upland areas. Data for 1970-71 were provided by Mr. John Taylor, Jr. of the Florida Division of Health.

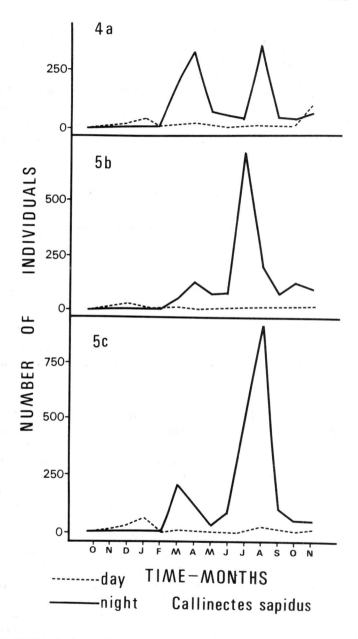

Figure 5. Field collections of blue crabs (*Callinectes sapidus*) taken in East Bay from March, 1975 to November, 1975. Collections were made during the day and at night at all three stations.

Pinfish studies

The avoidance responses of 30 pinfish to various concentrations of BKME are shown in Fig. 6a; mean levels of such behavior are given in Table 3. There was no significant avoidance at the lowest concentration. Over 70% of the fish avoided 0.1% BKME and there was significant avoidance for all fish tested at 1.0 and 10.0% BKME. With the exception of the response to 0.01% levels, tests of analysis of variance indicated that the reactions to BKME were significantly ($P <$ 0.01) different from controls. The avoidance threshold for $L.$ $rhomboides$ was estimated at 0.06% BKME (16, 17).

Table 3. Avoidance responses of $Lagodon$ $rhomboides$ to various concentrations of bleached kraft mill effluent. Avoidance is given as time spent in clean water as a percentage of total test time.

Concentration	Mean Response	95% Confidence Interval for Response
Control	50.4*	49.4-51.4*
0.01%	51.6*	49.6-53.6*
0.1%	60.3	58.6-62.0
1.0%	73.5	65.3-81.7
10.0%	80.8	78.8-82.8

*Indicates a non-significant response

In January, 1974, the Buckeye Cellulose Corporation instituted a pollution control program which resulted in a dramatic decrease in the B.O.D. levels of effluent leaving the plant (Fig. 6c). Prior to such a program, there were considerable differences in the numbers of pinfish taken at Fenholloway stations and their respective control stations (Table 4); recovery was seen at stations F12 and F16. Total numbers of pinfish taken at transect stations over a 4 year period are given in Fig. 6b while totals from the 3 inner stations of the Econfina and Fenholloway systems respectively are shown in Fig. 6c. These results indicate that numbers of pinfish taken in areas affected by BKME were considerably less than those taken in the control system. There appeared to be no response of this species to the improvements in the water quality in the area. Thus, although there was a clear-cut laboratory avoidance response of pinfish to relatively low levels of BKME, there was no sign of ecovery in the environment for over 1 year subsequent to a marked improvement in the water quality. Since the possibility exists that the pollution control effort was not totally effective in the elimination of effluents which could elicit an avoidance response, such avoidance could not be ruled out in the field.

Lagodon rhomboides

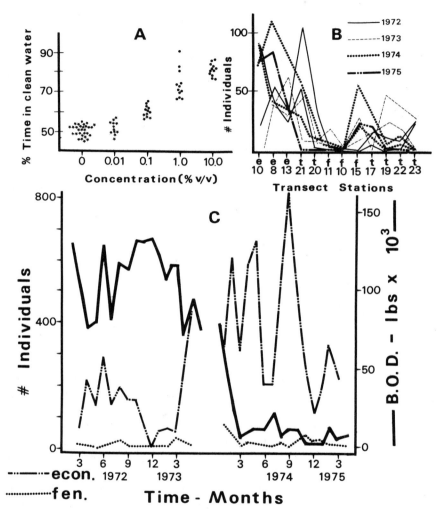

Figure 6. A. Experimental avoidance response of pinfish to BKME. Circles represent significant avoidance while stars represent no significance.

B. Transect data representing spring and fall collections of pinfish (number of individuals) taken along a transect in Apalachee Bay from 1972-1975.

C. Comparison of pinfish collections taken at 3 stations in the Econfina (E_7, E_8, E_{10}) and Fenholloway (F_9, F_{10}, F_{11}) systems prior to (1972-73) and after (1974-75) a pollution abatement program. B.O.D. figures (taken from an effluent canal leading into the Fenholloway River) were provided by Mr. J.H. Millican of the Buckeye Cellulose Corporation.

Table 4. Comparison of numbers of individuals, color and estimated BKME strength taken at comparable stations on the Econfina (E) and Fenholloway (F) rivers.

| Species | Stations | Individuals | Color | | Estimated BKME Strength (%) |
			Mean	Standard Deviation	
Lagodon	F5b	0	728	392	20.4
rhomboides	E4a	53	231	192	
	F9	61	283	277	7.3
	E7	760	105	122	
	F10	74	113	90	2.3
	E8	1059	56	62	
	F11	7	143	108	2.6
	E10	1006	79	77	
	F12	332	79	57	1.4
	E12	368	47	52	
	F14	30	134	90	2.2
	E11	621	81	84	
	F15	58	109	105	2.3
	E13	471	53	58	
	F16	164	62	59	--
	E9	300	42	62	

DISCUSSION

The laboratory avoidance data are consistent with previous studies; the pinfish response to BKME falls roughly within values observed by other investigators (10, 31). The pH range in Apalachicola Bay and the BKME concentrations in Apalachee Bay were within the range of levels which elicited marked laboratory avoidance in blue crabs and pinfish. Both species are capable of detecting such levels of effluents; however, there is a possibility that other factors in the field modify the relatively straight-forward laboratory behavioral reactions.

In both study areas, there is a dynamic situation with respect to environmental gradients which constantly undergo short- and long-term changes. Such variations are determined by several external factors. The inverse relationship of the field distribution of juvenile and adult blue crabs could indicate intraspecific predation patterns and/or developmental modifications of reactions to microhabitat displacement. Gradients of pH in the field are ephemeral and are usually in the order of hundreds of meters. This in itself presents a problem when attempting to apply laboratory results since all controlled tests were carried out with relatively steep gradients. In the Econfina-Fenholloway system, associated studies (38, 39) have shown that benthic macrophyte distribution and biomass were adversely affected by the dispersal of BKME in shallow offshore areas. A Pearson correlation analysis of water color (as an indicator of BKME) and pinfish numbers per station showed a low but significant relationship of these factors ($R = -0.2$.; $P < 0.02$, $N = 96$). A much higher correlation was made between number of fish and macrophyte biomass ($R = 0.47$; $P < 0.01$, $N = 96$). This plus the fact that pinfish did not return to previously polluted areas for over a year after there was significant improvement in water quality would indicate that other mechanisms could be involved such as habitat alteration (destruction of grassbeds), possible trophic responses, and other functions related to interspecific interactions. Contrary to the blue crab study, however, the field evidence in the Econfina-Fenholloway System does not directly conflict with the potential importance of avoidance behavior as a contributory factor to the observed distribution of pinfish. There are simply too few data to make a direct comparison. However, the data do indicate that other determinant functions are operational, and that the field distribution of this species could be determined by various environmental conditions other than water quality *per se*.

Both field studies are continuing, and the above data must be considered as preliminary. However, there is ample evidence that while laboratory results provide valuable insight into various problems involving specific biological reactions to aquatic pollutants, such studies without associated field information can lead to spurious conclusions.

ACKNOWLEDGEMENTS

The authors acknowledge the assistance of the Statistical Consulting Center of Florida State University for aid in the statistical analysis of the data. We also thank Dr. Tim S. Stuart and the Florida Department of Pollution Control for support in the field operations in East Bay. We are particularly indebted to Mr. J. H. Millican and his staff at the Buckeye Cellulose Corporation for their cooperation and support of both projects. We thank Mr. John Taylor, Jr. for his help with field data. Thanks are also due to Dr. Robert C. Harriss and his staff at the Edward Ball Marine Laboratory. Operations in the laboratory and the field were funded by grants from the NOAA Office of Sea Grant, Department of Commerce (Grant Number 04-3-158-43) and the Board of County Commissioners of Franklin County, Florida. Data analysis was carried out with a computer program developed with support from EPA Program Element #1 BA 025 under Grant Number R-803339.

LITERATURE CITED

1. Anderson, John M., and Margaret R. Peterson. 1969. DDT: Sublethal effects of brook trout nervous systems. Science 164: 440-441.
2. Cherry, D. S., K. L. Dickson, and J. Cairns, Jr. 1975. Temperatures selected and avoided by fish at various acclimation temperatures. J. Fish. Res. Bd. Canada 32: 485-491.

3. Grande, M. 1967. Effect of copper and zinc on salmonid fishes. *In* Advances in Water Pollution Research. Proc, 3rd Int. Conf. Munich, Germany 1: 75-95.
4. Hansen, D. J. 1969. Avoidance of pesticides by untrained sheepshead minnows. Trans. Amer. Fish. Soc. 98: 426-429.
5. _____. 1972. DDT and malathion: Effect of salinity selection by mosquitofish. Trans. Amer. Fish. Soc. 101: 346-350.
6. _____, E. Matthews, S. L. Nall, and D. P. Dumas. 1972. Avoidance of pesticides by untrained mosquitofish, *Gambusia affinis.* Bull. Environ. Contam. Toxicol. 8: 46-51.
7. _____, S.C. Schimmel and J.M. Keltner, Jr. 1973. Avoidance of pesticides by grass shrimp (*Palaemonetes pugio*) Bull. Environ. Contam. Toxicol. 9: 129-133.
8. Ishio, S. 1964. Behavior of fish exposed to toxic substances. Adv. Wat. Pol. Res., Proc. 2nd Conf. Tokyo; Editor O. Jaag. p. 19-40.
9. Jackson, D.A., J.M. Anderson and D.R. Gardner. 1970. Further investigations of the effect of DDT on learning in fish. Can. J. Zool. 48: 577-580.
10. Jones, B. F., C.E. Warren, C.E. Bond, and P. Doudoroff. 1956. Avoidance reactions of salmonid fishes to pulp mill effluents. Sewage Ind. Wastes 28: 1403-1413.
11. Jones, J.R.E. 1947. The reactions of *Pygosteus pungitius* L. to toxic solutions. J. Exp. Biol. 24: 110-122.
12. _____. 1948. A further study of the reactions of fish to toxic solutions. J. Exp. Biol. 25: 22-34.
13. _____. 1952. The reactions of fish to waters of low oxygen concentration. J. Exp. Biol. 29: 403-415.
14. Kleerekoper, H., G.F. Westlake, and J.H. Matis. 1972. Orientation of goldfish (*Carassius auratus*) in response to a shallow gradient of a sublethal concentration of copper in an open field. J. Fish. Res. Bd. Canada 29: 45-54.
15. Kynard, Boyd. 1974. Avoidance behavior of insecticide susceptible and resistant populations of mosquitofish to four insecticides. Trans. Amer. Fish. Soc. 103: 557-561.
16. Lewis, F. G., III. 1974. Avoidance reactions of two species of marine fishes to kraft pulp mill effluent. M.S. Thesis, Florida State University, Tallahassee. 79 p.
17. _____, and R. J. Livingston. (in press) Avoidance of bleached kraft mill effluents by two marine fishes. J. Fish. Res. Bd. Canada.
18. Livingston, R.J. 1974. Field and laboratory studies concerning the effects of various pollutants on estuarine and coastal organisms with application to the management of the Apalachicola Bay System (North Floirda, U.S.A.) Florida Sea Grant Report # R/EM-1: 574 p.
19. _____. 1975. Impact of pulp mill effluents on estuarine and coastal fishes in Apalachicola Bay, Florida U.S.A. Mar. Biol. 32 (1): 19-48.
20. _____. (in press) Diurnal and seasonal fluctuations of estuarine organisms in a north Florida estuary. Est. Coastal Mar. Sci.
21. _____, C.R. Cripe, C.C. Koenig, F.G. Lewis, and B.D. DeGrove. 1974. A system for the determination of chronic effects of pollutants on the physiology and behavior of marine organisms. Florida Sea Grant Prog., Rept. 4: 1-15.

22. _____, K.L. Heck, T.A. Hooks and M.S. Zimmerman. 1972. The ecological impact of pulp mill effluents on aquatic flora and fauna of north Florida: comparison of a polluted drainage system (Fenholloway) and an unpolluted one (Econfina). Report for Florida Coastal Coordinating Council (unpublished).

23. _____, R.L. Iverson, R.H. Estabrook, V.E. Keys, and J. Taylor, Jr. 1975. Major features of the Apalachicola Bay System: Physiography, biota, and resource management. Flor. Sci. 37: 245-271.

24. Meldrim, J.W., and J.J. Gift. 1971. Temperature preference, avoidance and shock experiments with estuarine fishes. Ichthyological Associates Bull. 7: 1-75.

25. Rehwoldt, R., and G. Bida. 1970. Fish avoidance reactions. Bull. Env. Contam. Tox. 5: 205-206.

26. Saunders, R.L., and J.B. Sprague. 1967. Effects of copper-zinc mining pollution on a spawning migration of Atlantic salmon (*Salmo salar*). Water Res. 1: 419-432.

27. Siegel, Sidney. 1956. Non parametric statistics for the behavioral sciences. McGraw-Hill, New York. 312 p.

28. Sprague, J.B. 1963. Avoidance of sublethal mining pollution by Atlantic salmon. Proc. Ont. Ind. Wastes Conf. Ont. Water Resour. Comm. 10: 221-236.

29. _____. 1964. Avoidance of copper zinc solutions by young salmon in the laboratory. J. Water Pollut. Contr. Fed. 36: 990-1004.

30. _____. 1968. Avoidance reactions of rainbow trout to zinc sulphate solutions. Water. Res. 2: 367-372.

31. _____, and D.E. Drury. 1969. Avoidance reactions of salmonid fish to representative pollutants. Adv. Wat. Pollut. Res., Proc. 4th Inter. Conf. Prague. Editor S. H. Jenkins. p. 169-179.

32. _____, P.F. Elson, and R.L. Saunders. 1965. Sublethal copper-zinc pollution in a salmon river—a field and laboratory study. Adv. Wat. Pollut. Res., Proc. 2nd Int. Conf., Tokyo, 1: 61-75.

33. _____, and B. A. Ramsey. 1965. Lethal levels of mixed copper zinc solutions for juvenile salmon. J. Fish. Res. Bd. Canada 22: 425-432.

34. Syanzuki, K. 1964. Studies on the toxic effects of industrial waste on fish and shellfish. J. Shimonoseki Coll. Fish 13: 157-211.

35. Timms, A. M., H. Kleerekoper, and J. Matis. 1972. Locomotor responses of goldfish, channel catfish, and largemouth bass to a "copper-polluted" mass of water in an open field. Water Resour. Res. 8: 1574-1580.

36. Westlake, G.F., and H. Kleerekoper. 1974. The locomotor response of goldfish to a steep gradient of copper ions. Water Resour. Res. 10:103-105.

37. Whitmore, C.M., C.E. Warren, and P. Doudoroff. 1960. Avoidance reactions of salmonid and centrarchid fishes to low oxygen concentrations. Trans. Amer. Fish. Soc. 89: 17-26.

38. Zimmerman, M.S. 1974. A comparison of the benthic macrophytes of a polluted drainage system (Fenholloway River) with an unpolluted drainage system (Econfina River). M.S. Thesis, Florida State University, Tallahassee.

39. _____, and R.J. Livingston. (in press) The effects of kraft mill effluents on benthic macrophyte assemblages in a shallow bay system (Apalachicola Bay, north Florida, U.S.A.) Mar. Biol.

PHYSIOLOGICAL AND BIOCHEMICAL ADAPTATIONS

Convened by:
F. John Vernberg
Belle W. Baruch Institute for Marine Biology and Coastal Research
University of South Carolina
Columbia, South Carolina 29208

One important aspect of estuarine science deals with the study of the functional responses of organisms to their environment. Knowledge of the physiological and biochemical capabilities of the estuarine biota is important for a number of reasons:

1) environmental impact assessment—one of the key features used in attempting to assess the possible effects of environmental alterations is the lethal and sublethal responses of organisms. Probably more proposed construction and environmental alterations have been halted based on knowledge of the physiological ecology of organisms than any other ecological subdiscipline.

2) ecosystem analysis—many of the functional attributes of estuarine ecosystems are based on the physiological and biochemical responses of its biota. For example, the bioenergetics of populations and communities is dependent upon respiratory, osmoregulatory, reproductive, and feeding and digestive processes.

3) adaptive mechanisms—a knowledge of the nature of the fundamental physiological and biochemical mechanisms which enable organisms to be adapted to their environment is basic to understanding life processes quite apart from any practical application.

Based partially on these three reasons, and because estuarine researchers and planners are concerned with the estuarine biota, a need was recognized for a review and evaluation session dealing with recent advances in physiology and biochemistry of estuarine organisms. The 1967 published volume entitled *Estuaries*, which was the written product of what is now called the First International Estuarine Research Conference, contained a session on the physiology of estuarine animals. Since then an ever-increasing research emphasis has been placed on this subject with the happy result that much new data and some new advances have been made.

Topics included in this session were of a general nature which would have appeal to many estuarine scientists irrespective of their specific subdiscipline. These topics were selected so that we would not duplicate functional studies discussed in other sessions, such as behavior and nutrient cycling.

RESPIRATORY ADAPTATION: FISHES

Donald E. Hoss and David S. Peters
National Marine Fisheries Service
Atlantic Estuarine Fisheries Center
Beaufort, North Carolina 28516

ABSTRACT: Changes in respiration rate (oxygen consumption) in relation to environmental alteration (both natural and man induced) may reflect how well, if at all, an organism has adjusted to a given situation. The purpose of this paper is to review the recent literature on fish respiration giving special emphasis to those papers that relate respiratory adaptation to certain environmental conditions. By adaptation we mean an adjustment of respiratory rate in responses to changed conditions. We have examined the utilitarian value of some, but not all, adaptations. In our discussion of respiratory adaptation, the experimental methodology was critically examined since the organism's response may be affected significantly by the experimental technique chosen.

Major topics of discussion will include: (1) the problems associated with methodology and defining the type of metabolism being measured, (2) the effects of temperature, salinity, oxygen tension, season, nutritional state, and developmental stage on oxygen consumption, and (3) the usefulness of oxygen consumption measurements in estuarine research. To stimulate discussion and research in those areas where we feel certain weaknesses exist, a critical position was taken on several subjects.

INTRODUCTION

This report, with a few exceptions, is a review of the recent literature on respiration of fish emphasizing those papers that show respiratory adaptation (i.e. adjustment) of fish to different temperature and oxygen concentrations. For a more general review of respiration we refer you to the excellent recent work of Fry (15) and Vernberg and Vernberg (33) and the classical work of Winberg (34).

Respiration is defined as metabolic activity which includes both oxygen utilization and carbon dioxide production. For several reasons, including the high alkalinity of estuarine water, CO_2 production by estuarine fish is more difficult

335

to measure than O_2 consumption; therefore, our discussion deals with oxygen consumption of the intact organism and the way in which these measurements have been used in recent estuarine research.

We agree with Prosser (30) that the word adaptation has been defined in so many ways that it is of questionable value. Rather than adding to the confusion we will follow Prosser's terminology where physiological adaptation refers to any state or rate function which favors biological activity in a specific but often stressful environment. Included within this definition are evolutionary adaptations connotating the genetic constitution of the organism and compensatory changes which occur within the existing genome such as seasonal changes in metabolic performances. In our discussion of adaptation, we will evaluate experimental methodology and indicate some areas which we feel need additional research.

TERMINOLOGY AND METHODOLOGY

Depending on conditions under which respiration is measured, several different levels of metabolism can be estimated. The three most commonly used terms for these levels in fish physiology are "standard" (= "resting"), "routine" and "active" metabolism. Standard metabolism is an approximation of the minimum rate for the intact organism determined at zero activity or extrapolated to zero activity from determinations at various levels of forced activity. Active metabolism is the maximum sustained rate for a fish swimming steadily (15). Routine metabolism falls between standard and active metabolism levels and has been defined as the oxygen consumed by fish whose only movements are spontaneous (2). A fourth term, "normal," connotes the metabolic rate under natural conditions.

Correct methodology is extremely important in the measurement of respiration as techniques may strongly influence the results. Winberg (34) and Beamish and Dickie (1) have reviewed the methodology and precautions to be observed; consequently, we will review only the main features and limitations of these methods. Beamish and Dickie (1) divide the methods of measuring respiration into three general groups: (1) sealed container, (2) flowing water, and (3) manometric. The choice of one method over another depends on the species, the size of the fish, and the purpose of the experiment.

The sealed container method is simple to set up, but the oxygen level in the container is constantly being reduced while the carbon dioxide level is increasing and at certain levels respiration will be affected. This method can yield useful information, but only during the short period before the gases reach critical concentrations.

Oxygen consumption of larval forms and eggs can best be measured using manometric methods. The system has some of the disadvantages of the sealed container method, i.e. decreasing oxygen content with time and the possibility that the smallness of the respiration flasks may either excite the fish or restrict spontaneous activity, thereby causing abnormal respiration. Despite these disadvantages the method seems to be the best one available for measuring oxygen consumed by individual larval fish.

In the flowing water method, water of constant oxygen content flows by the fish and the rate of oxygen consumption is determined by changes in the oxygen content of the water as it passes through the respiration chamber. This method also allows oxygen consumption to be measured over longer periods of time. When the rate of oxygen consumption is low there is an increase in the variability of the measurements, but for most studies some form of a flowing water system in which the various desired environmental variables can be maintained at constant levels is to be preferred over a closed system.

The techniques discussed above are useful in describing respiratory adaptation under laboratory conditions; however, the respiratory activity of a fish in nature may be vastly different than those described in the laboratory. In nature the fish will have to adjust to a dynamic environment: diurnal and tidal changes in temperature and salinity, food quantity and composition will probably be different, movement and behavior will also change. At the present time there is no generally acceptable way of measuring natural respiration, which for energy flow purposes is called "normal" respiration, although many investigators have estimated normal respiration by doubling either resting or routine metabolism. Such manipulations are too subjective and are of little value in studying natural respiratory adaptation.

Mishima and Odum (27) suggested that it might be possible to relate the elimination rate of a radioisotope by organisms under natural conditions to metabolism, as measured by oxygen consumption in the laboratory. This method is dependent on: (1) correlating the isotope elimination rate with the laboratory oxygen consumption rate; (2) assuming the same relationship exists in nature; (3) measuring the rate of elimination of the radionuclide from the organism in the field; and (4) estimating the oxygen requirement associated with the measured radioisotope loss.

Edwards (9) applied this method to plaice *(Pleuronectes platessa)* and concluded that the isotope loss was highly correlated with oxygen consumption and thus could be used to estimate normal oxygen consumption.

On the other hand, we have been unable to obtain a significant correlation between oxygen consumption and ^{65}Zn loss in either the pinfish *(Lagodon rhomboides)* or the black sea bass *(Centropristis striata)*, (unpublished data D.E. Hoss and D.S. Peters, NMFS, Beaufort, N.C.). Thus, we do not feel that the method has received sufficient testing to justify recommending its use for estimating normal metabolism in fish.

ADAPTATION TO TEMPERATURE

Although it is generally assumed that respiration in poikilotherms, such as fish, is temperature dependent, respiratory adaptation or partial temperature independence has been demonstrated. Respiratory compensation to temperature is generally assumed to be accomplished through kinetic changes in metabolic pathways, and that when it occurs, may be of considerable significance to poikilotherms. Complete compensation (i.e. homeostatic regulation) would allow their vital activities to proceed at steady rates, independent of the environmental temperature. The degree to which fish are able to adapt to temperature and the importance of compensation are not well understood.

Respiration rate-temperature (R-T) curves are frequently used to express the effect of temperature on metabolism. Vernberg and Vernberg (32) separate these curves into two types: (1) those determined at different temperatures using animals acclimated to each of these temperatures (acclimation curve) and (2) those based on the initial rate of the oxygen consumption determined at temperatures to which the animals have not been acclimated (acute curve). If a distinction is made between acclimation and acclimatization, a third type of curve would consist of those determined at different seasons using acclimatized animals, at seasonal temperatures. Acclimatized organisms (15) are those who have been exposed to the total environmental complex throughout their life up to the time of testing and acclimation refers to a "process of bringing the animal to a given steady state by setting one or more of the conditions to which it is exposed for an appropriate time before a given test."

R-T curves (acute, acclimation, and acclimatization) demonstrating adaptation exist for a number of fishes but differences in methodology make comparisons of questionable value. Most investigators working with fish have compared similar, but different species from different latitudes or species from one location under different conditions. However, there have been few comparisons made between geographically separated populations of the same species of estuarine fish.

A good recent example of adaptation between species at different latitudes is given in two related papers by Edwards, Blaxter, Gopalan, and Mathew (10) and Edwards, Blaxter, Gopalan, Mathew and Finlayson (11). These authors compare the effects of temperature on standard respiration of two tropical fish *(Cynoglossus* and *Halophryne)* and two ecologically equivalent temperate fishes *(Pleuronectes* and *Cottus)*. Respiration of tropical fish did not seem to be significantly higher than that of temperate counterparts when each is measured at its natural environmental temperature. The R-T curves (acclimated) for the temperate and tropical fish were similar, except that tropical species were temperature-independent in the extreme upper part of their temperature range. This data supports the concept of respiratory adaptation in poikilotherms at different latitudes.

An example of seasonal adaptation by a species is given by Parvatheswara Rao (29). Working with a tropical freshwater species *(Etroplus maculatus)* he found that oxygen consumption changed very little between seasons indicating a high degree of metabolic homeostasis. When respiration was measured at the same temperature (acute R-T curve) winter acclimatized fish consumed oxygen at a higher rate than summer acclimatized fish. He concluded that winter fish increased their oxygen consumption to compensate for depressing effects of low temperature and that summer acclimatized fish decreased their oxygen consumption to compensate stimulating effects of high temperature.

Another example of adaptation between seasons is given by Cameron (5) for the pinfish *(Lagodon rhomboides)* in Texas. Cameron in his discussion concluded that seasonal adaptation to cold occurs in pinfish since the acclimatized R-T curve is shifted substantially upward in winter from what would be expected on the basis of extrapolated summer data. Wohlschlag, Cameron and Cech (35) also working with pinfish in Texas concluded that seasonal cold adaptation existed.

On the other hand, Hoss (20) working on pinfish in North Carolina, did not find any significant amount of cold adaptation in acclimatized fish at different seasons. For comparison the calculated R-T curves for 30 g pinfish using the published metabolic equations of Cameron (5) and Hoss (20) are shown in Fig. 1. Major differences between the two sets of data are as follows: (1) the respiration for Texas pinfish is generally higher than that obtained for North Carolina

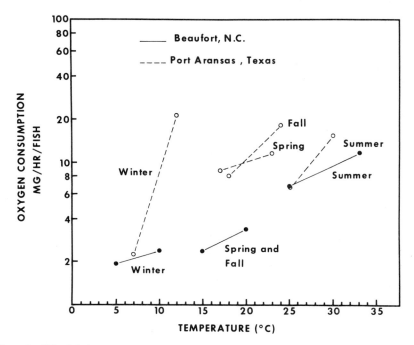

Figure 1. Calculated oxygen consumption of 30 g pinfish from Texas and North Carolina at seasonal temperatures. Open circles represent data of Cameron (5) and closed circles that of Hoss (20).

Table 1. Differences in methodology and procedures used to measure pinfish respiration in Texas (5) and in North Carolina (20).

Method and Procedure	Texas	North Carolina
Type of respirometer	modified sealed container	flowing water
Time of acclimation to respirometer	1 hr	24 hr
Ratio of chamber size to fish size	chamber size constant	chamber size approximately 10 x displacement volume of fish
Prior feeding history	unknown	not fed for 24 hr prior to being placed in respiration chamber

The "Location" spans Texas and North Carolina columns.

fish, (2) the response coefficient (Q_{10} = 22) for Texas fish at winter temperatures (7-12°C) was well above that for North Carolina pinfish (Q_{10} = 1.5) at winter temperatures (5-10°C), (3) at winter temperatures the Texas data indicates that large pinfish have a higher weight specific metabolic rate than small pinfish.

The difference in data from North Carolina and Texas could have a genetic basis; however, if cold adaptation has occurred in Texas pinfish one would expect it also in the more northern Carolina pinfish where its utility should be even greater. Since the Carolina fish do not show cold adaptation, we feel that the methodological differences shown in Table 1, could explain the difference in results. The first three factors would tend to increase respiration due to excitement of the fish, while the influence of food on respiration has been well documented and in some cases it has been suggested that respiration rate may be proportional to food intake (12, 13). Cameron (5) cited some of these factors as possible sources of error in his data.

In another recent paper concerned with adaptation to cold, Holeton (16) investigated eleven species of arctic fish and found that only one of these species, the polar cod, had a high oxygen uptake rate relative to temperate and tropical species at equivalent temperatures. He attributed the increased oxygen uptake of the polar cod to a high level of spontaneous activity in the respirometer rather than to cold adaptation. Holeton also states that much of the cold adaptation reported in the literature may be experimental artifact, and our data on pinfish supports Holeton's hypothesis.

ADAPTATION TO OXYGEN TENSION

In comparison to temperature, the influence of oxygen tension on respiration rate of estuarine fish has received relatively little study. The scanty data available provides no reason to expect different responses than those described in freshwater fishes. Active, routine and standard oxygen consumption rates are each affected differently by reduced oxygen concentration. As environmental oxygen declines, standard metabolism is relatively unaffected until extremely low levels of oxygen are reached, then higher ventilation rates can slightly increase the standard rate. Also random activity can either increase or decrease at low oxygen tension depending on the species of fish (23), making the relationship between routine respiration and oxygen concentration species dependent. The relationship between active metabolism and environmental oxygen is less species dependent, i.e. the scope for activity declines with reduced oxygen tension. When active and standard oxygen consumption are nearly equal, the oxygen concentration may be approaching a lethal level (15).

Fishes which encounter environments with reduced oxygen tensions can adjust in various ways, either behaviorally or physiologically. Behavioral compensation includes lowering the need for oxygen by decreasing movement (15) and avoidance of low oxygen tension, which may be accomplished by mullet through changes in their movements (26). Physiological adjustment may involve increasing the fish's ability to extract oxygen from the water or changing to metabolic pathways which require less oxygen. The extreme metabolic adaptation to low oxygen, a shift to anaerobic respiration, has been shown in laboratory fish by Blazka (3) and Mathur (25) and appears likely under some natural conditions (7). The relative frequency and ecological significance of these possible responses to hypoxic conditions has not been documented.

Fishes have evolved several strategies for increasing oxygen supply to their tissues when they are in a low oxygen environment. When mullet are exposed to hypoxic conditions they adapt by increasing their respiratory water flow (6). MacLeod and Smith (24) showed that fathead minnows exposed to an oxygen poor water increase the oxygen carrying capacity of their blood by increasing hematocrit levels. Experiments with freshwater trout indicate heart rate decreases and peripheral resistance to blood flow increases at low oxygen level; however, these factors which would tend to decrease blood flow may be offset by increased stroke volume (17). An increase in blood flow also would seem advantageous but apparently has not been demonstrated.

Another variation in adaptation to low oxygen tension involves maintenance of aerobic metabolism, with a shift toward more oxidation of the substrates which have higher oxycaloric coefficients (Q_{ox}, calories/mg oxygen used in aerobic metabolism). Since carbohydrates have a higher Q_{ox} than fats or proteins they would be the preferred oxidative substrate under low oxygen tensions. However, many fish eat and store relatively little carbohydrate; thus its metabolism may be of relatively little significance. Changes in the fats or proteins being oxidized may decrease the oxygen needed for a specific energy production. In general, fats have a higher Q_{ox} than proteins; however, considerable variation can occur, depending on the particular fat or protein being oxidized and the nitrogenous end product formed (14).

Nutritional ecology of estuarine fish has not been studied enough to say what percentage of their energy is normally derived from fat, carbohydrate or protein oxidation; however, the different oxycaloric coefficients for these substances indicate certain substrate changes could be advantageous under low oxygen conditions.

The frequency and importance of changes in the metabolic substrates oxidized by fish which are exposed to hypoxic conditions are poorly defined. Increases in the respiratory quotient (RQ = moles CO_2 produced/moles O_2 used) at low oxygen tensions (22, 23) may be partially due to a shift toward aerobic catabolism of materials requiring less oxygen, but other causes are possible, e.g. fat synthesis (22) or some anaerobic metabolism. Increases in ammonia excretion during hypoxia indicate increased protein catabolism in the mullet, *Rhinomugil corsula* (23) and could be interpreted as a shift toward increased oxidation of specific amino acids with higher oxycaloric coefficients. *Tilapia mossambica* (22) do not show an increase in hypoxic protein utilization (NH_3 excretion) but could still be altering their metabolism so that only proteins or fats with high oxycaloric coefficients are oxidized. Although it would appear advantageous to shift the substrates being oxidized aerobically, when hypoxic conditions are encountered, further research will be needed to demonstrate the occurrence and significance of such shifts.

EFFECT OF SIZE ON ADAPTATION

One of the most frequently cited metabolic relationships is between oxygen consumption and body size. Respiration rate is related to body weight by an exponential equation: $Q = aW^k$ where Q is rate of oxygen consumption and W is body weight. The relationship of respiration to body size is reflected in the exponent k. If k were equal to one, respiration would be proportional to body weight. When k is less than one, the usual case, weight specific respiration decreases with increasing fish size. Winberg (34) reviewed published data and concluded that (1) a depends on a variety of factors but often approximates 0.3 and (2) the weight exponent k is less variable and approximates 0.8. The exponent is not constant as some biologists have assumed, and as Fry (15) points out the value of 0.8 should not become a dogma. Numerous investigators have, in fact, shown that k is not the same in all cases. Hoss (19) found statistical differences in respiratory coefficients among five species of estuarine fish. Brett (4) found that the more the respiration measurement approaches standard metabolism the greater the rate-weight coefficient departs from unity. Morris (28) has demonstrated that the exponent will vary depending on thermal history, respiration temperature and the range of fish size considered. The important though often unrecognized effect of fish size or developmental stage on respiration rate is shown in Fig. 2.

It is obvious that the relationship between respiration rate and body size changes with developmental stage of pinfish. What portion of this change is attributable to methodology, size related changes in the fishes response to our

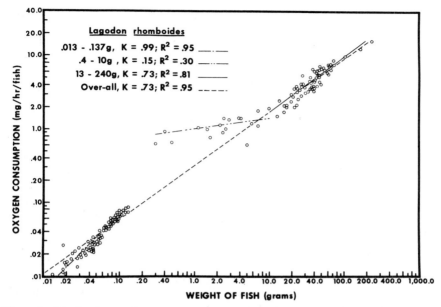

Figure 2. Calculated regression of oxygen consumption on weight for pinfish between 0.3 and 67 g. Regressions for 0.013-0.137 g, 0.4-10 g and 13-240 g fish are shown as well as overall regression equation. Each point represents data for one fish. (Unpublished data of Hoss).

techniques, or other factors is not clear. Further research is needed to show if this phenomenon applies to other fishes, why it occurs and if it has any adaptive value.

Most studies of respiratory adaptation have been conducted using juvenile-adult fish with little attention being paid to early developmental stages of marine and estuarine fish. Fish undergo metamorphic, metabolic and ecological change from larvae to adult; so, a concomitant change in respiratory adaptation should not be surprising even though it is generally undocumented. Determinations of Q_{10} values for herring eggs and larvae (18), Atlantic croaker larvae and pinfish larvae (20), and unpublished data, (D.E. Hoss, NMFS, Beaufort, N.C.) give no indication that respiratory response of these species to temperature is different than might be expected of more mature fish. Using juvenile fish Rao (31) showed that the effect of salinity on the weight exponent k changed with fish size; but this salinity effect has not been tested from egg to adult. DeSilva and Tytler (8) working with herring and plaice up to metamorphosis, have shown that dependence on environmental oxygen tensions does change with age. The early larval stages of both species showed some degree of "regulation" of oxygen uptake. Later larval stages showed "conformity," i.e. the oxygen uptake fell with the oxygen concentration of the surrounding water. After metamorphosis and the appearance of respiratory pigment a degree of "regulation" reappeared.

From the limited amount of available data, we are not able to say whether respiratory adaptability changes with developmental stage. It appears this would be a fruitful area for future research.

DISCUSSION

From an ecological standpoint, a knowledge of the energy of metabolism of fish is necessary if we are to understand important aspects of trophic interrelationships in marine and estuarine systems. In this context we feel two subject areas warrant further investigation. First, most respiration measurements have been obtained using single fish so there is little indication whether schooling alters oxygen consumption rates. Secondly, very little progress has been made toward measuring normal metabolism. Investigators are still forced to rely on various correction factors such as the value 2 proposed by Winberg (34) for increasing routine metabolism. Perhaps some sort of isotope loss experiment may prove useful for this purpose.

Most of the literature that we have reviewed shows that animals are adapted to the area in which they live, and as Fry (15) has stated the fish make this point themselves without recourse to respirometers. In spite of the numerous papers emphasizing the importance of methodology and proper handling of the fish prior to oxygen measurement, many investigators still proceed with inadequate methods and procedures, making it impossible to meaningfully compare their data with other data in the literature. Because of this, we agree with the statement by Holeton (16) that much "cold adaptation" may be an interpretational artifact due to the methods used. Furthermore he questions the underlying assumption that an elevated standard metabolic rate confers an advantage to a fish in cold water since it implies a high energy maintenance cost. The same problem is encountered in any discussion of adaptation. By definition a respiratory adaptation is favorable; we are therefore required to make a very subjective judgment about the utilitarian value of any respiratory change. Perhaps in the future we should discuss respiratory adjustments, rather than adaptations. This change in terminology would not require a subjective judgment and would therefore more accurately reflect the facts.

ACKNOWLEDGMENT

This research was supported under agreement AT(49-7)-5 between the Energy Research and Development Administration and the National Marine Fisheries Service.

REFERENCES

1. Beamish, F. W. H., and L. M. Dickie. 1967. Metabolism and biological production in fish, p. 215-242. In S. D. Gerking (ed.), The Biological Basis of Freshwater Fish Production, Blackwell, Oxford.
2. _____, and P. S. Mookherjii. 1964. Respiration of fish with special emphasis on standard oxygen consumption. I. Influence of weight and temperature

on respiration of goldfish, *Carassius auratus* L. Can. J. Zool. 42:161-175.

3. Blazka, P. 1958. Anaerobic metabolism in fish. Physiol. Zool. 31:117-128.

4. Brett, J. R. 1965. The relation of size to rate of oxygen consumption and sustained swimming speed of sockeye salmon *(Oncorhynchus nerka)*. J. Fish. Res. Board Canada. 22:1491-1501.

5. Cameron, J. N. 1969. Growth, respiratory metabolism and seasonal distribution of juvenile pinfish (*Lagodon rhomboides* Linnaeus) in Redfish Bay, Texas. Contr. Mar. Sci. Univ. Tex. 13:89-104.

6. Cech, J. J., Jr., and D. E. Wohlschlag. 1973. Respiratory responses of the striped mullet *Mugil cephalus* (L.) to hypoxic conditions. J. Fish Biol. 5:421-428.

7. Coulter, G. W. 1967. Low apparent oxygen requirements of deep-water fishes in Lake Tanganyika. Nature (Lond.). 215:317-318.

8. DeSilva, C. D., and P. Tytler. 1973. The influence of reduced environmental oxygen on the metabolism and survival of herring and plaice larvae. Neth. J. Sea Res. 7:345-362.

9. Edwards, R. R. C. 1967. Estimation of the respiratory rate of young plaice (*Pleuronectes platessa* L.) in natural conditions using zinc-65. Nature (Lond.). 216(5122):1335-1337.

10. Edwards, R. R. C., J. H. S. Blaxter, U. K. Gopalan, and C. V. Mathew. 1970. A comparison of standard oxygen consumption of temperate and tropical bottom-living marine fish. Comp. Biochem. Physiol. 34:491-495.

11. _____, _____, _____, _____, and D. M. Finlayson. 1971. Feeding, metabolism and growth of tropical flatfish. J. Exp. Mar. Biol. Ecol. 6:279-300.

12. _____, D. M. Finlayson, and J. H. Steele. 1969. The ecology of O-group plaice and common dabs in Loch Ewe. II. Experimental studies of metabolism. J. Exp. Mar. Biol. Ecol. 3:1-17.

13. _____, _____, _____. 1972. An experimental study of the oxygen consumption, growth, and metabolism of the cod (*Gadus morhua* L.). J. Exp. Mar. Biol. Ecol. 8:299-309.

14. Elliott, J. M. and W. Davison. 1975. Energy equivalents of oxygen consumption in animal energetics. Oecologia 19:195-201.

15. Fry, F. E. J. 1971. The effect of environmental factors on the physiology of fish, p. 1-98. *In* W. S. Hoar and D. J. Randall (eds.), Fish Physiology. Academic Press, New York.

16. Holeton, G. F. 1974. Metabolic cold adaptation of polar fish: fact or artefact? Physiol. Zool. 47(3):137-152.

17. _____, and D. J. Randall. 1967. Changes in blood pressures in the rainbow trout during hypoxia. J. Exp. Biol. 46:297-305.

18. Holliday, F. G. T., J. H. S. Blaxter, and R. Lasker. 1964. Oxygen uptake of developing eggs and larvae of the herring *(Clupea harengus)*. J. Mar. Biol. Assoc. U. K. 44:711-723.

19. Hoss, D. E. 1967. Rates of respiration of estuarine fish. Proc. 21st Annu. Conf. Southeast. Assoc. Game Fish Comm., p. 416-423.

20. _____. 1974. Energy requirements of the pinfish (*Lagodon rhomboides*) in the Newport River estuary, N.C. Ecology 55:848-855.

21. Klieber, M. 1961. The fire of life. Wiley, New York. 454p.

22. Kutty, M. N. 1972. Respiratory quotient and ammonia excretion in *T. mossambica*. Mar. Biol. (Berl.). 16(2):126-133.

23. _____, and M. P. Mohamed. 1975. Metabolic adaptations of mullet, *Rhinomugil corsula* (Hamilton) with special reference to energy utilization. Aquaculture 5:253-270.

24. MacLeod, J. C., and L. L. Smith, Jr. 1966. Effect of pulpwood fiber on oxygen consumption and swimming endurance of the fathead minnow, *Pimephales promelas.* Trans. Am. Fish. Soc. 95(1):71-84.
25. Mathur, G. B. 1967. Anaerobic respiration in a cyprinoid fish, *Rasbora daniconius.* Nature (Lond.). 214:318-319.
26. McFarland, W. N., and S. A. Moss. 1967. Internal behavior in fish schools. Science (Wash., DC). 156:260-262.
27. Mishima, J., and E. P. Odum. 1963. Excretion rate of [65]Zn by *Littorina irrorata* in relation to temperature and body size. Limnol. Oceanogr. 8:39-44.
28. Morris, R. W. 1965. Thermal acclimation of metabolism of the yellow bullhead, *Ictalurus natalis* (LeSueur). Physiol. Zool. 38(1):219-227.
29. Parvatheswara Rao, V. 1971. Compensatory metabolic regulation to seasonal thermal stress in the tropical freshwater fish, *Etroplus maculatus* (Bleeker). Indian J. Exp. Biol. 9:40-44.
30. Prosser, C. L. 1975. Physiological adaptations in animals, p. 3-18. *In* F. J. Vernberg (ed.), Physiological adaptation to the environment. Intext Educational Publishers, New York.
31. Rao, G. M. M. 1968. Oxygen consumption of rainbow trout *(Salmo gairdneri)* in relation to activity and salinity. Can. J. Zool. 46:781-785.
32. Vernberg, F. J., and W. B. Vernberg. 1970. The animal and the environment. Holt, Rinehart and Winston, New York. 398 p.
33. Vernberg, W. B., and F. J. Vernberg. 1972. Environmental physiology of marine animals. Springer-Verlag, New York, 346 p.
34. Winberg, G. G. 1956. Rate of metabolism and food requirements of fishes. (Transl. from Russian). Transl. Ser. 194. Fish. Res. Board Canada.
35. Wohlschlag, D. E., J. N. Cameron, and J. J. Cech, Jr. 1968. Seasonal changes in the respiratory metabolism of the pinfish *(Lagodon rhomboides).* Contr. Mar. Sci. Univ. Tex. 13:89-104.

RESPIRATORY ADAPTATIONS: INVERTEBRATES

C.S. Hammen
Department of Zoology
University of Rhode Island
Kingston, R.I. 02881, U.S.A.

ABSTRACT: The major recent advances are: (1) increased precision in describing the relation of oxygen consumption to oxygen concentration, temperature, and activity; (2) recognition of cessation of oxygen consumption and survival of prolonged anoxia as common features of invertebrate respiration; and (3) integration of data on properties of enzymes and pathways of metabolism into the developing story of respiration with and without oxygen. Variation in rates of uptake with oxygen content of the medium can be described by a quadratic equation with three parameters chosen to express departure from strict conformity. Over certain segments of the environmental range, the oxygen uptake of some species increases with temperature only 1.2 to 1.4 times for $10°$, rather than in accordance with the expected doubling of rate of chemical reactions. When animals survive very low oxygen tension for extended periods, they may be called "euryoxic", a new term analogous to euryhaline and eurythermal. The term "facultative anaerobe" is inappropriate because all animals are aerobes. At least five species of bivalve mollusks are highly euryoxic, surviving 21 to 51 days of oxygen lack at low temperatures. Since rates of enzyme activity utlimately determine rates of oxygen uptake and acid production, the kinetic properties of enzymes, together with endogenous concentrations of substrates, activators, and inhibitors, must eventually predict respiration rates.

INTRODUCTION

Respiration is taken here in the sense of oxygen consumption, carbon dioxide production, and the underlying cellular oxidations. Ventilatory mechanisms are not considered, nor is the transport of oxygen in body fluids. The major features of respiration in several of its senses were well described by Krogh (11) in "The comparative physiology of respiratory mechanisms." In recent years, the most noteworthy attempts to cope with the literature are: discussions of oxygen consumption in relation to environment by Vernberg & Vernberg (28), accounts of respiration in all senses in the treatise edited by Prosser (22), and an integration of metabolism with nutrition of marine animals by Pandian (19), who cited 912 references. This brief review is restricted to some of the more significant contributions of the last five years. The most interesting current topic, and a real challenge, is to connect respiration with intermediary metabolism, as deduced from isotope tracer work and properties of enzymes from different species.

OXYGEN CONSUMPTION RATES
Temperature effects

Superimposed on the general pattern of increased oxygen consumption with temperature are many instances of relative insensitivity. This usually means an increase of 1.2 to 1.4 times with a $10°$ rise, rather than the expected doubling. For example, a successful intertidal species, the barnacle *Pollicipes polymerus*, displayed a Q_{10} of only 1.3 over the range 20-27°C (21). The analysis was carried from intact animals through isolated tissue, homogenates, and mitochondria from *Littorina littorea*, and oxygen consumption remained relatively constant from 5° to 25°C at all levels of organization (18). Another demonstration that isolated tissues sometimes display the phenomenon was the finding that adductor muscle of *Crassostrea virginica* respired at the same rate over 16-24°C, while gill and mantle rates increased as expected (20). Of course, as the authors suggested, the adductor is critical to the defense of bivalves, and relative temperature independence represents an adaptive advantage. Newell (17) suggested that animals of the upper shore, more subject to temperature fluctuations, are more likely to display little change in rate of oxygen consumption over particular ranges of temperature, and that this may "represent a means by which energy is conserved."

The enzymic basis of insensitivity is unknown for invertebrates. From studies of enzymes of fishes, the hypothesis has been advanced that selection has produced enzymes with substrate affinity decreasing (K_m increasing) as temperature increases, a property called "positive thermal modulation" (9). A recent analysis suggests that substrate affinity is only one part of the explanation. The adenosine deaminases of four bivalves behaved as if "a temperature - independent process (the formation of the enzyme-substrate complex) dominates a temperature-dependent process (product release)" at low substrate concentrations (8). The underlying mechanisms for the phenomenon are probably multiple, varying with species and with enzymes within each species.

Oxygen Concentration

The relation between oxygen consumption and oxygen concentration has been described by calling animals either regulators or conformers. Few species fit exactly into either category, and new terms to describe all sorts of modified or intermediate categories would be a nuisance. It is therefore gratifying that a computer-assisted search was made for an equation to describe more precisely the respiratory responses of animals to variation in oxygen concentration (14). Several equations were tested for agreement with the data on oxygen consumption rate ($\overline{V}_{O_2}$) and partial pressure of oxygen (P_{O_2}), and the best equation in providing good fit, ease of computation, and parameters useful for comparing species (27) is:

$$y = B_0 + B_1 x + B_2 x^2,$$

where $y = \overline{V}_{O_2}$ and $x = P_{O_2}$.

The meanings of the parameters are: B_0 is the minimum rate of respiration found at very low partial pressures, near zero; it represents the imaginary intercept with the y axis in a graph of the quadratic polynomial equation, although y must equal zero when no O_2 is present; B_1 represents the linear effect of x on y, and varies from 0 to 22×10^3 in cases presented to date; B_2 represents the departure from linearity of the effect of x on y, varies from -11×10^{-5} to 2×10^{-5}, and is usually negative. A strict regulator, respiring at the same rate independently of O_2 concentration, would have $B_0 = 1$ on a relative scale, B_1 and B_2 equal to zero. A strict conformer, respiring in direct proportion to O_2 concentration, would have $B_0 = 0$, $B_1 = $ a constant > 0, and $B_2 = 0$.

Examples of equations for groups with different responses are: Bivalvia, $B_0 = -0.19$, $B_1 = 22.2 \times 10^3$, $B_2 = -10.5 \times 10^{-5}$, and Ophiuroidea, $B_0 = 0.30$, $B_1 = 0$, $B_2 = 2.38 \times 10^{-5}$ (values from 27). The curves of these equations are presented in Fig. 1, which shows that the bivalve tends to regulate at P_{O_2} above 80 mm Hg, and the ophiuroid at P_{O_2} below 80mm Hg. But it is much simpler and more precise to describe the responses by giving values of the three parameters. A recent paper (15) on vascular and coelomic hemoglobins in polychaetes used B_2 to describe differences between species, and even the same species under different conditions, such as ligation of gills, and presence of carbon monoxide. This advance in quantitative treatment of respiration data provides the analytical description that will permit attack on the problem of underlying control mechanisms.

ANAEROBIC PROCESSES
Cessation of Oxygen Consumption

In recent years, reports of marine invertebrates ceasing to respire have appeared. For example, isolated whole bodies of the barnacle *Chthamalus depressus* frequently consumed no oxygen at all for 40 to 80 minutes in a Warburg respirometer (6). Sixteen species from seven phyla were reported to stop respiring at low oxygen concentrations, generally 5-17 mm Hg, about one-tenth of full saturation (14). A sea urchin, a polychaete, and a sea anemone occasionally stopped respiring at relatively high oxygen levels, 31-55 mm Hg, 20-35% saturation, without exhausting the supply of oxygen available to them.

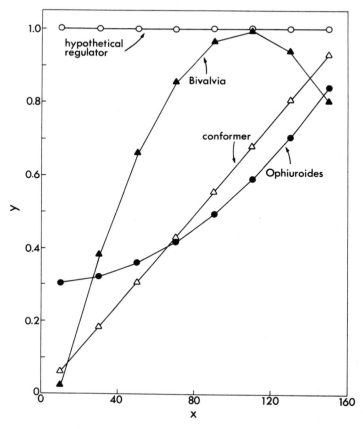

Figure 1. Relation between relative rate of oxygen consumption (y) and partial pressure of oxygen (x), for a hypothetical regulator (open circles), a conformer (open triangles), Ophiuroidea (solid circles) and Bivalvia (solid triangles). Partial pressure in mm Hg. Equations for real animals from Van Winkle & Mangum (27).

This represents a sort of "voluntary" shift to anaerobiosis, not previously suspected in fundamentally aerobic organisms.

Prolonged Survival of Anoxia

Many reports of tolerance to oxygen lack by marine invertebrates have appeared over the years. In a good recent survey, the survival of 14 species kept in sealed bottles nearly devoid of oxygen (<0.15ml O_2/1) was recorded at intervals (26). There were 7 mollusks, 2 echinoderms, 4 crustaceans, and a polychaete from the Baltic and the North Sea, and results were given as time for 50% mortality (LT_{50}). Four species of bivalve mollusks survived 21-51 days at

10°C, three species of crustaceans less than 14 hours, and the other seven species all tolerated anoxia for 1-6 days. Survival times in the literature are often maxima rather than LT_{50}. Some of these values are included in Fig. 2 for comparison with Theede's results. The conclusions are that (1) bivalve mollusks are most tolerant of anoxia, (2) a variety of marine invertebrates from at least 6 phyla tolerate anoxia quite well (1-6 days), and (3) most small crustaceans are intolerant of oxygen lack.

In a review of oyster metabolism (5), the term "facultative anaerobe" was borrowed from microbial metabolism to describe the ability of *Crassostrea virginica* to close its valves tightly and remain anaerobic for days. The oyster, although highly adapted to withstand anoxia, is not an anaerobe, and does not live indefinitely without oxygen. *C. virginica* survives about 35 days at 5°C and only 2 days at 25°C (3); thus the tolerance is near that of *Mytilus edulis*, but it is exceeded by *Cyprina islandica*. Even parasitic helminths and parasitic arthropods " . . . are all fundamentally aerobic organisms, but can tolerate lack of oxygen fairly well" (1). Unfortunately, the unsubtantiated assumption that there are

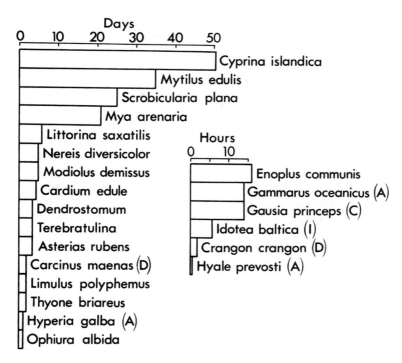

Figure 2. Time of survival of various species of marine invertebrates under conditions of anoxia or extreme hypoxia. Most values are LT_{50}, but some are maximum survival times. From Theede (26) and other sources.

facultative anaerobes among metazoans now seems firmly embedded in the literature. I suggest that animals capable of tolerating a wide range of oxygen concentrations, including zero, be called "euryoxic", and those with more limited tolerance be called stenoxic, in parallel with the terms euryhaline, eurythermal, etc. In this sense of the term, already in the literature (23), many estuarine invertebrates are euryoxic. This trait is valuable in poorly aerated benthic locations, and in intertidal situations where bivalves typically remain closed to avoid desiccation. A recent work on sea urchin respiration (29) suggests that the body wall is stenoxic and the internal tissues are euryoxic. At any rate, we now recognize anaerobic processes as of major importance in metabolism of marine invertebrates.

THE BALANCE BETWEEN AEROBIC AND ANAEROBIC METABOLISM
Anaerobic End-Products and Phosphorylation

If oxygen consumption sometimes can be kept from increasing with temperature, can be stopped entirely while oxygen is still available, and can be made unnecessary for survival over periods of days and weeks, then energy demand must be kept low and must be satisfied by anaerobic metabolism. In large, active invertebrates, the energy comes from glycolysis, lactic acid is formed, and its reoxidation is signalled by increases in respiration while the oxygen debt is repaid. In several bivalve mollusks, succinic acid is a major product and lactic acid a minor product (7). Succinic acid increased markedly in *Mytilus edulis* under anoxia, but aspartic acid remained the major product (13). In crustaceans, other products such as propionic acid and glycerol are formed (25). The probable pathways of metabolism that invertebrates use during anaerobiosis have been described as "simultaneous catabolism of carbohydrate and amino acids" (10). This idea is based on results with parasitic worms, especially *Ascaris* and *Fasciola*, and bivalve mollusks, especially *Crassostrea virginica* and *C. gigas*.

The actual method of production of ATP, which is needed for cellular work, remains largely hypothetical. One reaction, the conversion of fumarate to succinate, when catalyzed by extract of *Mytilus edulis* adductor, was stimulated by added ADP, and the reverse reaction was stimulated by ATP, indicating one site of anaerobic phosphorylation (7). Much work remains to be done, since the conditions for oxidative phosphorylation in marine invertebrate mitochondria are little known. In the case of the blue crab, *Callinectes sapidus*, media of more than 300 mosM were necessary for good activity (2).

The Change in Forms of Metabolism

The process of change from aerobic to anaerobic pathways is incompletely known. Based on studies of stimulation, inhibition and pH optima of two enzymes in oyster muscle, the hypothesis has been advanced that alanine accumulation and increased acidity can cause phosphoenolpyruvate (PEP) to be carboxylated and converted to succinate rather than to pyruvate (9). This scheme has been judged unlikely in *Mytilus* because of endogenous pool sizes, and because acidity could not increase until anaerobiosis was well advanced (12). Anaerobic pathways are probably in operation constantly in marine invertebrates, and changes are not abrupt, but rather involve a gradual deceleration of those pathways depending on oxygen. Many experiments with labelled substrates show that marine invertebrates carry on mixed fermentations in the presence of oxygen, and simply change the proportions of acid end-products when totally deprived of oxygen. On the whole—animal level, the shifting balance between aerobic and anaerobic metabolism might be studied by measuring heat production, which accompanies all energy transformations.

Rates of heat production by the marine gastropod *Nucella lapillus* were measured by Grainger (4), and rates of heat production of the shore crab *Carcinus maenas* were measured by Spaargaren (24), but for reasons other than investigating the anaerobiosis problem.

Evolution of Aerobic Ability

In evolution, aerobes followed anaerobes as oxygen from photosynthesis entered the atmosphere. The first eukaryotic cells were formed "when a pleiomorphic microbe, capable only of anaerobic fermentation of glucose to pyruvate, symbiotically harbored a smaller (aerobic) prokaryote" (16). The Metazoa then inherited both anaerobic and aerobic metabolism from their protozoan ancestors. On the argument that algae have contributed oxygen to the atmosphere for 2.7×10^9 years, and that the earliest metazoa had well-developed mitochondrial respiration, some authors (16) have suggested that the atmosphere was similar in composition to today's well before the Cambrian, about 0.6×10^9 years before the present. However, the oxygen requirement for calcification has led others (23) to the conclusion that oxygen was present at only one-tenth of present atmospheric concentration at the beginning of the Cambrian. Regardless of the actual time course of oxygen accumulation, it is probable that pressure to develop highly efficient aerobic respiration came with increased size, increased activity, thicker shells, and the consequent energy demands. Euryoxic animals, therefore, preceded highly aerobic forms in evolution. Pressure to retain ancestral ability to survive anoxia came from frequent exposure to hypoxic environments, such as those commonly encountered by estuarine invertebrates.

REFERENCES

1. Brand, T. von. 1973. Biochemistry of parasites. 2nd ed. Academic Press, New York.
2. Chen, C.H., and A.L. Lehninger. 1973. Respiration and phosphorylation by mitochondria from the hepatopancreas of the blue crab, *Callinectes sapidus*. Arch. Biochem. Biophys. 154:449-459.
3. Dunnington, E.A. 1968. Survival time of oysters after burial at various temperatures. Proc. Nat. Shellfish Assoc. 58:101-103.

4. Grainger, J.N.R. 1968. The relation between heat production, oxygen consumption and temperature in some poikilotherms, p. 86-89. *In* A. Locker (ed.), Quantitative Biology of Metabolism, Springer–Verlag, New York.
5. Hammen, C.S. 1969. Metabolism of the oyster, *Crassostrea virginica*. Amer. Zool. 9: 309-318.
6. _____. 1972. Lactate oxidation in the upper-shore barnacle, *Chthamalus depressus* (Poli). Comp. Biochem. Physiol. 43A: 435-441.
7. _____. 1975. Succinate and lactate oxidoreductases of bivalve mollusks. Comp. Biochem. Physiol. 50B: 407-412.
8. Harbison, G.R., and J.R. Fisher. 1974. Substrate-dependent apparent activation energies of the adenosine deaminases from bivalved molluscs. Comp. Biochem. Physiol. 47B: 27-32.
9. Hochachka, P.W., and G.N. Somero. 1973. Strategies of biochemical adaptation. Saunders, Philadelphia.
10. _____, J. Fields and T. Mustafa. 1973. Animal life without oxygen: basic biochemical mechanisms. Amer. Zool. 13: 543-555.
11. Krogh, A. 1941. The comparative physiology of respiratory mechanisms. Univ. of Pennsylvania Press, Philadelphia.
12. Livingstone, D.R., and B.L. Bayne. 1974. Pyruvate kinase from the mantle tissue of *Mytilus edulis* L. Comp. Biochem. Physiol. 48B: 481-497.
13. Loxton, J., and A.E. Chaplin. 1974. The metabolism of *Mytilus edulis* L. during facultative anaerobiosis. Biochem. Soc. Trans. 2: 419-421.
14. Mangum, C., and W. VanWinkle. 1973. Responses of aquatic invertebrates to declining oxygen conditions. Amer. Zool. 13: 529-541.
15. _____, B.R. Woodin, C. Bonaventura, J. Bonaventura, and B. Sullivan. 1975. The role of coelomic and vascular hemoglobin in the annelid family Terebellidae. Comp. Biochem. Physiol. 51A: 281-294.
16. Margulis, L. 1970. Origin of eukaryotic cells. Yale Univ. Press, New Haven.
17. Newell, R.C. 1973. Factors affecting the respiration of intertidal invertebrates. Amer. Zool. 13: 513-528.
18. _____, and V.I. Pye. 1971. Quantitative aspects of the relationship between metabolism and temperature in the winkle *Littorina littorea* (L.). Comp. Biochem. Physiol. 38B: 635-650.
19. Pandian, T.J. 1975. Mechanisms of heterotrophy. Chapter 3, p. 61-249. *In* O. Kinne (ed.), Marine Ecology. Volume 2. Part 1. Wiley, New York.
20. Percy, J.A., and F.A. Aldrich. 1971. Metabolic rate independent of temperature in mollusc tissue. Nature 231: 393-394.
21. Petersen, J.A., H.F. Fyhn, and K. Johansen. 1974. Eco-physiological studies of an intertidal crustacean, *Pollicipes polymerus* (Cirripedia, Lepadomorpha): aquatic and aerial respiration. J. Exptl. Biol. 61: 309-320.
22. Prosser, C.L. (ed.) 1973. Comparative animal physiology. 3rd edition, Saunders, Philadelphia.
23. Rhoads, D.C., and J.W. Morse. 1971. Evolutionary and ecologic significance of oxygen-deficient marine basins. Lethaia 4: 413-428.
24. Spaargaren, D.H. 1975. Heat production of the shore-crab *Carcinus maenas* (L.) and its relation to osmotic stress. Proc. 9th Europ. Mar. Biol. Symp. (H. Barnes, ed.) Aberdeen Univ. Press.
25. Thabrew, M.E., P.C. Poat and K.A. Munday. 1973. The effect of malonate on succinate metabolism in the excised gill of *Carcinus maenas*. Comp. Biochem. Physiol. 44B: 869-879.

26. Theede, H. 1973. Comparative studies on the influence of oxygen deficiency and hydrogen sulfide on marine bottom invertebrates. Neth. J. Sea Research 7: 244-252.

27. Van Winkle, W., and C. Mangum. 1975. Oxyconformers and oxyregulators: A quantitative index. J. Exp. Mar. Biol. Ecol. 17: 103-110.

28. Vernberg, Winona B., and F.J. Vernberg. 1972. Environmental physiology of marine animals. Springer-Verlag, New York.

29. Webster, S.K., and A.C. Giese. 1975. Oxygen consumption of the purple sea urchin with special reference to the reproductive cycle. Biol. Bull. 148: 165-180.

THE FUNCTION OF RESPIRATORY PIGMENTS IN ESTUARINE ANIMALS

Charlotte P. Mangum
Department of Biology
College of William and Mary
Williamsburg, VA 23185

ABSTRACT: Estuarine animals must either regulate the osmotic concentrations of their body fluids or conform to salinity changes, in which case they often regulate body volume. It is not clear, however, that either process involves changes in oxidative metabolism which are sufficient in magnitude to necessitate special adaptations of their respiratory pigments.

Unless the animal is a perfect regulator of the salt concentration of its blood, which is true only of the vertebrates, the changes in inorganic salts would be expected to have measureable effects on respiratory pigment oxygenation and, by implication, on oxygen consumption. In fact, these changes do not occur in the blue crab *Callinectes sapidus*, because the salt effect on hemocyanin oxygenation is opposed by a concomitant change in blood pH.

The pH change, which results in a stability of hemocyanin function in an unstable ionic environment, is important in osmotic as well as respiratory adaptation. It is believed to result from the altered level of ammonia production as intracellular fluids osmotically re-equilibrate to the blood after a salinity change. The process also provides a counterion to drive the salt pump at the gill.

INTRODUCTION

Studies of oxygen carrying pigments in aquatic animals have often emphasized the special problems of gas exchange and transport in the aquatic as opposed to the terrestrial medium. Adaptations have been related to physical features of the environment such as its density, oxygen permeability and oxygen capacity. The comparative analysis of respiration has progressed to the point where the questions, and hopefully the answers, may become more sophisticated, and we can begin to understand the subtler aspects of adaptation. Although they are all aquatic, the lacustrine, estuarine and marine environments are very different from one another and the differences must be reflected in the functioning of physiological and biochemical systems in the species found in each.

356

The physiology and biochemistry of estuarine animals have been investigated primarily to elucidate the tendency towards a homeostatic condition of the osmotic concentration of body fluids. This essay is focused upon respiratory rather than osmotic adaptations. I shall attempt to demonstrate the importance of understanding the oxygen transport system in relation to the unique features of the estuarine environment as it influences the operating conditions inside an estuarine animal as well as outside. I shall also try to explain the stability of a system that is not truly *homeostatic* but rather *enantiostatic*, in the sense that the standing condition results from a balance of opposing forces.

The concept of the estuary used implicitly below is Pritchard's (41) Type 1, because it is exemplified by the Chesapeake Bay where our original researches were conducted. The discussion, however, is also relevant to species found in other types of estuaries. Since our knowledge of it is so extensive, the chief example is the blue crab *Callinectes sapidus* Rathbun. Other species are discussed mainly to illustrate similarities and differences, and to supplement areas of ignorance. In this case the discussion is clearly relevant only to the decapod crustaceans; there is ample evidence to suggest that theirs is but one of many enantiostatic conditions in adaptation to estuarine life.

METABOLIC EFFECTS

Osmotic Adjustments to the External Medium

When an estuarine animal experiences dilution of the external medium, it either conforms to the change by permitting an equivalent dilution of its body fluids, or else it compensates for the change, in part or in full, by actively absorbing salts. Osmoconformers often regulate body volume, but the ubiquity of the response has not been established and its mechanisms are not understood. Therefore the metabolic demands of osmoconformity on oxygen transport cannot be predicted. Responses of oxygen consumption in dilute media include increases, decreases and no change (42).

Osmoregulating species are believed to fuel the process of salt absorption by aerobic pathways, which may result in increased demands on the oxygen transport system. In fact, this expectation is not always realized in experimental data. Calculations of the small amount of physical work entailed in osmoregulation are often cited to resolve the apparent paradox, but these calculations are not really critical to the present problem, whether or not their numerical assumptions are correct. The efficiency of biological systems in converting metabolic energy into physical work is so low (59) that the actual change in metabolism may be measureable when a calculation would suggest otherwise. The actual response of systems performing ion transport is more pertinent to the question of understanding metabolic demands. The transport of ions across the membranes of gill tissue excised from the blue crab *Callinectes sapidus* entails an increase in oxygen consumption of at least 22% even when the crab osmotically conforms to the medium (37). The metabolic expenditure involved in moving ions across the membranes of a mammalian red blood cell is similar (4). Moreover, the increase in oxygen consumption of blue crab gill tissue placed in various dilutions of seawater is proportional to the osmotic change (14); it is no less in dilutions to which the intact animal would conform, than those in which the animal begins to regulate. The sodium pump in the gill is quite active even when no net salt absorption takes place; its activity is merely lower (by about 80%) at 35 than at $5^\circ/oo$ salinity (56). There are several alternative interpretations of these findings, but none permits a clear prediction of the response of oxygen consumption in intact animals.

On morphological grounds alone, a change in oxygen consumption with salinity would be expected in some osmoregulators but not in others. The arthropods and the vertebrates largely avoid the respiratory problems of living in an unstable ionic environment by covering most of their tissue with impermeable substances and restricting exchange with the environment to a very small fraction of total biomass. If the metabolic expenditure of salt absorption is 20-35% of the total and if the fraction of the animal that performs this process is less than 10% of the total, then a metabolic change in that small fraction is unlikely to be detected in measurements on intact animals. Laird and Haefner (27) concluded that acclimation of adult blue crabs to 10 and 30 $^\circ/oo$ has no effect on total oxygen consumption. Other groups of estuarine animals expose large fractions of their body tissue directly to the external medium, devoid of protective coverings. The site of salt absorption must be less restricted in a vermiform osmoregulator such as the annelid *Nereis succinea*, and a large fraction of the animal must engage in active salt absorption. Recent studies by Oglesby (39) clearly demonstrate the increase in aerobic metabolism accompanying the onset of osmoregulation.

Osmotic Adjustments to the Internal Medium

Only among the vertebrates do regulatory mechanisms compensate fully by maintaining a constant osmotic concentration of body fluids throughout the range of variation in the external environment. Members of other animal groups are imperfect, though often quite strong, osmoregulators. In both imperfect osmoregulators and in osmoconformers, the osmotic concentration of body fluids changes in dilute media. Intracellular fluids re-establish equilibrium with extracellular fluids by reduction of the size of a labile pool of osmotically active particles. These labile constituents include small organic molecules such as amino acids, taurine, betaine and other compounds whose identity is unknown. During initial acclimation to a new salinity, changes in the size of the free amino acid pool may be brought about by anaerobic pathways that terminate in substances such as alanine and proline (2). This response would necessitate no change in aerobic metabolism or its demands on the oxygen transport system. After intracellular and extracellular fluids equilibrate, however, different amino acids predominate and the size of the pool is believed to be maintained by oxidative processes. When the body fluids of the blue crab reach osmotic equilibrium with the external medium, the size of the free amino acid pool in muscle and nerve cells is believed to be controlled by catabolic processes. Changes in oxygen consumption of isolated axons exposed to relatively small dilutions of the medium were reported by Gilles (19). Since the aerobic metabolism of deep tissue in crustaceans must be supplied by blood, this change, if it occurs *in vivo*, might influence the function of the oxygen transport system. The magnitude of such a change, and the possibility of detecting it in intact animals, however, would be difficult to predict from the response of *in vitro* preparations.

The amino acids are not the only organic substances that participate in the osmotic adjustment of cells to a dilute body fluid. Betaine and taurine play a larger role than free amino acids in several molluscs, for example (48). Betaine is metabolized by oxidative processes; taurine is formed oxidatively but its fate is unknown—it may simply be excreted during acclimation to a dilution. In the osmotic adjustment of intracellular fluids in arthropods such as the horseshoe crab *Limulus polyphemus* (7, 46) and the giant barnacle *Pollicipes polymerus* (17), well over half of the change in intracellular fluids is mediated by unknown substances other than amino acids and taurine. Therefore, the changes in oxygen requirements of their deep tissues cannot be predicted at present.

In summary, the active transport of ions across biological membranes is an oxidative process, but it is not clear whether hyperosmotic regulation of body fluids in dilute media entails a measureable increase in oxygen consumption of intact animals. The process may make no special demands on the oxygen transport system of estuarine animals.

Osmotic conformity and imperfect regulation both result in the dilution of extracellular fluids, which is believed to stimulate oxidative catabolism of osmotically active particles inside cells. These processes are quantitatively important in the low salinity acclimation of decapod crustaceans but not in several species of molluscs, other groups of arthropods and coelenterates. Thus the only estuarine animals in which the oxygen transport system might clearly be stressed by osmotic adaptation *per se* are the decapod crustaceans, and the magnitude of this stress cannot be gauged from available information.

DIRECT EFFECTS
The Ion Sensitivity of Respiratory Pigments

There is, in fact, a more cogent reason to expect specific respiratory adaptations in estuarine animals. This reason is a corollary of fundamental properties of the respiratory pigments. Lower levels of the structure and hence the function of complex proteins are usually sensitive to the ionic environment of the protein. As the peptide chains bind or release protons and the ions of salts, weak bonds are formed or broken and the physical configuration of the molecule is altered. In some cases the peptide subunits of a complex macromolecule may fall apart or aggregate, depending on the concentration of ions in solution. In this way the microenvironment around the reactive site of an enzyme is altered, and its affinity for substrate is either enhanced or reduced.

The specialized carrier molecules that transport much of the oxygen consumed by large animals are proteins. They can be classified according to their sensitivity to physiological changes in ion concentrations, as follows:

1) No sensitivity to changes in inorganic salts or to H^+ (e.g., the intracellular respiratory pigments of invertebrates: the intracellular hemoglobins and, possibly, the hemerythrins). The intracellular hemoglobins are relatively simple molecules, often composed of a single peptide chain, as in many annelids, or two peptides, as in the lamellibranch molluscs. Somewhat larger oligomers occur in glycerid bloodworms, but they respond very little, if at all, to the addition of salts or H^+ (21, 61). Sipunculid hemerythrins are usually composed of eight subunits, so the structural prerequisite for ion sensitivity seems to be present. However, the absence of a H^+ effect has been reported several times, and the effects of salts on oxygen binding have not been studied. The chloride ion modifies electrophoretic mobility and spectral properties (25), which may or may not be related to oxygen-binding.

2) Oxygen affinity is depressed by both H^+ and salts (e.g. the intracellular hemoglobins of most vertebrates, which are tetrameric complexes of two different kinds of subunits). The salts that have received the most attention are small organic molecules that are not directly involved in osmotic adaptation, but the addition of chloride ions also lowers oxygen affinity.

3) Oxygen affinity is decreased by H^+ and increased by salts (Fig. 1A) (e.g. the intracellular hemoglobins of elasmobranchs, the extracellular heme proteins hemoglobin and chlorocruorin, and crustacean and cephalopod hemocyanins). The structure of the elasmobranch hemoglobins differs only in small ways from that of mammalian hemoglobin, but the extracellular compounds are high molecular weight $(0.5 - > 3 \times 10^6$ daltons) complexes of at least two different peptides. The magnitude of the ionic response of the heme proteins is quite variable (1, 5, 6, 16).

The effective ions which have been emphasized in the literature are Ca^{+2} and Mg^{+2}, which in some species change very little with salinity and are thus unlikely to have the dramatic effects demonstrated in many *in vitro* experiments. Truchot (58), however, has shown that physiological changes in these two cations do induce changes in oxygen affinity of the hemocyanin of the portunid crab *Carcinus maenas*, and the change is relatively large. P_{50} varies from 14 mm Hg for the hemocyanin in blood taken from animals acclimated to 11.7 $°/oo$, to 10 mm Hg for 36 $°/oo$ (15° C; pH 7.75). Of the variables in annelid blood, however, the physiological changes in Na^+ and Cl^- would have larger effects on oxygen affinity of *Arenicola marina* hemoglobin than Ca^{+2}, Mg^{+2} and K^+ combined (16).

4) Oxygen affinity is increased by H^+ and decreased by salts (Fig. 1B) (e.g. the hemocyanins of gastropods and xiphosurans). Molluscan hemocyanins, whose

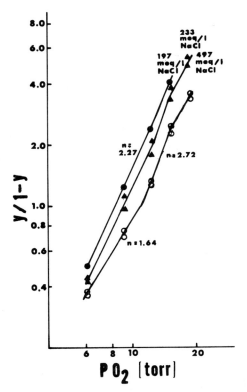

Figure 1. Hill plot of the oxygen equilibrium of hemocyanin from *Busycon canaliculatum* at 23° C in NaCl buffered with Tris HCl (ionic strength 0.1, pH 7.843-7.854). Hemocyanin prepared by centrifugation of blood at 45×10^3 rpm, followed by resuspension of pellet in buffered saline. Salt concentration and pH of solutions measured with Radiometer Corp. electrodes; % oxygenation (Y) measured in diffusion chamber, as described by Mangum, Lykkeboe and Johansen (34). Data collected by the author in collaboration with G. Lykkeboe. Positions of fitted regression lines ($r^2 = 0.991$-0.999) different at P = .05.

structure is more completely known, are very large molecules (up to 9×10^6 daltons) composed of two different kinds of subunits of about 50×10^3 each. When certain cations (such as Mg^{+2}) are absent, these giant molecules dissociate to much smaller particles and their oxygen affinity increases (10, 63). It is not clear, however, that the concentration of Mg^{+2} in animals ever changes enough to bring about detectable alterations of oxygenation. On the other hand, when the NaCl concentration is reduced from the level in the blood of the large conch *Busycon canaliculatum* at 35 °/oo salinity to that at 20 °/oo, the lowest salinity in which I have found the species, the oxygen affinity of its hemocyanin is significantly (P<.001) raised (Fig. 2). Sullivan et al. (52) have reported a large effect of Cl⁻ on the hemocyanin of the horseshoe crab *Limulus polyphemus*; the effect is significant at physiological levels.

The structural basis of the ion sensitivity of mammalian hemoglobin is partly understood (26), but the opposite effect of H^+ and salts on the other respiratory pigments (categories 3 and 4) is obscure. The classification is merely correlative, and exceptions are not precluded by present understanding of their molecular basis. For adaptive reasons, however, the correlation makes good sense.

In addition to the response of oxygen affinity, the property known as cooperativity of oxygen binding (the slope of an oxygen equilibrium curve), or the Hill constant *n*, may respond to H^+ and salts, and thus induce further modifications

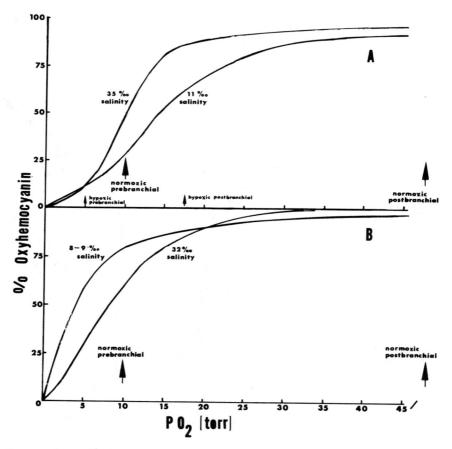

Figure 2. Oxygen equilibrium of respiratory pigments in categories 3 (A, *Carcinus maenas*, 15° C) and 4 (B, *Limulus polyphemus*, 22 °C) at different salt concentrations. Reconstructed from data shown by Truchot (57, 58) and Sullivan *et al.* (52).

of oxygenation *in vivo*. The available information on this property defies generalization about its physiological importance, however, and its response cannot be conveniently categorized. The discussion below is focused upon oxygen affinity, even though ion effects on cooperativity are included in the quantitative examples for a particular species.

Physiological and Ecological Implications

Animals with oxygen carrying pigments belonging to categories 1) and 2) avoid the respiratory problems associated with an unstable ionic environment in body fluids, but only at some cost. Insensitivity to ionic change seems to accompany osmoconformity, a response that allows the species to colonize the polyhaline regions of the estuary (> 18 °/oo salinity). But osmoconformers with intracellular respiratory pigments are not found in more mesohaline and in oligohaline waters. Below about 8 °/oo the cells containing low molecular weight hemoglobins and hemerythrins begin to lyse. Ion insensitivity also means that the function of these molecules cannot be modulated, as it is in vertebrates, by controlled changes in H^+ and salts. Thus, one of the major advantages of the vertebrate red blood cell was not exploited in its earlier version (30).

High ionic sensitivity of the oxygen carrying molecule, accompanied by homeostatic regulation of the ions in blood, is the ultimate solution. Among the several classes of vertebrates that have successfully colonized estuarine habitats, few tolerate large ionic changes in blood. The function of hemoglobin in *Rana cancrivora* (20), in which blood chloride varies from 98 meq/l in freshwater to 227 meq/l in 80% seawater (about 28 °/oo salinity), should be an especially interesting subject. Considerably smaller changes occur in other amphibians (11) and in several species of fish (22). If the chloride sensitivity of their hemoglobins is similar to that of mammalian hemoglobins (47), and if the ionic instability of serum is paralleled inside the red blood cell, then these changes should have measureable effects on oxygen transport. Although blood chloride levels in elasmobranchs change very little with salinity (22), the chloride sensitivity of their hemoglobins is so great that oxygen affinity might vary from 5 to 8 mm Hg, which is physiologically significant. Other than a few exceptions, various species of vertebrates strongly regulate both the salt concentration and the acid-base status of their blood. In birds and mammals the regulation of blood pH is accomplished largely by the kidney, which shows an allometric increase in size when compared, for example, with the antennal gland of decapod crustaceans. It is highly probable that this increase is reflected in its metabolism, and that homeostatic ion regulation in vertebrates is expensive in the overall energy budget of the animal.

None of the species with oxygen carrying pigments belonging to categories 3) and 4) are perfect regulators of the ion concentrations in their body fluids, and yet many of them are more euryhaline than any vertebrate. The blue crab *Callinectes sapidus* is found in freshwater (31) and in hypersaline lagoons with salinities as high as 75 °/oo (13). Its osmotic response to very high salinities is not known; it essentially conforms to salinities in the range 28 to 35 °/oo, but below 28 °/oo it is a strong regulator (29; Fig. 3A). A species such as the annelid polychaete *Nereis succinea* may be even more interesting. Its distribution is almost as euryhaline as that of the blue crab, but the changes in blood salt concentration are four times greater (38).

If the dilution of blood salt is not opposed by compensatory responses of other systems, the oxygenation of the respiratory pigments in category 3) should *decrease* and the oxygenation of substances in category 4) should *increase* as the animal moves upstream (Fig. 1). For the moment I am assuming that the pigments deliver about half of their oxygen when the animals are acclimated to high salinity waters (a figure representative of many aquatic species), and that blood PO_2 does not change at low salinity. Oxygen transport by pigments in category 3) should *increase* in low salinities because more oxygen is unloaded to respiring tissues. Oxygen transport by those in category 4) should *decrease* because less oxygen is unloaded to respiring tissues (Fig. 1). Unless other respiratory activities such as circulation or ventilation compensate, the rate of oxygen consumption in decapod crustaceans and in annelids with extracellular respiratory pigments would be elevated in dilute waters and that in xiphosurans and hemocyanin containing molluscs would be reduced. *This postulated change in oxygen consumption would be unrelated to osmotic work; it would be a passive consequence of the altered ionic environment in which the respiratory pigment operates.*

The hypothetical change in oxygen consumption would be relatively great. In the decapod crustaceans hemocyanin (category 3; Fig. 1A) transports 80-95% of the oxygen consumed at summer temperatures (32, 44, 51); the increase in Fig. 1A would be about 15%. Hemocyanin would take up somewhat less oxygen at the gills (about 5%), but it would deliver considerably more to the tissues (about 20%). The hemocyanin of *Limulus polyphemus* (category 4; Fig. 1B) is less important in respiration under these conditions (24, 33, 35). At high *in vivo* concentrations, it transports about 75% of the oxygen consumed, and the decrease in Fig. 1B would be about 20%. Oxygenation at the gills would not change appreciably, but the hemocyanin would deliver 20% less oxygen to the tissues.

In other species the change might be extremely small, if the ion sensitivity of the respiratory pigment were diminished. Thus this property can account for measureable increases or decreases in the oxygen consumption of intact animals, and it may also explain the absence of a detectable response.

The relationship between *in vitro* oxygen equilibrium data and the respiratory response of an intact animal, as described above, is a gross oversimplification. For example, the oxygen consumption of blue crabs may not change until the animals are exposed to salinities lower than 10 °/oo (27; Table 1). And yet the oxygen equilibrium data do not suggest a nonlinear response.

The certainty of a prediction of oxygen consumption from data on *in vitro* oxygen binding of respiratory pigments is small because the prediction must be made on the very tenuous condition that other respiratory activities do not

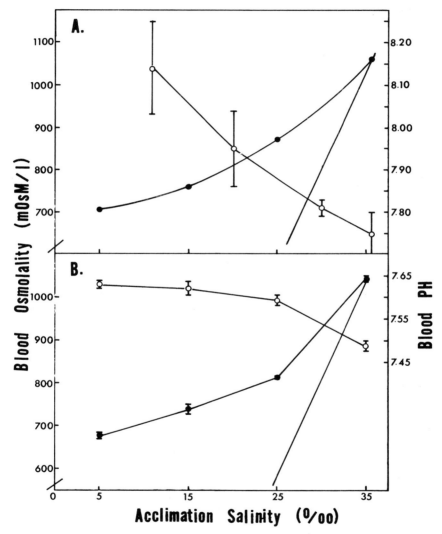

Figure 3. Relationship of blood osmotic concentration (•) and pH (o) in the portunid crabs *Carcinus maenas* (A, 15° C) and *Callinectes sapidus* (B, 22° C). Data in A (mean ± 95% C.I.) from Truchot (57) and Theede (54); data in B (mean ± S.E. for 10 winter crabs) from Weiland and Mangum (62). Solid diagonal at right is isosmotic line.

Table 1. Rate of oxygen consumption (mean VO_2 ± S.E.) of adult male *Callinectes sapidus* before and two days after change in acclimation salinity. $\dot{V}O_2$ calculated from record of oxygen depletion (Yellow Springs Instrument Co. 5420-A polarographic electrode) in closed container from 159 to 100 torr. 19°C. Summary of unpublished data collected by the author.

Wt. (gm)	$\dot{V}O_2$ (μl/gm-hr)		% change
	25 °/oo	5 °/oo	
A. Animals transferred from 5 to 25 °/oo			
203	97 (±9)	71 (±3)	27
100	123 (±6)	92 (±3)	25
149	70 (±6)	58 (±4)	17
B. Animals transferred from 25 to 5 °/oo			
111	52 (±5)	75 (±2)	44
114	100 (±3)	110 (±2)	10
157	77 (±1)	99 (±3)	29

change. As concluded above, the available information does not suggest that cell respiration in most of the crab's deep tissues changes enough with acclimation to salinity to have appreciable effects on deoxygenation of the blood. Increases in aerobic metabolism of isolated gill (37) and nerve (19) are known, but these tissues comprise a very small fraction of the total body mass. More convincingly, the PO_2 of blood returning from the tissues changes very little, and it actually goes *up* at low salinity and not *down* (60). Blood PO_2 is determined by many respiratory activities in addition to mitochondrial metabolism, such as the rates of circulation and ventilation, of which we have little knowledge. Thus the assumption that other respiratory activities do not change with salinity is partly a heuristic device to emphasize the importance of oxygen transport, but it is also supported by present knowledge and necessitated by areas of ignorance. In any event, *in vitro* oxygen equilibrium data are useful primarily in understanding the influence of the oxygen transport system on metabolism, and only if the pH and PO_2 of body fluids are known.

Effect of Acclimation Salinity on Body Fluid pH and PO_2

Naively, one might expect the pH of circulating body fluids to vary directly with the salt concentration of the medium. If an increase in osmotic work entails an increase in aerobic metabolism, then the additional CO_2 in the blood could generate more H^+. This relationship has been demonstrated in the barnacle *Pollicipes polymerus* (17). As already indicated, however, oxygen consumption in many species does not increase in low salinity and no metabolic effect on the bicarbonate buffer system would be expected. Moreover, an extremely important phenomenon was discovered in the green crab *Carcinus maenas*, a member of the same decapod family as *Callinectes sapidus* (57). In this species oxygen consumption increases as the blood becomes hyperosmotic to the medium, in salinities only slightly lower than full strength seawater (53, 54; Fig. 3A). Although there is a small increase in PCO_2 as the salinity is reduced from 36 to about 30 $°/oo$, PCO_2 then returns to basal levels and there is no significant difference in the values at 36, 26, 20 and 11.7 $°/oo$ (57). Contrary to naive expectation, blood pH varies *inversely* with acclimation salinity; the overall change is quite large (-0.4 in the range 11.7-36 $°/oo$), and it cannot be brought about by the end products of aerobic metabolism. It is great enough to *reverse* the predicted effects of salt reduction on hemocyanin oxygen affinity. Instead of decreasing as the crab invades dilute waters, hemocyanin oxygen affinity increases from 10 to 7.3 mm Hg (15° C), because blood pH increases at the same time from 7.75 to 8.15 (57). In some crabs this pH change would be sufficient to dissociate most of the hemocyanin molecules into smaller subunits with a higher oxygen affinity, but the aggregation of *Carcinus* hemocyanin does not change in this range (9). This difference may represent a structural adaptation of the molecule to the unstable ionic environment in which it works. In any event, the increase in oxygen consumption (53) occurs when oxygen transport by hemocyanin would be expected to be *impaired* (58) not enhanced, if the *in vitro* data are interpreted simply.

The pH change occurs in the blue crab, also a portunid, although the form of the response is different. Unlike *Carcinus maenas*, *Callinectes sapidus* makes a sharp transition from osmoconformity to osmoregulation. If the acclimation salinity 27-28 $°/oo$ had been used in the experiment illustrated in Fig. 3B, the line describing changes in blood osmolality would exactly parallel the isosmotic line (29) to that point, and then it would depart. Below 27-28 $°/oo$ salinity, blood osmolality and chloride change very little (Fig. 3B; 29, 62). The percent oxygenation of blue crab hemocyanin is reduced at low salt levels (32), and the effective ions include Cl^- as well as divalent cations (Bonaventuras and Sullivan, pers. commun.). However, the percent oxygenation is increased at low H^+ levels, and the effect is very large (log P_{50}/ pH = -1.3). The net result is a change in the oxygen transport function of hemocyanin only at the lowest experimental salinity (5 $°/oo$), and the change may not occur in natural waters. In our experiments the dilutions were prepared with distilled water, to maintain constant ion ratios for the different measurements (62). Consequently, blood Ca^{+2} was probably lower than it is in crabs found in natural fresh waters, where the oxygen transport system may remain unchanged.

THE RELATIONSHIP BETWEEN BODY FLUID pH
AND ENVIRONMENTAL SALINITY
Its Adaptive Significance

In neither species of portunid crab does the overall change in ionic composition of the blood promote an enhanced oxygen transport function of hemocyanin in low salinity. Indeed, the *in vitro* data suggest that hemocyanin may play a smaller respiratory role in *Carcinus maenas*, which is the less euryhaline

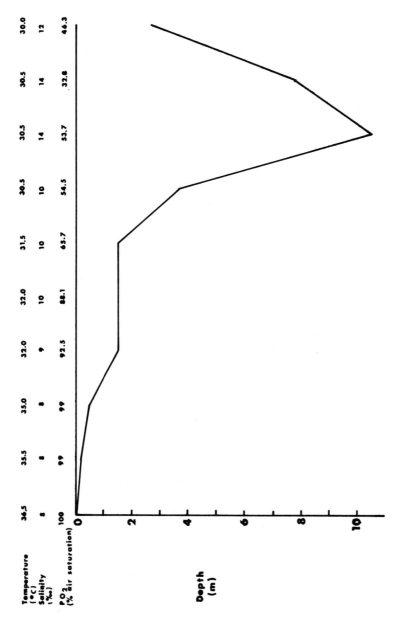

Figure 4. An oxygen profile (% air saturation) of the bottom layer of the York River estuary at Sycamore Landing, Virginia on August 15, 1975 at mid-flood tide (1400-1500 hr EDT). PO_2 in surface water ≥ 92.5% air saturation. Temperature and oxygen content measured with Yellow Springs Instrument Co. Model 5419 oxygen probe; salinity with refractometer (Atago) calibrated with distilled and Copenhagen Normal waters.

species. This seemingly unadaptive response is an outcome of the surprising change in blood pH. Why does it occur?

All of the respiratory measurements on intact blue crabs were made with the animals carefully held in continuously aerated water; the PO_2 of the water entering the branchial chamber was never lower than 138.5 mm Hg, or about 87% air saturation (62). All of the predictions have been formulated on the assumption that neither the oxygen demand nor the environmental oxygen concentration is altered when the animal enters the estuary. And yet the animal must increase its locomotor activity in order to make the estuarine migration and oxygen conditions may be drastically reduced in the deeper waters below the halocline (8). In highly stratified waters blue crabs may experience severe hypoxia; in more common situations they encounter oxygen regimes such as that shown in Fig. 4.

If the oxygen affinity of hemocyanin were lowered, due to a reduction in salts, at the same time that blood PO_2 fell in response to increased oxygen demand or low oxygen levels in the environment, the role of hemocyanin would be less in a habitat such as the York River than in more saline and also better oxygenated waters of the Chesapeake Bay (Fig. 1A). The observed stability of hemocyanin function, due the counterbalancing changes in H^+ and salts, may represent the best compromise between the demands of aerobic metabolism at high levels of locomotor and osmotic activity, and the likelihood of encountering oxygen depleted waters during the migration upstream.

The mechanism of the pH change may also provide an insight into its evolutionary and adaptive significance.

The Relationship Between Ammonia Production, Blood pH and Environmental Salinity

Although there are alternative explanations that must be kept in mind (see below), we believe that the increase in blood pH with dilution of the medium is brought about by ammonia produced in catabolism of the free amino acids that maintain intracellular osmolality (62). The intimate connection between blood pH and osmotic adaptation was apparent in the different responses of the two species of portunid crabs and the barnacle *Pollicipes polymerus* (17). Since the osmotic balance between cells and blood of *P. polymerus* is not maintained by changes in the size of an amino acid pool, there is no reason to suppose that ammonia production changes with acclimation salinity. The elevation of aerobic metabolism yields more CO_2 at low salinity, and blood pH goes down. But in the portunid crabs, free amino acids contribute as much as 70% of the total intracellular osmolality (18). The catabolism of these compounds seems to be a critical mechanism of osmotic adjustment, and the ammonia in blood rises at low salinity. At 11 $^\circ$/oo, the pool in *Carcinus maenas* is reduced to less than half its initial size, and the blood ammonia level rises by about 0.7 mM/1 (49). In *Callinectes sapidus* the concentration of amino acids in muscle decreases to 32%

of its initial value when crabs are moved from 32-34 $^\circ/oo$ to 16-17 $^\circ/oo$; blood ammonia rises by about 0.1 mM/l (18). The changes are somewhat smaller in blue crabs, even when a correction is made for the smaller dilution of the medium. The change in blood pH is also smaller (Fig. 3), but its effect on hemocyanin oxygenation is greater.

The pattern of the responses of blood pH and osmolality suggests that the two changes are closely related. The shape of each curve is somewhat different in the two species, but in both cases the pH curve appears to be a mirror image of the osmolality curve (Fig. 3). In *Carcinus maenas* there is a gradual transition from osmotic conformity to hyperosmoticity and a similarly monotonic change in pH as the external medium is diluted. In contrast, blood pH, osmolality and ammonia all change sharply in blue crabs when the acclimation salinity is reduced from 35 to 25 $^\circ/oo$, but very little below 25 $^\circ/oo$ (36). Seasonal differences in blue crabs also support the trend; winter crabs (shown in Fig. 3B) have significantly higher blood osmotic concentrations and lower values of blood pH than summer crabs acclimated to the same salinity (62).

The Origin and Movement of Ammonia

The ammonia excreted by the mammalian kidney appears to originate in kidney cells from precursors transported from other sites (40). A peripheral site of deamination is necessary to the hypothesis that ammonia crosses cell membranes in the molecular form. The ammonia buffer system favors increasing ionization of the NH_3 molecules to NH_4^+ as pH is decreased from pK (9.15). Thus a pH difference across a membrane separating two aqueous solutions such as blood and intracellular fluid (or intracellular fluid and urine) results in the diffusion of NH_3, which moves freely, but not the charged particle NH_4^+, which is considerably less permeant. NH_3 moves in the direction of increasing H^+ concentration, because the higher H^+ concentration induces more protonation and thus leaves less NH_3. Thus molecular diffusion can account for the movement of the ammonia found in crab blood into the urine, as illustrated in Fig. 5.

Unlike mammals, considerable quantities of ammonia exist in the blood of crabs. It is highly probable that at least a fraction of the ammonia excreted does not originate at the site of output. The movement of ammonia from blood into the epithelial cells at the gill can be interpreted as a process of "downhill" diffusion, since blood H^+ is believed to be lower than intracellular H^+ concentration. Some of the ammonia leaving the gill is probably not derived from the blood; it may be produced inside the cells that line the gill epithelium. Silverthorn (50 and pers. commun.) has detected activity in the blue crab gill of glutamic dehydrogenase, an enzyme that catalyzes deamination, and she has also shown that this activity increases at low salinity. Her findings are consistent with the idea that the size of the free amino acid pool in tissue is maintained catabolically, but it is unlikely that the gill is the major site of deamination in the animal. There is little evidence of amino acid transport in blood, and salinity induced changes in the amino acid pool inside gill cells are the smallest of those in the several tissues examined (18, 19, 49). When a blue crab is moved from water of about 32-34 °/oo to 16-17 °/oo, the total amino acid concentration in gill cells is reduced by less than 8 mM/l, whereas the net Na^+ flux changes from a somewhat negative value to about 100 mM/l positive in blood. The role of the gill (relative to other tissue) in producing ammonia cannot be quantitatively evaluated without knowledge of the kinetics of amino acid metabolism. Static

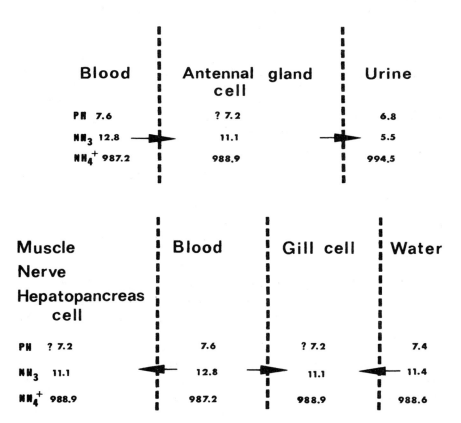

Figure 5. The molecular diffusion of NH_3 across membranes of the blue crab *Callinectes sapidus* at measured and expected (?) pH. The pH of urine in this species is not known; the value given is for *Limulus polyphemus* (35) at the same temperature (22° C).

At equilibrium a total of 1000 particles of NH_3 and NH_4^+ would be distributed as shown if pK for the ammonia buffer system is the same (9.15) as that in mammalian blood at 37°C. The temperature sensitivity of pK is not unequivocally known and the absolute NH_3 values would be different at 22°C, at which the other data on blue crabs were collected. The sign of the diffusion gradients and the direction of molecular movement (arrows) would not change, however.

measurements suggest that other tissues such as the hepatopancreas, muscle and nerve are more important sources of ammonia. We do not know whether the process of deamination is centralized, but the changes in blood ammonia levels strongly suggest that it is not confined to the gill.

The H^+ gradient dictates that NH_3 must diffuse from blood into gill cells, but not from there into water, until the salinity becomes very low (Fig. 5). Excretion of ammonia at the gill must take place by means of active transport. This conclusion is supported not only by the *a priori* consideration of the H^+ gradient, but also by the finding that ammonia is still excreted when the medium is so overloaded with $NH_4 Cl$ that the NH_3 gradient would block molecular diffusion (36).

Another difficult problem is the site of protonation of the ammonia and how it affects blood pH. If intracellular pH is lower than blood pH, then ammonia will not diffuse from muscle, for example, into blood (Fig. 5). A similar problem exists in accounting for the ammonia in renal veins of mammals, where it is supposed that "flow rather than pH is the major determinant of transport" (40, p. 464). This statement implies that the driving force for molecular diffusion—a downhill gradient—is maintained by the rapid removal of NH_3 in flowing blood, or that blood flow is faster than NH_3 permeation and H^+ release, so that equilibrium between the molecular and ionized forms in cells and in blood is not established in time to reverse the diffusion gradient. This idea is apparently based on intuitive considerations rather than knowledge of the kinetics of these processes. The possibility that NH_4^+ crosses the cell membrane, and that intracellular protonation causes the observed change in blood pH must also be considered.

The Relationship Between Nitrogen Excretion and Osmoregulation

The change in blood NH_4^+ concentration may have a very important effect on the osmotic balance between blood and water as well as that between blood and intracellular fluids.

A hyperosmotic condition of the blood in estuarine crustaceans is generated by a "salt pump" at the gill; as in the lower vertebrates, the crab "kidneys" (called antennal or maxillary glands, depending on the location of their openings to the environment) play a very small role, if any. The pump consists of carrier molecules bound to the gill membrane that actively move ions across the barrier. These carriers are enzymes whose oxidative activity is often monitored as the breakdown of ATP; in many biological systems (e.g., squid axons, red blood cells) they exchange three Na^+ for two K^+ and hence they are called $Na^+ + K^+$ ATPases. In fact they are always active at the crab gill, but the loss of Na^+ by some route equals the intake at salinities above 27-28 $^{\circ}/oo$. At low salinity their activity is greater (56), and Na^+ loss is believed to be reduced.

Clearly a salt pump which functions to maintain a hyperosmotic condition in blood would not work if the net exchange were one K^+ for one Na^+, and yet some cation must be secreted each time a Na^+ is absorbed, to maintain electro-chemical balance. A good counterion would be one that is produced inside the animal in excess, and the excess should be proportional to the net absorption of Na^+. There is a growing body of evidence that a counterion of importance in the blue crab is NH_4^+, which meets many of the requisite conditions.

Towle (56, 57) has studied the properties of the carrier enzyme in micro-somes prepared from the gill of blue crabs. When he substituted lower concentra-tions of NH_4^+ for K^+ in the incubation medium, there was no significant reduc-tion in enzyme activity (Fig. 6). This finding suggests that the production of excess ammonia in low salinity, while playing a critical role in respiratory adjust-ment, is also important in salt absorption at the gill. There is, however, reason to

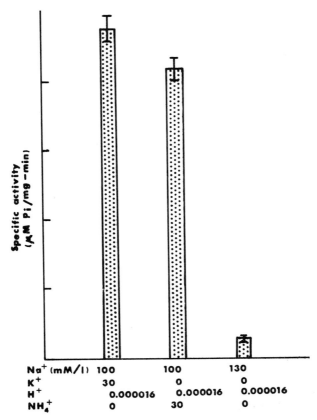

Figure 6. ATPase activity in microsomal preparations of gills from male adult individuals of *Callinectes sapidus* acclimated to 5°/₀₀ salinity. Microsomes incubated at 25° C in 20 mM imidazole-HC1 (pH 7.8), 5 mM $MgCl_2$, 5 mM disodium ATP, plus NaCl, KCl and/or NH_4Cl, as indicated. Data from Towle and Harris (56).

believe that the stoichiometry of cation exchange does not take the simple form of one NH_4^+ for one Na^+. Even though the enzyme activity is unchanged by the substitution of NH_4^+ for K^+ in equal concentrations, *in vitro* kinetic data suggest that the transport enzyme may move far more K^+ than NH_4^+ at physiological concentrations (ca. 67 mM Na^+, 1.65 mM K^+ and 0.14 mM NH_4^+ at 5 °/oo salinity).

In order to understand how the salt pump maintains a hyperosmotic blood without continuously increasing the electrical potential across the gill membrane, we must learn whether a NH_4^+ accompanies the enzyme catalyzed exchange of two K^+ for three Na^+, and also the net flux of K^+. Regardless of the possibile participation of K^+ in Na^+ uptake, one would not expect a simple exchange of one NH_4^+ for one Na^+; the role of molecular diffusion in ammonia excretion may have been overemphasized in the past, but the process must occur when it is driven by an acid gradient. When the crab migrates from high to low salinity, it also migrates from a medium which is alkaline to the blood and would therefore block NH_3 output by diffusion, to an acid medium in which NH_3 diffusion would occur, regardless of the state of equilibrium across the gill membrane. Some fraction of the increase in ammonia excretion at low salinity must be passive. We do not believe that the fraction is very large, because ammonia output can be greatly reduced by the compound ouabain, which specifically inhibits active transport[1]. Nonetheless, even a small change in the diffusion component would obscure the expected stoichiometry.

A $Na^+ + H^+$ Exchange?

When the level of NaCl in mammalian blood is changed, the concentration of H^+ changes in the opposite direction (43). One might suppose that the same trend would occur in crabs, and that the pH change is unrelated to the osmotic response of intracellular fluids and the level of ammonia in the blood. Although the evidence is indirect, the secretion of H^+ across membranes in the mammalian kidney is believed to occur, and it is linked with Na^+ absorption (43). The nature of the relationship is not entirely clear, but one possibility is that H^+ functions as a counterion for Na^+ in carrier mediated exchange.

This hypothesis has been invoked many times in the study of osmoregulation in teleost fish. In a provocative review, Evans (15) has even concluded that the adaptive significance of Na^+ uptake is the maintenance of a stable acid-base status of the blood and the efficient extrusion of wastes (both H^+ and NH_4^+), rather than a constant osmotic concentration. The methodological difficulty in testing the notion of H^+ secretion is to distinguish between the movement of that ion *per se* and the movement of another substance that generates a new H^+ in the

[1] For technical reasons, this experiment was done on the annelid *Nereis succinea* and not the blue crab.

medium; a truly definitive experiment has not been performed. When the reduction of the buffering capacity of the medium is considered, the net acid output by blue crabs is not clearly related to either blood H^+ concentration or osmoregulation (36). This item of negative evidence, however, cannot decide the question.

The hypothesis of a H^+ for Na^+ exchange in crustaceans lacks a more crucial supporting fact. There is no evidence of a carrier enzyme which is capable of performing the exchange. The lack of evidence is not due to the formidable difficulties in experimental methodology. For the present purposes, the only necessary evidence would be a demonstration of substantial enzyme activity in the presence of physiological concentrations of H^+ and Na^+. The data for the crab gill enzymes seem unambiguous on this point, if not conclusive on the roles of other ions. When both K^+ and NH_4^+ are removed from the incubation medium, leaving behind physiological concentrations of H^+ in the buffer, enzyme activity is reduced so much that it is barely detectable (Fig. 6). Although other kinds of carriers may be discovered, the hypothesis of a $Na^+ + H^+$ exchange at the crab gill does not seem to be consistent with the available information on transport enzymes. These enzymes have been studied in the partially purified form for only a few years, however, and it is possible that further advances will include purification of a molecule that exchanges Na^+ for H^+.

The Relationship of Blood NH_4^+ Concentration to NH_4^+ Excretion

The notion that the osmotic and excretory responses are coupled has another feature that may seem paradoxical if one believes that all animals are like mammals. When mammalian blood is depleted of salts, its acid-base status shifts in the alkaline direction, which is interpreted as evidence of increased H^+ secretion by the kidney (43). Here I am reversing the logic to suppose that a low concentration of a potential counterion (H^+) in crab blood argues *against* its importance in driving the salt pump, and that the production of a high concentration of another ion (NH_4^+) makes it a good candidate. Does the elevation of blood NH_4^+ have little to do with the adaptive response of intracellular amino acids? Does it reflect only a decrease in the excretion of that ion, along with others whose permeation is reduced at low salinity?

The answer to the latter question is most convincing (Table 2). There is no significant change in NH_4^+ excretion when blue crabs are transferred from a higher (35 $°/\text{oo}$) to a lower (28 $°/\text{oo}$) salinity, so long as the crab osmotically conforms to the change. However, the level of ammonia in blood goes up. In contrast, when crabs are moved from 28 $°/\text{oo}$ to a much lower salinity (5 $°/\text{oo}$), NH_4^+ output is more than doubled (Table 2). During this transition, the crab ceases to conform and its blood becomes hypersaline to the medium. These results, although far from definitive in locating the site of the increased NH_4^+ output, clearly demonstrate the parallelism of the osmotic and excretory responses; they suggest that NH_4^+ output changes only when salt absorption at the gill takes place, and not merely when excess counterion is present in blood. They also show that blood alkalosis is not a simple outcome of kidney function. The same phenomenon occurs in *Nereis succinea*, where it has been shown to be sensitive to the compound ouabain which inhibits the activity of the salt pump (Mangum, unpublished).

Table 2. Rates of ammonia excretion and net acid output of blue crabs (*Callinectes sapidus*) at three acclimation salinities. 22° C. Data from Mangum et al. (36).

	NH_4^+ output (μM/gm-hr) (paired observations)			Net acid output (nM/gm-hr) (unpaired observations)		
	Mean	S.E.	N	Mean	S.E.	N
Salinity						
35 °/oo	1.10	0.06	8	1.08	0.13	7
28 °/oo	1.09	0.06	8	0.83	0.09	7
% change	-0.01			-23		
5 °/oo	2.45	0.37	6	1.06	0.05	9
% change	135			-2		

THE INTEGRATION OF RESPIRATORY, OSMOTIC AND EXCRETORY RESPONSES TO LOW SALINITY
Evolution

It is highly probable that the sensitivity of the H^+ concentration of body fluids to environmental salinity *preceded* the origin of respiratory pigments. The simplest of animals—turbellarian flatworms (23) and scyphozoan polyps (60)—show small but significant changes in the intracellular pool of free amino acids when subjected to a change in salinity. Thus, the selection pressure for the osmotic response at the cellular level was neither the counterbalancing effect on oxygen transport nor the requirement for a superfluous counterion to balance the absorption of salts into blood. In the sense of a circulating body fluid, blood does not exist in the cnidarians and platyhelminths, and they are osmoconformers.

However, the quantitative aspects of these concomitant changes appear to vary enormously in different species, and the balance may very well be a product of evolutionary adaptation to particular combinations of environmental factors. For example, the free amino acids of *Limulus polyphemus* change with acclimation salinity, but they contribute only about 10% of the total intracellular osmolality (7, 46). The pH change in blood is almost imperceptible (35). The changes in free amino acids and body fluid pH in nereid polychaetes are relatively large, but the H^+ sensitivity of their hemoglobins is very small (12; C.P. Mangum and L.C. Oglesby, in prep.).

Distribution

The integrated effects of ionic change on respiratory pigment function may be quite different in various groups of animals, and the balance may be influenced by other variables accompanying the ionic change. There is no reason to suppose that the findings for decapod crustaceans are general. The response of oxygen uptake and transport by respiratory pigments to the environmental conditions in estuarine habitats depends upon complex interactions of the respiratory, osmotic and excretory systems, including aspects such as ventilation and circulation which are virtually unknown. The basis of the enantiostatic response of an intact animal is so complex that it cannot be described in qualitative terms and it cannot be predicted from data on different properties measured separately. At the present level of understanding it is clear only that no one of the traditionally defined physiological functions of an estuarine animal, and no one of the environmental variables in the estuary can be understood in isolation.

ACKNOWLEDGMENTS

I am grateful to J. and C. Bonaventura, S.U. Silverthorn, B. Sullivan and D.W. Towle for sharing with me their unpublished results. Much of the original work discussed above was supported by research grants GB-43488 and PCM 74-09345 A02 from the National Science Foundation (Regulatory Biology).

LITERATURE CITED

1. Antonini, E., A. Rossi-Fanelli and A. Caputo. 1962. Studies on chlorocruorin. I. The oxygen equilibrium of Spirographis chlorocruorin. Arch. Biochem. Biophys. 97: 336-342.
2. Baginski, R.M., and S.K. Pierce. 1975. Anaerobiosis: a possible source of osmotic solute for high salinity acclimation in marine molluscs. J. Exp. Biol. 62:589-598.
3. Ballard, B.S., and W. Abbott. 1969. Osmotic accommodation in *Callinectes sapidus* Rathbun. Comp. Biochem. Physiol. 29: 671-687.
4. Bernstein, R.E. 1953. Rates of glycolysis in human red cells in relation to energy requirements for cation transport. Nature 172: 911-912.
5. Bonaventura, J., C. Bonaventura and B. Sullivan. 1974. Urea tolerance as a molecular adaptation of elasmobranch hemoglobins. Science 186: 57-59.
6. Bonaventura, C., B. Sullivan, J. Bonaventura and S. Bourne. 1974. CO binding by hemocyanins of *Limulus polyphemus, Busycon carica* and *Callinectes sapidus*. Biochem. 13: 4784-4789.
7. Bricteux-Grégoire, S., Gh. Dûchateau-Bosson, Ch. Jeuniaux and M. Florkin. 1966. Les constituants osmotiquement actifs des muscles et leur contribution à la régulation isosmotique intracellulaire chez *Limulus polyphemus*. Comp. Biochem. Physiol. 19: 729-736.
8. Carpenter, J.H., and D.G. Cargo. 1957. Oxygen requirement and mortality of the blue crab in the Chesapeake Bay. Ches. Bay Inst. Tech. Rep. 13, 1-22.

9. Chantler, E.N., R.R. Harris and W.N. Bannister. 1973. Oxygenation and aggregation properties of haemocyanin from *Carcinus mediterraneus* and *Potamon edulis*. Comp. Biochem. Physiol. 46A: 333-344.

10. DePhillips, H.A., K.W. Nickerson, M. Johnson and K.E. van Holde. 1969. Physical studies of hemocyanins. IV. Oxygen-linked dissociation of *Loligo pealei* hemocyanin. Biochemistry 8: 3665-3672.

11. Deyrup, I.J. 1964. Water balance and kidney, p. 251-328. *In* J.H. Moore (ed.) Physiology of the Amphibia, Academic Press. New York.

12. Economides, A.P., and R.M.G. Wells. 1975. The respiratory function of the blood of *Neanthes* (= *Nereis*) *virens* (Sars) (Polychaeta: Nereidae). Comp. Biochem. Physiol. 51A: 219-224.

13. Emery, K.O., R.E. Stevenson and J.W. Hedgpeth. 1957. Estuaries and lagoons, p. 673-750. *In* J.W. Hedgpeth (ed.) Treatise on Marine Ecology and Paleoecology, Vol. 1. Mem. 67, Geol. Soc. Amer., Washington, D.C.

14. Engel, D.W., and L.D. Eggert. 1974. The effect of salinity and sex on the respiration rates of excised gills of the blue crab, *Callinectes sapidus*. Comp. Biochem. Physiol. 47A: 1005-1013.

15. Evans, D.H. 1975. Ion exchange mechanisms in fish gills. Comp. Biochem. Physiol. 51A: 491-496.

16. Everaarts, J.A., and R.E. Weber. 1974. Effects of inorganic anions and cations on oxygen-binding of haemoglobin from *Arenicola marina* (Polychaeta). Comp. Biochem. Physiol. 48A: 507-520.

17. Fyhn, H.J., J.A. Petersen and K. Johansen. 1972. Eco-physiological studies of an intertidal crustacean, *Pollicipes polymerus* (Cirripedia, Lepadomorpha). J. Exp. Biol. 57: 83-102.

18. Gerard, J-F., and R. Gilles. 1972. The free amino acid pool in *Callinectes sapidus* (Rathbun) tissues and its role in the osmotic intracellular regulation. J. Exp. Mar. Biol. Ecol. 10: 125-136.

19. Gilles, R. 1973. Oxygen consumption as related to the amino-acid metabolism during osmoregulation of the blue crab *Callinectes sapidus*. Neth. J. Sea Res. 7: 280-289.

20. Gordon, M.S., K. Schmidt-Nielsen and H.M. Kelly. 1961. Osmotic regulation in the crab-eating frog (*Rana cancrivora*). J. Exp. Biol. 38: 659-678.

21. Hoffmann, R.J., and C.P. Mangum. 1970. The function of coelmic cell hemoglobin in the polychaete *Glycera dibranchiata*. Comp. Biochem. Physiol. 36: 211-228.

22. Holmes, W.N., and E.N. Donaldson. 1969. Body compartments and distribution of electrolytes, p. 1-90. *In* W.S. Hoar and D.J. Randall (eds.), Fish Physiology, Vol. 1, Academic Press. New York.

23. Johannes, R.E., S.J. Coward and K.L. Webb. 1969. Are dissolved amino acids an energy source for marine invertebrates? Comp. Biochem. Physiol. 29: 283-288.

24. Johansen, K., and J.A. Petersen. 1975. Respiratory adaptations in *Limulus polyphemus* (L.), p. 129-146. *In* F.J. Vernberg (ed.) Physiological ecology of estuarine organisms, Univ. So. Carolina Press. Columbia, S.C.

25. Keresztes-Nagy, S., and I.M. Klotz. 1967. Cooperative interactions of bound ions and sulfhydryl groups in hemerythrin. J. Polymer Sci. 16C: 561-570.

26. Kilmartin, J.V., and L. Rossi-Bernardi. 1973. Interaction of hemoglobin with hydrogen ions, carbon dioxide and organic phosphates. Physiol. Rev. 53: 836-890.

27. Laird, C.M., and P.A. Haefner. 1976. The effects of intrinsic and environ-

mental factors on the oxygen consumption of the blue crab, *Callinectes sapidus* Rathbun. J. Exp. Mar. Biol. Ecol. (in press).

28. Larimer, J.L., and A.F. Riggs. 1964. Properties of hemocyanins. I. The effect of calcium ions on the oxygen equilibrium of crayfish hemocyanin. Comp. Biochem. Physiol. 13: 35-46.

29. Lynch, M.P., K.L. Webb and W.A. Van Engel. 1973. Variations in serum constituents of the blue crab *Callinectes sapidus*: chloride and osmotic concentration. Comp. Biochem. Physiol. 44A: 719-734.

30. Mangum, C.P. 1976. Primitive respiratory adaptations. *In* R.C. Newell, (ed.) Adaptations to the Environment: Physiology of Marine Animals, Butterworth's & Co., Ltd. London (in press).

31. _____, and L.M. Amende. 1972. Blood osmotic concentration of blue crabs (*Callinectes sapidus* Rathbun) found in fresh water. Chesapeake Sci. 13: 318-320.

32. _____, and A.L. Weiland. 1975. The function of hemocyanin in respiration of the blue crab *Callinectes sapidus*. J. Exp. Zool. 193:257-264.

33. _____, M.A. Freadman and K. Johansen. 1975. The quantitative role of hemocyanin in aerobic respiration of *Limulus polyphemus*. J. Exp. Zool. 191: 279-285.

34. _____, G. Lykkeboe and K. Johansen. 1975. Oxygen uptake and the role of hemoglobin in the East African swampworm *Alma emini*. Comp. Biochem. Physiol. 52A:477-482.

35. _____, C.E. Booth, P.L. DeFur, N.A. Heckel, R.P. Henry, L.C. Oglesby and G. Polites. 1976. The ionic environment of hemocyanin in *Limulus polyphemus*. Biol. Bull. 150 (in press).

36. _____, S.U. Silverthorn, J.L. Harris, D.W. Towle and A.R. Krall. 1976. The relationship between blood pH, ammonia excretion and adaptation to low salinity in the blue crab *Callinectes sapidus*. J. Exp. Zool. 195: 129-136.

37. Mantel, L.H. 1967. Asymmetry potentials, metabolism and sodium fluxes in gills of the blue crab *Callinectes sapidus*. Comp. Biochem. Physiol. 20: 743-754.

38. Oglesby, L.C. 1965. Steady-state parameters of water and chloride regulation in estuarine nereid polychaetes. Comp. Biochem. Physiol. 14: 621-640.

39. _____. 1975. Effects of salinity changes on metabolism of estuarine nereid polychaetes. Amer. Zool. 15:795.

40. Pitts, R.F. 1973. Production and excretion of ammonia in relation to acid-base regulation, p. 455-496. *In* J. Orloff and R.W. Berliner (eds.) Handbook of Physiology, Sect. 8, Amer. Physiol. Soc. Washington, D.C.

41. Pritchard, D.W. 1967. What is an estuary: physical viewpoint, p. 3-6. *In* G.H. Lauff (ed.) Estuaries, Amer. Assoc. Adv. Sci. Publ. No. 83. Washington, D.C.

42. Prosser, C.L. 1973. Comparative Animal Physiology (3rd edition). W.B. Saunders Co. Philadelphia. 966 p.

43. Rector, F.C. 1973. Acidification of the urine, p. 431-454. *In* J. Orloff and R.W. Berliner (eds.) Handbook of Physiology, Sect. 8, Amer. Physiol. Soc., Washington, D.C.

44. Redmond, J.R. 1968. Transport of oxygen by the blood of the land crab, *Gecarcinus lateralis*. Amer. Zool. 8: 471-479.

45. Robertson, J.D. 1960. Ionic regulation in the crab *Carcinus maenas* (L.) in relation to the moulting cycle. Comp. Biochem. Physiol. 1: 183-212.

380   C. P. MANGUM

46. _____. 1970. Osmotic and ionic regulation in the horseshoe crab *Limulus polyphemus* (Linnaeus). Biol. Bull. 138: 157-183.
47. Rossi-Fanelli, A., E. Antonini and A. Caputo. 1961. Studies on the relation between molecular and functional properties of hemoglobin. II. The effects of salts on the oxygen equilibrium of human hemoglobin. J. Biol. Chem. 236: 397-400.
48. Schoffeniels, E., and R. Gilles. 1972. Ionregulation and osmoregulation in molluscs, p. 393-420. *In* M. Florkin and B.T. Scheer (eds.) Chemical Zoology, Vol. 7, Academic Press, New York.
49. Siebers, D., C. Lucu, K-R. Sperling and K. Eberlein. 1972. Kinetics of osmoregulation in the crab *Carcinus maenas*. Mar. Biol. 17: 291-303.
50. Silverthorn, S.U., and A.R. Krall. 1975. Ammonia excretion and low salinity acclimation in *Callinectes sapidus*. Amer. Zool. 14: 795.
51. Spoek, G.L. 1962. Verslag van onderzoekingen gedaan in het "Stazione Zoologica" te Naples. Koninkl. Nederl. Akad. Wetens., Amsterdam 71: 29-34.
52. Sullivan, B., J. Bonaventura and C. Bonaventura. 1974. Functional differences in the multiple hemocyanins of the horseshoe crab, *Limulus polyphemus* (L.). Proc. Natl. Acad. Sci. 71: 2558-2562.
53. Theede, H. 1964. Physiologische Unterscheide bei der Strandkrabbe *Carcinus maenas* L. aus Nord- und Ostsee. Kieler Meeresforsch. 20: 179-191.
54. _____. 1969. Einige neue Aspekte bei der Osmoregulation von *Carcinus maenas*. Mar. Biol. 2: 114-120.
55. Towle, D.W. 1976. Effect of NH_4^+ and K^+ on Na^+ - transport ATPase activity of blue crab gill. Amer. Zool. 16 (in press).
56. _____, and J.L. Harris. 1976. Role of gill $Na^+ + K^+$-dependent ATPases in acclimation of blue crabs to low salinity. J. Exp. Zool. (in press).
57. Truchot, J-P. 1973. Fixation et transport de l'oxygene par le sang de *Carcinus maenas:* variations en rapport avec diverses conditions de température et de salinité. Neth. J. Sea Res. 7: 482-495.
58. _____. 1975. Factors controlling the *in vitro* and *in vivo* oxygen affinity of the hemocyanin in the crab *Carcinus maenas* (L.) Resp. Physiol. 24: 173-190.
59. Tucker, V.A. 1975. The energetic cost of moving about. Amer. Scient. 63: 413-419.
60. Webb, K.L., A.L. Schimpff and J. Olmon. 1972. Free amino acid composition of scyphozoan polyps of *Aurelia aurita, Chrysaora quinquecirrha* and *Cyanea capillata* at various salinities. Comp. Biochem. Physiol. 43B: 653-663.
61. Weber, R.E. 1973. Functional and molecular properties of corpuscular haemoglobin from the bloodworm *Glycera gigantea*. Neth. J. Sea Res. 7: 316-327.
62. Weiland, A.L., and C.P. Mangum. 1975. The influence of environmental salinity on hemocyanin function in the blue crab *Callinectes sapidus*. J. Exp. Zool. 193:265-274.
63. Wood, E.J., and D.G. Dalgleish. 1973. *Murex trunculus* haemocyanin. II. The oxygenation reaction and circular dichroism. Eur. J. Biochem. 35: 421-427.

ACTIVITY OF BACTERIA IN THE ESTUARINE ENVIRONMENT[1]

L. Harold Stevenson and Carl W. Erkenbrecher
Department of Biology
and
Belle W. Baruch Institute for Marine
Biology and Coastal Research
University of South Carolina
Columbia, South Carolina

ABSTRACT: Water in near-shore and estuarine environments characteristically contains a large number of bacteria and a generally high "activity" or heterotrophic potential. The bacteria in near-shore waters do not represent a uniform population and the density probably does not reflect growth and multiplication. The population consists of those bacteria that are adapted to function in the environment together with those that are simply surviving. Studies with pure cultures are often suspect and there are also inherent problems involved with metabolic studies of mixed populations. The enhanced metabolic activity in estuarine areas may be a simple reflection of the favorable nutritional status of the water, however, it may also represent increased activities of freshwater and marine forms in an attempt to compensate for the adverse impact of some perturbation. No single factor can account for the demise or survival of bacteria in these environments. A variety of both structural and physiological explanations may be put forth to explain how bacteria have adapted to the near-shore areas. The structural adaptations may include modification of the cell surface, membrane permiability and transport, and enzyme structure. The physiological responses may include changes in metabolic rate, metabolite pools, proteins and nucleic acid synthesis, and growth patterns.

INTRODUCTION

Water in the near-shore and estuarine environments characteristically contains a larger number of bacteria per unit volume than can be recovered from oceanic

---

[1] Contribution No. 131 of the Belle W. Baruch Institute for Marine Biology and Coastal Research.

and some inland waters; likewise, the "activity" of bacteria in estuarine systems is reported to be higher than observed in most other aquatic systems. The bacteria inhabiting these areas must be capable of a variety of physiological adaptations in response to fluctuating physical and chemical conditions. The ideal would be to correlate the physiological and metabolic properties of estuarine bacteria with their capacity to survive and function in this environment. The state-of-the-art, however, allows one to do this only to a limited extent. The objectives of this review are, therefore, limited in scope. We will attempt to point to some of the approaches that can be used to investigate the metabolism of estuarine bacteria, to outline the information available on the physiological properties of bacteria found in the estuary, and to consider some of the physiological properties conducive to survival of microbes in an estuarine system. We have limited our treatment to heterotrophic bacteria in the water column.

BACTERIAL METABOLISM

The study of bacterial metabolism has moved through several periods. The phases have included metabolic pathway determination, mechanisms and control of enzyme action, genetic control of macromolecular synthesis and gene expression, and finally the control of cell growth and differentiation (68). With some notable exceptions, most of the work in bacterial metabolism has been laboratory oriented using pure cultures. This approach would limit the scope of investigations of estuarine bacteria to that small percentage of bacteria that we can grow on laboratory media (34). Studies with pure cultures of estuarine bacteria are further limited because the origin of the isolate is difficult to determine. Bacteria in coastal waters clearly do not represent a uniform population. The populations include bacteria uniquely adapted to function in the environment together with those that are simply surviving (31). The latter group may consist of freshwater organisms derived from lakes and streams, those from salt marsh sediment, and those from terrestrial run-off. There is really no clear-cut definition of an estuarine bacterium similar to the one proposed for marine bacteria by MacLeod (43).

The suitability of laboratory cultures to the study of the metabolism and physiology of bacteria from natural systems has been questioned (1). The prevailing opinion in the last decade has been that the metabolism of bacteria should be measured *in situ*. The point of focus for the study of the physiology of estuarine bacteria should be the environment where the bacteria grow and affect their surroundings and not the "hot house" conditions imposed on the isolate by laboratory cultivation (1). Consequently, efforts have been made to adopt laboratory concepts and methodology to *in situ* investigations (See 59, 66, 70).

The adaptations of laboratory techniques to the study of natural systems have included, among others, investigations of the production and activity of cell-free enzymes (39, 42, 64, 65), nitrogen fixation as measured by acetylene reduction (4, 19, 47, 76), growth of bacteria employing the chemostat (32, 33), and heterotrophic metabolism (see below). The study of heterotrophic activity has received major emphasis in several laboratories and will be considered here in some detail. Several techniques are currently being used to study heterotrophic activity in nature. These methods are oxygen uptake, radiorespirometry, ^{14}C-labeled substrate uptake, tritiated substrate uptake, autoradiography, and microcalorimetry.

Respirometric techniques in which oxygen uptake is measured by oxygen electrode or manometric procedures have been useful as tools to indicate rates of microbial respiration (55). These methods, however, provide only rough estimates of the separate respiration rates of plankton and bacteria and cannot distinguish heterotrophic from autotrophic activity. Respiration rates, therefore, usually accompany biomass and radioactive substrate uptake measurements (28).

Radiorespirometry is one of the more sensitive techniques for estimating mineralization rates of organic substances in aquatic environments. This procedure measures the rate of evolution (as turnover time) of radioactive carbon dioxide resulting from the oxidation of a ^{14}C-labeled substrate (37). Interference due to sample storage, "wall effect", and substrate induced activity has been minimized by reducing incubation time and by maintaining low substrate concentrations (80). Using this technique, typical turnover times for glucose in the English Channel were observed to range from 10 to 66 days.

Perhaps the most widely used technique for measuring rates of heterotrophic activity is the uptake of specific ^{14}C-labeled substrates. This method has been very successful in monitoring activities of bacteria in pure cultures since they appear to exhibit typical first-order kinetics as described by Michaelis and Menten. The application of this technique to the natural environment was first proposed by Parsons and Strickland (54) and is based on several key assumptions. (1) The uptake of organic substrates by a heterogeneous microbial population in nature can be described by Michaelis-Menten enzyme kinetics. (2) All heterotrophs present in the sample should respond in the same way to varying substrate concentrations. (3) Bacterial growth and induction of transport systems should not occur during the measurement period. (4) Substrate concentrations in the sample should be known. (5) The substrate concentration should not change significantly. (6) The kinetics are only applicable to entry of the substrate into the cell and then only if no significant transport occurs in the opposite direction (84). Since the natural substrate concentrations are extremely low (a few μ g/liter), they are very difficult to measure quantitatively. The Hanes modification of the Lineweaver-Burk kinetic equation has been used most frequenlty to calculate graphically the kinetic parameters (14). By plotting the ratio of the incubation time (t) over the fraction of isotope taken up (f) versus the amount of substrate added (A), the following parameters can be calculated: maximum uptake velocity (V_{max}), uptake constant ($K+S_n$), and turnover time (T_t) (See 85 for a discussion of notation).

The maximum potential uptake velocity, V_{max}, is the most commonly used indicator of heterotrophic activity (87). Typical values for V_{max} measured in various aquatic environments are illustrated in Table 1. Values for estuarine systems were several orders of magnitude higher than those in open-ocean systems and normally higher than those observed in lakes. Although it has been shown to be proprotional to biomass in pure culture (22, 83), V_{max} per cell varies sufficiently to caution its use in estimating bacterial biomass (24). The interpretation of V_{max} as an indicator of heterotrophic activity must be done with caution. In cases where the natural substrate concentration (S_n) has been determined for certain amino acids, the natural uptake velocity represented only 1 to 10 percent of the V_{max}, and varied both monthly and among substrates (10).

The value of the ($K+S_n$) parameter is difficult to interpret and is best understood if K and S_n are considered separately. Theoretically K is more than the transport constant for the uptake of a particular substrate or substrates, and should be more correctly referred to as "the constant" (Hamilton, see p. 179 in 10). Typically bacterial uptake is characterized by low values (high affinity transport systems) and K values are usually close to S_n in natural waters (10). S_n, the natural substrate concentration, is difficult to measure in nature but can be used to calculate actual velocities of uptake, v_n, when it is known (25).

The third parameter is the turnover time (T_t), the time required by the bacteria to utilize all of the existing substrate. Turnover rates have been observed to range from 0.5 hr in a polluted pond to greater than 5,000 hr in an oligotrophic lake (25), with typical estuarine values of 7 to 93 hrs for specific amino acids (27). Turnover times for mineralization have been shown to be about double those for assimilation (85).

An important modification of the basic uptake procedure was developed to account for a loss of radioactivity due to respiration as $^{14}CO_2$ (22, 26). Values obtained by methods prior to this modification underestimated the total uptake. Losses due to respiration ranged from a low of 4 percent to upwards of 60 percent of the amount taken up. Although both assimilated and respired substrate give total substrate transported into the cell, the loss of substrate in dissolved form has not been accounted for. An additional advantage to the measurement of respiration is the ability to calculate mineralization rates (37, 80, 85).

As is the case with most methods, various problems have arisen that question the validity of the technique as applied to natural systems and several papers, each dealing with specific problems, have been published (20, 72, 79, 84). Some of the problems relevant to the mechanics of measurement are the following. (1) What type of blank should be used? A blank should theoretically detect all abiological effects especially where low levels of uptake are encountered. (2) Should acidified or nonacidified treatment be employed to terminate the uptake reaction (20, 72)? Because both methods are currently in use, caution should be

Table 1.   Ranges of typical V_{max} values for various aquatic environments.

Source	Substrate	V_{max} ($\mu gCl^{-1}hr^{-1}$)	Reference
ESTUARINE			
Pamlico River Estuary	Glucose	0.06 to 9.64	11
	Various Amino Acids	0.04 to 69.42	11
Nanaimo River Estuary	Glucose	0.004 to 62.5*	60
Toronto Harbor Lake Ontario	Glucose	0.199 to 0.811	82
Woods Hole Harbor	Glycine	$\sim$0.01 to 0.05*	73
Yaquina Bay	Glutamate	0.12 to 0.52	20
OCEAN			
Atlantic	Glucose	0.001 to 0.006*	74
North Atlantic	Glucose	0.0002 to 0.0012	10
Western North Atlantic	Glucose	0.0005 to 0.006	28
Western Sargasso Sea	Glucose	0.003	28
Tropical Pacific	Glucose	0.00056 to .0540	23
Nanaimo, (British Columbia, Canada)	Glucose	0.007 to 0.03*	54
	Acetate	0.0031 to 0.02*	
FRESHWATER			
Dairy Pond, N.C.	Aspartic Acid	0.072	26
Swedish Lake (oligotrophic)	Glucose	0.00032	29
Gravelly Pond (mesotrophic)	Glycolate Acetate Glycine Glucose	0.2 to 0.9	86
Lake Erken (eutrophic)	Glucose	0.0024 to 0.04	29
	Glucose	0.023 to 0.18*	87
	Acetate	0.019 to 0.19*	
Upper Klamath Lake (eutrophic)	Glucose	0.08 to 1.29	86
	Acetate	0.08 to 2.29	
	Glycolate	0.04 to 1.54	
	Glycine	0.002 to 0.96	
Upper Klamath Lake (algal bloom)	Various Amino Acids	0.592 to 5.958	7
Lotsjon Lake (highly polluted)	Glucose	6.00	2

*no correction of V_{max} for loss of $^{14}CO_2$ by respiration.

taken when comparing data from different sources. (3) Is a correction for filtration error necessary? The construction of a filtration correction curve has been recommended although the exact nature of the reason for the error is unclear (72). (4) What is the effect of agitation during incubation? This factor has not yet been tested as a source of variability in the technique (84). (5) Is there competitive inhibition between groups of substrates in the natural environment? This factor has prompted a reevaluation of the uptake technique as it applies to natural environments (6, 84).

A seldom used technique that eliminates some of the problems associated with ^{14}C-uptake is the use of tritiated substrates (3). This method provides a direct determination of the uptake and, because the tritiated substrates used have higher specific activities than ^{14}C-labeled substrates, problems with substrate activation of uptake are reduced.

BACTERIAL SURVIVAL

The presence of large numbers of heterotrophic bacteria in estuarine systems suggests that the organisms are well adapted to cope with the broad range of fluctuations in environmental conditions (1). Considering the diversity of the bacterial community and the many types of environmental conditions, it would be fruitless to propose that only a few common properties would be responsible for the fitness of the organisms to inhabit the estuary. Several of the general properties of estuarine bacteria that may contribute to their continued presence in the system (fitness traits) include tolerance to or dependence on saline conditions and heavy metals, competitive advantages in utilization of organic nutrients, capacity for attachment to particulate material and subsequent colonization, and, perhaps most important of all, dormancy.

Bacteria in the estuarine environment must tolerate intrusion by seawater and its accompanying salts and heavy metals. Most of the information relative to the influence of seawater ions on bacteria has been gleaned from studies of marine bacteria (44), enteric bacteria exposed to seawater (36), and extreme halophiles (41). Three reviews of the ionic requirements of marine bacteria have been published (43, 44, 56). Based on the information accumulated thus far, Na^+ appears to play a major role in membrane transport (15, 16, 45, 46, 67) and cell wall integrity (12, 17, 18).

Fluctuations in salinity have been shown to influence the physiology of marine bacteria. Ishida et al. (30) were able to demonstrate that the uptake of amino acids by a marine pseudomonad was generally greater when the organism was incubated in less than full-strength seawater. The growth rate of the organism, however, was maximum in media containing a larger percentage of seawater. The influence of salinity on the amino acid pool size and composition as well as the release of amino acids has been reported by Brown and Stanley (5) and Stanley and Brown (69). Generally, pool contents increased with increases in NaCl concentrations and "pulse" increases in salinity resulted in rapid synthesis of glutamate and proline. On the other hand, there was a marked decrease in amino acid pool size when salinity decreased. Griffiths and Morita (21) examined the effect of salinity on the catabolism of a marine bacterium and found significant shifts in uptake and respiration with increasing salinities. The observed patterns of CO_2 evolution indicated that the hexose monophosphate pathway (the catabolic system employed by the organism) was greatly altered by changes in salinity. The apparent shutdown of the pathway coincided with the minimum salinity for growth.

The second fitness trait proposed for estuarine bacteria concerns the ability of bacteria to compete for organic substrates. Although there is no uniminity of opinion, it appears that bacteria are more efficient than algae at substrate uptake at low concentrations (84, 87). Wright and Hobbie (87) employed kinetic analysis of heterotrophic uptake and found that a mixed culture of bacteria and algae closely resembled natural populations in uptake kinetics with bacteria accounting for most of the activity. Williams (78), using differential filtration to study the size of the organisms responsible for heterotrophic uptake, found that about 50 percent of the substrates were assimilated by organisms smaller than 1 μm and about 30 percent was incorporated by organisms in the size class of protozoa and algae (or attached bacteria). The involvement of bacteria in uptake of nutrients was also noted by Robinson et al. (58). They demonstrated an excellent correlation (r = 0.98) between bacterial populations and the uptake of six substrates. Using autoradiographic techniques, Munro and Brock (51) also implicated bacteria in heterotrophic uptake. Jannasch (31) has demonstrated the growth of marine bacteria at very low substrate concentrations using the chemostat. However, a blanket statement concerning the utilization of all substrates cannot be made. Remsen et al. (57) reported that up to 86 percent of the urea decomposing activity could be removed from coastal waters by filtration through a 10–20 μm net. They concluded that bacteria are relatively unimportant in the uptake and decomposition of urea in aquatic environments. The competitive advantage afforded by efficient nutrient uptake would primarily benefit those bacteria that are free-living.

The attachment of bacteria to surfaces may represent a third important fitness trait for estuarine bacteria. The ability of microbes to attach to and colonize particulate material may afford them with a micro-environment higher in nutrient concentration than the surrounding water. The attachment of bacteria to solid objects submerged in aquatic systems was reported in the late 1930's and a possible role of inert materials in the concentration of nutrients from natural waters was suggested in the early 1940's (See Corpe, 9, for a historical review.) ZoBell (89) was able to show that materials such as glass, porcelain, and sand stimulated bacterial growth in seawater. Meadows and Anderson (50) described the colonization of marine sand grains by various microorganisms and Munro and Brock (51) also demonstrated the same process.

A variety of conditions has been demonstrated to influence the attachment of bacteria to surfaces. These include the production of exocellular materials (48, 49), surface charge (52), surface tension (13), as well as fimbriae and pili (53).

Dexter et al. (13) used a variety of man-made materials of varying surface tension (18 to 45 dynes per cm) to show that wetability directly influenced the attachment of marine bacteria. Generally, bacteria absorbed more readily to those surfaces with the highest critical surface tension. Utilizing the *Limulus* lysate method to quantify the lipopolysaccharide present, Dexter and coworkers were able to show that glass (46 dynes/cm) was the most heavily colonized with as many as 10^9 cells per cm^2 and Teflon (18 dynes/cm) supported the least number of attached bacteria ($5 \times 10^5/cm^2$) after 7 days of submergence.

Microbial populations have also been recovered from the surface of marine plants and animals (8, 35, 63, 77, 88). Sieburth et al. (63) employed the scanning electron microscope to study the attachment of microorganisms to a wide variety of materials of biological origin. They were able to show that living-plant surfaces appeared to have a markedly reduced diversity in comparison to the organisms attaching to glass slides. The colonization appeared much reduced during periods of active plant growth; both density and diversity increased during dormancy and senescence. Sieburth (62) showed bacteria attached to unicellular organisms like diatoms and foraminifera as well as the surfaces of leafy green, brown and red algae. Sea grasses, such as eel, turtle, or manatee, supported extensive mircrobial populations. The surfaces of some animal forms, i.e., lobster larvae and amphipods, were heavily colonized, while others were free of bacteria. Man-made items like plastic and nylon line as well as processed wood were also colonized.

The fourth, and possibly the most important, physiological adaptation contributing to the survival of estuarine bacteria may be dormancy. Dormancy has been defined by Sussman and Halvorson (71) as "any rest period or reversible interruption of the phenotypic development of an organism." They further delineated a dichotomy in the types of dormancy-constitutive and exogenous. It is to exogenous dormancy, "a condition in which development is delayed because of unfavorable chemical or physical conditions of the environment," (71) that we refer. The possible role of dormancy in the distribution of bacteria was proposed by Lamanna and Malette (40) when they pointed out that dormancy may be involved if the fluctuations in the natural environment are too severe to permit continuous growth. This would permit survival so that proliferation could ensue upon the reappearance of favorable conditions.

Every student of microbiology knows that when bacteria are inoculated into a suitable medium and placed in a compatable chemical-physical environment, laboratory cultures exhibit standard growth patterns or phases, namely, lag, log, stationary, and death. Two generally accepted reasons for the termination of active metabolism and multiplication are the accumulation of toxic products and the depletion of an essential nutrient (68). The batch culture system, as outlined above, has been criticized as a poor substitute for the natural environment. The chemostat or continuous culture system has been recommended as a better choice to study the growth of bacteria in nature. The continuous culture technique has been used in some excellent and sophisticated studies of the growth of bacteria in natural waters (31, 32, 33) and these studies have yielded important information. In some systems there may be a stable environment with a continuous input of nutrients resulting in a "continuous" growth of the bacteria therein. However, stable conditions are manifestly not the hallmark of estuarine ecosystems. Physical, chemical, nutritional, and biological conditions change markedly through time in estuarine and coastal systems. The only thing constant is change itself. There is no reason to assume that all of the bacteria in the system can function over the entire range of variables; some must experience periods of inactivity.

There are three lines of evidence suggesting periods of dormancy for estuarine bacteria. The first is derived from studies of population structure. Extensive literature showing seasonal fluctuations in total bacterial numbers is available. However, most of these reports deal with "total heterotrophs" or "total microbial biomass" rather than with fluctuations in specific types. There are some notable exceptions like the work of Sieburth (61), and Kaneko and Colwell (38). Sieburth's work was based on the fluctuations of physiological types relative to seasonal temperature. He demonstrated that different groups of organisms were dominant during various times of the year. Kaneko and Colwell studied the conditions influencing the distribution of a single species of *Vibrio*. They were able to show that the organism was not found during the winter and reappeared during the summer months. A second indication of bacterial dormancy is provided by Jannasch (31). Using the continuous culture technique, he was able to show that cultures of marine bacteria would not divide in seawater containing limiting substrate concentrations. He suggested that below certain levels of substrate, cultures were surviving but "inactive." The third line of evidence is derived from heterotrophic uptake experiments. Williams and Gray (81) reported two distinct increases in the rates of uptake of amino acids by marine populations. The first was immediately after substrate addition and the second was observed 20 to 35 hours after addition of substrate. The latter could have been the result of activation of an inactive cell or to microbial growth. Vaccaro and Jannasch (75) reported that natural bacterial populations in seawater respond rapidly to organic enrichment, showing a good kinetic response within 12 hours. Since the response did not allow time for multiplication, Wright (84) suggested that some members of the community were functionally "turned off" or dormant.

REFERENCES

1. Alexander, M. 1971. Biochemical ecology of microorganisms. Ann. Rev. Microbiol. 25:361-392.
2. Allen, H.L. 1967. Chemo-organotrophic utilization of dissolved organic compounds by planktic algae and bacteria in a pond. M.S. thesis, Michigan State University.
3. Azam, F., and O. Holm-Hansen. 1973. Use of tritiated substrates in the study of heterotrophy in seawater. Marine Biol. 23:191-196.
4. Brooks, R.H., P.L. Brezonik, H.D. Putnam, and M.A. Keirn. 1971. Nitrogen

fixation in an estuarine environment: The Waccasassa on the Florida Gulf Coast. Limnol. Oceanogr. 16:701-710.

5. Brown, C.M., and S.O. Stanley. 1972. Environment-mediated changes in the cellular content of the 'pool' constituents and their associated changes in cell physiology. *In* A.C.R. Dean, S.J. Pirt, and D.W. Tempest (eds.), Environmental Control of Cell Synthesis and Structure. Academic Press, London.

6. Burnison, B.K., and R.Y. Morita. 1973. Competitive inhibition for amino acid uptake by the indigenous microflora of Upper Klamath Lake. Appl. Microbiol. 25:103-106.

7. _____, and _____. 1974. Heterotrophic potential for amino acid uptake in a naturally eutrophic lake. Appl. Microbiol. 27:488-495.

8. Chan, E.C.S., and E.A. McManus. 1967. Development of a method for the total count of marine bacteria on algae. Can. J. Microbiol. 13:295-301.

9. Corpe, W.A. 1974. Periphytic marine bacteria and the formation of microbial films on solid surfaces, p. 397-417. *In* R.R. Colwell, and R.Y. Morita (eds.), Effect of the ocean environment on microbial activities. University Park Press, Baltimore.

10. Crawford, C.C., J.E. Hobbie, and K.L. Webb. 1973. Utilization of dissolved organic compounds by microorganisms in an estuary, p. 169-180. *In* L.H. Stevenson and R.R. Colwell (eds.), Belle W. Baruch library in marine science. No. 1. Estuarine microbial ecology. Univ. South Carolina Press, Columbia.

11. _____, _____, and _____. 1974. The utilization of dissolved free amino acids by estuarine microorganisms. Ecology 55:551-563.

12. DeVoe, I.W., and E.L. Oginsky. 1969. Antagonistic effect of monovalent cations in maintenance of cellular integrity of a marine bacterium. J. Bacteriol. 98:1355-1367.

13. Dexter, S.C., J.D. Sullivan, J. Williams, and S.W. Watson. 1975. Influence of substrate wettability on the attachment of marine bacteria to various surfaces. Appl. Microbiol. 30:298-308.

14. Dixon, M., and E.C. Webb. 1964. Enzymes, 2nd ed., Academic Press, Inc., New York.

15. Drapeau, G.R., and R.A. MacLeod. 1963. Na$^+$-dependent active transport of α-aminoisobutyric acid into cells of a marine pseudomonad. Biochem. Biophys. Res. Commun. 12:111-115.

16. Drapeau, G.R., T.I. Matula, and R.A. MacLeod. 1966. Nutrition and metabolism of marine bacteria. XV. Relation of Na$^+$-activated transport to the Na$^+$ requirement of a marine pseudomonad for growth. J. Bacteriol. 92:63-71.

17. Forsberg, C.W., J.W. Costerton, and R.A. MacLeod. 1970. Separation and localization of cell wall layers of a gram-negative bacterium. J. Bacteriol. 104:1338-1353.

18. _____, _____, and _____. 1970. Quantitation, chemical characteristics, and ultrastructure of the three outer cell wall layers of a gram-negative bacterium. J. Bacteriol. 104:1354-1368.

19. Goering, J.J., R.C. Dugdale, and D.W. Menzel. 1966. Estimates of *in situ* rates of nitrogen uptake by *Trichodesmium* sp. in the tropical Atlantic Ocean. Limnol. Oceanogr. 11:614-620.

20. Griffiths, R.P., F.J. Hanus, and R.Y. Morita. 1974. The effects of various water-sample treatments on the apparent uptake of glutamic acid by

natural marine microbial populations. Can. J. Microbiol. 20:1261-1266.
21. _____, and R.Y. Morita. 1973. Salinity effects on glucose uptake and catabolism in the obligately psychrophilic marine bacterium *Vibrio marinus*. Mar. Biol. 23:177-182.
22. Hamilton, R.D., and K.E. Austin. 1967. Assay of relative heterotrophic potential in the sea: the use of specifically labelled glucose. Can. J. Microbiol. 13:1165-1173.
23. _____, and J.E. Preslan. 1970. Observations on heterotrophic activity in the eastern tropical Pacific. Limnol. Oceanogr. 15:395-401.
24. _____, K.M. Morgan, and J.D.H. Strickland. 1966. The glucose uptake kinetics of some marine bacteria. Can. J. Microbiol. 12:995-1003.
25. Hobbie, J.E. 1973. Using kinetic analyses of uptake of carbon-14 to measure rates of movement of individual organic compounds into aquatic bacteria. Bull. Ecol. Res. Comm. 17:207-214.
26. _____, and C.C. Crawford. 1969. Respiration corrections for bacterial uptake of dissolved organic compounds in natural waters. Limnol. Oceanogr. 14:528-532.
27. _____, _____, and K.L. Webb. 1968. Amino acid flux in an estuary. Science 159:1463-1464.
28. _____, O. Holm-Hansen, T.T. Packard, L.R. Pomeroy, R.W. Sheldon, J.P. Thomas, and W.J. Wiebe. 1972. A study of the distribution and activity of microorganisms in ocean water. Limnol. Oceanogr. 17:544-555.
29. _____, and R.T. Wright. 1968. A new method for the study of bacteria in lakes: description and results. Mitt. Internat. Verein. Limnol. 14:64-71.
30. Ishida, Y., A. Nakayama, and H. Kadota. 1974. Temperature-salinity effects upon the growth of marine bacteria, p. 80-91. *In* R.R. Colwell, and R.Y. Morita (eds.), Effect of the ocean environment on microbial activities. University Park Press, Baltimore.
31. Jannasch, H.W. 1967. Growth of marine bacteria at limiting concentrations of organic carbon in seawater. Limnol. Oceanogr. 12:264-271.
32. _____. 1968. Competitive elimination of Enterobacteriaceae from seawater. Appl. Microbiol. 16:1616-1618.
33. _____. 1969. Estimations of bacterial growth rates in natural waters. J. Bacteriol. 99:156-160.
34. _____, and G.E. Jones. 1959. Bacterial populations in seawater as determined by different methods of enumeration. Limnol. Oceanogr. 4:128-139.
35. Jones, G.E. 1958. Attachment of marine bacteria in zooplankton. U.S. Fish Wildl. Serv. Spec. Sci. Rep. Fish. 279:77-78.
36. _____. 1971. The fate of freshwater bacteria in the sea. Devel. Ind. Microbiol. 12:141-151.
37. Kadota, H.Y., Y. Hata, and H. Miyoshi. 1966. A new method for estimating the mineralization activity of lakewater and sediment. Mem. Res. Inst. Food Sci. Kyoto Univ. 27:28-30.
38. Kaneko, T., and R.R. Colwell. 1973. Ecology of *Vibrio parahaemolyticus* and related organisms in Chesapeake Bay. J. Bacteriol. 113:24-32.
39. Kim, J., and C.E. ZoBell. 1974. Occurrence and activities of cellfree enzymes in oceanic environments, p. 368-385. *In* R.R. Colwell, and R.Y. Morita (eds.), Effect of the ocean environment on microbial activities. University Park Press, Baltimore.
40. Lamanna, C., and M.F. Mallette. 1965. Basic bacteriology its biological and

chemical background, 3rd ed. The Williams and Wilkins Co., Baltimore.

41. Larsen, H. 1967. Biochemical aspects of extreme halophilism. Adv. Microbial. Physiol. 1:97-132.

42. Litchfield, C.D. 1973. Interactions of amino acids and marine bacteria, p. 145-168. *In* L.H. Stevenson, and R.R. Colwell (eds.) Belle W. Baruch library in marine science. No. 1. Estuarine microbial ecology. Univ. South Carolina Press, Columbia.

43. MacLeod, R.A. 1965. The question of the existence of specific marine bacteria. Bacteriol. Rev. 29:9-23.

44. _____. 1968. On the role of inorganic ions in the physiology of marine bacteria, p. 95-126. *In* M.R. Droop and E.J. Ferguson Wood (eds.), Advances in microbiology of the sea, Vol. 1. Academic Press, New York.

45. _____, and A. Hori. 1960. Nutrition and metabolism of marine bacteria. VIII. Tricarboxylic acid enzymes in a marine bacterium and their response to inorganic salts. J. Bacteriol. 80:464-471.

46. _____, and E. Onofrey. 1957. Nutrition and metabolism of marine bacteria. III. The relation of sodium and potassium to growth. J. Cell Comp. Physiol. 50:389-402.

47. Maruyama, Y., T. Suzuki, and K. Otobe. 1974. Nitrogen fixation in the marine environment: the effect of organic substrates on acetylene reduction, p. 341-353. *In* R.R. Colwell, and R.Y. Morita (eds.), Effect of the ocean environment on microbial activities. University Park Press, Baltimore.

48. Marshall, K.C., R. Stout, and R. Mitchell. 1971. Mechanism of the initial events in the sorption of marine bacteria to surfaces. J. Gen. Microbiol. 68:337-348.

49. _____, _____, and _____. 1971. Selective sorption of bacteria from sea water. Can. J. Microbiol. 17:1413-1416.

50. Meadows, P.S., and I.G. Anderson. 1968. Microorganisms attached to marine sand grains. J. Mar. Biol. Ass. U. K. 48:161-175.

51. Munro, A.L.S., and T.D. Brock. 1967. Distinction between bacterial and algal utilization of soluble substances in the sea. J. Gen. Microbiol. 51:35-42.

52. Neihoh, R., and G. Loeb. 1972. Molecular fouling of surfaces in sea water, p. 710-718. *In* Proc. Third Int. Congr. on marine corrosion and fouling. National Bureau of Standards, Gaithersburg.

53. Ottow, J.C.G. 1975. Ecology, physiology, and genetics of fimbriae and pili. Ann. Rev. Microbiol. 29:79-108.

54. Parsons, T.R., and J.D.H. Strickland. 1962. On the production of particulate organic carbon by heterotrophic processes in sea water. Deep-Sea Res. 8:211-222.

55. Pomeroy, L.E., and R.E. Johannes. 1968. Occurrence and respiration of ultraplankton in the upper 500 meters of the ocean. Deep-Sea Res. 15:381-391.

56. Pratt, D. 1974. Salt requirements for growth and function of marine bacteria, p. 3-15. *In* R.R. Colwell, and R.Y. Morita (eds.), Effect of the ocean environment on microbial activities. University Park Press, Baltimore.

57. Remsen, C.C., E.J. Carpenter, and B.W. Schroeder. 1974. The role of urea in marine microbial ecology, p. 286-304. *In* R.R. Colwell, and R.Y. Morita (eds.), Effect of the ocean environment on microbial activities. University Park Press, Baltimore.

58. Robinson, G.G.C., L.L. Hendzel, and D.C. Gillespie. 1973. A relationship between heterotrophic utilization of organic acids and bacterial populations in West Blue Lake, Manitoba. Limnol. Oceanogr. 18:264-269.
59. Rosswall, T. 1972. Modern methods in the study of microbial ecology. Bulletin No. 17, Swedish National Research Council, IBP Symposium, Rotobeckman, Stockholm, Sweden.
60. Seki, H., K.V. Stephens, and T.R. Parsons. 1969. The contribution of allochtonous bacteria and organic materials from a small river into a semi-enclosed sea. Arch. Hydrobiol. 66:37-47.
61. Sieburth, J. McN. 1967. Seasonal selection of estuarine bacteria by water temperature. J. Exp. Mar. Biol. Ecol. 1:98-121.
62. _____. 1975. Microbial seascapes a pictorial essay on marine microorganisms and their environment. University Park Press, Baltimore.
63. _____, R.D. Brooks, R.V. Gessner, C.D. Thomas, and J.L. Tootle. 1974. Microbial colonization of marine plant surfaces as observed by scanning electron microscopy, p. 418-432. In R.R. Colwell, and R.Y. Morita (eds.), Effect of the ocean environment on microbial activities. University Park Press, Baltimore.
64. Sizemore, R. K., and L.H. Stevenson. 1974. Environmental factors associated with proteolytic activity of estuarine bacteria. Life Sci. 15:1425-1432.
65. _____, _____, and B.H. Hebeler. 1973. Distribution and activity of proteolytic bacteria in estuarine sediments, p. 133-143. In L.H. Stevenson and R.R. Colwell (eds.) Belle W. Baruch Library in marine science. No. 1. Estuarine microbial ecology. Univ. South Carolina Press, Columbia.
66. Sorokin, Y.I., and H. Kadota. 1972. Techniques for the assessment of microbial production and decomposition in fresh waters. IBP Handbook No. 23. Blackwell Scientific Publications, London, England.
67. Sprott, G.O., J.P. Drozdowski, E.L. Martin, and R.A. MacLeod. 1975. Kinetics of Na$^+$-dependent amino acid transport using cells and membrane vesicles of a marine pseudomonad. Can. J. Microbiol. 21:43-50.
68. Stanier, R.Y., M. Doudoroff, and E.A. Adelberg. 1970. The microbial world. 3rd ed., Prentice-Hall, Inc., Englewood Cliffs.
69. Stanley, S.O., and C.M. Brown. 1974. Influence of temperature and salinity on the amino-acid pools of some marine pseudomonads, p. 92-103. In R.R. Colwell, and R.Y. Morita (eds.), Effect of the ocean environment on microbial activities. University Park Press, Baltimore.
70. Stevenson, L.H., and R.R. Colwell. 1973. Belle W. Baruch library in marine science. No. 1. Estuarine microbial ecology. Univ. South Carolina Press, Columbia.
71. Sussman, A.S., and H.O. Halvorson. 1966. Spores, their dormancy and germination. Harper and Row, New York.
72. Thompson, B., and R.D. Hamilton. 1974. Some problems with heterotrophic uptake methodology, p. 566-575. In R.R. Colwell, and R.Y. Morita (eds.), Effect of the ocean environment on microbial activities. University Park Press, Baltimore.
73. Vaccaro, R.F. 1969. The response of natural microbial populations in seawater to organic enrichment. Limnol. Oceanogr. 14:726-735.
74. _____, S.E. Hicks, H.W. Jannasch, and F.G. Carey. 1968. The occurrence and role of glucose in seawater. Limnol. Oceanogr. 13:356-360.
75. _____, and H.W. Jannasch. 1966. Studies on heterotrophic activity in seawater based on glucose assimilation. Limnol. Oceanogr. 11:596-607.

76. Whitney, D.E., G.M. Woodwell, and R.W. Howarth. 1975. Nitrogen fixation in Flax Pond: a Long Island salt marsh. Limnol. Oceanogr. 20:640-643.

77. Wiebe, W.J., and L.R. Pomeroy. 1972. Microorganisms and their association with aggregates and detritus in the sea: a microscopic study, p. 325-352. In U. Melchiorri-Santolini, and J.W. Hopton (eds.), Detritus and its role in aquatic ecosystems. Proceedings of the IBP-UNESCO Symposium. Mem. 1st. Ital. Idrobiol., 29 Suppl. Pallanza, Italy.

78. Williams, P.J. LeB. 1970. Heterotrophic utilization of dissolved organic compounds in the sea. I. Size distribution of population and relationship between respiration and incorporation of growth substrates. J. Mar. Biol. Ass. U.K. 50:859-870.

79. _____. 1973. The validity of the application of simple kinetic analysis to heterogeneous microbial populations. Limnol. Oceanogr. 18:159-165.

80. _____, and C. Askew. 1968. A method of measuring the mineralization by micro-organisms of organic compounds in sea-water. Deep-Sea Res. 15:365-375.

81. _____, and R.W. Gray. 1970. Heterotrophic utilization of dissolved organic compounds in the sea. II. Observations on the response of heterotrophic marine populations to abrupt increases in amino acid concentration. J. Mar. Biol. Ass. U.K. 50:871-881.

82. Wood, L.W. 1973. Monosaccharide and disaccharide interactions on uptake and catabolism of carbohydrates by mixed microbial communities, p. 181-197. In L.H. Stevenson, and R.R. Colwell (eds.), Belle W. Baruch library in marine science. No. 1. Estuarine microbial ecology.. Univ. South Carolina Press, Columbia.

83. Wright, R.T. 1970. Glycollic acid uptake by planktonic bacteria, p. 521-536. In D.W. Hood (ed.), Organic matter in natural waters. Institute of Marine Science, Occassional Publication No. 1. College, Alaska: University of Alaska.

84. _____. 1973. Some difficulties in using ^{14}C-organic solutes to measure heterotrophic bacterial activity, p. 199-217. In L.H. Stevenson, and R.R. Colwell (eds.), Belle W. Baruch library in marine science. No. 1. Estuarine microbial ecology. Univ. South Carolina Press, Columbia.

85. _____. 1974. Mineralization of organic solutes by heterotrophic bacteria, p. 546-565. In R.R. Colwell, and R.Y. Morita (eds.), The effect of the ocean environment on microbial activities. University Park Press, Baltimore.

86. _____. 1975. Studies on glycollic acid metabolism by fresh water bacteria. Limnol. Oceanogr. 20:626-633.

87. _____, and J.E. Hobbie. 1966. Use of glucose and acetate by bacteria and algae in aquatic ecosystems. Ecology 47:447-464.

88. ZoBell, C.E. 1970. Substratum as an environmental factor for aquatic bacteria, fungi and blue green algae. In O. Kinne (ed.), Marine ecology. Vol. 1. Environmental factors. John Wiley, New York.

89. _____. 1943. The effect of solid surfaces on bacterial activity. J. Bacteriol. 46:38-59.

OSMOREGULATORY RESPONSES TO ESTUARINE CONDITIONS:

CHRONIC OSMOTIC STRESS AND COMPETITION

Pierre Lasserre
Université de Bordeaux
Institut de Biologie Marine
33120 Arcachon, France

ABSTRACT: Estuarine systems confront potential colonizing organisms with severe osmotic problems and so come to support highly adapted communities. Osmoregulation has a central rôle to play in this context. Many important physiological mechanisms have been extensively studied, and yet many patterns involved do not fit into a unifying scheme. In euryhaline fish, sodium transport involving $Na^+ - K^+$ ATPase pools have been described in a large variety of effector organs. In fish gills, this ATPase activity has been related to Na^+ extrusion and there is now some evidence that Na^+/K^+ exchange carrier is linked with a Cl^- pump. When considering the final balance of osmotic pressure that must be attained by tissues, not only inorganic ions but also non-protein nitrogenous compounds, notably free amino acids, can be considered as effectors in the osmoregulation of many estuarine organisms and even in the so-called "homeosmotic regulators." As regards hormonal and neurosecretory effects on osmoregulation, studies on mechanisms must wait on the elucidation of the enzymatic pathways involved.

Osmotic and ionic stress is an inescapable part of life in estuaries, coastal lagoons, and mangroves. These ecotones are largely utilized for extensive aquaculture in temperate and tropical latitudes. An outline of some osmotic and ionic problems which have become apparent in the traditional and more advanced forms of coastal aquafarming are reviewed. The hypo-osmotic or hyperosmotic stress is, in general, mediated through a highly adapted osmotic behaviour. Isolation of the temporary immigrants in coastal enclosures can alter the *ad libitum* migratory tendencies and trigger osmoregulatory disfunction. Even in the absence of mortality, the osmotic stress becomes chronic and some of its consequences (growth retardation, infertility) seem frequently to have been underestimated.

396   P. LASSERRE

Comparative studies of the osmotic behaviour of all the species to be farmed with respect to their differential penetration of estuarine ecotones are urgently needed. Some osmoregulatory properties of highly tolerant estuarine inhabitants (e. g. meiofauna) are considered. Their competitive advantages over stressed macrofaunal assemblages might alter the organic nutrient supply in these extremely productive but unstable habitats.

INTRODUCTION

Estuarine waters fall into the mixohaline category, which may range from oligohaline to hypersaline concentrations. The very existence of any organism will depend on its capacity to adjust to daily and seasonal changes in salinity. The question of euryhalinity has, therefore, a more central than peripheral rôle to rôle to play in this context. Each euryhaline organism does not regulate by any single pattern but is resourceful in utilizing a wide range of regulatory mechanisms. The osmoregulators are able to regulate the composition of their body fluids by passive or active mechanisms which maintain a steady state of ionic and water flux between external and internal media. In osmoconforming species, fluctuations in the external environment are paralleled by similar variations in the concentration of the body fluids. All these adaptive properties imply different specializations. A considerable amount of work has been done during the last decades on osmoregulation, and excellent reviews of much of the literature that is pertinent to the physiology of euryhaline plants, bacteria, invertebrates and vertebrates are available. Other papers in this volume also contain relevant information. For these reasons, a comprehensive review of the literature is not given in this paper, nor are all physiological mechanisms possessed by estuarine organisms for dealing with the hypo-osmotic and hyperosmotic environment described. Instead, my basic idea has been to concentrate on some effects of hypo- and hyperosmotic stress and some of its consequences which are not immediately obvious. There is, at the present, widespread interest in rearing and harvesting organisms of nutritional value in those estuaries, those coastal lagoons and mangroves that are extremely productive and are, nevertheless, in a state of borderline excess of eutrophication. These ecotones receive temporary immigration from the sea and, to a lesser extent, from the land-based waters. While many species may actively seek the most favorable environment by migration others, with limited locomotor ability or which are herded in enclosed or semi-enclosed volumes, must deal with the environment.[1] These traditional forms of extensive aquaculture give yields which vary greatly and are not as good as expected. Among other aspects, competition for food and space, predation, silting, and

[1] In general, the technique consists of constructing impoundments in the intertidal region or in shallow coastal lagoons and regulating the inflow and outflow of sea water so that mollusc larvae, juvenile and post-juvenile fishes and crustaceans can only enter and not leave the impoundments. Frequently, juveniles and adults are captured near the shore and stocked in the rearing ponds.

stress are processes that shape and limit the communities. Such processes may also affect more advanced forms of extensive estuarine aquafarming. From the ecophysiologist's point of view, immigrants are confronted with severe osmotic problems and eventually they have to compete with more characteristic inhabitants. The latter have attained a remarkable degree of independence from, or tolerance to, the high-stress environment. Even though they do not necessarily involve the death of the animal, the long-term effects of osmotic stress are of importance and I shall attempt to outline some osmotic and ionic problems which have become apparent in extensive estuarine aquafarming. Although data essential to the understanding of the "normal" (experimental) osmoregulatory physiology have been obtained, amplitude and long term effects of osmotic stress, under both field and laboratory conditions are poorly understood. The value and practicability of this to controlling and preventing osmotic stress by appropriate environmental manipulation are yet to be evaluated. As a basis for this discussion, I shall outline the spectrum of salinity responses and osmotic stress evident among cultured fish. When the osmotic stress is maintained, growth and gametogenesis may frequently be impaired as well. Death, usually preceeded by infectious diseases, eventually occurs, sometimes after a considerable delay (up to several months). Another neglected aspect which is considered here is the degree of euryhalinity and the competitive advantages of highly adapted benthic communities of euryhaline micro-metazoans i.e., the meiofauna.

OSMOREGULATION OF EURYHALINE FISH IN ESTUARINE HABITATS

Hypo-osmotic and hyperosmotic regulators

Water balance and electrolyte transfers

It is well known that euryhaline fish are hetero-osmotic regulators. They have the capacity to control osmotically and ionically their body fluids. They maintain their plasma hyperosmotic in fresh water and hypo-osmotic in sea water by regulating ion levels, predominantly sodium and chloride. In sea water, the problems of osmotic "desiccation" and salt loading have been recognized for many years as well as the broad outlines of their solution. To maintain its water balance, the fish actively drinks salt water and much of the medium is absorbed by the gut. The main solutes are then excreted—Na^+ and Cl^- through the gills, Mg^{++} and $So_4^{=}$ by the kidney. In fresh water, water tends to move osmotically into the body. The excess of water is excreted by the kidney. Since the fish drinks little, it is able to maintain its salt balance during prolonged periods without food. Furthermore, the renal Na^+ and Cl^- loss and the passive loss of these electrolytes along the concentration gradient across the external boundaries is compensated for by an active uptake of NaCl through the gills. Therefore, gills are of fundamental importance in these homeostatic mechanisms, for it is at this level that the major movements of ions against the concentration gradients take place (59). The urinary bladder and the skin play also a not negligible rôle (3, 31). Studies of electrolyte transfer across gills have shown functional differences in the net absorption of Na^+ and Cl^- in fresh water and extrusion of these ions in sea water. At present, the mechanisms underlying NaCl fluxes are reasonably well known but, many fundamental aspects remain unclear. In fresh water, a coupling between endogenous NH_4^+, HCO_3^- and H^+ excretion and Na^+, Cl^- absorption is observed. In sea water there must be an active transport of Cl^-. It is excreted against the electrochemical gradient which favours inward movement of the anion. Net chloride extrusion means that there must be an outwardly oriented active transport mechanism. The situation is more complex for Na^+. Motais et al. (66) observed that in the flounder Na^+ efflux varies as a function of external Na^+ according to Michaelis-Menten curves. They suggested the presence of a Na^+ pump associated with an exchange diffusion mechanism. Maetz (58) found that a fraction of the Na^+ efflux in the flounder also varies with external K^+. He suggested that a sodium pump operated by means of a Na/K exchange carrier with a high affinity for K^+ and that the Na^+/Na^+ exchange resulted from a competition between external Na^+ and K^+. He proposed that a branchial $Na^+ - K^+$ ATPase was responsible for the Na^+/K^+ exchange. These hypotheses have been challenged by Potts and Eddy (75) and by Kirschner and his coworkers (45). They concluded that the transfer of Na^+ across the gill is a passive phenomenon and that the external Na^+ and K^+-dependent Na^+ effluxes were accounted for by potential changes across the gills. The very recent experiments of Maetz and Pic (62) on the mullet, *Liza ramada*, are in favour of a possible linkage between the Cl^- pump and the Na^+/K^+ exchange mechanism proposed for the eel by Epstein et al. (18).

Hormonal control of osmoregulation in teleosts has led to a very large literature which has been recently surveyed (3, 38). According to many authors, the primary osmoregulatory hormones in teleosts are prolactin and cortisol. Cortisol is considered as favouring sea water adaptation and prolactin as favouring fresh water adaptation. The question is complex and the answer not yet clear since measurements of the blood levels of these hormones using heterologous radioimmunoassay systems have been subject to recent criticisms (67).

According to Evans (21), the limiting factor in euryhalinity of marine immigrant fish "... is not the absence of the necessary ion (NaCl) uptake mechanisms. They are always present as secondary couplers to needed ionic extrusion (HCO_3^-, H^+, NH_4^+). The entrance into fresh water is limited by the relation between these uptake mechanisms and the ionic permeability of a given species. ... This ionic permeability may be under the control of intrinsic hormones (prolactin) and external calcium concentration."

Biochemical aspects

Some recent and important developments relate to the biochemical mechanisms of osmoregulation in euryhaline fish. The possible cellular and molecular processes, principally at the gill level, in relation to the osmoregulatory needs of euryhaline fish have been considered by different authors (reviews 11, 60). Many aspects remain speculative. One interesting discovery concerns the presence of cation and anion ATPase in gill microsomal preparations.

Cation ATPases. The presence of $Na^+ - K^+$ ATPase has been demonstrated on a wide variety of teleosts (see e.g., 13, 17, 35, 41, 63, 90, 91). It has been shown that the amount of branchial $Na^+ - K^+$ ATPase is greater in fish adapted to sea water than those adapted to fresh water. This difference has been attributed to the increased load of Na^+ in sea water which must be pumped across the gills from the blood into the external environment. The finding (39) that eel chloride cells, in the gill, contain the bulk of the $Na^+ - K^+$ ATPase has been recently confirmed (77). At the same time, succinic dehydrogenase has been found at higher specific activities in sea water chloride cells than in fresh water cells (78). It is, therefore, well established that adaptation of the eel to sea water is accompanied not only by the production of greater amounts of $Na^+ - K^+$ ATPase but also by increased quantities of ATP to drive these reactions.

Nevertheless, the extent to which measurements of the above properties may be used to assess the degree to which euryhaline fish are adapted to either sea water or fresh water is a complex question. The sodium-potassium exchange has not been clearly established to be, in general, ATP-linked even though there is some evidence to believe that this happens. In the eel, a parallel evolution of Na^+ transport and $Na^+ - K^+$ ATPase has been shown (6).

Furthermore, Lasserre (50) found a reverse pattern of enzyme induction. Branchial $Na^+ - K^+$ ATPase (ouabain sensitive) increases when sea water grey mullet and sea bass have adapted to fresh water. Yet, Zaugg and Wagner (91) state that gill $Na^+ - K^+$ ATPase is elevated in juvenile fresh water smolts of *Salmo gairdneri* exhibiting migratory tendencies. In these particular cases, fresh water is an adequate stimulus for the induction of $Na^+ - K^+$ ATPase synthesis.

There is the possibility that more than one functional form of $Na^+ - K^+$ ATPase develops in the gills of fish under different environmental conditions. Sargent and Thomson (77) have demonstrated that $Na^+ - K^+$ ATPases from the gills of sea water and fresh water adapted eels are kinetically and structurally indistinguishable. The two protein sub-units of the enzyme are present in both sea water and fresh water chloride cells, more enzyme being present in sea water cells than in fresh water cells. These results invalidate the hypothesis of two $Na^+ - K^+$ ATPase pools in eel gill (65). There is, however, some further evidence that ATPase with markedly different characteristics may be involved in fresh- and salt water media. The enzyme activity in gills of fresh water adapted rainbow trout is activated by Na^+ but does not require K^+. This Na^+ ATPase is only partially inhibited by ouabain at concentrations below $10^{-3}M$. The enzyme from salt water fish requires both Na^+ and K^+ for maximal activation (71). Moreover, kinetic differences in $Na^+ - K^+$ ATPase from gills of sea water and fresh water adapted grey mullet have been recently described (28). NH_4^+ has the same activation effect in both sea water and fresh water enzymes (same apparent Km) but one part of the $Na^+ - K^+$ ATPase is strictly K^+-dependent. Activation by K^+ gives two successive plateaus in the fresh water enzyme. These data support the hypothesis of two gill $Na^+ - K^+$ ATPase pools in fresh water adapted mullet.

The functional significance of high amounts of $Na^+ - K^+$ ATPase in fresh water remains obscure. The possibility that the enzyme could be involved in a branchial sodium absorption pump seems unlikely. One possibility exists, however, if we suppose the presence of a Na^+/K^+ exchange located at the internal-facing membrane of the chloride cells in the fresh water adapted fish, as suggested for rainbow trout (43). Further investigations are necessary to determine the status of the one or several forms of cation ATPases in the gills of euryhaline fish in relation to Na^+ transport.

Anion ATPase. There is also ample reason to postulate the presence of an active carrier for a separate chloride pump. The carrier is blocked by low concentrations of thiocyanate—which inhibit the efflux of chloride in sea water. An anion-stimulated ATPase specifically inhibited by SCN^- has been studied in relation to the energetics of gill Cl^- uptake. This HCO_3^- activated, thiocyanate inhibited ATPase was found in the gills of rainbow trout (42), the eel and goldfish (61), and recently the mullet (Lasserre, unpublished). The connection between the anion-activated ATPase and Cl^- transport remains to be elucidated because no clear-cut activation by Cl^- was demonstrated. Furthermore, no significant difference in the amount of enzyme was found between sea water and fresh water. There is no evidence, at present, that an anion ATPase is involved in gill Cl^- transport. This possibility, however, remains, the Cl^- active transport being an energy requiring step.

Tissue water and intracellular osmoregulation

Euryhaline teleosts are classically considered as perfect homeosmotic animals; however, several estuarine species regulate their internal milieu within quite wide limits, e.g. the flounders (46, 55), eels (34), mullets and sea bass (54, 55), and several other intertidal species (2, 19, 10). The tissues of these species have, therefore, to withstand an osmotic stress. A cellular volume regulation has been postulated since only slight changes in muscle water contents were found during the adaptation of the fishes to varying salinities (34, 46, 55). Lange and Fugelli (46) first suggested the involvement of trimethylamine oxide and the total free nitrogenous compounds in the osmotic adjustment of teleost tissues. The detailed analysis of Huggins and Colley on *Anguilla anguilla,* Lasserre and Gilles on *Paralichthys lethostigma* and *Chelon labrosus* have shown that free pools of taurine and non-essential amino acid pools are involved in intracellular osmotic regulation (34, 55). Whereas involvement of the non-essential amino acids in osmotic adaptation was common to mullet, flounder, and eel, other nitrogenous constituents such as taurine or carnosine showed variable contribution—taurine, for example, being quantitatively an important osmotic effector in mullet, flounder (55) and the intertidal fish *Agonus cataphractus* (10) but not in the eel

(34). These studies give support to an involvement of free pools of nitrogenous constituents, predominantly amino acids, in the biochemical changes associated with the adaptation of euryhaline teleosts to their surroundings. This is also true of most euryhaline invertebrates (30, 79).

Differential penetration of estuarine habitats and osmotic stress

Each year, both catadromous and anadromous fishes make their way into the open waters of estuaries and lagoons. They are continually adjusting to the fluctuating environment so that their reaction always depends on their life history. The many mechanisms involved in the migration of amphihaline fish have been extensively reviewed by M. Fontaine (25).

Migratory tendencies and ATPase adjustments

In young salmonids, the physiological changes that are preparatory to seaward migration are of a rhythmical nature. They regress when smolts are retained in fresh water. The seasonal temporal patterns in salinity responses of *Salmo* and *Oncorhynchus* have been studied extensively (review 32). Zaugg and his coworkers traced seasonal changes including a spring rise in the gill $Na^+ - K^+$ ATPase of coho salmon and a sharp decline in this activity during the summer if the fish remained in fresh water (90). A decrease in salinity resistance was noted during the summer months. It was concluded that the smolts of all salmonids experience physiological changes which re-adapt them to the fresh water environment if they do not reach the sea. The changes in gill ATPase indicate one of the changes in physiology that are preparatory to the catadromous migration of the smolt.

Control of ATPase levels in fish gills is probably mediated through some hormonal and neuronal stimulation. ACTH restores the reduced gill $Na^+ - K^+$ ATPase following hypophysectomy (63). Very significant increases in the level of $Na^+ - K^+$ ATPase have been obtained in gills of fresh water eels after cortisol and ACTH treatments (16,27,40,63). Data obtained on *Anguilla rostrata* indicate that a "burst of cortisol secretion occurs shortly after eels make their first contact with sea water in riverine estuaries" (16), and that the increase in circulating hormone is accompanied by a rise in the amount of branchial $Na^+ - K^+$ ATPase. Epstein et al. (16) concluded from their experiments on fresh water adapted *A. rostrata* that "cortisol induces a series of changes in freshwater eels, including a rise in $Na^+ - K^+$ ATPase in gills and intestine, that successfully prepares these euryhaline telosts to combat the osmotic stress of migrating to sea water."

Chronic osmotic stress

Salinity is probably, with temperature, one of the major factors influencing the differential penetration of mixohaline lagoons. The natural distribution of the temporary immigrants, such as grey mullets, in lagoonal ponds at Arcachon shows a good correlation with their respective capabilities for osmoregulation in hypotonic media (54). Both the thick- and thin-lipped grey mullet tolerate a wide range of salinity, but *Liza ramada* (thin-lipped mullet), which controls plasma osmolality throughout the salinity gradient, colonizes oligohaline and fresh water ponds. In contrast, *Chelon labrosus* (thick-lipped mullet), which is unable to develop a long term osmoregulation in fresh water media, is distributed mostly in polyhaline and mesohaline ponds. Similar patterns have been described in other temporary estuarine immigrants. It is reasonable to propose that the fishes are restricted in their long-term upstream penetration, not because they lack ion uptake mechanisms, but because the mechanisms are not sufficient enough to balance, on a long term basis, the salt losses by net diffusional loss.

A number of species are commonly captured during their seaward or landward migrations and are often cultured in estuarine impoundments. The species, mostly of marine origin, include salmonids, eels, flounders, bass, breams, etc. Land-locked living conditions may modify normal endogenous cyclic adjustments. Confinement to fresh water can give an unequivocal decrease in plama osmolality and chief electrolytes and many problems of survival, altered growth rates, and infertility could result as a consequence of chronic osmotic stress. These osmoregulatory perturbations could be related, at least in part, to a stress-mediated increase in the circulating catecholamines and notably epinephrine. In the mullet, *Liza ramada*, epinephrine injections increase diffusional water permeability in both fresh and salt water (73). Moreover, the laboratory diuresis observed after the handling of migrating salmonids in fresh water also reflects the increased osmotic permeability of the gills (33). In the trout, stress also increases the diffusional water permeability (20).

In the mullet, *Chelon labrosus,* fresh water survival is considerably longer if the external calcium concentration is raised. Physiologically, calcium is known to have a permissive action of the active transport of Na^+ in mullet (72). Normalization of osmoregulatory functions could be obtained by an appropriate level of calcium in the environment. Increasing the Ca^{++} content of fresh water to 100 ppm, partially alleviated hypo-osmoticity in grey mullets (Lasserre, unpublished data).

The amount of $Na^+ - K^+$ ATPase in gill of *C. labrosus* and *Dicentrarchus labrax* (bass) rises when the fishes enter oligohaline and fresh water lagoons (50, 54). Such enzymatic stimulation might be a secondary response to stress-mediated physiological changes, and not an adaptation per se. The high amount of gill $Na^+ - K^+$ ATPase once attained (10 days) remained unchanged by continuous residence in oligohaline water. Conversely, in *Liza ramada* (a better osmoregulator), the enzyme activity was increased after 10 days of fresh water adaptation but more prolonged adaptation was marked by a reduction in ATPase level. After one month, no significant differences occured between fresh and salt water adapted fish (5, and unpublished data).

In these temporary immigrant fish, the transition from the sea to mixohaline waters under naturally occuring seasonal migratory or experimental conditions could involve physiological changes mediated by hypophyseal or interrenal activation which in turn could enhance the induction of ion transport ATPase systems. Synergistic prolactinic and corticotropic hyperactivity has been noted in grey mullets and sea bass invading oligohaline and fresh waters (4, 5, 70, 83). Moreover, interrenal activity is greater in mullet from fresh water than in those from sea water (37). These histometric observations need to be correlated with measurements of circulating hormones (76). It is probable, nevertheless, that endocrine secretions could trigger the increase of ion-stimulated ATPases and other physiological mechanisms in such a way as to promote rapid seaward migration.

Some possible effects of chronic osmotic stress.

Altered growth rates. In Arcachon Bay, grey mullets which are herded in semi-enclosed lagoonal ponds have lower growth rates than populations which live freely outside the fish ponds either in the intertidal flats or in the ocean. In the coastal lagoons of Guerrero (Mexico), the depression of growth of *Mugil curema* has been reported during the dry period (March to May), which brings about marked evaporation with a resultant hypersalinity (87). Impaired growth and decreased gill ATPase of the Pacific salmon can result by increasing the NaCl content of diets or the external salinity at an unappropriate stage (89). Unpublished and recent data on grey mullets and sea bass herded in the lagoonal ponds at Arcachon showed high specific growth rates in autumn coincident with seaward migration. This phenomenon could be attributed to an increase in protein metabolism which is associated with osmoregulation and which may stimulate, as a secondary effect, the specific growth rates and reproduction. After the seaward migratory period (September-October), the growth rates of the remaining fishes become slow. Moreover, salt-water salmonid populations have increased growth rates relative to those in fresh water (22). A very recent investigation on nitrogen metabolism in fresh water steelhead trout (82) has shown a close association between rapid spring growth, salt-water tolerance, and increased protein metabolism.

Infertility. There is evidence that many estuarine species can live but not reproduce in land-locked saline enclosures. The lack of gamete production in grey mullet confined to olighaline and fresh water lagoons seems to be limited to the female fish, spermatogenesis being normal in both habitats (1, 5, 8). Moreover, recent results on grey mullets and sea bass from the lagoonal ponds at Arcachon (5, 83) confirm the findings of Abraham and his coworkers (1, 4) that there is an inverse relationship between ovarian maturity and the levels of activity of prolactin and corticotropin producing cells. The ovocyte development is blocked at stage III, and probably further development in ovarian maturity is incompatible with the simultaneous high levels of prolactin and corticotropin-producing cells found in both mullet and bass herded in brackish and fresh waters. The 11-ketotestosterone, a potent androgen for grey mullet, has been found in high concentrations in the ovarian tissue of *Liza ramada* confined to fresh water (14), while small or unmeasurable concentrations of the steroid were present in ovaries of mullets caught in the open sea (15). The accumulation of 11-ketotestosterone could inhibit the release of ovulating hormones, or activity (14). Recently, spawning was obtained in oligohaline media by progesterone treatment in *Chelon labrosus* (8).

Hyperosmotic stress and changes in the non-protein nitrogenous compounds. Many forms living in low salinity waters occur in hypersaline media. The most abundant and widely distributed mullets in tropical lagoons are *Mugil cephalus* and *M. curema*. The striped mullet penetrates fresh water and can occur also up to at least 80 ppt. The white mullet (*M. curema*) can tolerate an overall range of salinity between 0 and 50 ppt (64). High salinities are generally well accepted by mullets. When the thick-lipped mullet *(C. labrosus)* goes from normal sea water (33 ppt) to 70 ppt, there is a very significant increase (50 to 80%) in the muscle amino acids, which is apparent after 10 days of adaptation. The pool size of the non-essential amino acids, glycine, alanine, aspartic acid, glutamic acid, serine and proline follows the salinity. The amount of taurine present in muscle is also clearly related to salinity (55). Such mechanisms can explain, in part, the ability of grey mullets to tolerate high internal osmolalities and, therefore, to colonize hypersaline habitats. The above changes are reversible: after a hypo-osmotic stress (direct transfer from sea water to water of 5 ppt), there is a 42 to 58% decrease in the same non-essential free amino acid pools, as was observed in muscle after four days (Lasserre, unpublished). The low levels of nitrogenous osmotic effectors in mullet herded in fresh water and oligohaline enclosures could be related to the fact that the species is unable to develop a long term osmoregulation in very diluted media. An interesting parallel may be made with the Atlantic salmon (*Salmo salar*). Cowey et al. (12) found no significant change in muscle or plasma amino acid levels in both fresh and salt water. Fontaine and Marchelidon (26) have found, however, that smolts in fresh water exhibit a decrease in muscle free amino acids, during their catadromous migration. The largest significant changes involved glycine and taurine. This adaptation pattern has been related to some "preparatory changes in the physiology of the smolting."

Concluding remarks. The nature of the so-called alimental upstream migration of many catadromous fish conflicts somewhat with their limited aptitude to grow, reproduce and eventually to survive in coastal lagoons, mangroves and estuaries. It is noteworthy that normal upstream migrations are most often temporary, and going to lower (or higher) salinities seems to physiologically precondition the fish to immediate or short delayed seaward migration. Impaired growth, infertility and considerably delayed mortalities preceeded frequently by infectious diseases are often the physiological consequences of farming in too much confined estuarine impoundments. These phenomena are also influenced by temperature (9) and probably many other altered environmental factors. In the case of benthophageous macrofauna (grey mullets, crustaceans etc.), trophic problems with highly adapted communities of euryhaline microfauna and meiofauna are uneligible problems of competition.

COMPETITIVE ADVANTAGES OF EURYHALINE MICROFAUNA AND MEIOFAUNA IN ESTUARINE HABITATS

There is now much evidence to indicate that estuarine microfauna and meiofauna have a remarkable degree of tolerance to their environment. Recent developments on their physiological ecology have been discussed (53, 56). Their adaptations include tolerance to wide ranges in temperature, how oxygen tension, and changes in salinity.

Salinity tolerance and osmoregulation

Most estuarine microfauna and meiofauna are very tolerant to a wide range of salinities (24, 36, 53, 85). Osmoregulation in the meiobenthic oligochaete *Marionina achaeta* has been studied in detail (47, 48, 49, 52). This broadly distributed species (49, 51) can adapt to a wide range of salinities, from fresh water to concentrated sea water (48). Changes in body volume have been measured both during and after equilibration of the animals in media of altered salinity. *M. achaeta* clearly shows volume regulation after a hypo-osmotic stress (52). The time course of these volume regulations are rapid and of the same magnitude as those found in a few polychaetes (69, 80). The rates of respiration have been found to be constant over a salinity range of 3-25 ppt (48). The species is able to maintain its oxygen uptake relatively constant despite large salinity changes. The time course of this metabolic compensation is immediate, or at least very rapid (one or two hours after the salinity change). At both very high and very low (1 ppt) salinities (and in fresh water), respiration was significantly elevated, and these elevated rates have been related to the increased metabolic cost of attempting to maintain normal ionic concentrations (47, 52). The stimulating effects of Na^+ on respiration, and the inhibitory action of ouabain on sodium-activated oxygen uptake, suggest a close relationship between a sodium transport mechanism and respiration. A significant percentage (20 to 30%) of total respiration is most probably utilized by some active process of ionic regulation. The rate of these active uptakes might well be controlled, in some way, by the concentration of body fluids. In view of this, only preliminary experiments using microtechniques have been pursued. The uptake processes probably involve active transport of solute against an electrochemical potential gradient. It is conceivable that nephridia exhibit, at least to some extent, a major role in these phenomena (52). Similar adaptations have been found in some other groups of estuarine meiofauna (53, and unpublished data).

From the above, it appears that meiofaunal species have mechanisms for active transport which are probably used to maintain the body volume and to stabilize the ionic composition of body fluids. This elaborate strategy of osmoregulation is energetically quite costly, particularly at very low and high environmental salinities.

Competition

The microfaunal and meiofaunal communities are very successful in reproducing and growing in estuarine conditions. They have short generation times and their production : biomass ratios tend to be considerably higher than for macrofauna (see e.g., 24, 29). Euryhaline aerobic microfauna and meiofauna might be expected to use a significant amount of metabolic energy for osmoregulation. They must therefore, require a high level of food consumption to complete their life cycle and it may be anticipated that their protein metabolism is elevated. Euryhalinity and high rates of both growth and reproduction are probably related to a general high protein turnover in microfauna and meiofauna. In this way, they can have a competitive advantage over macrofaunal assemblages.

The presence of microfaunal (ciliates) and meiofaunal communities with extremely high numbers of individuals in estuarine habitats is well documented (24, 29, 57). Estimates of the meiofaunal contribution to total community respiration in the lagoonal fish ponds at Arcachon have recently been made from accurate individual determinations on the principal taxa (57). In view of the very high specific oxygen consumption (53, 84), the euryhaline meiofauna (mainly composed of nematodes and copepods) contributed from 9% to up to 58% of total community respiration (57). Similar (but lower) percentages have been estimated indirectly in other shallow-water areas (81). In this context, a significant part of the benthic primary production which is converted to meiofaunal tissue might not be passed upwards to higher trophic levels. Spring and summer growth retardation of grey mullets and other benthophageous macrofauna could therefore, also, be explained in terms of a trophic competition with such highly adapted small benthic metazoans.

GENERAL CONCLUSIONS

Osmotic and ionic stress is an inescapable part of life in estuarine habitats. The salinity effects are complicated by the interaction of other environmental factors (temperature, oxygen, photoperiod, food, etc.) (86), and are mediated, generally, through an osmotic behaviour which is highly adapted in nature. Isolation of temporary immigrants in coastal enclosures can alter the *ad libitum* migratory tendencies and trigger osmoregulatory disfunctions. Some competitive advantages of highly tolerant estuarine meiofauna over stressed macrofaunal assemblages have often been underestimated. As illustrated by meiofauna (52, 53), euryhaline fish (23, 68), and also many other animals (7, 44), the amount of energy that is required to osmoregulate is expressed by appreciable changes in resting oxygen consumption. The high-energy requirement of osmoregulatory functions, first suggested by Schlieper, has been criticized by many authors, principally on a thermodynamic argument (74). Kirschner (44) was correct in pointing out that this argument was not valid. It should be noted, however, that the effects of salinity on metabolism are often masked, especially in intact animals, by many other factors. Moreover, the reaction of the animal need not be strictly dependent on thermodynamic conditions but also on the kinetic properties of enzymes.

One important feature in inorganic and organic osmoregulatory systems is that ATP appears to be one of the most immediate energizers. In many estuarine osmoregulators, periods of increased osmoregulatory activity lead to a rise in enzyme activity (ATPase, succinic dehydrogenase, enzymes implicated in amino acid metabolism, etc.). These changes can certainly affect the ion gradient dependent uptakes of various cations, anions, amino acids and transmitter substances. Moreover, it is very probable that periods of prolonged neuronal and hormonal activity, often related to prolonged osmotic stress, may influence the synthesis of the different carriers (e.g., cation and anion activated enzymes). This may produce some shortage of energy demands for growth or maturity, which is reinforced at a certain moment by some generalized reduction of metabolic energy (reduction of ATP production), as in the case of low temperature in winter, accidental anoxia due to eutrophication in summer, hypoglycemia due to starvation, or to some unbalanced condition of nutrient availability.

Besides the efforts of estuarine aquaculturists, an eventual improvement in production will depend on a fundamental understanding of the natural osmotic behaviour, which demands, in turn, the differential penetration of many species in estuarine areas. This aim will allow the migratory tendencies to be taken into account and also allow the various demands for the cultured organisms to be better separated and to be related as regards survival, growth and reproduction. Apart from the naturally occuring stress, many polluting substances can alter the osmoregulatory mechanisms. Low concentrations of many specific cations or anions (e.g., copper, lead, thiocyanate) present in estuarine waters in the form of various complexes, might poison cation and anion pumps and inhibit enzyme activities (18, 88).

Osmotic balances as derived from a study of blood chemistry and tissue composition can indicate a "healthy" or "unhealthy" state relative to given environmental conditions. Gill ATPase of amphihaline fish can give some significant indication on their migratory tendencies. Physiological periodicities, temporal shifts due to localized abiotic or biotic effects, and variation due to reproductive conditions must be taken into consideration and all this is beset by both methodological and interpretative problems. Only in a few instances is there sufficient information of this nature available. As a consequence, the physiological ecologist will have a two-fold task: to determine the effects of environmental change, and to separate and distinguish normal response patterns of the organisms. Yet, in any cataclysmic event, it will be possible to partition the effect due to a pollution factor from that of a normal environmental stress.

ACKNOWLEDGMENTS

I wish to express my gratitude to Dr. Martin L. Wiley, University of Maryland, for his linguistic revision of the manuscript. I am indebted, also, to Dr. Harold Barnes, Dunstaffnage Laboratory, Oban, for his critical reading of the revised manuscript. Special thanks are given to Dr. F. John Vernberg, University of South Carolina, for the opportunity to participate in the conference. It is a pleasure to acknowledge support from the following organizations: Estuarine Research Federation, Belle Baruch Coastal Research Institute, Centre National de la Recherche Scientifique.

REFERENCES

1. Abraham M., N. Blanc, and A. Yashouv. 1966. Oogenesis in five species of grey mullets (Teleostei, Mugilidae) from natural and landlocked habitats. Israel J. Zool. 15:155-172.
2. Ahokas, R.A., and F.G. Duerr. 1975. Tissue water and intracellular osmoregulation in two species of euryhaline telosts, *Culea inconstans* and *Fundulus diaphanus*. Comp. Biochem. Physiol. 52A:449-454.
3. Bern, H.A., 1975. Prolactin and osmoregulation. Amer. Zool. 15:937-948.
4. Blanc-Livni, N., and M. Abraham. 1970. The influence of environmental salinity on the prolactin and gonadotropin-secreting regions in the pituitary of *Mugil* (Teleostei). Gen. Comp. Endocrinol. 14:184-197.
5. Boisseau, J., P. Lasserre, J.L. Gallis and P. Cassifour. 1975. Aspects écophysiologiques de l'osmorégulation et de l'évolution génitale de poissons mugilidés en milieu lagunaire. J. Physiol. (Paris) 70:669-670.
6. Bornancin, M., and G. DeRenzis. 1972. Evolution of the branchial sodium outflux and its components, especially the Na^+/K^+ exchange and the Na^+/K^+ dependent ATPase activity during adaptation to sea water in *Anguilla anguilla*. Comp. Biochem. Physiol. 43A:577-591.
7. Burton, R.F. 1973. The significance of ionic concentrations in the internal media of animals. Biol. Rev. 48:195-231.
8. Cassifour, P., and P. Chambolle. 1975. Induction de la ponte par injection de progestérone chez *Crenimugil labrosus* (Risso) poisson téléostéen, en milieu lagunaire. J. Physiol. (Paris) 70:565-570.
9. Cech, J.J., and D.E. Wohlschlag. 1975. Summer growth depression in the striped mullet, *Mugil cephalus* L. Contrib. Mar. Sci. 19:91-100.
10. Colley, L., F.R. Fox, and A.K. Huggins. 1974. The effect of changes in external salinity on the non-protein nitrogenous constituents of parietal muscle from *Agonus cataphractus*. Comp. Biochem. Physiol. 48A:757-763.
11. Conte, F.P. 1969. Salt secretion, p. 241-292. *In* W.S. Hoar and D.J. Randall (eds.), Fish Physiology, vol. 1. Academic Press, New York and London.
12. Cowey, C.B., K.W. Dailey and G. Parry. 1962. Study of amino acids, free or as components of protein, and some B vitamines in the tissues of Atlantic

salmon, *Salmo salar,* during spawning migration. Comp. Biochem. Physiol. 7:29-38.
13. Dharmamba, M., M. Bornancin and J. Maetz. 1975. Environmental salinity and sodium and chloride exchanges across the gill of *Tilapia mossambica.* J. Physiol. (Paris), 70:627-636.
14. Eckstein, B. 1975. Possible reasons for the infertility of grey mullets confined to fresh water. Aquaculture 5:9-17.
15. _____, and V. Eylath. 1970. The occurence and biosynthesis *in vitro* of 11-ketotestosterone in ovarian tissue of the mullet derived from two biotopes. Gen. Comp. Endocrinol. 14:396-403.
16. Epstein, F.H., M. Cynamon, and W. McKay. 1971. Endocrine control of Na-K ATPase and sea water adaptation in *Anguilla rostrata.* Gen. Comp. Endocrinol. 16:323-328.
17. _____, A.I. Katz, and G.E. Pickford, 1967. Sodium- and potassium- activated adenosine triphosphatase of gills: role in adaptation of teleosts to salt water. Science 156:1245-1247.
18. _____, J. Maetz, and G. DeRenzis. 1973. Active transport of chloride by the teleost gills: inhibition by thiocyanate. Amer. J. Physiol. 224:1295-1299.
19. Evans, D.H. 1967. Sodium, chloride and water balance of the intertidal teleost, *Xiphister atropurpureus.* I. Regulation of plasma concentration and water content. J. exp. Biol., 47:513-517.
20. _____. 1969. Studies on the permeability of selected marine, freshwater and euryhaline teleosts. J. exp. Biol. 50:698-703.
21. _____. 1975. Ionic exchange mechanisms in fish gills. Comp. Biochem. Physiol. 51A:491-596.
22. Falk, K. 1969. Experiments on fattening trout on coastal and inland waters. Fischereiforschung 6(1):93-98.
23. Farmer, G.J., and F.W.H. Beamish. 1969. Oxygen consumption in *Tilapia nilotica* in relation to swimming speed and salinity. J. Fish. Res. Bd. Canada. 26:2807-2821.
24. Fenchel, T. 1969. The ecology of marine microbenthos. Ophelia 6:1-182.
25. Fontaine, M. 1975. Physiological mechanisms in the migration of marine and amphihaline fish, p. 241-355. *In* F.S. Russel and M. Yonge (eds.), Advances in Marine Biology, vol. 13, Academic Press, New York and London.
26. _____, and J. Marchelidon. 1971. Amino acid contents of the brain and the muscle of young salmon (*Salmo salar* L.) at parr and smolt stages. Comp. Biochem. Physiol. 40A:127-134.
27. Forrest, J.N., A.D. Cohen, D.A. Schon, and F.H. Epstein. 1973. Na transport and Na-K-ATPase in gills during adaptation to sea water: effects of cortisol. Amer. J. Physiol. 224:709-713.
28. Gallis, J.L. 1976. Some kinetic properties of the $Na^+ - K^+$ ATPase in gills of euryhaline teleost, *Chelon labrosus* (Risso) adapted to fresh water and sea water. Biochimie (in press).
29. Gerlach, S.A. 1971. On the importance of marine meiofauna for benthos communities. Oecologia (Berlin) 6:176-190.
30. Gilles, R. 1975. Mechanisms of ion and osmoregulation, p. 259-347. *In* O. Kinne (ed.), Marine Ecology, vol. 2(1). Wiley, Melbourne and London.
31. Hirano, T., D.W. Johnson, H.A. Bern and S. Utida. 1973. Studies on water and ion movements in the isolated urinary bladder of selected freshwater, marine and euryhaline teleosts. Comp. Biochem. Physiol. 45A:529-540.

32. Hoar, W.S. 1976. Smolt transformation. Evolution, behaviour, and physiology. J. Fish. Res. Bd. Canada 33:1233-1252.

33. Holmes, R.M. 1961. Kidney function in migrating salmonids. Rep. Challenger Soc. 13:23.

34. Huggins, A.K., and L. Colley. 1971. The changes in the non-protein nitrogenous constituents of muscles during the adaptation of the eel *Anguilla anguilla* L. from fresh water to sea water. Comp. Biochem. Physiol. 38B:537-541.

35. Jampol, L.M., and F.H. Epstein. 1970. Sodium-potassium-activated adenosine triphosphatase and osmotic regulation by fishes. Amer. J. Physiol. 218:607-611.

36. Jansson, B.O. 1971. The "Umwelt" of the interstitial fauna. Smithson. Contr. Zool. 76:129-140.

37. Johnson, D.W. 1972. Variations in the interrenal and corpuscles of Stannius of *Mugil cephalus* from the Colorado River and its estuary. Gen. Comp. Endocrinol. 19:7-25.

38. _____. 1973. Endocrine control of hydromineral balance in teleosts. Amer. Zool. 13:799-818.

39. Kamiya, M. 1972. Sodium-potassium-activated adenosine triphosphatase in isolated chloride cells from eel gills. Comp. Biochem. Physiol. 43B:611-617.

40. _____. 1972. Hormonal effect on Na-K-ATPase activity in the gill of Japanese eel, *Anguilla japonica*, with special reference to seawater adaptation. Endocrinol. Japon. 19:489-493.

41. _____, and S. Utida. 1968. Changes in activity of sodium-potassium activated adenosine triphosphatase in gills during adaptation of the Japanese eel to sea water. Comp. Biochem. Physiol. 26:675-685.

42. Kerstetter, T.H., and L. B. Kirschner. 1974. HCO_3^--dependent ATPase activity in the gills of rainbow trout *(Salmo gairdneri)*. Comp. Biochem. Physiol. 48B:581-589.

43. _____, L.B. Kirschner, and D.D. Rafuse. 1970. On the mechanisms of sodium ion transport by the irrigated gills of rainbow trout *(Salmo gairdneri)*. J. Gen. Physiol. 56:342-359.

44. Kirschner, L.B. 1967. Comparative physiology: invertebrates excretory organs. Ann. Rev. Physiol. 29:169-196.

45. Kirschner, L.B., L. Greenwald, and L. Sanders. 1974. On the mechanism of sodium extrusion across the irrigated gill of sea water adapted rainbow trout *(Salmo gairdneri)*. J. gen. Physiol. 64:148-165.

46. Lange, R., and K. Fugelli. 1965. The osmotic adjustment in the euryhaline teleosts, the flounder, *Pleuronectes flesus* L. and the three-spined stickleback, *Gasterosteus aculeatus* L. Comp. Biochem. Physiol. 15:283-292.

47. Lasserre, P. 1969. Relations énergétiques entre le métabolisme respiratoire et la régulation ionique chez une Annélide oligochète euryhaline, *Marionina achaeta* (Hagen). C. R. Acad. Sci. Paris 268:1541-1544.

48. _____. 1970. Action des variations de salinité sur le métabolisme respiratoire d'oligochètes euryhalins du genre *Marionina* Michaelsen. J. Exp. Mar. Biol. Ecol. 4:150-155.

49. _____. 1971. Données écophysiologiques sur la répartition des oligochètes marins meiobenthiques. Vie Milieu 22:523-540.

50. _____. 1971. Increase of the $(Na^+ + K^+)$-dependent ATPase activity in gills and kidneys of two euryhaline marine teleosts *Crenimugil labrosus* (Risso,

1826) and *Dicentrarchus labrax* (Linnaeus, 1758) during adaptation to freshwater. Life Sciences 10(11):113-119.

51. _____. 1975. Clitellata, p. 215-275. *In* A.C. Giese and J.S. Pearse (eds), Reproduction of Marine Invertebrates, vol. III: Annelids and Echiurans. Academic Press, New York and London.

52. _____. 1975. Mécanismes osmorégulateurs chez une Annélide oligochète de la meiofaune. Cahiers Biol. Mar. 16:765-798.

53. _____. 1976. Metabolic activities of benthic microfauna and meiofauna: Recent advances and review of suitable methods of analysis, (in press). *In* I.N. McCave (ed.), The Benthic Boundary Layer, Plenum Press (NATO Science Conf., Les Arcs, France, 3-10 Nov. 1974).

54. _____, and J.L. Gallis. 1975. Osmoregulation and differential penetration of two grey mullets, *Chelon labrosus* (Risso) and *Liza ramada* (Risso) in estuarine fish ponds. Aquaculture 5:323-344.

55. _____, and R. Gilles. 1971. Modification of the amino acid pool in the parietal muscle of two euryhaline teleosts during osmotic adjustment. Experientia 27:1434-1435.

56. _____, and J. Renaud-Mornant (eds.). 1975. Aspects of meiofauna research. Proc. Intern. Meeting on Meiofauna Physiological Ecology, Arcachon, 25-29 Sept. 1974. Cah. Biol. Mar. 16:593-798.

57. _____, J. Renaud-Mornant and J. Castel. 1976. Metabolic activities of meiofaunal communities in a semi-enclosed lagoon. Possibilities of trophic competition between meiofauna and mugilid fish. 10th European Symposium on Marine Biology, Ostend, Belgium, Sept. 17-23, 1975 (in press).

58. Maetz, J. 1969. Sea water teleosts: evidence for a sodium-potassium exchange in a branchial sodium-excreting pump. Science 166:613-615.

59. _____. 1971. Fish gills: mechanisms of salt transfer in fresh water and sea water. Phil. Trans. Roy. Soc. Lond. B. 262:209-249.

60. _____, and M. Bornancin. 1975. Biochemical and biophysical aspects of salt secretion by chloride cells in teleosts. Fortschr. Zool. 23:322-362.

61. _____, and F. Garcia-Romeu. 1964. The mechanism of sodium and chloride uptake by the gills of a freshwater fish, *Carassius auratus*. II. Evidences for NH_4^+/ Na^+ and HCO_3^-/ Cl^- exchanges. J. gen. Physiol. 47:1209-1227.

62. _____, and P. Pic. 1975. New evidence for a Na/K and Na/Na exchange carrier linked with the Cl^- pump in the gill of *Mugil capito* in sea water. J. Comp. Physiol. 102:85-100.

63. Milne, K.P., J.N. Ball and I. Chester Jones. 1971. Effects of salinity, hypophysectomy and corticotrophin on branchial Na- and K-activated ATPase in the eel, *Anguilla anguilla* L. J. Endocrinol. 49:177-178.

64. Moore, R.H. 1974. General ecology, distribution and relative abundance of *Mugil cephalus* and *Mugil curema* on the South Texas coast. Contrib. Mar. Sci. 18:241-255.

65. Motais R, and F. Garcia-Romeu. 1972. Transport mechanisms in the teleostean gill and amphibian skin. Ann. Rev. Physiol. 34:141-176.

66. _____, _____, and J. Maetz. 1966. Exchange-diffusion effect and euryhalinity in teleosts. J. gen. Physiol. 50:391-422.

67. Nicoll, C.S.1975. Radioimmunoassay and radio receptor assays for prolactin and growth hormone. Amer. Zool. 15:881-903.

68. Nordlie, F.G., and C.W. Leffler. 1975. Ionic regulation and the energetics of osmoregulation in *Mugil cephalus* Lin. Comp. Biochem. Physiol. 51A:125-131.

69. Oglesby, L.C. 1969. Inorganic components and metabolism; ionic and osmotic regulation: Annelida, Sipuncula, and Echiura, p. 211-310. *In* M. Florkin and B.T. Scheer (eds.), Chemical Zoology, vol. 4, Academic Press, New York and London

70. Olivereau, M. 1968. Etude cytologique de l'hypophyse du muge, en particulier en relation avec la salinité extérieure. Z. Zellforsch. Mikrosk. Anat. Abt. Histochem. 87:545-561.

71. Pfeiler, E., and L.B. Kirschner. 1972. Studies on gill ATPase of rainbow trout *(Salmo gairdneri)*. Biochem. Biophys. Acta. 282:301-310.

72. Pic, P., and J. Maetz. 1975. Différences de potentiel trans-branchial et flux ioniques chez *Mugil capito* adapté à l'eau de mer. Importance de l'ion Ca^{++}. C. R. Acad. Sci. Paris 280:983-986.

73. _____, N. Mayer-Gostan, and J. Maetz. 1975. Branchial effects of epinephrine in the sea-water adapted mullet. I. Water permeability. Amer. J. Physiol. 226:698-702.

74. Potts, W.T.W. 1954. The energetics of osmotic regulation in brackish- and freshwater animals. J. exp. Biol. 31:618-630.

75. _____, and F.B. Eddy. 1973. Gill potentials and sodium fluxes in the flounder *Platichthys flesus*. J. Comp. Physiol. 87:29-48.

76. Sage, M. 1973. The relationship between the pituitary content of prolactin and blood sodium levels in mullet *(Mugil cephalus)* transferred from sea water to fresh water. Contrib. Mar. Sci. 17:163-167.

77. Sargent, J.R., and A.J. Thomson. 1974. The nature and properties of the inducible $(Na^+ + K^+)$ – dependent adenosine triphosphatase in the gills of eels *(Anguilla anguilla)* adapted to fresh water and sea water. Biochem. J. 144:69-75.

78. _____, _____, and M. Bornancin. 1975. Activities and localization of succinic dehydrogenase and Na^+/K^+-activated adenosine triphosphatase in the gills of fresh water and sea water eels *(Anguilla anguilla)*. Comp. Biochem. Physiol. 51B:75-79.

79. Schoffeniels, E., and R. Gilles. 1970. Osmoregulation in aquatic arthropods, p. 255-286. *In* M. Florkin and B.T. Scheer (eds.), Chemical Zoology, vol. 5. Academic Press, New York and London.

80. Skaer, H. Le B. 1974. The water balance of a serpulid polychaete, *Mercierella enigmatica* (Fauvel). I. Osmotic concentration and volume regulation. J. exp. Biol. 60:321-330.

81. Smith, K.L. 1973. Respiration of a sublittoral community. Ecology 54:1064-1075.

82. Smith, M.A.K., and A. Thorpe. 1976. Nitrogen metabolism and trophic input in relation to growth in freshwater and saltwater *Salmo gairdneri*. Biol. Bull. 150:139-151.

83. Stequert, B. 1972. Contribution à l'étude du Bar des réservoirs à poissons de la région d'Arcachon. Thèse 3° cycle, Univ. Bordeaux.

84. Vernberg, W.B., and B.C. Coull. 1974. Respiration of an interstitial ciliate and benthic energy relationships. Oecologia (Berlin) 16:259-264.

85. _____, and _____. 1975. Multiple factor effects of environmental parameters on the physiology, ecology and distribution of some marine meiofauna. Cah. Biol. Mar. 16:721-732.

86. _____, and F.J. Vernberg. 1972. Environmental physiology of marine animals. Springer Verlag, Berlin and New York. 346 p.

87. Yañez, L.A. 1976. Observaciones sobre *Mugil curema* Valenciennes en areas

naturales de crianza, Mexico. Alimentación, crecimiento, madurez y relaciones ecologicas. Ann. Centro. Cienc. del Mar Y Limnol. Univ. Nat. Autón. México (in press).
88. Zaugg, W.S. 1968. Copper-sensitive, EDTA activated ATPase in gill microsomes. Fed. Proc. 27:806.
89. _____, and L.R. McLain. 1969. Inorganic salt effects on growth, salt water adaptation, and gill ATPase of Pacific salmon, p. 293-306. *In* O.W. Neuhaus and J.E. Halver (eds.), Fish in Research, Academic Press, New York and London.
90. _____, and _____. 1970. Adenosine triphosphatase activity in gills of salmonids: seasonal variations and salt water influence in coho salmon *(Oncorhynchus kisutch)*. Comp. Biochem. Physiol. 35:587-596.
91. _____, and H.H. Wagner. 1973. Gill ATPase activity related to parr-smolt transformation and migration in steelhead trout *(Salmo gairdneri):* influence of photoperiod and temperature. Comp. Biochem. Physiol. 45:955-965.

NITROGEN METABOLISM AND EXCRETION: REGULATION OF

INTRACELLULAR AMINO ACID CONCENTRATIONS

Stephen H. Bishop
Marrs McLean Department of Biochemistry
Baylor College of Medicine
Houston, Texas 77025

ABSTRACT: Amino acids and taurine are used by osmoconforming and osmo-regulating estuarine invertebrate animals to control the intracellular osmotic pressure. These amino acids are readily derived from intermediates of glycolysis and the tricarboxylic acid cycle. Isolated molluscan hearts adapted to low and high salinities *in vitro* show adaptive changes in amino acid levels which correlate with changes in mechanical properties. Transaminases are at high levels in all tissues of all organisms studied. High levels of the primary ammonia forming-fixing enzyme (GDH) have been found only in crustaceans. Ammonia and small amounts of free amino acids are the major nitrogen excretory products. These are released through the gills, body wall, and gut. Animals adapted to low salinities show increased ammonia production. In animals with low GDH activi-ties (flatworms, molluscs, annelids), the serine cycle and purine nucleotide cycle have been suggested as primary ammonia forming mechanisms. The enzyme responsible for ammonia fixation as these animals adapt to high salinity has not been described.

INTRODUCTION

Free amino acids and amines comprise the bulk of the organic solute used by estuarine and marine invertebrates for regulation of the cellular osmotic pres-sure. The amino acids most commonly used are the nonessential amino acids, alanine, glycine, aspartate, glutamate, proline, and taurine (28). In full sea water, each species uses only a few of these amino acids for intracellular osmoregula-tion. For instance, glycine can account for one-half to two-thirds of the amino acid pool in *Arenicola marina* (13), *Rangia cuneata* (1), and *Crangon crangon* (48); alanine glycine, proline and taurine account for close to 90% of the free amino acid pool in *Mya arenaria* (47), *Mytilus edulis* (9, 33), *Modiolus demissus*

414

(38) and *Callinectes sapidus* (20); and taurine, glutamate and aspartate make up 60% of the amino acid pool in *Nassarius obsoletus* (32). Most estuarine animals experience large changes in the intracellular concentrations of these amino acids with adaptation to diluted sea water. The regulation of the turnover of these amino acids then becomes an important aspect of their overall metabolic activity. This discussion will focus on amino acid metabolism and nitrogen excretion as it relates to the abilities of these organisms to withstand salinity changes in the estuarine environment.

EXCRETION AND RESPONSE TO DECREASE IN SALINITY

Aquatic invertebrate animals tend to excrete ammonia as the major end product of nitrogen metabolism. With oysters, 65-70% of the excreted nitrogen was ammonia; uric acid, urea, amino acids and various other unidentified nitrogen compounds accounted for about 30% of the nitrogen (31). Similar studies with other marine invertebrates yield essentially the same results (37).

Volume regulation by loss of salt and organic solute generally took a few hours to a few days after the animals were placed in diluted sea water. In the few species tested, such as the bivalve and polychaete in Fig. 1, there was an increased ammonia elaboration to the media (15). Similar experiments have been carried out with euryhaline crustaceans (28). Ammonia production was greatest just after transfer to diluted sea water, then declined as the intracellular amino acid concentrations declined. When ribbed mussels (3) (Fig. 2) were adapted to full sea water and placed in one-third sea water, ammonia excretion did not rise markedly until twelve hours after transfer whereas blood ammonia and amino acid levels rose immediately and the amino acid levels in the ventricle declined immediately. Free amino acids were not released to the bathing media. This change of intracellular amino acid concentrations in the ventricles and other tissues was not detected in mantle tissue. Bartberger and Pierce (3) suggest that amino acids are released to the blood by most tissues and that the lack of change in the mantle was because the mantle accumulated and deaminated these amino acids.

In some other species, ammonia production did not change markedly when the animals were placed in diluted sea water (14, 15). For instance, *Thais laminosa* (15) showed little change in ammonia production and only small changes in the intracellular amino acid content. In *Thais*, however, the intracellular amino acids accounted for less than 20% of the osmotic pressure. In some sea stars (18), changes in the amino acid pool size throughout the year varied as a function their feeding and reproductive cycle and not with changes in environmental salinity. Similarly, concentrations of tissue amino acids in the shore crab, *Carcinus maenas*, were influenced markedly by the diet; fed animals responded better than starved animals (12). Florkin and Schoffeniels (17) have suggested that species showing little or no increased ammonia production with dilution of the sea water, probably do not show dramatic changes in intracellular amino acid levels and show modest volume regulation. The erratic ammonia production by a number of polychaetes transferred to low salinity (13) may indicate a lack of control of volume regulation (36).

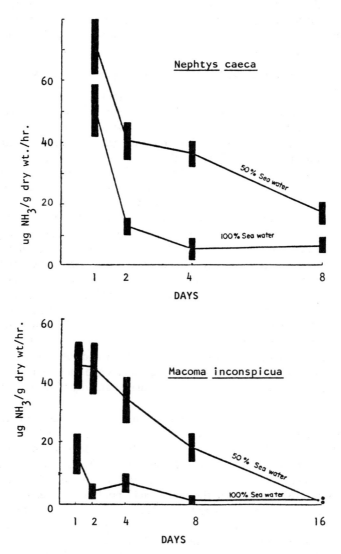

Figure 1. Ammonia excretion of *Macoma inconspicua* and *Nephtys caeca* as modified by salinity and time. Values are based on dry weight of soft parts of animals for *M. inconspicua* and dry weight of whole animals for *N. caeca* (15).

The role of amino acid metabolism in volume regulation has been studied in isolated tissues from a number of estuarine invertebrates because the responses seen in individual tissues seem similar to those seen in whole animals. The isolated tissue preparations serve as model systems for a wide variety of physiological and metabolic investigations which are not possible with whole animals.

Nerve tissue from Chinese crabs, previously acclimated to sea water, was transferred to half strength sea water and the loss of amino acids observed (30) (Table 1). The loss or decline in concentration of amino acids in the tissue was accompanied by an increase in the amino acids and ammonia in the bathing media. The increased ammonia production accounted for less than ten percent of the loss of amino acids from the tissue. Most of the amino acid nitrogen lost from the tissue appeared as amino acid nitrogen in the bathing media. There appeared to be no extensive catabolism of the amino acids. However, a quantitative assessment of the changes within the cellular compartment relative to the bathing media was not possible.

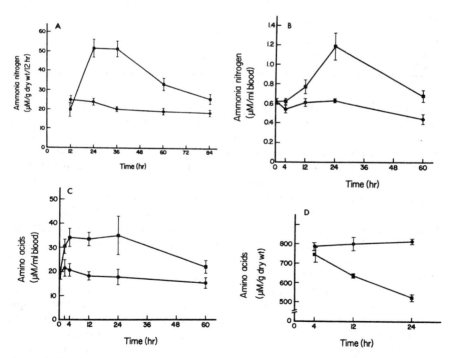

Figure 2. Time course changes in ammonia excretion (A), haemolymph ammonia (B), haemolymph amino acids (C), and ventricle amino acids (D) of propped open *Modiolus demissus* during low salinity acclimation (510 mOsm). Mussels were previously acclimated to full strength sea water (1020 mOsm). (■–■) Response at 510 mOsm. (●–●) Control response at 1020 mOsm (3).

Table 1. *Eriocheir sinensis.* Effect of osmotic shocks on the amino-acid concentration in isolated axons and ambient media when placed in 50% sea water media. All values expressed in μM 100 mg^{-1} dry weight (30).

Amino acid	Axon Hypo-osmotic stress		Ambient Media Hypo-osmotic shock	
	Experimentals	Controls	Experimentals	Controls
Taurine	5.38	7.03	0.83	0.70
Asp	10.25	23.65	11.17	11.11
Thr	0.87	1.04	0.18	0.22
Ser	3.10	3.34	0.83	0.88
Glu	2.62	3.03	1.43	2.43
Pro	6.94	17.78	19.34	7.93
Gly	3.24	3.83	1.25	1.09
Ala	8.07	12.16	5.99	6.06
Cys	traces	traces	traces	traces
Val	0.52	0.58	0.38	0.29
Met	traces	traces	traces	traces
Ilu	0.30	0.50	0.30	0.15
Leu	0.32	0.45	0.36	0.20
Tyr	0.21	0.30	0.22	0.15
Phe	0.24	0.30	0.27	0.19
Lys	0.42	0.39	——	——
His	0.49	0.51	——	——
Arg	1.45	3.03	——	——
Ammonia	2.19	10.00	17.39	8.75

In preliminary experiments with blue crab axons pre-loaded with ^{14}C-alanine (21), there was an increased efflux of radioactivity when nerves were exposed to either hypo- or hyperosmotic conditions. Although identification of the radioactive material leaving the axon was not reported, it would appear that the rate of exit of free amino acids from the isolated tissues increased with media dilution. In more recent experiments, the rate of uptake of alanine by the axon was inhibited when the axons were placed in diluted sea water media (19). Both Ca^{++} and Na^+ seem to be involved in regulating influx and efflux (19). When axons from blue crabs in sea water were removed and transferred to half strength sea water containing a variety of ^{14}C-amino acids (25), the increased catabolic rate of the amino acids was less than two fold and accounted for only a small fraction of the amino acid loss.

In other experiments, ventricles isolated from ribbed mussels acclimated to sea water were placed in muscle baths containing diluted sea water and changes in mechanical properties and amino acid levels were measured (38) (Fig. 3). Amino acid levels in the tissue declined rapidly as volume regulation neared completion (Fig. 3A) and there was a loss of amino acids from the ventricle to the bathing media (Fig. 3B). No ammonia was produced and all amino acid nitrogen lost by the tissue was accounted for as amino acid nitrogen in the bathing media (Fig. 3B). The ventricle stopped beating with initial transfer, then resumed beating in one to three hours as volume regulation neared completion.

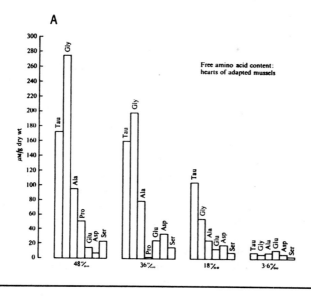

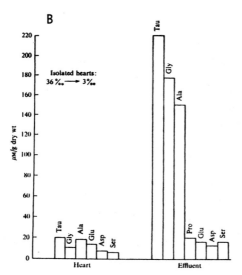

Figure 3. A. Intracellular free amino acid concentrations in ventricles taken from *Modiolus demissus granosissimus* acclimated to various salinities (38).

B. Composition of the NPS efflux from isolated *Modiolus demissus granosissimus* ventricles in 3°/₀₀ sea water. The heart histogram indicates the concentrations of free amino acids left in the tissues. The effluent histogram shows the free amino acids which diffused into the bath water (38).

Volume regulation by solute elimination was somehow involved in proper function of the tissue. The effect of the ions on volume regulation and heart rate with osmotic stress was investigated (39) because both required proper membrane function. Loss of amino acids from ventricles placed in diluted sea water was reduced by high concentrations of Ca^{++}, Mg^{++} or by La^{+++} (Ca^{++} antagonist) but not by changes in Na^+, Cl^-, or K^+. As long as Ca^{++} and Mg^{++} ion concentrations were high and the osmotic pressure was maintained by Tris and propionate, leakage of amino acids from the tissue was minimal.

One can conclude then that in the mussel hearts, crab nerve tissue, and most isolated tissues, volume regulation by reduction in tissue amino acid levels with sea water dilution may be controlled by membrane permeability rather than by extensive tissue metabolism. Membrane permeability is related closely to the levels of Ca^{++}, Mg^{++}, and Na^+ ions. The lack of apparent volume regulation in the mantle tissue of *Modiolus demissus* (3) may be a function of the Ca^{++}-Mg^{++} binding capacity of the tissue and an altered membrane permeability to amino acids. The mantle lays down the $CaCO_3$ shell and has unusual Ca^{++}-Mg^{++} binding properties. The poor volume regulation by some prosobranchs (15) and sea stars (18) may be related to peculiar Ca^{++}-Mg^{++} binding activity in their tissues.

With reduction of the extracellular fluid osmolality, tissues of most volume regulators release amino acids to the blood or haemolymph. However, the animals excrete ammonia rather than amino acids. Ammonia production from these amino acids may occur in only a few, as yet, undefined tissues. Several theories relating amino acid metabolism to ammonia formation during salinity adaptation have been put forward and three will be discussed. None have been proven but all have some merit and may co-exist. In general, these mechanisms suppose that there is some integrated control of amino acid degradation so that ammonia production will be related to the metabolic rate.

The first scheme (Fig. 4A) (16) extends the transdeaminase model of Braunstein (7). In general, amino groups are brought to α-ketoglutarate for transamination to glutamate. Ammonia is then released from glutamate by action of glutamate dehydrogenase (GDH) and α-ketoglutarate is made available for another transamination. The transaminases are found in high concentrations in all tissues examined (12, 23, 28) and are not markedly influenced by changes in ionic strength. GDH, however, has only been found in measureable amounts in a few crustaceans (12, 27). In molluscs (10, 11) and polychaete annelids (6), the GDH levels measured in the ammonia forming direction were below detectable limits. In *Homarus, Eriochier,* and *Helix* (land snail) the enzyme is 10-30 times more active in the direction of glutamate synthesis than ammonia formation (16, 43). Although GDH is found in some crab muscles and gills in levels comparable to mammalian skeletal muscle (12, 27), the activity in the ammonia forming direction is essentially non-existant. Pierce and coworkers (3, 39) have suggested that GDH linked mechanisms do not function in volume regulation.

PRIMARY AMMONIA FORMING MECHANISMS:

A. GLUTAMATE DEHYDROGENASE - TRANSAMINASE SCHEME

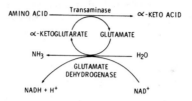

Stoichiometry: Amino Acid + H_2O + NAD^+ $\rightleftharpoons$ α-Keto Acid + NADH + H^+ + NH_3

B. ADENINE NUCLEOTIDE - TRANSAMINASE SCHEME

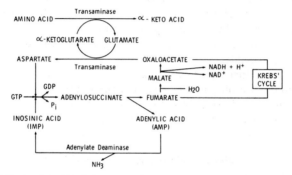

Stoichiometry: Amino Acid + H_2O + GTP + NAD^+ $\longrightarrow$ α-Keto Acid + NADH + GDP + P_i + H^+ + NH_3

C. SERINE CYCLE

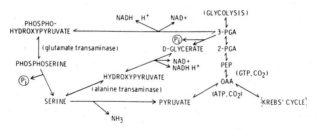

Stoichiometry: Amino Acid + GTP + ATP + NAD^+ + $3H_2O$ $\longrightarrow$ α-Keto Acid + GDP + ADP + $2P_i$ + NADH + H^+ + NH_3

Figure 4. Primary ammonia mechanisms:
  a. Glutamate dehydrogenase-transaminase scheme (6).
  b. Adenine nucleotide-transaminase scheme (6).
  c. The serine cycle.

An alternate mechanism employing the purine nucleotide cycle and adenylate deaminase (Fig. 4B) as the ammonia forming step is energetically more advantageous (6, 7, 34) than the GDH linked mechanism. Adenylate deaminase has a wide distribution in free living marine invertebrate species (6). However, the levels of the enzymes responsible for IMP to AMP conversion have not been evaluated.

A third mechanism (Fig. 4C) uses serine dehydrase as the ammonia forming step. As originally proposed by Gilles (23), the intermediates would not cycle and catalyze ammonia production. This scheme (Fig. 4C) makes a cycle out of the pathway. Serine dehydrase has a fairly wide distribution and seems to occur in all organisms examined (5). Because the enzyme is inhibited by high salt, it would be more active at reduced salinities (23). Additionally, the pathway involves a CO_2 fixation step. In the few studies with ^{14}C-CO_2 fixation where the specific radioactivity of serine has been measured, it has an extremely high specific radioactivity which seems inappropriate for its normally low concentration (see 5). Again, data on serine biosynthesis and turnover in invertebrate animals are not available.

To summarize, then, in this vast group of ammonotelic invertebrates, the primary ammonia forming mechanisms and the relationships between ammonia formation, amino acid metabolism and the metabolic rate are not understood. These relationships must be established if we are to understand how the amino acid pool sizes are regulated.

RESPONSE TO INCREASED SALINITIES

The mechanism for increases in cellular amino acid content with transfer of animals from low to high salinity requires knowledge of the origins of both the amino group and the carbon skeletons. The amino group can be derived by ammonia fixation or amino acid uptake and subsequent transamination or by protein degradation and subsequent transamination. Because the amino acids used as solute are non-essential, the carbon skeletons can be synthesized at varying rates from glucose, acetate, and other amino acids (12, 24, 28). Two schemes (Fig. 5) and the evidence for and against these are discussed.

The first model (Fig. 5A) (17) is essentially the reverse of the deamination scheme in Fig. 4A. The carbon skeletons are derived from intermediates of polysaccharide and fat metabolism. The amino group is derived from ammonia by ammonia fixation into glutamate by GDH. The specific amino acids which accumulate then are derived by transamination from glutamate.

An alternative scheme is described in Fig. 5B. The carbon skeleton of the amino acids can originate from fat, polysaccharide or protein. The amino group, however, originates from amino acids rather than from ammonia. These amino acids can be derived by uptake of free amino acids from outside the cell or organism or by protein degration. After entering the cellular pool, these amino acids contribute directly or transfer their amino group to the appropriate TCA cycle or glycolytic intermediate for synthesis of the specific amino acids in the amino acid pool. The keto acids, formed from the carbon skeletons of these amino acids after amino transfer, enter the metabolic pool for energy production or conversion to intermediates for specific amino acid synthesis. In this scheme, there is a net gain in the number of solute molecules.

AMINO ACID GENERATING SCHEMES

A. GLUTAMATE DEHYDROGENASE PATHWAY

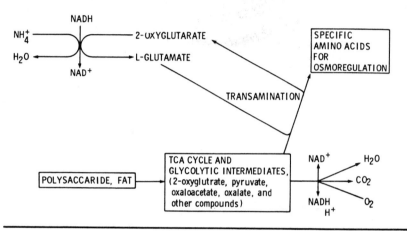

B. PROTEIN TURNOVER-AMINO ACID UPTAKE PATHWAY

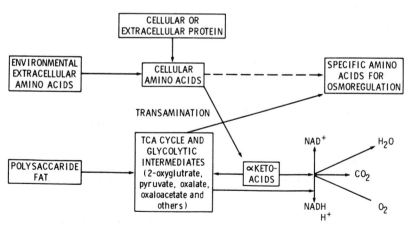

Figure 5. Amino acid generating scheme:
         a. Glutamate dehyrogenase pathway.
         b. Protein turnover—amino acid uptake.

Any discussion of these models requires some knowledge of the relative rates of adaptation and of the availability of substrate.

Recently, Baginski and Pierce (2) have determined the rates of change of all the amino acids in the pool after ribbed mussels were transferred from low to high salinity. After acclimation at $12°/_{oo}$, mussels were transferred to $36°/_{oo}$ and the ventricles removed at intervals after transfer. The total amino acid pool size in the ventricles reaches a maximum and stabilizes after five days (Fig. 6). At this time, alanine, aspartate, glutamate, glycine, proline, serine, threonine, and taurine account for 90% of the amino acid pool. In the initial phase of adaptation (0-5 days), the alanine concentration reaches a peak then begins to decline. Glycine increases slowly then stabilizes after 60 days and taurine increases very slowly throughout the experimental period. Glutamate concentrations do not change. Proline follows the same trend as glycine, then disappears from the pool after 65 days.

To evaluate this adaptation at the tissue level (2), ventricles were removed from mussels adapted at $12 °/_{oo}$ and transferred to muscle baths containing artificial sea water at $36°/_{oo}$. After nine hours, there was a striking increase in the alanine and proline content, a modest increase in the glycine content and no change in the taurine content (Table 2). These increases parallel those seen with

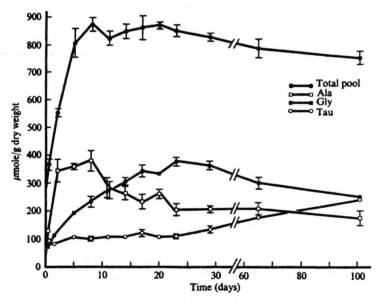

Figure 6. Time course of intracellular amino acid concentration increases in ventricles removed from *M. demissus* during acclimation to high salinity. At time 0 animals were transferred from 12 to $36°/_{oo}$. Error bars are standard errors. For those points without error bars, standard errors are less than the size of the point (3).

Table 2. Changes in intracellular free amino acid concentrations in isolated ventricles of low salinity acclimated *M. d. demissus* following exposure to increased salinity (2).

Salinity change ($^{\circ}/_{\circ\circ}$)	Amino acid concentrations (μmole/g dry wt)[1]			
	Alanine	Glycine	Proline	Taurine
12-12	30 (7.5)	33 (4.3)	0	103 (5.1)
12-36	231 (14.3)	53 (1.9)	61 (0.9)	105 (8.1)

[1] S.E.M. shown in parenthesis.

ventricles taken from whole animals nine hours after transfer (Fig. 6). Alanine was used as the primary organic solute in the initial phases of isosmotic adaptation in both the isolated tissue and the whole animal. Baginski and Pierce (2) suggest that glycogenolysis and some degree of anaerobic metabolism is involved in alanine production. The rapid rise and fall of alanine and the slow accumulation of taurine and glycine indicate that the carbon skeletons of these amino acids may be derived from metabolic pools with very different turnover rates. The changes in the concentration of individual components with prolonged incubation at high salinities suggest a re-organization and stabilization of the amino acid pool by an accumulation of amino acids which may have a slow turnover compared to alanine. This stabilization would be aided by the decreased efflux of amino acids seen at high salinities (2, 19, 45, 46).

These experiments (2) offer an explanation for the lack of ^{14}C incorporation into the amino acids which are most active in osmoregulation in well adapted animals (12, 28). For instance, in short term ^{14}C-glucose tracer experiments with *Arenicola marina* (50), most of the radioactivity was in alanine, aspartate, glutamate and a variety of TCA cycle and glycolytic intermediates; little or no radioactivity was detected in glycine which is the most abundant amino acid in the amino acid pool. Additionally, investigators in different laboratories using the same animal species but different or unknown adaptation times often obtain data which differ quantitatively with regard to the concentrations of specific amino acids. For instance, glycine was the most abundant tissue amino acid in populations of *Nereis succinea* from Alamitos Bay (California), whereas with Salton Sea (California) population, total amino acid levels were higher and alanine and glycine were present in equal amounts (35). If the kinetics of glycine and alanine accumulation in *N. succinea* (35) resemble those in *Modiolus* (2), possibly the Salton Sea worm population had been subjected to an increase in environmental salinity a few days before analysis. Genetic differences between the populations may be minimal.

Support for the model in Fig. 5A is based on the presence of GDH and a reduced oxidative metabolism at high salinities in a variety of crustaceans. Higher NADH levels and low O_2 consumption are found in mitochondia isolated from a number of crabs adapted to high salinity compared to those adapted to low salinity (26, 29). GDH from *Eriochier* leg muscle is activated by increasing NaCl concentrations at high NADH and α-ketoglutarate concentrations and inhibited by high NaCl concentrations at low NADH and α-ketoglutarate concentrations (27). Cellular NADH concentrations are generally in the high range (0.1-0.3mM). Therefore, GDH activity should increase with increasing ionic strength, favoring NH_3 fixation. However, as discussed above, the level of GDH is below detectable levels in most other marine animals. If NH_4^+ ions were available, they would themselves contribute to the solute pool and raise the osmotic pressure; transfer to a carbon skeleton to form an amino acid would not increase the number of solute particles. Nitrogen may be stored in the tissues in some non-polymeric form, such as with uric acid in the land crab *Cardiosoma* (22). When *Cardiosoma* enters sea water, some of the uric acid is degraded to NH_4^+ and released. No such nitrogen store has been found in most estuarine invertebrates and the NH_4^+ content of the tissues is low at all salinities (28). Until there is more information on NH_4^+ availability and GDH distribution, this model (Fig. 5A) seems untenable.

With the second model (Fig. 5B), the amino group may be derived from amino acids in the environment. Most marine animals can absorb low molecular weight, dissolved organic matter directly through the body wall or through soft body parts for general nutrition (46). Amino acid uptake follows Michaelis-Menten or saturation kinetics against a 10^2 to 10^5 concentration gradient. Apparent K_m's (K_t's for transport) for amino acids are 10^{-4} M to 10^{-6} M or 10 to 1000 times the concentration found in sea water or estuarine waters (44). The V_{max}'s for uptake are calculated as a function of surface area/g but generally fall in the range of 10^{-8} to 10^{-5} moles/g-hr at 10-25°C.

In deciding whether or not dissolved amino acids serve as a source of amino nitrogen for cellular osmotic pressure adjustment, one must look at uptake rates at high salinities. With isolated *Mytilus californianus* gill tissue (49), uptake of cycloleucine shows a V_{max} of 30 μmoles/g-hr and a K_m of 0.25mM. At amino acid concentrations in the bathing media of 10^{-5} or 10^{-6}M, then uptake rates are 1.15 μmoles/g-hr and 0.124 μmoles/g-hr, respectively. In the transfer of *Mytilus edulis* from half to full sea water, Bricteaux-Gregoire (9) found an increase of 84 μmoles of amino acids/g body water. At amino acid concentrations of 10^{-5} or 10^{-6}M, it would take 12 days or 113 days, respectively, for this average tissue to come to equilibrium after transfer from half to full sea water assuming that this transport rate limits accumulation. The 10^{-6}M amino acid concentration is a high value for normal sea water and 10^{-5}M is rarely attained (46). In isolated tissue experiments, no amino acids are added to the artificial sea water bathing media (2). In practice, volume regulation takes from a few hours to a few days (2, 48). The high K_m's for transport and low environmental concentrations of amino acids in experimental preparations where adaptation occurs rapidly all argue against the contention that uptake of free amino acids makes a significant contribution to the solute used for rapid volume regulation.

The most reasonable source of the amino group for these amino acids appears to be protein (Fig. 5B). Amino acids generated by increased protein catabolism and reduced oxidative metabolism would be transaminated to a common acceptor, such as pyruvate with subsequent alanine accumulation. The fate of the variety of α-keto acids which would be generated in this scheme has never been evaluated. Most should end up as glucogenic or ketogenic precursors. Using the data of Baginski and Pierce (2) (Fig. 6) with mussel ventricles, there is a net increase of about 500 μmoles of free amino acid/g dry-wt after 5 days. Assuming that a g-dry wt. is equivalent to about 5 g-wet wt., then about 0.2 g of the dry wt. is salt. Approximately 50% of the remaining dry weight should be protein. Taking 100 as an average amino acid residue weight in the protein, then 500 μmoles of amino acid residue is equivalent to about 50 mg of dry protein or about 13% of the protein in the preparation. If one considers the ventricle preparation described in Table 2 using the same assumptions, the increase in alanine plus glycine and proline would be equivalent to 28 mg or 7% of the tissue protein.

One can conclude that small decreases in tissue protein content can account for very large changes in the free amino acid pool sizes. Arguments against the view that protein catabolism is involved in regulation of amino acid pool sizes, stem from the inability of investigators to measure significant changes in protein content when animals are adapted at various salinities (16, 17, 28). However, if the required changes are as small as supposed, then they could be overlooked. Protein determinations based on the Lowry method, biuret, or dry weight measurements are accurate only to 10-15%.

There have been only a few investigations of protein turnover as a function of salinity. An increase or decrease in total protein of 10-15% should be sufficient to account for the observed changes in the amino acid pool sizes. With ^{14}C-alanine incubations (4) using foot tissue from the snail *Melanopsis trifasciata*, the amount of ^{14}C in protein increased with transfer from high to low salinity and decreased with transfer from low to high salinity. The total amount of nitrogen in the system was essentially unchanged. With the crayfish *Orconectes limosus* (40), transfer to $26^\circ/_{oo}$ in the absence of food caused a great increase in the free non-essential amino acids and a smaller increase in free essential amino acids. The loss in body protein could be accounted for as individual free amino acids. With transfer of *Carcinus maenas* from low to high salinity (41), serum protein decreased from 4.0 to 2.8 g/100 ml, suggesting that serum proteins may serve as a source of amino acids for osmoregulation. More recently, Siebers (42) injected ^{14}C-labelled *Carcinus* plasma protein into *Carcinus*, then transferred them from $11^\circ/_{oo}$ to $38^\circ/_{oo}$ sea water. When compared to controls held at $11^\circ/_{oo}$ there was a 5% increase in ^{14}C in both the low molecular weight solute pool and CO_2 and about a 12% drop in ^{14}C in the haemolymph protein. Because of the small changes, the significance is questionable; however, only small changes are anticipated.

More data on protein turnover with salinity change is required to substantiate the model in Fig. 5B. At least this model has none of the flaws of the GDH linked models (Fig. 5A). The fate of the amino group in these processes seems to be the key to understanding this regulatory process. Experiments with ^{15}N-NH$_3$ and ^{15}N amino acids are required to properly evaluate the pathways for regulating the changes in tissue amino acid concentration with changes in the estuarine salinity.

SUMMARY

In addition to dietary considerations, the amount of nitrogen excreted by most estuarine invertebrates is closely related to the regulation of the intracellular amino acid pool size. At low salinities, most tissues volume regulate by increasing the efflux of amino acids to the blood or haemolymph. Ammonia, however, is the major nitrogenous excretory product. The mechanisms for ammonia production are not well defined. When these organisms are transferred to high salinities, the intracellular amino acid pool size increases so as to be in osmotic equilibrium with the media or blood. It is postulated that this increase results from slight alterations in protein turnover and anaerobic metabolism in combination with decreased amino acid efflux. The models presented suggest that independent pathways are used for increasing and decreasing the amino acid pool sizes.

ACKNOWLEDGMENTS

I wish to thank K.L. Gibbs for reading and commenting on the manuscript, Dr. S.K. Pierce for providing a copy of his manuscript prior to publication, and the NIH, NSF, and Robert A. Welch Foundation for grant support.

REFERENCES

1. Allen, K. 1961. The effect of salinity on the amino acid concentration in *Rangia cuneata* (Pelecypoda). Biol. Bull. 121:419-424.
2. Baginski, R.M., and S.K. Pierce, Jr. 1975. Anaerobiosis: A possible source of osmotic solute for high-salinity acclimation in marine molluscs. J. Exp. Biol. 62:589-598.
3. Bartberger, C.A., and S.K. Pierce, Jr. 1976. Relationship between ammonia excretion rates and hemolymph nitrogenous compounds of a euryhaline bivalve during low salinity acclimation. Biol. Bull. 150:1-14.
4. Bedford, J.J. 1971. Osmoregulation in *Melanopsis trifasiata*, IV. The possible control of intracellular isosmotic regulation. Comp. Biochem. Physiol. 40A: 1015-1028.
5. Bishop, S.H. 1975. Ammonia formation and amino acid excretion by *Gyrocotyle fimbriata* (Cestoidea). J. Parasiol. 61:79-88.
6. ____, and L.B. Barnes. 1971. Ammonia forming mechanisms: Deamination of 5'-adenylic acid (AMP) by some polychaete annelids. Comp. Biochem. Physiol. 40B:407-422.
7. Braunstein, A.E. 1957. Les voies principales de l'assumulation de dissimilation de l'azote chez les animaux. Adv. Enzymol. 19:335-389.

8. Bricteux-Gregoire, S., G. Duchateau-Bosson, C. Jeuniaux, and M. Florkin. 1962. Constituants osmotiquements actifs des muscles du crab chenois *Eriocheir sinensis*, adoptee a l'eau douce ou a l'eau de mer. Arch. Int. Physiol. Biochim. 70:273-286.

9. _____, _____, _____, and _____. 1964. Constituants osmotiquements actifs des muscles adducteurs des *Mytilus edulis* adaptee a l'eau de mer ou a l'eau saumatre. Archs. Int. Physiol. Biochim. 72:116-123.

10. Campbell, J.W., and S.H. Bishop. 1970. Nitrogen metabolism in molluscs, p. 103-206. *In* J.W. Campbell (ed.), Comparative Biochemistry of Nitrogen Metabolism. Vol. 1. The Invertebrates. Academic Press, Inc., N.Y.

11. _____, R.B. Drotman, J.A. MacDonald, and Tramell. 1972. Nitrogen metabolism in terrestrial invertebrates, p. 1-54. *In* J.W. Campbell and L. Goldstein (eds.), Nitrogen Metabolism and the Environment. Academic Press, Inc., N.Y.

12. Chaplin, A.E., A.K. Huggins, and K.A. Munday. 1970. The effect of salinity on the metabolism of nitrogen containing compounds by *Carcinus maenas*. Int. J. Biochem. 1:385-400.

13. Clark, M.E. 1968. A survey of the effect of osmotic dilution on free amino acids of various polychaetes. Biol. Bull. 134:252-260.

14. Duerr, F.G. 1968. Excretion of ammonia and urea in seven species of marine prosobranch snails. Comp. Biochem. Physiol. 26:1051-1059.

15. Emerson, D.N. 1969. Influence of salinity on ammonia excretion rates and tissue constituents of euryhaline invertebrates. Comp. Biochem. Physiol. 29:1115-1133.

16. Florkin, M., and E. Schoffeniels. 1965. Euryhalinity and the concept of physiological radiation, p. 6-40. *In* K.A. Munday (ed.), Studies in Comparative Biochemistry. Pergamon Press, N.Y.

17. _____, and _____. 1969. Molecular appraoches to ecology. Academic Press, Inc., N.Y.

18. Furgeson, J. 1975. The role of free amino acids in nitrogen storage during the annual cycle of a starfish. Comp. Biochem. Physiol. 51A:341-350.

19. Gerard, J.F. 1975. Volume regulation and alanine transport. Response of isolated axons of *Callinectes sapidus* Rathburn to hypo-osmotic conditions. Comp. Biochem. Physiol. 51A:225-229.

20. _____, and R. Gilles. 1972. The free amino acid pool in *Callinectes sapidus* tissues and its role in the osmotic intracellular regulation. J. Exp. Mar. Biol. Ecol. 10:125-136.

21. _____, and _____. 1972. Modification of amino acid efflux during the osmotic adjustment of isolated axons of *Callinectes sapidus*. Experientia 28:863-864.

22. Gifford, C.A. 1968. Accumulation of uric acid in the land crab *Cardiosoma guanhami*. Am. Zool. 8:521-528.

23. Gilles, R. 1969. Effect of various salts on the activity of enzymes implicated in amino acid metabolism. Arch. Int. Physiol. Biochim 77:441-464.

24. _____. 1970. Intermediary metabolism and energy production in some invertebrates. Arch. Int. Physiol. Biochim. 78:313-326.

25. _____. 1972. Amino acid metabolism and isosmotic intracellular regulation in isolated surviving axons in the blue crab *Callinectes sapidus*. Life Sci. II:565-572.

26. _____. 1972. Oxygen consumption as related to amino acid metabolism during osmoregulation in the blue crab *Callinectes sapidus*. Neth. J. Sea Res. 7:280-289.

27. _____. 1974. Studies on the effect of NaCl on the activity of *Eriocheir sinensis* glutamate dehydrogenase. Int. J. Biochem. 5:623-628.

28. _____. 1975. Mechanisms of ion and osmoregulation, p. 259-347. *In* O. Kinne (ed.), Marine Ecology Vol. 11. J. Wiley and Sons, N.Y.

29. _____, and F.F. Jobsis. 1972. Isosmotic intracellular regulation and redox changes in the respiratory chain components of *Callinectes sapidus* isolated muscle fibers. Life Sci. II:877-886.

30. _____, and E. Schoffeniels. 1969. Isomotic regulation in isolated surviving nerves of *Eriocheir sinensis* Milne Edwards. Comp. Biochem. Physiol. 31:927-939.

31. Hammen, C.A. 1969. Metabolism of the oyster *Crassostria virginica*. Am. Zool. 9:309-318.

32. Kasschau, M.R. 1975. The relationship of free amino acids to salinity changes and temperature—salinity interactions in the mud-flat snail, *Nassarius obsoletus*. Comp. Biochem. Physiol. 51A:301-308.

33. Lange, R. 1963. Osmotic function of amino acids and taurine in the mussel *Mytilus edulis*. Comp. Biochem. Physiol. 10:173-179.

34. Lowenstein, J.M. 1972. Ammonia production in muscle and other tissues: the purine nucleotide cycle. Physiol. Rev. 52:382-414.

35. Mearns, A.J., and D.J. Reish. 1969. A comparison of the free amino acids in two populations of the polychaetous annelids *Neanthes succinea*. Bull. So. Calif. Acad. Sci. 68:43-53.

36. Oglesby, L.C. 1973. Salt and water balance in lugworms (Polychaeta: Arenicolidae), with particular reference to *Abarenicola pacifica* in Coos Bay. Oregon Biol. Bull. 133:643-658.

37. Pandian, T.J. 1975. Mechanisms of heterotrophy, p. 61-249. *In* O. Kinne (ed.), Marine Ecology, Vol. 11, pt. 1. John Wiley and Sons, N.Y.

38. Pierce, S.K., Jr., and M.J. Greenberg. 1972. The nature of cellular volume regulation in marine bivalves. J. Exp. Biol. 57:681-792.

39. _____, and _____. 1973. The initiation and control of free amino acid regulation of cell volume control in salinity stressed marine bivalves. J. Exp. Biol. 59:435-446.

40. Siebers, D. 1972. Mechanismen der intrazellularen isosmotischen Regulation der Amino-saurekonzentration bei dem Flußkrebs *Orconectes limosus*. Z. Vergl. Physiol. 76:97-114.

41. _____. 1974. Mechanisms of intracellular isosmotic regulation: Fate of ^{14}C-labelled serum proteins in the shore crab *Carcinus maenas* after changed environmental salinity. Helgolander. Wiss. Meers. 26:375-381.

42. _____, K.R. Sperling, and K. Eberkin. 1972. Kinetics of osmoregulation in the crab *Carcinus maenas*. Mar. Biol. 17:291-303.

43. Sollock, R.L. 1975. Amino acid catabolism in gastropod molluscs. Ph.D. thesis Rice University, Houston, Texas, 92 p.

44. Southward, A.J., and E.L. Southward. 1972. Observations on the role of dissolved organic compounds in the nutrition of benthic invertebrates, III. Uptake in relation to organic content of the habitat. Sarsia 50:29-46.

45. Stephens, G.C. 1964. Uptake of organic material by aquatic invertebrates. III Uptake of glycine by brackish water annelids. Biol. Bull. 126:150-162.

46. _____. 1972. Amino acid accumulation and assimilation in marine organisms, p. 155-184. *In* J.W. Campbell and L. Goldstein (eds.), Nitrogen Metabolism and the Environment. Acadmic Press, Inc., N.Y.

47. Virkar, R.A., and K.L. Webb. 1970. Free amino acid composition of the

soft-shell clam *Mya arenaria* in relation to salinity of the medium. Comp. Biochem. Physiol. 32:775-784.

48. Weber, R.E., and W.J.A. van Marrewijk. 1972. Free amino acids in the shrimp *Crangon crangon* and their osmoregulatory significance. Neth. J. Sea Res. 6:391-415.

49. Wright, S.H., T.L. Johnson, and J.H. Crowe. 1975. Transport of amino acids by isolated gills of the mussel *Mytilus californianus* Conrad. J. Exp. Biol. 62:313-325.

50. Zebe, E. 1975. *In vivo*—Untersuchungen uber den Glucose-Abbau bei *Arenicola marina* (Annelida, Polychaeta). J. Comp. Physiol. 101:133-145.

ASPECTS OF REPRODUCTION IN BIVALVE MOLLUSCS

B.L. Bayne
Institute for Marine Environmental Research
67 Citadel Road
Plymouth PL1 3DH, England

ABSTRACT: There exists a large literature on the components of the environment that effect the phasing of the production, growth and release of gametes by estuarine invertebrates; interactions between temperature, salinity, ration and day-length are particularly important. Linked with these cycles of gametogenesis are a variety of biochemical processes, including the synthesis, storage and utilisation of specific metabolites, which affect the efficiency of the gametogenic events. Endocrinological processes probably provide the integrated control between the biochemical and gametogenic cycles, and the phasing, in time, of these cycles in turn affects the physiological responses of the organism to environmental change. The efficiency of gametogenesis is reflected in the fecundity of the animal, and the vigour of the larvae, which are major components of population fitness. Both of these components are influenced by the physiological condition of the adult.

These points are all discussed, and relevant literature reviewed, within the context of the ecology and physiology of estuarine populations.

INTRODUCTION

Reproduction includes gametogenesis, larval development and metamorphosis, all of which are energy-consuming processes. Intuitively, we may imagine physiological and biochemical adaptation to these demands for energy taking the form of answers to certain questions: How much of the available resource should be allocated to gametogenesis? What form should this allocation take? When should it be made? Once such an allocation has been made, how may it be protected from potentially deleterious environmental change? And finally, what are the energy demands of the larvae, and how can they be met? Studies on bivalve molluscs have suggested certain answers to these questions. The answers may, or may not, be unique to estuarine conditions, but taken together they can help to illustrate some aspects of the reproductive process that are relevant to estuarine studies.

432

THE ALLOCATION OF ENERGY TO GAMETOGENESIS

Most bivalve molluscs, and all of the species considered in this paper, reproduce by means of a planktotrophic larval development, i.e. by the development of a swimming larva that feeds on suspended particulate matter whilst temporarily resident in the plankton. The species considered also all reproduce annually. For some species, all the energy available for production, once the demands of maintenance have been met, is eventually utilised in gametogenesis; these species have no somatic growth once they are adult (for example, *Macoma balthica* in Massachusetts 32). Even in species in which somatic growth continues throughout life, the caloric demands of gametogenesis may be very high. In Table 1 some published data for various species of bivalve mollusc are summarised. Disregarding the lowest values recorded for *Tellina* and *Patinopecten*, the mean of the published values is 39%. This value is likely to increase with age (29). Fugi and Hashizume (26) calculated that gamete production was 3% of total production in 1-year old *Patinopecten*, increasing to 38% in 3-year old individuals. Bayne (unpublished data) estimated the following relationships from observations on *Mytilus edulis*:

Age in years	Gamete production as a percentage of total production
1	8
2	45
3	72
4	94

In a study on *Tellina tenuis*, Trevallion (54) showed that gamete production can be extremely variable (9-62% of total production) within a single population

Table 1. Gamete production in calories (L) as a proportion of total production (L + G) in eight species of bivalve mollusc.

Species	$\left[\dfrac{L}{(L+G)}\right].100$	Reference and comments
Modiolus demissus	17	(40)
Mercenaria mercenaria	52	(8)
Chione cancellata	50	(45)
Dosinia elegans	47	(46)
Scrobicularia plana	24-52	(37)
Tellina tenuis	9-62	(54); highly variable over three years of observation
Patinopecten yessoensis	3-38	(26); gamete production increased with age
Mytilus edulis	8-94	(11) and unpublished; derived from laboratory observations

over a period of years. These values for gamete production in bivalves are high, but are consistent with a planktotrophic "strategy" (55, 56) which is characterised by a high fecundity and a high metabolic cost.

THE GAMETOGENIC CYCLE

It is commonplace to observe that for most marine and estuarine invertebrates gametogenesis is cyclical. The cycles may occur with an annual, monthly or lunar periodicity, and are found in polar, temperate and in tropical species (30, 44). Clarke (19) has pointed to the dichotomy of interest between the ecologist, who is concerned largely with the synchronisation of the reproductive behaviour of the individual, or the population, with external events, and the physiologist, who is more concerned with the co-ordination of events within the individual animal. The reviews of the reproductive biology of polychaetes by Clarke (19) and by Clarke and Olive (20) demonstrate that a full understanding of breeding can result only when this dichotomy is breached.

In seeking global trends in the timing of spawning by marine invertebrates, Moore (44) re-stated the view that in polar waters summer spawning predominates; in temperate waters spring spawning is common, whereas in the tropics there is a return shift towards spawning in the summer. In temperate and in sub-tropical regions, species with a northerly distribution tend to breed in winter while the warmer-water species breed in summer. For example, Seed (51) has demonstrated that the boreo-arctic bivalve *Mytilus edulis* spawns earlier in the year in British waters than does the Mediterranean species *M. galloprovincialis* in the same locality. *M. edulis* on the south coast initiate spawning earlier than mussels of the same species further north.

Wilson and Hodgkin (62) correlated the season and the duration of reproductive activity in four species of bivalve from Western Australia with their geographical distribution and with temperature. However, some of the finer details of the cycle, and particularly the restriction of spawning to a part of the potential reproductive season, suggested the operation of factors other than temperature in the synchronisation of breeding.

Studies of this kind have led to the conclusion that temperature, either in terms of an absolute value, or in its rate of change, is a major factor in the synchronisation of the breeding cycle (48, 18, 11). Carriker (17) found that *Mercenaria mercenaria* in a mixed estuary generally spawned at or just after low tide, which coincided with periods of maximum temperature as a result of heat brought down the estuary by the ebb tide. However, other factors, such as the abundance of phytoplankton and salinity, can also be important in stimulating spawning. Wilson (60, 61) concluded that salinity changes, and not temperature, controlled spawning by the estuarine mussel *Xenostrobus securis*. In this species the potential breeding temperature range is wide relative to ambient environmental temperatures and the mussels are able to breed opportunistically according to the very variable salinities. The estuarine clam *Rangia cuneata* is also stimulated to spawn by a change in salinity (Anderson, pers. commun.).

It is often possible to recognise that spawning occurs under conditions that are advantageous for development of the larvae. *Xenostrobus* spawns when salinity reaches a critical lower limit for larval development (61). Spring spawners in temperate climates ensure some synchrony between larval development and maximum production of the phytoplankton. When food is abundant for long periods, as in some estuaries, spawning might be timed to prevent the flushing of the larvae out to sea, e.g. in the summer months, at flood tide, when fresh water run-off is at a minimum.

Similar attempts to correlate breeding cycles with environmental factors are legion, because of the obvious advantages to a species of synchronisation between the sexes in the release of gametes, and of the adaptive value of matching embryogenesis and larval growth to optimal environmental conditions. Three states in the breeding cycle are amenable to environmental synchronisation (or entrainment), viz. the onset of proliferation of the gametes from the germinal epithelium, maturation of the gametes, and the initiation of spawning. Clarke (19) discussed some consequences of synchronisation operating at the initiation of gametogenesis compared with the synchronisation of spawning. In the first case, variable rates of oocyte growth (and vitellogenesis), resulting from short-term variation in food supply, might result in a prolonged spawning period. If maturation and spawning are synchronised, ova of variable age but of similar maturity, can be released within a brief spawning period. There is no inherent reason why synchronisation cannot occur both at the initiation of gamete growth and at spawning, although most 'ecological' studies have concentrated on looking for the latter. Any serious search for entrainment earlier in the breeding cycle must include the very first stages of gametogonal proliferation, requiring the use of cytological and endocrinological techniques (20). There is a need for research of this kind on estuarine invertebrates. Equally, the growth of the oocytes and vitellogenesis may be controlled so as to reduce the variability in time of the production of mature gametes. Some of the ways in which this may be achieved are discussed below.

ENERGY (=NUTRIENT) RESERVES

The energy derived from food and destined for use in gametogenesis may be stored and used at a later date, resulting in a 'biochemical cycle' that is related to the gametogenic cycle. Such cycles of storage and utilisation of reserves have long been recognised and documented (30, 31). For example, in a series of papers Ansell (1, 3, 4, 5) [see also Ansell and Trevallion, (7); Trevallion, (54)] has recorded the seasonal gross biochemical changes in bivalves from the Clyde Sea area. Walne (57) reviewed some of the literature on biochemical content of oysters. Williams (59), De Zwaan and Zandee (25), Gabbot and Bayne (28) and Dare and Edwards (23) have recently described the biochemical cycle in *Mytilus edulis*.

Certain common features emerge from these studies. Nutrient reserves are accumulated when food is abundant; these nutrients may subsequenlty be utilised, in part, to meet maintenance energy demands when food is scarce, but they are also used in the production of gametes. Vitellogenesis occurs at the expense of the energy store, and in some, possibly all, species the growth of the oocytes and sperm does not begin until some nutrient reserve has been accumulated (28, 49, 24). Also, the number of ova produced is dependent on the size of the nutrient store (42, 11). The rate of gametogenesis is therefore placed at one remove from a direct dependence on an uncertain food supply, providing the possibility for some control over the variability of oocyte growth.

The timing of the co-ordination between the cycles of nutrient storage and gamete production varies between species, however. In *Pecten maximus* (21), *Macoma balthica* (24) and in *Mytilus edulis* a nutrient store is accumulated in the summer and utilised for vitellogenesis in the autumn and winter. In *Tellina tenuis* (54), *Abra alba* (3), *Chlamys septemradiata* (4) and in *Cardium edule* (Newell, unpublished data) the two cycles are more nearly coincident in time. The detailed relationship between these cycles can be complex, involving differences between tissues and in the fate of the reserve. Examples from three bivalves are given in illustration.

In *Tellina tenuis* (54) carbohydrate reserves are accumulated in the late summer (August) and utilised for maintenance metabolism in the late autumn and winter (November to January). During the winter gametogenesis is dormant. In the spring (April and May) new reserves of carbohydrate are accumulated and at the same time there is rapid growth and gonad proliferation. Both somatic and gamete growth continue in the summer (June to August); some spawning occurs throughout the summer, but is synchronised within the population. Trevallion suggests that temperature may have a direct effect on growth and reproduction, and also an indirect effect *via* the food supply.

In *Macoma balthica* from the Wadden Sea (24), gametogenesis is initiated in the late summer (August) and continues through the autumn and winter until spawning over a short period in April and May. Growth occurs from March until June after which high temperatures inhibit feeding, growth ceases and the animals become emaciated through utilisation of reserves for maintenance. During much of this period of emaciation (June to September) gametogenesis is dormant. In October and November falling temperatures allow renewed feeding to take place; reserves are accumulated and these are used in gametogenesis during the winter. Thus, in both *Tellina* and *Macoma* there are periods when food is abundant and growth and gametogenesis occur simultaneously, and other periods when reserves are utilised only in maintenance. The species differ in that *Macoma* accumulates reserves in the autumn which are utilised for gametogenesis in the winter. However, just as *Tellina* shows differences in detail between different populations in one year and in a single population between years (7, 54), *Macoma* may also show differences between populations (15, 41, 32).

Mytilus edulis has a seasonal cycle which is similar to *Macoma*, but two tissues are involved in the storage and mobilisation of reserves (28, 53, 23). In

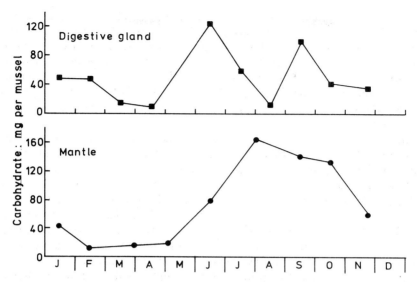

Figure 1. Seasonal changes in the carbohydrate content of mantle tissue (●) and digestive gland (■) of *Mytilus edulis*. From (25, 28, 53).

the summer after spawning (July, August) gametogenesis is dormant and glycogen accumulation takes place in the Leydig cells in the mantle (Fig. 1). In the weeks just preceding spawning there is a period of accumulation of glycogen in the digestive gland. After spawning, the glycogen content of the digestive gland declines as it is transferred to the mantle tissues. There follows another period of glycogen accumulation in the digestive gland in September, and the material is utilised for maintenance metabolism during the winter when food may be scarce. The glycogen content of the mantle tissue is also at a maximum in the summer, but this store is utilised in the autumn and winter for gametogenesis. Gametogenesis is initiated in the autumn (September) and proceeds through the winter until spawning in the spring and summer. Therefore, in this species also there is a period when glycogen is mobilised (from the digestive gland) to 'fuel' gametogenesis at a time when somatic growth is also taking place. In the winter, growth ceases and gametogenesis takes place at the expense of an energy reserve.

Another common feature of these cycles in the bivalves concerns the chemical nature of the components involved. Whereas storage, either in the mantle, or the digestive gland (or the adductor muscle in some species) mainly involves glycogen, the developing oocytes are rich in lipid; the implication is that glycogen is converted into the lipid of the ova. Gabbott (27) has interpreted these events as a "storage cycle" (38) distinguishable from a "metabolic cycle" in that the different steps (glycogen synthesis, glycogen degradation, lipid synthesis) are separated in time and, in this case, occur also in different tissues. Gabbott considers control of this storage cycle in terms of the scheme illustrated in Fig. 2. In *Mytilus*, a high level of blood glucose in the summer, after spawning, helps to trigger glycogen synthesis. In the autumn, neurosecretory activity provides a hormonal control which switches the cycle to the degradation of glycogen and the initiation of gametogenesis. Once initiated, gametogenesis proceeds through the winter until spawning, when control *via* the digestive gland is re-established.

The proposed function of hormonal regulators in this storage cycle (Fig. 2) is speculative. However, experimental work by Lubet and his colleagues in recent years suggests a neurosecretory basis for the integrated control of glycogen synthesis and degradation, and the gametogenic cycle. Houtteville and Lubet (36) maintained the mantle tissue of male *Mytilus edulis* in culture, both in isolation and in association with cerebral and visceral ganglia. They concluded, in accordance with earlier work (43), that secretions from the visceral ganglia control the accumulation of glycogen reserves in the mantle tissue, whereas secretions from the cerebral ganglia encourage the liberation of these reserves and the development of the gametes. When coupled with a control *via* blood glucose this hormonal regulation suggests a scheme similar to the control of glycogen metabolism in vertebrate systems (27, 47). In other bivalves, (e.g. *Aequipecten irradians* (49)) control of gametogenesis by a single neurosecretory cycle has been proposed. More research within the context of a storage cycle as proposed by Gabbott (27) is clearly warranted.

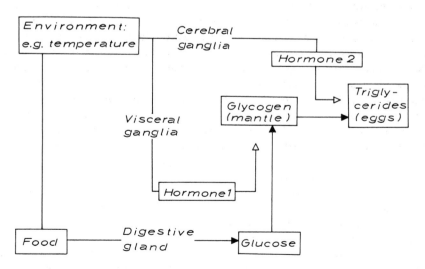

Figure 2. A scheme for the regulation of glycogen metabolism and gametogenesis in bivalves. Modified from (27).

OTHER SEASONAL PHYSIOLOGICAL CHANGES

In view of the considerable metabolic changes that characterise the seasonal cycles of nutrient storage and breeding, it is not surprising to find that other, related physiological processes are also seasonally variable. Experiments with *Mytilus edulis* (58, 13) have shown that the rates of acclimation of oxygen consumption and feeding are sufficiently high to keep these processes virtually independent of temperature during the year, since the seasonal rate of change of temperature is less than the animal's capacity to acclimate. However, even when the effects of temperature and of animal size are excluded from determinations of oxygen consumption rates, a seasonal pattern to the values remains. Bruce (14) reduced his data on oxygen consumption by *M. edulis* to standard values at 10 C (Fig. 3a) and still recorded high values in winter and spring and low values in the summer. He concluded that an increasing proportion of gonad material in the body through late summer and autumn increased the demand for oxygen. Kruger (39) measured oxygen consumption by the same species at constant temperature (15 C) and recorded a similar seasonal pattern (Fig. 3b). Bayne (10) also reported a seasonal pattern of high respiration rates in the winter and lower rates in the summer (Fig. 3c), and interpreted this pattern as reflecting the high energy demands of gametogenesis in the winter.

In other bivalves, such as *Donax vittatus* (6), in which gametogenesis occurs in the spring and summer but is dormant in the winter, oxygen consumption is depressed in the winter and is high in the summer. In addition, *Donax* does not acclimate to temperature change to any marked degree (2). This, coupled with the differences between the seasonal storage cycles of *Donax* and *Mytilus*, has important ecological consequences. In *Mytilus*, although the rate of gametogenesis increases with rise in temperature, total aerobic metabolism is regulated by the acclimation response. The demands on the energy store for processes of activity and growth are controlled (13), conserving the store to meet the requirements of gametogenesis. During starvation there is a decline in metabolic rate, resulting in further conservation of the nutrient reserve. This would seem to be adaptive under conditions where, for reasons of insurance against food limitation in the larval stages, spawning is synchronised to occur early in the year, following a winter in which temperatures are low and food for the adult might be scarce. However, the price paid for this degree of 'buffering' between gamete production and environmental changes may be a reduced capacity to benefit from unusually favourable conditions. If phytoplankton production is unusually high in the spring, *Mytilus* may be unable to respond by an increase in fecundity.

In *Donax vittatus*, metabolic rate increases with rise in temperature. Although, by virtue of close coupling between feeding and gametogenesis, fecundity can probably be increased when food is abundant, lack of temperature acclimation makes the individual vulnerable to prolonged stress, especially if the food reserve is limited. Experiments by Sastry (48, 49) on *Aequipecten*

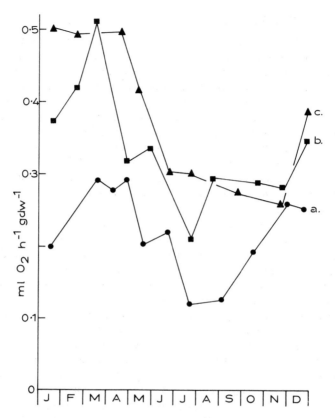

Figure 3. Seasonal changes in the rates of oxygen consumption by *Mytilus edulis*, reduced to constant temperature, and for animals of standard size: a) from Bruce (14), b) from Kruger (39), c) from Bayne (10). 'gdw' is grams dry flesh weight.

irradians indicate that as long as a certain threshold of nutrient reserve is present in the animal, gametogenesis, once triggered, will go to completion in spite of adverse food conditions. Nevertheless, if high temperatures persist at a time when food reserves are low, and the animal is unable to acclimate, mortality is likely. Ansell and Sivadas (6) suggest that such mortalities do occasionally occur in populations of *Donax*.

Recent unpublished studies by Newell on the euryhaline bivalve *Cardium* (= *Cerastoderma*) *edule* demonstrate a pattern of events intermediate between *Mytilus* and *Donax*. In *Cardium* (Fig. 4) carbohydrate reserves are accumulated in the spring and summer coincidentally with the main period of gametogenesis. After spawning in the summer, there is a gradual decline in the carbohydrate level to minimal values in the winter. This cycle resembles the storage cycle of *Donax*. However, unlike *Donax*, *Cardium* can acclimate its rate of oxygen

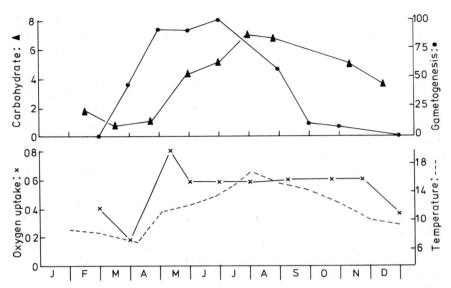

Figure 4. The seasonal cycles of the rate of oxygen consumption (X; ml O_2 h^{-1} per $W^{0.44}$), water temperature (dashed line; degrees C), total carbohydrate (▲; percent of flesh weight) and stage of gametogenesis (●; percent of the population with mature gametes in the germinal follicles) for *Cardium* (= *Cerastoderma*) *edule* in a Plymouth population. Unpublished data due to Roger Newell.

consumption in response to change in temperature, in a manner apparently similar to *Mytilus*. During the summer, in spite of rising water temperatures, metabolic rate is at a constant value; in the winter, during the sexually dormant period, metabolic rate is reduced. The outcome is a cycle of physiological changes similar to *Donax* but with the important difference that, by virtue of a "temperature-independent" metabolism in summer, there is some conservation of its carbohydrate reserve for gametogenesis. Intuitively, these three patterns of physiological change (*Donax*, *Cardium*, *Mytilus*) seem to represent an increasing degree of regulation of gametogenesis and control over the balance of resources made available to gamete production. All species are iteroparous, although they may reflect different degrees of "opportunism" in their reproduction. Calow (16) has discussed the advantages that accrue from " . . . increased feedback intensity on reproduction *via* the physiological pathways" that regulate the allocation of energy resources to fecundity. Future studies should examine the reproductive patterns in bivalves with these considerations in mind.

RELATIONSHIPS BETWEEN ADULT CONDITION AND LARVAL VIGOUR

As discussed earlier, the storage cycle in bivalves involves the accumulation of carbohydrate, mainly glycogen, in the adult, and its conversion to lipid for storage in the ova. The advantage to the adult of storing glycogen is that it represents an easily accessible store of energy (*via* glucose) in a relatively inert form. The limitations on the storage of glycogen are those of space, although bivalves may accumulate up to 40% of their dry weight. The advantages to the developing larva of storing lipids, however, are two: they have a higher energy content per unit weight than do carbohydrates, and they confer buoyancy, so reducing the energy required to maintain vertical position in the water column (22). In both lecithotrophic and planktotrophic larval development, embryogenesis is dependent on these reserves of lipid. In the former case the reserves must suffice until metamorphosis; in the latter the reserves must meet the energy demands of the developing embryo until the larval feeding organs are functional. Experiments with bivalves have shown that both the amounts and the character of these reserves are vulnerable to stress in the adult, and these studies also illustrate some of the functions of the reserves in the developing larva.

It is useful to view planktotrophic larval development as comprising three stages. Stage I covers the period from fertilisation to the first functioning of the larval feeding organs. This is a time of intense morphogenetic activity, during which there is a complete dependence on the stored energy reserves in the egg. Stage II covers a period of rapid growth, when there is less reliance on stored reserves since the larva is now able to feed. In bivalves, considerable lipid reserves are accumulated during this stage (35). At the end of stage II the larva is, for the first time, capable of metamorphosis. Stage III is the "delay" stage (50), during which the larva will metamorphose if a suitable substrate is encountered. There is some growth at this time, and some development of new tissue, but this is primarily the stage of settlement from the plankton, and of "swimming-crawling" behaviour (17) which culminates in metamorphosis.

Helm, Holland and Stephenson (33) showed that the rate of growth of the larvae of the European oyster, *Ostrea edulis*, was less when the larvae were released from adults kept at a low ration in the laboratory, than in larvae from adults kept at a higher ration. The viability (or vigour) of the larvae also declined as the length of time which the adults were held in the laboratory increased. These losses of larval vigour were correlated with loss of "physiological condition" of the adults. Also, in larval cultures derived from stressed adults there was a decrease in the yield of post-larvae, relative to the control cultures. Bayne (10) found that abnormal embryonic development occurred in eggs which had been spawned by adult *Mytilus edulis* subjected to high temperatures and low ration. He predicted that these effects of adult stress would be most evident during periods of active morphogenesis, i.e., in stage I of larval development and during metamorphosis. Bayne, Gabbott and Widdows (12) recorded a decline in the rate of growth of larvae from stressed adult mussels.

When these effects of adult condition are evident at stage I of larval development, it is the size of the embryo, rather than the duration of the stage, which is altered. This results in a smaller larva entering stage II, and such a larva is probably disadvantaged through being limited to a narrower variety of food species (52) and a reduced capacity to survive starvation. When the effects of adult condition impinge on stage II, it is the rate of growth which might be affected; and since the size at which the larvae enter stage III is remarkably constant [at least for *Mytilus edulis* (9)], the duration of the stage would be altered. Any increase in the duration of larval life increases the probability of predation, and of "excessive dispersal" (22). The study by Helm et al. (33) also suggested that a reduced rate of growth in stage II results in a decreased capacity for successful metamorphosis.

Helm, Holland and Stephenson (33) correlated larval growth rates with the amount of neutral lipid in the newly-liberated larvae of *Ostrea*. Bayne et al. (12) found that eggs with a high phospho-lipid content developed into larger embryos than eggs low in phospho-lipid. Studies on bivalve larvae therefore hold some promise that relationships between adult condition and larval development can be measured in biochemical terms, possibly as a function of lipid metabolism. The importance of lipids in the economy of these planktotrophic larvae is illustrated by experimental studies. Holland and Spencer (35) found that during the growth of *Ostrea edulis* larvae, the level of neutral lipid rose from 8.8 to 23.2% by weight of total organic matter, and then fell dramatically during metamorphosis to about 9% (Fig. 5). During short periods of starvation both protein and lipid level declined, with little change in carbohydrate; but the caloric contribution was greater from lipid (0.9×10^{-3} cals day^{-1} per larva) than from protein (0.5×10^{-3} cals day^{-1} per larva). Lipid is the main energy reserve of the larva and is

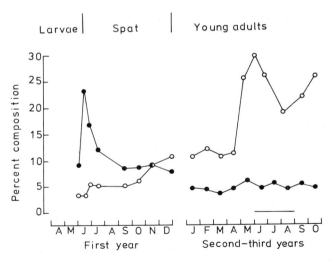

Figure 5. Changes in the percentage of lipid and carbohydrate during development of *Ostrea edulis* (redrawn from Holland and Hannant, (34); after Gabbott, (27). •, neutral lipid; ○, glycogen. The horizontal bar indicates the spawning season.

utilised both during starvation and during metamorphosis when, due to re-orientation of the organ systems in the mantle cavity, the larva is unable to feed. Bayne et al. (12), working with *Mytilus* larvae, found that the lipid level fell during the seven days after fertilisation, and that phospho-lipids made a significant contribution to energy metabolism during larval stage I.

The importance of lipids to these larvae, and the apparent vulnerability of the glycogen-to-lipid conversion in the adult, are apparent from these studies. In addition to a qualitative effect of adult stress on embryogenesis, there is also a quantitative effect, seen as a reduction in fecundity (11, 12). It is instructive to compare the magnitude of these effects on germinal tissue with the effects on somatic tissue during stress. Table 2 compares the components of an energy budget for two individual mussels, one of which is fed a ration greater than its maintenance requirement, the other of which is starved. In spite of a very considerable loss of body reserves during starvation, the amount of energy devoted to vitellogenesis is only halved. Starvation may also result in a less vigorous larva, but the buffering of the germinal processes from the full effects of the stress provides a further example of the "protection" of gametogenesis by physiological adaptation. The studies reported by Gabbott and Bayne (28) and Sastry (49) indicate that this protection breaks down only when the conditions suitable for spawning are absent at a time that mature gametes are present in the germinal tissues; gamete resorption may then occur.

Table 2. The energy budgets of two *Mytilus edulis*, one fed above maintenance ration, the other starved, for 30 days. The symbols are according to IBP terminology, where G = C-(F+U+R+L). Based on Bayne (11) and unpublished data.

Components of the energy budget		Fed individual	Starved individual
1. Calories consumed	(C)	3000	0
2. Calories lost as faeces	(F)	450	0
3. Calories lost as urine	(U)	175	290
4. Calories lost as heat	(R)	1725	730
5. Calories produced as gametes	(L)	150	80
6. Somatic production (calories)	(G)	+ 500	-1100

CONCLUSIONS

The species of bivalve molluscs that have been considered in this discussion allocate a large proportion of their energy resources to reproduction. This allocation takes the form of the storage and utilisation of glycogen, coupled with an annual gametogenic cycle that is probably synchronised by a variety of environmental factors. The detailed sequencing of these events varies between species, but all involve the accumulation of an energy reserve when food is abundant, and the utilisation of this reserve for maintenance and for gametogenesis. The regulation of the allocation of energy resources between maintenance, growth and gametogenesis is complex. Recent studies suggest, however, that a mutual control of the storage cycle between glycogen synthesis in the adult and lipid synthesis in the developing oocyte may exist, based on neurosecretory hormones and the level of blood glucose. Various physiological factors operate to buffer the gametogenic events from environmental disturbance; these include the ability to acclimate the metabolic rate in response to changes in temperature, and a switch mechanism that allows gametogenesis to proceed to completion once a certain level of nutrient reserve has been accumulated, regardless of subsequent short-term deterioration in the environment. This buffering is not wholly efficient, however, and stress on the adult affects both fecundity and the vigour of the larvae that subsequently develop. Lipid reserves in the egg are necessary for successful embryogenesis to the stage when the larva is able to feed, and also seem to play a part (not presently understood) in the growth of the larvae. The feeding larva accumulates further lipid reserves which are then utilised at metamorphosis.

ACKNOWLEDGMENTS

I am grateful to Roger Newell for permission to use his unpublished data on *Cardium*, and to Peter Gabbott for many discussions on the subject of this paper. This work forms part of the experimental ecology research programme of the Institute for Marine Environmental Research; it was supported by the Natural Environment Research Council and commissioned in part by the Department of the Environment.

REFERENCES

1. Ansell, A.D. 1972. Distribution, growth and seasonal changes in biochemical composition for the bivalve *Donax vittatus* (da Costa) from Kames Bay, Millport. J. exp. mar. Biol. Ecol. 10:137-150.
2. _____. 1973. Oxygen consumption by the bivalve *Donax vittatus* (da Costa). Ibid 11:311-328.
3. _____. 1974. Seasonal changes in biochemical composition of the bivalve *Abra alba* from the Clyde Sea area. Mar. Biol. 25:13-20.
4. _____. 1974. Seasonal changes in biochemical composition of the bivalve *Chlamys septemradiata* from the Clyde Sea area. Ibid 25:85-99.
5. _____. 1974. Seasonal changes in biochemical composition of the bivalve *Nucula sulcata* from the Clyde Sea area. Ibid 25:101-108.
6. _____, and P. Sivadas. 1973. Some effects of temperature and starvation on the bivalve *Donax vittatus* (da Costa) in experimental laboratory populations. J. exp. mar. Biol. Ecol. 13:229-262.
7. _____, and A. Trevallion. 1967. Studies of *Tellina tenuis* (da Costa). I. Seasonal growth and biochemical cycle. Ibid 1:220-235.
8. _____, F.A. Loosmore and K.F. Lander. 1964. Studies on the hard-shell clam, *Venus mercenaria*, in British waters. II. Seasonal cycle in condition and biochemical composition. J. appl. Ecol. 1:83-95.
9. Bayne, B.L. 1965. Growth and the delay of metamorphosis of the larvae of *Mytilus edulis* (L). Ophelia 2:1-47.

10. _____. 1972. Some effects of stress in the adult on the larval development of *Mytilus edulis*. Nature, Lond. 237:459.

11. _____. 1975. Reproduction in bivalve molluscs under environmental stress, p. 259-277. *In* F.J. Vernberg (ed.), Physiological Ecology of Estuarine Organisms. University of South Carolina Press, Columbia.

12. _____, P.A. Gabbott and J. Widdows. 1975. Some effects of stress in the adult on the eggs and larvae of *Mytilus edulis* (L). J. mar. biol. Ass. U.K. 55:675-690.

13. _____, R.J. Thompson and J. Widdows. 1973. Some effects of temperature and food on the rate of oxygen consumption by *Mytilus edulis* (L)., p. 181-193. *In* W. Wieser (ed.), Effects of Temperature on Ectothermic Organisms. Springer-Verlag, Berlin.

14. Bruce, J.R. 1926. The respiratory exchange of the mussel (*Mytilus edulis* L.). Biochem. J. 20:829-846.

15. Caddy, J.F. 1967. Maturation of gametes and spawning in *Macoma balthica* (L). Can J. Zool. 45:955-965.

16. Calow, P. 1973. The relationship between fecundity, phenology and longevity: A systems approach. Amer. Natur. 107:559-574.

17. Carriker, M.R. 1961. Interrelation of functional morphology, behaviour and autecology in early stages of the bivalve *Mercenaria mercenaria*. J. Elisha Mitch. Sci. Soc. 77:168-241.

18. _____. 1967. Ecology of estuarine benthic invertebrates: A perspective, p. 442-487. *In* G.H. Lauff (ed.), Estuaries. American Association for the Advancement of Science, Washington.

19. Clarke, R.B. 1965. Endocrinology and the reproductive biology of polychaetes. Oceanogr. mar. biol. Ann. Rev. 3:211-255.

20. _____, and P.J.W. Olive. 1973. Recent advances in polychaete endocrinology and reproductive biology. Ibid 11:175-222.

21. Comely, C.A. 1974. Seasonal variations in the flesh weights and biochemical content of the scallop *Pecten maximus* (L.) in the Clyde Sea area. J. Cons. int. Explor. Mer. 35:281-295.

22. Crisp, D.J. 1975. The role of the pelagic larva. (In press.) *In* P. Spencer Davies (ed.), Perspectives in Exerpimental Biology. Pergamon Press, Oxford.

23. Dare, P.J., and D.B. Edwards. 1975. Seasonal changes in flesh weight and biochemical composition of mussels (*Mytilus edulis* L.) in the Conway estuary, North Wales. J. exp. mar. Biol. Ecol. 18:89-97.

24. De Wilde, P.A.W. 1975. Influence of temperature on behaviour, energy metabolism and growth of *Macoma balthica* (L.), p. 239-256. *In* H. Barnes (ed.), Proc. 9th Europ. mar. biol. Symp. Aberdeen University Press, Aberdeen.

25. De Zwaan, A. and D.I. Zandee. 1972. Body distribution and seasonal changes in the glycogen content of the common sea mussel, *Mytilus edulis*. Comp. Biochem. Physiol. 43A:53-58.

26. Fuji, A., and M. Hashizume. 1974. Energy budget for a Japanese common scallop *Patinopecten yessoensis* (Jay) in Mutsu Bay. Bull. Fac. Fish. Hokkaido Univ. 25:7-19.

27. Gabbott, P.A. 1975. Storage cycles in marine bivalve molluscs: A hypothesis concerning the relationship between glycogen metabolism and gametogenesis, p. 191-211. *In* H. Barnes (ed.), Proc. 9th Europ. mar. biol. Symp. Aberdeen University Press, Aberdeen.

28. _____, and B.L. Bayne. 1973. Biochemical effects of temperature and nutritive stress on *Mytilus edulis* (L). J. mar. biol. Ass. U.K. 53:269-286.
29. Gadgil, M., and W.H. Bossert. 1970. Life historical consequences of natural selection. Amer. Natur. 104:1-24.
30. Giese, A.C. 1959. Comparative physiology: Annual reproductive cycles of marine invertebrates. Ann. rev. Physiol. 21:547-576.
31. _____. 1967. Some methods for study of biochemical constitution of marine invertebrates. Oceanogr. mar. Biol. Ann. Rev. 5:253-288.
32. Gilbert, M.A. 1973. Growth rate, longevity and maximum size of *Macoma balthica* (L.). Biol. Bull. 145:119-126.
33. Helm, M.M., D.L. Holland and R.R. Stephenson. 1973. The effect of supplementary algal feeding of a hatchery breeding stock of *Ostrea edulis* (L.) on larval vigour. J. mar. biol. Ass. U.K. 53:673-684.
34. Holland, D.L., and P.J. Hannant. 1974. Biochemical changes during growth of the spat of the oyster, *Ostrea edulis* (L.). Ibid 54:1007-1016.
35. _____, and B.E. Spencer. 1973. Biochemical changes in fed and starved oysters, *Ostrea edulis* (L.) during larval development, metamorphosis and early spat growth. Ibid 53:287-298.
36. Houtteville, P., and P. Lubet. 1974. Analyse experimentale, en culture organotypique, de l'action des ganglions cerebropteureux et viscereux sur le manteau de la moule male, *Mytilus edulis* (L). (Mollusque, Pelecypode). Compt. rend. hebd. seances de l'Acad. sciences. Ser. D. 278:2469-2472.
37. Hughes, R.N. 1970. An energy budget for a tidal flat population of the bivalve *Scrobicularia plana* (da Costa). J. Anim. Ecol. 39:333-356.
38. Krebs, H.A. 1970. Some aspects of the regulation of fuel supply in omnivorous animals, p. 397-420. *In* G. Weber (ed.), Advances in Enzyme Regulation. Pergamon Press, Oxford.
39. Kruger, F. 1960. Zur Frage der Grossenabhangigkeit des Sauerstoffverbranchs von *Mytilus edulis* (L). Helgol. wissen. Meeres. 7:125-148.
40. Kuenzler, E.J. 1961. Structure and energy flow of a mussel population in a Georgia salt marsh. Limnol. Oceanogr. 6:191-204.
41. Lammens, J.J. 1967. Growth and reproduction in a tidal flat population of *Macoma balthica* (L). Neth. J. sea. res. 3:315-382.
42. Loosanoff, U.L. 1965. Gonad development and discharge of spawn in oysters of Long Island Sound. Biol. Bull. 129:546-561.
43. Lubet, P. 1966. Essai d'analyse experimentale des perturbations produites par les ablations de ganglions nerveux chez *Mytilus edulis* (L.) et *Mytilus galloprovincialis* Lmk. (Mollusques: Lamellibranches). Ann. d'Endocrin. 27:353-365.
44. Moore, H.B. 1972. Aspects of stress in the tropical environment. Adv. mar. Biol. 10:217-269.
45. _____, and N.N. Lopez. 1969. The ecology of *Chione cancellata*. Bull. mar. sci. 19:131-148.
46. _____, and _____. 1970. A contribution to the ecology of the lamellibranch *Dosinia elegans*. Ibid 20:980-986.
47. Newsholme, E.A., and C. Start. 1973. Regulation in Metabolism. John Wiley, London, 349 p.
48. Sastry, A.N. 1963. Reproduction of the bay scallop, *Aequipecten irradians* Lamarck. Influence of temperature on maturation and spawning. Biol. Bull. 125:146-153.

49. _____. 1975. Physiology and ecology of reproduction in marine inverte-brates, p. 279-299. *In* F. J. Vernberg (ed.), Physiological Ecology of Estu-arine Organisms. University of South Carolina Press, Columbia.

50. Scheltema, R. 1967. The relationship of temperature to the larval develop-ment of *Nassarius obsoletus* (Gastropoda). Biol. Bull. 132:253-265.

51. Seed, R. 1976. Reproduction in *Mytilus* (Mollusca: Bivalvia) in European waters. In press, Proc. 8th Europ. mar. biol. Symp.

52. Strathmann, R. 1975. Larval feeding in echinoderms. Amer. Zool. 15:717-730.

53. Thompson, R.J., N.A. Ratcliffe and B.L. Bayne. 1974. Effects of starvation on structure and function in the digestive gland of the mussel (*Mytilus edulis* L.). J. mar. biol. Ass. U.K. 54:699-712.

54. Trevallion, A. 1971. Studies on *Tellina tenuis* (da Costa). III. Aspects of general biology and energy flow. J. exp. mar. Biol. Ecol. 7:95-122.

55. Vance, R.R. 1973. On reproductive strategies in marine benthic inverte-brates. Amer. Natur. 107:339-352.

56. _____. 1973. More on reproductive strategies in marine benthic inverte-brates. Amer. Natur. 107:353-361.

57. Walne, P.R. 1970. The seasonal variation of meat and glycogen content of seven populations of oysters *Ostrea edulis* (L.) and a review of the litera-ture. Fish. Invest., Min. Ag. Fish and Food, Ser. 2. 26:1-35.

58. Widdows, J., and B.L. Bayne. 1971. Temperature acclimation of *Mytilus edulis* (L.) with reference to its energy budget. J. mar. biol. Ass. U.K. 51:827-843.

59. Williams, C.S. 1969. The effect of *Mytilicola intestinalis* on the biochemical composition of mussels. Ibid 49:161-173.

60. Wilson, B.R. 1968. Survival and reproduction of the mussel *Xenostrobus securis* (Lamarck) (Mollusca, Bivalvia, Mytilidae) in the Swan Estuary, Western Australia. Part I: Salinity tolerance. J. nat. Hist. 2:307-328.

61. _____. 1969. Survival and reproduction of the mussel *Xenostrobus securis* (Lamarck) (Mollusca, Bivalvia, Mytilidae) in a Western Australian estuary. Pt. II: Reproduction, growth and longevity. Ibid 3:93-120.

62. _____, and E.P. Hodgkin. 1967. A comparative account of the reproductive cycles of five species of marine mussels (Bivalvia: Mytilidae) in the vicinity of Freemantle, Western Australia. Austr. J. mar. fresh. Res. 18:175-203.

ENDOCRINE MECHANISMS IN ESTUARINE ANIMALS:

INVERTEBRATES AND FISHES

Milton Fingerman
Department of Biology
Tulane University
New Orleans, Louisiana 70118

ABSTRACT: Estuarine crustaceans and fishes have highly evolved endocrine systems. Nevertheless, considerable endocrine involvement has been demonstrated in other estuarine invertebrates, particularly annelids and mollusks. Estuarine organisms face a wide variety of environmental problems, such as changes of salinity and temperature, that endanger them. Consequently, their endocrine systems have an important role in integrating the functions necessary to assure survival. Particularly important for survival of animals in an estuary, which is subject to wide salinity fluctuations, are the hormones that provide for salt and water balance, enabling the animals to regulate their "internal environment" within fairly narrow limits. Color changes also have adaptive significance, for example, by providing protective coloration. They are regulated by hormones in all crustaceans, and with few possible exceptions in all fishes. Furthermore, although hormonal regulation of growth and reproduction has been studied extensively in fishes and crustaceans, endocrine control of these two functions in other invertebrates also occurs. Other processes such as heart rate and carbohydrate metabolism are also hormonally controlled in at least some estuarine animals.

INTRODUCTION

Few environments expose their inhabitants to as many ecological variables, and of such large magnitude, as does an estuary. Estuarine organisms are confronted, for example, by wide ranging salinity changes due to the tidal ebb and flow and the inflow of fresh water, this fresh water inflow tending to be restricted as the tide rises. The physiological responses of an animal living in an estuary to physical ecological stresses are adaptive, and the endocrine system has a large

449

role in coordinating these responses to assure the animal's survival. Much significant information relating to endocrine mechanisms in strictly fresh water and marine invertebrates and fishes must, because of space limitations, be omitted from the following discussion.

SALT AND WATER BALANCE

Hormones that function in salt and water balance are particularly important for survival of estuarine organisms. Estuaries are characterized by fluctuating salinities. Yet the "internal environment" maintains a fairly constant salt concentration.

Platyhelminths

Platyhelminths are an example of animals with an endocrine system that functions without a circulatory system. Their endocrine system consists of neurosecretory cells distributed in the brain and longitudinal nerve cords, and appears to produce an antidiuretic hormone (136).

Nemertines

Cerebral neurosecretory cells appear to be involved in salt and water balance in *Lineus ruber* and *Lineus viridis* (83). When an intact nemertine is exposed to diluted sea water it at first gains weight and its cerebral neurosecretory cells become hyperactive. It then returns to its original weight as these neurosecretory cells discharge their product. The cerebral organ, a structure composed of glandular and nervous elements, also seems to have a role in osmoregulation. Both the brain and cerebral organ must be present or the worm is unable to osmoregulate as well as does an intact one. However, whether the cerebral organ actually secretes a hormone or has some other role in the osmoregulatory process is not known.

Annelids

There is indirect evidence for neuroendocrine control of salt and water balance in the estuarine annelid, *Neanthes virens* (71). When this annelid was taken from 100% sea water and placed in 25% sea water the number of aldehyde-fuchsin-positive cells in the brain increased from 30 to 423.

Mollusks

An antidiuretic hormone appears present in the pleural ganglia of the opisthobranch, *Aplysia rosea* (138).

Crustaceans

A substance in the sinus glands of the shore crab, *Carcinus maenas*, reduced water uptake at ecdysis (27). It appeared to be neither the molt-inhibiting nor gonad-inhibiting hormone. A substance, presumably of neurosecretory origin, in the cephalothorax of the prawn, *Pandalus jordani*, decreased the outflux of sodium ions from the crab, *Hemigrapsus nudus*, but eyestalk extracts were inactive (118). The thoracic ganglion of the crab, *Metopograpsus messor*, contains a substance that increases the permeability of the body surfaces to water, whereas the eyestalk has a substance that either inhibits the release of the thoracic ganglion factor, is a direct antagonist of it, or acts directly on the body surfaces decreasing their permeability to water (72). Removal of both eyestalks from the crab, *Eriocheir sinensis*, results in an increased urine output (38), probably the result of loss of this same eyestalk factor.

Two factors in the nervous system of the crab, *Thalamita crenata*, affect water balance (135). One, which is acetone-insoluble, decreased ^3H_2O influx whereas the second, which is acetone-soluble, increased the influx rate. Neither affected sodium fluxes. Eyestalks of the fiddler crab, *Uca pugilator*, have a substance that promotes the uptake of sodium ions when it is in a hyposmotic medium (60).

Fishes

Arginine vasotocin appears to be the active neurohypophyseal principle involved in hydromineral regulation in teleosts. It increases sodium uptake by the teleost gill in fresh water but increases sodium efflux by this route when the teleost is in the sea (97). There is no evidence that neurohypophyseal factors affect the salt or water balance of elasmobranchs.

Certain euryhaline teleosts, mainly cyprinodonts, are unable to survive more than a few days in fresh water after hypophysectomy (22) because of the need for prolactin (115). Prolactin reduces the extrarenal loss of sodium by decreasing the activity of the sodium transport enzyme, Na^+-K^+-ATPase, in the gills (the main site of sodium excretion) and increasing the activity of this enzyme in the kidney, thus enhancing sodium resorption, but has no effect on this enzyme in the intestine (114). Prolactin also stimulates sodium resorption from the urinary bladder of the flounder (69). Release of prolactin appears to be controlled by an inhibitory hormone from the hypothalamus (101, 111), direct stimulation of the prolactin cells by decreasing blood osmotic pressure, and negative feedback by either prolactin or increased blood sodium (98, 112). Prolactin may be essential for resistance of fishes to thermal stress, possibly counteracting the osmoregulatory imbalance induced by high temperatures (67). Hypophysectomy results in hypocalcemia in *Fundulus heteroclitus* in calcium-deficient sea water whereas pituitary gland extracts injected into hypophysectomized *Fundulus* produced hypercalcemia due to either prolactin, ACTH, or both (106).

As in higher vertebrates, the adrenocorticosteroids of teleost fishes have an important role in salt and water balance. Cortisol seems to be the most important mineralocorticoid in fishes (68). It increases the rate of water uptake by the intestine (61), and raises the Na^+-K^+-ATPase activity in the gills and intestine

(44). In adrenalectomized eels in fresh water there is a reduction in urine production (30). Cortisol reduced the sodium concentration in the rectal gland fluid secreted by the lip shark, *Hemiscyllium plagiosum* (29). In teleosts, at least, adrenocorticotropic hormone release is stimulated by a hormone from the hypothalamus (101, 111).

Extracts of the corpuscles of Stannius, minute endocrine glands associated with the kidneys of teleosts, produce a strong hypocalcemic response (51). The name "hypocalcin" was recently suggested for this substance (105).

The ultimobranchial glands, important in calcium regulation in higher vertebrates, also occur in fishes but apparently have little or no effect on their calcium levels (104). However, fish ultimobranchial glands are large and contain much calcitonin, which suggests that calcitonin has an important function even if it may not be in calcium regulation. In fishes, calcitonin may regulate the total osmotic concentration of the blood, as in the eel where it causes a drop in serum osmolality (102).

Another structure that seems to be involved in salt and water balance in teleosts is the caudal neurosecretory system from which at least six different active factors have been extracted with three of them appearing to be involved in salt and water balance. One increases the sodium influx across the gills (93). A second increases the glomerular filtration rate (34) and causes bladder contraction (84). The bladder effect may help to eliminate the large amount of water teleosts excrete when in fresh water. A third substance appears to be arginine vasotocin (81). Bern (9) suggested that the teleost caudal neurosecretory system serves only as a fine control over salt and water balance with other endocrine centers exerting coarse control.

GROWTH
Coelenterates

Neurosecretory cells of hydras secrete a growth hormone (24, 123). A substance, named the neck-inducing factor, found in the Chesapeake Bay sea nettle, initiates strobilation (89). This factor not only appears essential for the continued development of the animal that produces it, but also diffuses into the water to cause strobilation in other members of the species as well. Whether this factor is the same as the hydra growth hormone has not been determined.

Annelids

Normal and regenerative growth in nereids requires a hormone from the anterior end of the worm, most likely produced in the brain, but possibly in the cells of the infracerebral gland, an enigmatic structure closely associated with the brain (54, 55, 140).

Mollusks

Growth ceases or is slowed considerably in the gastropod, *Crepidula fornicata*, after its cerebral ganglia are removed (90).

Crustaceans

Crustaceans produce a molt-inhibiting hormone (18). In higher crustaceans the medulla terminalis X-organ-sinus gland complex in the eyestalks is the main source of this hormone (109). Crustaceans that never have eyestalks also secrete this neurohormone (37, 96). It appears to exert its action mainly by inhibiting the release of a molting hormone or prohormone from a pair of ecdysial glands, called Y-organs, in the cephalothorax. The fact that eyestalk removal results in activation of the Y-organs is evidenced by an increased rate of RNA synthesis in these glands in eyestalkless individuals (126). The Y-organs of bigger crabs are large and can be readily identified. However, in other crustaceans they are apparently not so obvious. In fact, it seems that what have been called the Y-organs in macrurans by some recent investigators are actually the mandibular organs of Le Roux (87) which are apparently not molting glands. More recently, however, Le Roux (88) described the locations of both the Y-organs and mandibular glands in some macrurans, which should eliminate, future confusion of these organs. A gland at the base of each antenna of *Orconectes propinquus* has been reported to contain molting hormone but these glands are not the Y-organs (28). It remains to be seen whether these antennary glands occur in other crustaceans and how they might cooperate with the Y-organs in the control of molting.

An ecdysone appears to be the normal molting hormone of all arthropods (79). α-Ecdysone is the normal precursor of 20-hydroxyecdysone (=crustecdysone, β-ecdysone, ecdysterone), both of which have been found in crustaceans. The role of the Y-organs in ecdysone metabolism has still not been fully elucidated. Y-organs of late intermolt crabs, *Hemigrapsus nudus*, took up 650 times more labeled cholesterol per unit weight than did the rest of the crab (131). α-Ecdysone and 20-hydroxyecdysone were tentatively identified in the Y-organs, but the small amounts of the compounds in the Y-organs prohibited absolute identification.

Both α-ecdysone and 20-hydroxyecdysone induce precocious molting activity when administered to intermolt crustaceans. However, differences in the responses to these steroids have been observed. Every fiddler crab, *Uca pugilator*, that received 20-hydroxyecdysone completed proecdysis precociously and began ecdysis but all died during ecdysis (120). In contrast, α-ecdysone induced precocious molting in only up to 80% of the crabs but each of these crabs that began ecdysis precociously survived ecdysis. Similar differences have been observed with fourth stage larval lobsters (121). 20-Hydroxyecdysone also increased the rate of molting of a barnacle (134).

If 20-hydroxyecdysone is administered to an isopod or amphipod during a specific stage of proecdysis, the stage varying with the species, ecdysis will be delayed (13, 94). To help explain this delay the existence of an "ecdysis factor" was postulated. This factor is supposed to bring on ecdysis, but only when 20-hydroxyecdysone is absent or in low concentration in the blood;

20-hydroxyecdysone apparently inducing the proecdysial events, but not ecdysis. 20-Hydroxyecdysone induced proecdysis in intermolt, Y-organless amphipods, but ecdysis did not occur, indicating that the Y-organs are the source of the ecdysis factor also (14). Interestingly, the Y-organs of the crab, *Varuna litterata*, have two cell types which raises the possibility that this organ is the source of two hormones (92). The need for such an ecdysis factor being secreted could explain why no fiddler crabs (120) and lobsters (121) successfully completed ecdysis after being induced to enter proecdysis precociously by administration of 20-hydroxyecdysone. However, because all of the crabs and some of the lobsters that entered proecdysis precociously after receiving α-ecdysone successfully underwent proecdysis, α-ecdysone may not only be the precursor of 20-hydroxyecdysone but also of the ecdysis factor.

An interesting aspect of molting in the crustaceans is the fact that removal of 6-8 walking legs from otherwise intact individuals results in precocious molting, providing the opportunity for regenerating appendages to become unfolded and functional. However, removal of even one limb from eyestalkless fiddler crabs, *Uca pugilator*, has an inhibitory effect on the rate of molting of the eyestalkless individuals (48). Also, removal of the Y-organs alone from intermolt crabs, *Pachygrapsus marmoratus*, causes the crabs, as expected, to remain permanently in intermolt, but when eight limbs were removed along with the Y-organs the crabs not only entered proecdysis, but completed it (31). However, all of the crabs died during ecdysis. Y-organs are, therefore, not always needed for the onset of proecdysis, but the fact that none of these Y-organless crabs was able to complete ecdysis successfully is another indication of not only the existence of an ecdysis factor but also that the Y-organs produce it. Presumably, in the absence of the organs some other, as yet unidentified, structure produced a molting hormone.

Fishes

Growth hormone is the adenhypophyseal factor that promotes overall growth even in the absence of all other pituitary hormones. Although growth hormone from pig, sheep, beef, monkey, and man will stimulate growth in hypophysectomized *Fundulus heteroclitus*, teleost pituitary extracts will not promote growth in rats (53, 116). However, elasmobranch growth hormone is effective in rats (59). Comparable experiments with agnathans do not seem to have been performed. Release of fish growth hormone appears to be regulated by a release-stimulating hormone from the hypothalamus (101, 111).

The thyroid gland may also have a role in stimulating growth of fishes, but there is no clear evidence that it does. Some investigators using teleosts obtained growth stimulation (*e.g.*, 117) whereas others found growth was retarded (*e.g.*, 36). Thyrotropic hormone release is regulated by a release-inhibiting hormone from the hypothalamus (101, 111).

REPRODUCTION

Coelenterates

In hydras as in many other animals somatic growth and reproduction are antagonistic processes. The growth hormone of hydras seems to inhibit gonadal development while stimulating growth (23). Gonadal development occurs only in the complete or near absence of the growth hormone. There is no evidence that such is the situation in estuarine coelenterates, but study of the possible endocrine control of reproduction in them would be worthwhile.

Nemertines

Brains of male and female *Lineus ruber* contain a hormone that inhibits the onset of sexual maturity (11). Furthermore, males have an androgenic hormone, most likely produced in the testes, that is responsible for differentiation of the accessory sexual structures and secondary sexual characteristics.

Annelids

The hormonal control of reproduction has been studied with nereids in more detail than with any other family of annelids. The brain-infracerebral gland complex produces a hormone that inhibits precocious gamete maturation (41). However, the nereid brain-infracerebral gland complex appears to also have a tropic action on small oocytes because if this complex is removed while the oocytes are still very small, they degenerate instead of developing precociously (58, 124). Removal of this complex also brings on precocious epitokal metamorphosis. A single hormone can account for these tropic and inhibitory actions on reproduction and sexual maturation as well as growth stimulation in nereids (55). A substance has been obtained from immature oocytes of *Perinereis cultrifera*, injection of which not only resulted in epitoky but also inhibition of caudal regeneration (26). This substance presumably acts by inhibiting release of the appropriate hormone or hormones from the brain complex. In contrast to nereids, the brain of the lugworm, *Arenicola marina*, contains a hormone that is required for gamete maturation (64).

Mollusks

Egg-laying in the gastropod, *Aplysia californica*, is stimulated by a hormone produced by the neurosecretory bag cells in the abdominal ganglion (80). This hormone is a 6000 dalton polypeptide that is formed when a 25,000 dalton precursor is enzymatically cleaved (5). Development of the male reproductive system and sperm production in protandric gastropods appears due to an androgenic hormone produced by the nervous system. The subsequent sex reversal to a female also appears to depend upon a neurohormone that causes dedifferentiation of the male reproductive system (85, 132).

The optic glands of the octopus and squid produce a gonadotropic hormone in both sexes (139). The ultrastructure of these glands suggests that they have a neural origin but they are not neurohemal organs nor do they contain classical neurosecretory cells (12).

Crustaceans

The sinus gland contains a gonad-inhibiting hormone (107) while central nervous elements in the body proper contains a gonad-stimulating hormone (3). Also, males have an epithelial androgenic gland attached to each sperm duct (32). This pair of glands is responsible for the differentiation of the testes and male accessory reproductive structures. The testes do not secrete a hormone. Adult females lack a counterpart to the androgenic gland. Instead their ovaries secrete a hormone that is necessary for differentiation of secondary sexual structures such as a brood pouch (86). The gonad-inhibiting and gonad-stimulating hormones appear to act directly on the ovaries but only indirectly on the testes via the androgenic glands. The androgenic gland hormone appears to be a lipoidal substance, about 200-250 daltons (10). There is an annual cycle in the level of gonad-inhibiting hormone in the eyestalks of female shrimp, *Crangon crangon*, with levels highest after the breeding season and lowest in the early part of the breeding season (74).

Fishes

The adenohypophysis regulates both gametogenesis and gonadal steroidogenesis (39, 40). In tetrapods the gonadotropins, follicle-stimulating hormone (FSH) and luteinizing hormone (LH), are distinct substances whereas fishes appear to produce only a single gonadotropin (25). When fish pituitary extracts are injected into tetrapods, both FSH- and LH-like effects are frequently produced (143). On the other hand, injections of mammalian FSH and LH into fishes revealed that this FSH was inactive in fishes but the LH usually elicited both gametogenesis and steroidogenesis (62). The release of fish gonadotropic hormone appears to be regulated by a stimulating hormone from the hypothalamus (17).

The androgens, testosterone and 11-ketotestosterone, occur in the testes and blood of salmon (65). Estradiol-17β, estrone, and estriol are estrogens found in several fishes (57). Progesterone has also been found in two species of the electric ray, *Torpedo*, but whether this substance is actually a hormone in fishes or only a precursor of other steroids remains to be determined (35).

The pineal gland may also be involved in regulating reproductive function of some fishes. For example, pinealectomy slightly accelerates sexual maturation in *Poecilia reticulata* (113). On the other hand, sexual maturation in pinealectomized *Fundulus heteroclitus* was delayed (103). Such apparently conflicting results may be due to species or seasonal differences in the response to the pinealectomy.

The caudal neurosecretory system is also suspected of having a role in reproduction of fishes, extracts causing smooth muscle contraction in the sperm ducts and the oviducts of teleosts (7).

PIGMENTARY EFFECTORS
Crustaceans

Chromatophores and color change. Chromatophores of crustaceans are never innervated. From several species there is clear evidence for pigment-dispersing and pigment-concentrating neurohormones that affect the chromatophores to produce changes in the color of the animal's integument (20, 21, 46).

Recently, the structure (pGlu-Leu-Asn-Phe-Ser-Pro-Gly-Trp-NH_2) of the red pigment-concentrating hormone from the eyestalks of the prawn, *Pandalus borealis*, was determined and some synthetic hormone was prepared (45). The red pigment-concentrating hormones of the fiddler crab, *Uca pugilator*, and the prawn, *Palaemonetes vulgaris,* have either the same or nearly the same molecular size as that of the synthetic hormone (49). This synthetic hormone caused strong aggregation of the pigment in the red chromatophores of *Uca pugilator* and the dwarf crayfish, *Cambarellus shufeldti*, but at the concentration used had no effect on melanophores of the crab or the white chromatophores of both test species (47). Fractionation of the eyestalks of *Crangon crangon, Rhithropanopeus harrisi,* and *Pandalus jordani* showed that their red and white pigment-concentrating hormones are distinct substances (127, 128, 129). However, Josefsson (70) reported that synthetic *Pandalus* red pigment-concentrating hormone also caused white pigment concentration in the prawn, *Palaemon adspersus*, and concluded that red and white pigment concentration in the latter species "seems to be regulated by only one hormone." However, the total response of the red chromatophores of *Palaemon* was extremely large, lasting 190 minutes, with the red pigment becoming maximally concentrated (stage 1 of the Hogben-Slome index) whereas the response of the white pigment lasted only 31 minutes, with the white pigment never reaching the maximally aggregated condition (only stage 2.5). Although the synthetic red pigment-concentrating hormone appears to have some inherent ability to concentrate the white pigment of *Palaemon adspersus*, because of the small amount of white pigment concentration that occurred in contrast to the extremely large response of the red chromatophores it is likely that a distinct white pigment-concentrating hormone is also present in this prawn. Furthermore, the dose administered was quite likely more than occurs in the blood of this prawn at any one time, so that it is possible that at physiological concentrations there is never enough red pigment-concentrating hormone in the blood to affect its white chromatophores.

Retinal pigments. The distal and reflecting retinal pigments are clearly hormonally regulated. In some species there is also evidence for endocrine control of the proximal pigment, in others it may be an independent effector. These pigments regulate the amount of light striking the light-sensitive portion of each facet in the eye. The distal pigment is clearly regulated by light-adapting (76) and dark-adapting hormones (19, 50). For the proximal and reflecting pigments the existence of light-adapting and dark-adapting hormones has also been reported. But only the light-adapting or dark-adapting hormone has been reported in any one species for the proximal and reflecting pigments, dual control not having been demonstrated as yet.

Fishes

Melanophores are the main type of chromatophore in fishes. The control of fish color changes shows considerable diversity, ranging from what may be strictly nervous control through a combination of nervous and hormonal mechanisms to hormones alone. In lampreys hormonal control alone exists, hypophysectomy resulting in permanent blanching due to removal of the source of intermedin (melanocyte-stimulating hormone), the darkening hormone (144). The ammocoete larva of *Geotria australis* blanches when pineal glands are implanted or melatonin is injected (42). Darkening of elasmobranchs is due to intermedin alone (91). However, there are three main theories to account for the blanching when elasmobranchs are placed on a white background which may be due to species differences. One is that the melanophores are innervated by melanin-concentrating nerve fibers (108). A second is that the adenohypophysis produces a blanching hormone (63). The third theory is that blanching is simply due to the absence of intermedin from the blood (1). More recently the pineal has been implicated in the blanching that occurs when elasmobranchs are placed in darkness, this gland acting either directly by producing a blanching hormone or indirectly by inhibiting intermedin release (141). Pinealectomy did not affect the responses to black and to white backgrounds, but abolished the paling response that normally occurs in darkness.

Among teleosts there is greater diversity of the control of color changes than in other fishes. While *Fundulus* possesses intermedin, it is probably only used to help produce the extreme darkening of the skin that results from keeping fish on a black background for a long period of time. Intermedin apparently has no effect on rapid color changes because hypophysectomized *Fundulus* respond to different shades of background as fast as intact specimens (75). In practically all other teleosts intermedin is involved in rapid darkening of the skin. There is also evidence in some teleosts that the melanophores are innervated by melanin-dispersing nerve fibers. Evidence obtained from a variety of fishes favors an inhibitory control of intermedin release by the hypothalamus with monoamines somehow involved (95, 111, 142).

All teleost melanophores receive pigment-concentrating innervation. In addition, a melanin-concentrating factor from the pituitary gland of some teleosts has been reported, most recently by Baker and Ball (6). Epinephrine from the adrenal medulla is also a potent melanin-concentrator (as in excitement blanching) of most teleosts, but this hormone does darken rather than lighten a few teleosts and has no effect in still others (52). Thyroid hormone causes silvering of the skin, especially in salmonids, apparently due to guanine deposition (56).

HEART RATE
Mollusks

A cardioexcitatory substance having a molecular weight of about 1300 has been extracted from the vena cava neurosecretory system of cephalopods (15). It increases both the frequency and amplitude of the hearts beat.

Crustaceans

Extracts of the neurohemal pericardial organs always increase the amplitude of the heart beat and usually the frequency also (4). The hormone seems to be released into the blood as a free peptide (8).

METABOLISM
Annelids

Neurosecretory cells in the central nervous system of cold-acclimated earth-worms were found to contain an increased level of neurosecretory material (119). This increase was the apparent basis for a rise in the rate of production of a hormone that elevates the metabolism of cold-acclimated worms. Perhaps estuarine annelids are similarly adapted to produce such a hormone in times of cold stress.

Crustaceans

Blood sugar. The eyestalks contain a hyperglycemic hormone that is a heat-labile protein with a molecular weight of about 6200 (2, 78). It shows some species specificity of action (77). The integumentary tissue and muscles appear to be the most important targets of this hormone (73).

Lipid metabolism. The eyestalks of some crustaceans have been reported to contain a substance that inhibits lipid synthesis in the midgut gland, the level rising after eyestalk removal (100). This increase is not controlled by the Y-organs. However, sinus gland removal resulted in a decrease of the lipid content in the shore crab, *Hemigrapsus nudus* (99), and removal of the eyestalks from the crab, *Pachygrapsus marmoratus*, resulted in a fall in the lipid level of its midgut gland (82).

Enzyme synthesis. Evidence for two eyestalk hormones, one a stimulator and the second an inhibitor of digestive enzyme synthesis in the midgut gland of the prawn, *Palaemon serratus*, has been obtained (137). The relative amounts of these hormones vary according to the time of year.

Oxidative metabolism. The oxygen consumption of eyestalkless fiddler crabs, *Uca pugnax*, is greater than that of intact specimens (125). The investigator (125) suggested that the difference was due to a hormone that directly regulates respiration. However, many previous investigators who observed a similar rise in

460   M. FINGERMAN

oxygen consumption after eyestalk removal (*e.g.*, 16) suggested that the rise was not due to such a hormone but rather to the absence of the molt-inhibiting hormone and consequent onset of proecdysial activities. The main reason for Silverthorn's (125) conclusion is that whereas intact crabs showed a seasonal variation (higher in winter at 25°C) in their oxygen consumption, eyestalkless crabs did not.

Two inhibitors and one stimulator of succinate oxidation in intact rat liver mitochondria have been isolated by gel filtration from shrimp eyestalks (130). One inhibitor was found in the zone containing the hormone that disperses the pigment in melanophores (and may be the same substance) and the second inhibitor was present in the region where the red pigment-concentrating hormone is eluted from the column, but synthetic *Pandalus* red pigment-concentrating hormone had no effect on succinate oxidation. The stimulator was eluted from the column along with white pigment-concentrating hormone, but this stimulator could be separated from the white pigment concentrator.

Fishes

Blood sugar, carbohydrate metabolism, and lipid metabolism. The islet tissue of all fishes produces insulin, but only the islets of jawed fishes synthesize glucagon also. Insulin has a hypoglycemic action in fishes and decreases their liver and muscle lipids (43, 66). It also increases or decreases liver glycogen (depending on the species) while muscle glycogen is unaffected or increases. The role of glucagon, while hyperglycemic in tetrapods, is not clear in fishes as this hormone has either no effect or produces only a slight hyperglycemia. Adrenalin is hyperglycemic in fishes (133) while glucocorticoids promote gluconeogenesis in fishes (33). Because the hormones which are so effective on carbohydrate metabolism in tetrapods have minimal effects in elasmobranchs, Patent (110) suggested that at least in the elasmobranchs that he worked on these hormones are primarily concerned with the control of lipid metabolism rather than carbohydrates.

Oxidative metabolism. Although most of the evidence (56) indicates that thyroid hormone does not affect the rate of oxygen consumption of fishes, Ruhland (122) has reported a stimulatory effect of this hormone on teleost respiration.

ACKNOWLEDGMENT

The research of the author is being supported by Grant No. BMS75-03029 from the National Science Foundation.

REFERENCES

1. Abramowitz, A. A. 1939. The pituitary control of chromatophores in the dogfish. Am. Nat., 73: 208-218.
2. _____, F. L. Hisaw, and D. N. Papandrea. 1944. The occurrence of a diabetogenic factor in the eyestalk of crustaceans. Biol. Bull. 86:1-5.

3. Adiyodi, K. G., and R. G. Adiyodi. 1970. Endocrine control of reproduction in decapod Crustacea. Biol. Rev., 45: 121-165.
4. Alexandrowicz, J. S., and D. B. Carlisle. 1953. Some experiments on the function of the pericardial organs in Crustacea. J. Mar. Biol. Assoc. U. K., 32: 175-192.
5. Arch, S. 1972. Biosynthesis of the egg-laying hormone (ELH) in the bag cell neurons of *Aplysia californica*. J. Gen. Physiol., 60: 102-119.
6. Baker, B. I., and J. N. Ball. 1975. Evidence for a dual pituitary control of teleost melanophores. Gen. Comp. Endocrinol., 25: 147-152.
7. Berlind, A. 1973. Caudal neurosectory system: a physiologist's view. Am. Zoologist, 13: 759-770.
8. _____, and I. M. Cooke. 1970. Release of a neurosecretory hormone as peptide by electrical stimulation of crab pericardial organs. J. Exp. Biol., 53: 679-686.
9. Bern, H. A. 1969. Urophysis and caudal neurosecretory system. In W. S. Hoar and D. J. Randall (eds.), Fish Physiology, Vol. II, Academic Press, New York.
10. Berreur-Bonnenfant, J., and J. Meusy, J. P. Ferezou, M. Devys, A. Quesneau-Thierry, and M. Barbier. 1973. Recherches sur la sécrétion de la glande androgène des Crustacés Malacostracés: purification d'une substance à activité androgène. C. R. Acad. Sci. Paris, 277d:971-974.
11. Bierne, J. 1973. Contrôle neuroendocrinien de la puberté chez le mâle de *Lineus ruber* Müller (Hétéronémerte). C. R. Acad. Sci. Paris, 176D: 363-366.
12. Björkman, N. 1963. On the ultrastructure of the optic gland in *Octopus*. J. Ultrastruct. Res. 8: 195.
13. Blanchet, M. F. 1972. Effets sur la mue et sur la vitellogenèse de la β-ecdysone introduite aux étapes A et D_2 du cycle d' intermue chez *Orchestia gammarella* Pallas (Crustacé Amphipode). Comparaison avec les effets de la β et de l'α-ecdysone aux autres étapes de l'intermue. C. R. Acad. Sci. Paris, 274D: 3015-3018.
14. _____. 1974. Etude du contrôle hormonal du cycle d'intermue et de l'exuviation chez *Orchestia gammarella* par microcautérisation des organes Y suivie d'introduction d'ecdystérone. C. R. Acad. Sci. Paris, 278D:509-512.
15. Blanchi, D., L. Noviello, and M. Libonati. 1973. A neurohormone of cephalopods with cardioexcitatory activity. Gen. Comp. Endocrinol., 21:267-277.
16. Bliss, D. E. 1953. Endocrine control of metabolism in the land crab *Gecarcinus lateralis* (Fréminville). I. Differences in the respiratory metabolism of sinusglandless and eyestalkless crabs. Biol. Bull., 104: 275-296.
17. Breton, B., B. Jalabert, R. Billard, and C. Weil. 1971. Stimulation *in vitro* de la libération d'hormone gonadotrope hypophysaire par un facteur hypothalamique chez la Carpe *Cyprinus carpio* L. C. R. Acad. Sci. Paris, 273B:2591-2594.
18. Brown, F. A., Jr., and O. Cunningham. 1939. Influence of the sinusgland of crustaceans on normal viability and ecdysis. Biol. Bull., 77: 104-114.
19. _____, M. N. Hines, and M. Fingerman. 1952. Hormonal regulation of the distal retinal pigment of *Palaemonetes*. Biol. Bull., 102: 212-225.
20. _____, and I. M. Klotz. 1947. Separation of two mutually antagonistic chromatophorotropins from the tritocerebral commissure of *Crago*. Proc. Soc. Exp. Biol. Med., 64:310-313.

21. _____. H. M. Webb, and M. I. Sandeen. 1952. The action of two hormones regulating the red chromatophores of *Palaemonetes*. J. Exp. Zool., 120: 391-420.

22. Burden, C. E., 1956. The failure of hypophysectomized *Fundulus heteroclitus* to survive in fresh water. Biol. Bull., 110: 8-28.

23. Burnett, A. L., and N. A. Diehl. 1964. The nervous system of *Hydra*. III. The initiation of sexuality with special reference to the nervous system. J. Exp. Zool., 157: 237-250.

24. _____, _____, and F. Diehl. 1964. The nervous system of *Hydra*. II. Control of growth and regeneration by neurosecretory cells. J. Exp. Zool., 157: 227-236.

25. Burzawa-Gérard, E. and Y.A. Fontaine. 1965. Activités biologiques d'un facteur hypophysaire gonadotrope purifé de Poisson téléstéen. Gen. Comp. Endocrinol., 5:87-95.

26. Cardon, C., and M. Porchet. 1973. Purification partielle d'une substance responsable de la rétroaction génitale sur l'activité endocrine cérébrale chez es Nereidae (Annélides Polychètes). C. R. Acad. Sci. Paris, 277D: 1761-1764.

27. Carlisle, D. B. 1955. On the hormonal control of water balance in *Carcinus*. Publ. Staz. Zool. Napoli, 27: 227-231.

28. _____, and R.O. Connick. 1973. Crustecdysone (20-hydroxyecdysone): site of storage in the crayfish *Orconectes propinquus*. Can. J. Zool., 51: 417-420.

29. Chan, D. K. O., J. G. Phillips, and I. Chester Jones. 1967. Studies on electrolyte changes in the lip-shark, *Hemiscyllium plagiosum* (Bennett), with special reference to hormonal influence on the rectal gland. Comp. Biochem. Physiol. 23: 185-198.

30. _____, J. C. Rankin, and I. Chester Jones. 1969. Influences of the adrenal cortex and the corpuscles of Stannius on osmoregulation in the European eel (*Anguilla anguilla* L.), adapted to freshwater. Gen. Comp. Endocrinol., Suppl. 2: 342-353.

31. Charmantier-Daures, M. and G. Vernet. 1974. Nouvelles données sur le rôle de l'organe Y dans le déroulement de la mue chez *Pachygrapsus marmoratus* (Décapode, Grapsidé). Influence de la régénération intensive. C. R. Acad. Sci. Paris, 278D: 3367-3370.

32. Charniaux-Cotton, H. 1954. Découverte chez un Crustacé Amphipode (*Orchestia gammarella*) d'une glande endocrine responsable de la différenciation des caractères sexuels primaires et secondaires mâles. C. R. Acad. Sci. Paris. 239: 780-782.

33. Chester Jones, I., D. K. O. Chan, I. W. Henderson, and J. N. Ball. 1969. The adrenocortical steroids, adrenocorticotropin and the corpuscles of Stannius. *In* W.S. Hoar and D.J. Randall (eds.), Fish Physiology, Vol. II, Academic Press, New York.

34. _____, _____, and J. C. Rankin. 1969. Renal function in the European eel (*Anguilla anguilla* L.): effects of the caudal neurosecretory system, corpuscles of Stannius, neurohypophysial peptides and vasoactive substances. J. Endocrinol., 43: 21-31.

35. Chieffi, G. 1961. La luteogenesi nei selaci ovovivipari. Ricerche istologiche e istochimiche in *Torpedo marmorata* e *Torpedo ocellata*. Pubbl. Staz. Zool. Napoli, 32:145-166.

36. Dales, S., and W. S. Hoar. 1954. Effects of thyroxine and thiourea on the

early development of Chum salmon (*Oncorhynchus keta*). Can. J. Zool., 32:244-254.

37. Davis, C. W., and J. D. Costlow. 1974. Evidence for a molt inhibiting hormone in the barnacle *Balanus improvisus* (Crustacea, Cirripedia). J. Comp. Physiol., 93: 85-91.

38. De Leersnyder, M. 1967. Le milieu interieur d'*Eriocheir sinensis* H. Milne-Edwards et ses variations. II. Etude expérimentale. Cah. Biol. Mar., 8: 295-321.

39. de Vlaming, V. L. 1974. Environmental and endocrine control of teleost reproduction. *In* C. B. Schreck (ed.), Control of Sex in Fishes, Virginia Polytechnic Institute, Blacksburg.

40. Donaldson, E. M. 1973. Reproductive endocrinology of fishes. Am. Zoologist, 13: 909-927.

41. Durchon, M. 1948. Épitoquie expérimentale chez deux Polychètes: *Perinereis cultrifera* Grube et *Nereis irrorata* Malmgren. C. R. Acad. Sci. Paris, 227: 157.

42. Eddy, J. M. P., and R. Strahan. 1968. The role of the pineal complex in the pigmentary effector system of the lampreys, *Mordacia mordax* (Richardson) and *Geotria australis* Gray. Gen. Comp. Endocrinol., 11: 528-534.

43. Epple, A. 1969. The endocrine pancreas. *In* W. S. Hoar and D. J. Randall (eds.), Fish Physiology, Vol. II, Academic Press, New York.

44. Epstein, F. H., M. Cynamon, and W. McKay. 1971. Endocrine control of Na-K-ATPase and seawater adaptation in *Anguilla rostrata*. Gen. Comp. Endocrinol., 16: 323-328.

45. Fernlund, P., and L. Josefsson. 1972. Crustacean color-change hormone: amino acid sequence and chemical synthesis. Science, 177:173-175.

46. Fingerman, M. 1970. Dual control of the leucophores in the prawn, *Palaemonetes vulgaris*, by pigment-dispersing and pigment-concentrating substances. Biol. Bull., 138: 26-34.

47. _____, 1973. Behavior of chromatophores of the fiddler crab *Uca pugilator* and the dwarf crayfish *Cambarellus shufeldti* in response to synthetic *Pandalus* red pigment-concentrating hormone. Gen. Comp. Endocrinol., 20: 589-592.

48. _____, and S. W. Fingerman. 1974. The effects of limb removal on the rates of ecdysis of eyed and eyestalkless fiddler crabs, *Uca pugilator*, Zool. Jb. Physiol., 78:301-309.

49. _____, _____, and R. D. Hammond. 1974. Comparison of red pigment-concentrating hormones from the eyestalks of the fiddler crab, *Uca pugilator*, and the prawn, *Palaemonetes vulgaris*, with synthetic red pigment-concentrating hormone of *Pandalus borealis*. Gen. Comp. Endocrinol., 23: 124-126.

50. _____, M. E. Lowe, and B. I. Sundararaj. 1959. Dark-adapting and light-adapting hormones controlling the distal retinal pigment of the prawn *Palaemonetes vulgaris*. Biol. Bull., 116: 30-36.

51. Fontaine, M. 1964. Corpuscles de Stannius et régulation ionique. (Ca, K, Na) du milieu intérieur de l'Anguille (*Anguilla anguilla* L.). C. R. Acad. Sci. Paris, 259: 875-878.

52. Fujii, R. 1969. Chromatophores and pigments. *In* W. S. Hoar and D. J. Randall (eds.), Fish Physiology, Vol. III, Academic Press, New York.

53. Geschwind, I. I. 1967. Molecular variation and possible lines of evolution of peptide and protein hormones. Am. Zoologist, 7: 89-108.
54. Golding, D. W. 1967. Regeneration and growth control in *Nereis*. II. An axial gradient in growth potentiality. J. Embryol. Exp. Morph., 18: 79-90.
55. _____. 1972. Studies in the comparative neuroendocrinology of polychaete reproduction. Gen. Comp. Endocrinol, Suppl. 3: 580-590.
56. Gorbman, A. 1969. Thyroid function and its control in fishes. *In* W. S. Hoar and D. J. Randall (eds.), Fish Physiology, Vol. II, Academic Press, New York.
57. Gottfried, H., S. V. Hunt, T. H. Simpson, and R. S. Wright. 1962. Sex hormones in fish. The oestrogens of cod (*Gadus callarias*). J. Endocrinol., 24: 425-430.
58. Hauenschild, C. 1966. Der hormanale Einfluss des Gehirns auf die sexuelle Entwicklung bei dem Polychaeten *Platynereis dumerilii*. Gen. Comp. Endocrinol., 6: 26-73.
59. Hayashida, T. 1973. Biological and immunochemical studies with growth hormone in pituitary extracts of elasmobranchs. Gen. Comp. Endocrinol., 20: 377-385.
60. Heit, M., and M. Fingerman. 1975. The role of an eyestalk hormone in the regulation of the sodium concentration of the blood of the fiddler crab, *Uca pugilator*. Comp. Biochem. Physiol., 50A: 277-280.
61. Hirano, T., and S. Utida. 1968. Effects of ACTH and cortisol on water movement in isolated intestine of the eel, *Anguilla japonica*. Gen. Comp. Endocrinol., 11: 373-380.
62. Hoar, W. S. 1969. Reproduction. *In* W. S. Hoar and D. J. Randall (eds.), Fish Physiology, Vol. III, Academic Press, New York.
63. Hogben, L. T. 1936. The pigmentary effector system. VII. The chromatic function in elasmobranch fishes. Proc. Roy. Soc. London, 120B: 142-158.
64. Howie, D. I. D. 1966. Further data relating to the maturation hormone and its site of secretion in *Arenicola marina* L. Gen. Comp. Endocrinol., 6: 347-361.
65. Idler, D. R., P. J. Schmidt, and A. P. Ronald. 1960. Isolation and identification of 11-ketotestosterone in salmon plasma. Can. J. Biochem. Physiol., 38: 1053-1057.
66. Ince, B. W., and A. Thorpe. 1974. Effects of insulin and of metabolite loading on blood metabolites in the European silver eel (*Anguilla anguilla* L.) Gen. Comp. Endocrinol, 23: 460-471.
67. Johansen, P. H. 1967. The role of the pituitary in the resistance of the goldfish (*Carassius auratus* L.) to a high temperature. Can. J. Zool., 45: 329-345.
68. Johnson, D. W. 1973. Endocrine control of hydromineral balance in teleosts. Am. Zoologist, 13: 799-818.
69. _____, T. Hirano, H. A. Bern, and F. P. Conte. 1973. Hormonal control of water and sodium movements in the urinary bladder of the starry founder, *Platichthyes stellatus*. Gen. Comp. Endocrinol., 19: 115-128.
70. Josefsson, L. 1975. Structure and function of crustacean chromatophorotropins. Gen. Comp. Endocrinol, 25: 199-202.
71. Kamemoto, F. J., K. N. Kato, and L. E. Tucker. 1966. Neurosecretion and salt and water balance in the Annelida and Crustacea. Am. Zoologist. 6: 213-219.

72. Kato, K. N., and F. I. Kamemoto. 1969. Neuroendocrine involvement in osmoregulation in the grapsid crab *Metopograpsus messor*. Comp. Biochem. Physiol., 28: 665-674.
73. Keller, R., and E. M. Andrew. 1973. The site of action of the crustacean hyperglycemic hormone. Gen. Comp. Endocrinol: 20: 572-578.
74. Klek-Kawińska, E., and A. Bomirski. 1975. Ovary-inhibiting hormone activity in shrimp (*Crangon crangon*) eyestalks during the annual reproductive cycle. Gen. Comp. Endocrinol., 25: 9-13.
75. Kleinholz, L. H. 1935. The melanophore-dispersing principle in the hypophysis of *Fundulus heteroclitus*. Biol. Bull. 69:379-390.
76. ———. 1936. Crustacean eye-stalk hormone and retinal pigment migration. Biol. Bull. 70: 159-184.
77. ———, and R. Keller. 1973. Comparative studies in crustacean neurosecretory hyperglycemic hormones. I. The initial survey. Gen. Comp. Endocrinol., 21: 554-564.
78. ———, F. Kimball, and M. McGarvey. 1967. Initial characterization and separation of hyperglycemic (diabetogenic) hormone from the crustacean eyestalk. Gen. Comp. Endocrinol., 8: 75-81.
79. Krishnakumaran, A., and H. A. Schneiderman. 1970. Control of molting in mandibulate and chelicerate arthropods by ecdysones. Biol. Bull., 139: 520-538.
80. Kupfermann, I. 1967. Stimulation of egg laying: possible neuroendocrine function of bag cells of abdominal ganglion of *Aplysia californica*. Nature, 216: 814-815.
81. Lacanilao, F. 1972. The urophysial hydrosmotic factor of fishes. II. Chromatographic and pharmacologic indications of similarity to arginine vasotocin. Gen. Comp. Endocrinol., 19: 413-420.
82. Lautier, J., and G. Vernet. 1972. Comparaison du métabolisme lipidique de l'hépatopancréas de *Pachygrapsus marmoratus* Fabricius (Décapode Brachyoure) chez des animaux témoins et opérés des pédoncles oculaires en fonction du cycle d'intermue. C. R. Acad. Sci. Paris, 275D: 1899-1902.
83. Lechenault, H. 1965. Neurosécrétion et osmorégulation chez les Lineidae (Hétéronémertes). C. R. Acad. Sci. Paris. 261: 4868-4871.
84. Lederis, K. 1970. Teleost urophysis: I. Biossay of an active urophysial principle on the isolated urinary bladder of the rainbow trout, *Salmo gairdnerii*. Gen. Comp. Endocrinol., 14: 417-426.
85. Le Gall, S. 1974. Synchronisme entre l'activité sécrétrice des neurones de la jonction cérébropleurale et l'activité dédifférenciatrice de cette zone chez *Crepidula fornicata* Phil. en phase femelle. C. R. Acad. Sci. Paris, 278D: 939-942.
86. Legrand, J. J. 1955. Rôle endocrinien de l'ovaire dans la differenciation des oostégites chez les Crustacés Isopodes terrestres. C. R. Acad. Sci. Paris, 241: 1083-1085.
87. Le Roux, A. 1968. Description d'organes mandibulaires nouveaux chez les Crustacés Décapodes. C. R. Acad. Sci. Paris, 266D:1414-1417.
88. ———. 1974. Mise au point à propos de la distinction entre l'organe Y et l'organe mandibulaire chez les Crustacés eucarides. C. R. Acad. Sci. Paris, 278D: 1261-1264.
89. Loeb, M. J. 1974. Strobilation in the Chesapeake Bay sea nettle *Chrysaora quinquecirrha*—III. Dissociation of the neck-inducing factor from strobilating polyps. Comp. Biochem. Physiol., 49A: 423-432.

90. Lubet, P., and M. Silberzahn. 1971. Recherches sur les effets de l'ablation bilatérale des ganglions cerebroïdes chez la Crepidule (*Crepidula fornicata* Phil., Mollusque gastéropode): effets somatrotrope et gonadotrope. C. R. Soc. Biol., 165: 590-594.

91. Lundstrom, H. M., and P. Bard. 1932. Hypophysial control of cutaneous pigmentation in an elasmobranch fish. Biol. Bull., 62: 1-9.

92. Madhyastha, M. N., and P. V. Rangneker. 1972. Y-organ of the crab, *Varuna litterata* (Fabricius). Experientia, 28:580.

93. Maetz, J., J. Bourguet, and B. Lahlouh. 1964. Urophyse et osmorégulation chez *Carassius auratus*. Gen. Comp. Endocrinol., 4: 401-414.

94. Maissiat, J., and F. Graf. 1973. Action de l'ecdystérone sur l'apolysis et l'ecdysis de divers Crustacés Isopodes. J. Insect Physiol., 19: 1265-1276.

95. Meurling, P., and M. Fremberg. 1974. Effects of partial denervations of the neuro-intermediate lobe in the skate, *Raja radiata*. Acta Zool., 55: 7-15.

96. Mocquard, J. P., G. Besse, P. Juchault, J. J. Legrand, J. Maissiat, and C. Noulin. 1971. Contribution à l'analyse du controle neurohumoral de la croissance, de la mue et de la physiologie sexuelle mâle et femelle chez l'Oniscoïde *Ligia oceanica* L. (Crustacé, Isopode). Annal. Embryol. Morphogen., 4: 45-63.

97. Motais, R., and J. Maetz. 1964. Action des hormones neurohypophysaires sur les échanges de sodium (Mesurés à l'aide du radio-sodium Na^{24}) chez un téléostéen euryhalin: *Platichthys flesus* L. Gen. Comp. Endocrinol., 4: 210-224.

98. Nagahama, Y., R. S. Nishioka, H. A. Bern, and R. L. Gunther. 1975. Control of prolactin secretion in teleosts, with special reference to *Gillichthys mirabilis* and *Tilapia mossambica*. Gen. Comp. Endocrinol., 25: 166-188.

99. Neiland, K. A., and B. T. Scheer. 1953. The influence of fasting and of sinus gland removal on body composition of *Hemigrapsus nudus*. Physiol. Comp. Oecol., 3: 321-326.

100. O'Connor, J.D., and L. I. Gilbert. 1969. Alterations in lipid metabolism associated with premolt activity in a land crab and fresh-water crayfish. Comp. Biochem. Physiol., 29: 889-904.

101. Olivereau, M., and J. N. Ball. 1966. Histological study of functional ectopic pituitary transplants in a teleost fish (*Poecilia formosa*). Proc. Roy. Soc. London, 164B: 106-129.

102. Orimo, H., M. Ohata, M. Yoshikawa, J. Abe, S. Watanabe, M. Kotani, and T. Higashi. 1971. Ultimobranchial calcitonin. II. Physiological role of calcitonin. Igaku No Aymi, 79: 480.

103. Pang, P. K. T. 1967. The effect of pinealectomy on the adult male killifish, *Fundulus heteroclitus*. Am. Zoologist, 7: 715.

104. ———. 1973. Endocrine control of calcium metabolism in teleosts. Am. Zoologist, 13: 775-792.

105. ———, R. K. Pang, and W. H. Sawyer. 1974. Environmental calcium and the sensitivity of killifish (*Fundulus heteroclitus*) in bioassays for the hypocalcemic response to Stannius corpuscles from killifish and cod (*Gadus morhua*). Endocrinology, 94: 548-555.

106. ———, M. P. Schreibman, and R. W. Griffith. 1973. Pituitary regulation of serum calcium levels in the killifish, *Fundulus heteroclitus* L. Gen. Comp. Endocrinol., 21: 536-542.

107. Panouse, J. B. 1943. Influence de l'ablation du pédoncle oculaire sur la croissance de l'ovaire chez la Crevette *Leander serratus*. C. R. Acad. Sci. Paris, 217: 553-555.
108. Parker, G. H., and H. Porter. 1934. The control of the dermal melanophores in elasmobranch fishes. Biol. Bull., 66:30-37.
109. Passano, L. M. 1953. Neurosecretory control of molting in crabs by the Y-organ sinus gland complex. Physiol. Comp. Oecol., 3: 155-189.
110. Patent, G. 1970. Comparison of some hormonal effects on carbohydrate metabolism in an elasmobranch (*Squalus acanthias*) and a holocephalan (*Hydrolagus colliei*). Gen. Comp. Endocrinol., 14: 215-242.
111. Peter, R. E. 1973. Neuroendocrinology of teleosts. Am. Zoologist., 13: 743-755.
112. _____, and B. A. McKeown. 1975. Hypothalamic control of prolactin and thyrotropin secretion in teleosts, with special reference to recent studies on the goldfish. Gen. Comp. Endocrinol., 25: 153-165.
113. Pflugfelder, O. 1954. Wirkungen partieller Zerstörungen der Parietalregion von *Lebistes reticulatus*. Arch. Entwicklungsmech. Organ., 147:42-60.
114. Pickford, G. E., R. W. Griffith, J. Torretti, E. Hendler, and F. H. Epstein. 1970. Branchial reduction and renal stimulation of (Na$^+$, K$^+$)-ATPase by prolactin in hypophysectomized killifish in fresh water. Nature, 228: 378-379.
115. _____, and J. G. Phillips. 1959. Prolactin, a factor in promoting survival of hypophysectomized killifish in fresh water. Science, 130: 454-455.
116. _____, A. E. Wilhelmi, and N. Nussbaum. 1959. Comparative studies of the response of hypophysectomized killifish, *Fundulus heteroclitus*, to growth hormone preparations. Anat. Rec., 134: 624-625.
117. Piggins, D. J. 1962. Thyroid feeding of salmon parr. Nature, 195: 1017-1018.
118. Ramamurthi, R., and B. T. Scheer. 1967. A factor influencing sodium regulation in crustaceans. Life Sci., 6: 2171-2175.
119. Rao, K. P. 1962. Physiology of acclimation to low temperature in poikilotherms. Science, 137: 682-683.
120. _____, M. Fingerman, and C. Hays. 1972. Comparison of the abilities of α-ecdysone and 20-hydroxyecdysone to induce precocious proecdysis and ecdysis in the fiddler crab, *Uca pugilator*. Z. Vergl. Physiol., 76: 270-280.
121. Rao, K. R., S. W. Fingerman, and M. Fingerman. 1973. Effects of exogenous ecdysones on the molt cycles of fourth and fifth stage American lobsters, *Homarus americanus*. Comp. Biochem. Physiol., 44A:1105-1120.
122. Ruhland, M. L. 1971. La radiothyroïdectomie et son effet sur la consommation d'oxygène chez les cichlidés *Aequidans latifrons*. Can. J. Zool., 49: 423-425.
123. Schaller, H.C., and A. Gierer. 1973. Distribution of the head-activating substance in *Hydra* and its localization in membranous particles in nerve cells. J. Embryol. Exp. Morphol., 29: 39-52.
124. Schroeder, P. C. 1971. Studies on oogenesis in the polychaete annelid *Nereis grubei* (Kinberg). Gen. Comp. Endocrinol., 16: 312-322.
125. Silverthorn, S. U. 1973. Respiration in eyestalkless *Uca* (Crustacea: Decapoda) acclimated to two temperatures. Comp. Biochem. Physiol., 45A: 417-420.
126. Simione, F. P., Jr., and D. L. Hoffman. 1975. Some effects of eyestalk removal on the Y-organs of *Cancer irroratus* Say. Biol. Bull. 148: 440-447.

127. Skorkowski, E. F. 1971. Isolation of three chromatophorotropic hormones from the eyestalk of the shrimp *Crangon crangon*. Mar. Biol. 8: 220-223.

128. _____, 1972. Separation of three chromatophorotropic hormones from the eyestalk of the crab *Rhithropanopeus harrisi* (Gould). Gen. Comp. Endocrinol., 18: 329-334.

129. _____, and L. H. Kleinholz. 1973. Comparison of white pigment concentrating hormone from *Crangon* and *Pandalus*. Gen. Comp. Endocrinol., 20: 595-597.

130. _____. J. Świerczyński, Z. Aleksandrowicz, and L. H. Kleinholz. 1974. Effect of partially purified crustacean eyestalk factor on mitochondrial respiration. Comp. Biochem. Physiol., 49B: 627-630.

131. Spaziani, E., and S. B. Kater. 1973. Uptake and turnover of cholesterol-^{14}C in Y-organs of the crab *Hemigrapsus* as a function of the molt cycle. Gen. Comp. Endocrinol., 20:534-549.

132. Streiff, W. 1970. Analyse experimentale du déterminisme de la morphogenese et du cycle du tractus génital des Prosobranches. Bull. Soc. Zool. Fr., 95: 451-460.

133. Thorpe, A., and B. W. Ince. 1974. The effects of pancreatic hormones, catecholamines, and glucose loading on blood metabolites in the northern pike (*Esox lucius* L.). Gen. Comp. Endocrinol., 23: 29-44.

134. Tighe-Ford, D. J., and D. C. Vaile. 1972. The action of crustecdysone on the cirripede *Balanus balanoides* (L.). J. Exp. Mar. Biol. Ecol., 9:19-28.

135. Tullis, R. E., and F. I. Kamemoto. 1974. Separation and biological effects of CNS factors affecting water balance in the decapod crustacean *Thalamita crenata*. Gen. Comp. Endocrinol., 23: 19-28.

136. Ude, J. 1964. Untersuchungen zur Neurosekretion bei *Dendrocoelum lacteum* Oerst. (Platyhelminthes-Turbellaria). Z. Wiss. Zool, 170: 223-255.

137. Van Wormhoudt, A. 1974. Variations of the level of the digestive enzymes during the intermolt cycle of *Palaemon serratus*: influence of the season and effect of the eyestalk ablation. Comp. Biochem. Physiol., 49A: 707-715.

138. Vicente, N. 1969. Contribution a l'étude des Gastéropodes Opisthobranches du Golfe de Marseille. II—Histophysiologie du système nerveux. Étude des phénomènes neurosécrétoires. Recueil Traveux Stat. Mar. Endoume, 46, Fasc. 62, 13-121.

139. Wells, M. J., and J. Wells. 1959. Hormonal control of sexual maturity in *Octopus*. J. Exp. Biol., 36: 1-33.

140. Whittle, A. C., and D. W. Golding. 1974. The infracerebral gland and cerebral neurosecretory system—a probable neuroendocrine complex in phyllodocid polychaetes. Gen. Comp. Endocrinol., 24: 87-98.

141. Wilson, J.F., and J.M. Dodd. 1973. The role of the pineal complex and lateral eyes in the colour change response of the dogfish, *Scyliorhinus canicula* L. J. Endocrinol. 58: 591-598.

142. _____, and _____. 1973. Effects of pharmacological agents on the *in vivo* release of melanophore-stimulating hormone in the dogfish, *Scyliorhinus canicula*. Gen. Comp. Endocrinol., 20: 556-566.

143. Witschi, E. 1955. Vertebrate gonadotropins in comparative physiology of reproduction. Mem. Soc. Endocrinol., 4: 149-165.

144. Young, J. Z. 1937. The photoreceptors of lampreys. II. The functions of the pineal complex. J. Exp. Biol. 12: 254-270.

THE DETRITUS PROBLEM AND THE

FEEDING AND DIGESTION OF AN ESTUARINE ORGANISM

William W. Kirby-Smith
Duke University Marine Laboratory
Beaufort, North Carolina 28516

ABSTRACT: A review of past and present research on ecological aspects of food supplied to and used by estuarine organisms indicates that one major problem encountered is a lack of specific knowledge concerning pathways and rates of transfer of carbon and nitrogen from the producers to the consumers. This is especially true for suspension and deposit feeding animals which tend to be nonselective in their feeding habits. Recent papers which relate to these problems describe (1) salt marsh systems which have a net import instead of export of POC, PON, and nutrients; (2) suspended organic detritus which is synthesized by bacteria and not the result of bacterial decomposition; (3) suspended detritus with a low C/N ratio which is refractory to digestion by bacteria and suspension feeders. Physiological techniques of assessing digestive capabilities of estuarine organisms have not been widely used in ecological studies; rather, there commonly has been a usage of terms (herbivore, detritivore, suspension or deposit feeder) which is based upon ingestion or gut contents but not based upon digestion, net assimilation, or growth rates. For example, natural populations of the bay scallop, *Argopecten irradians*, ingest a mixture of POC which is 80% detritus and microheterotrophs and 20% phytoplankton. However, in laboratory experiments, the growth rate of scallops is greatest on a diet of mixed phytoplankton and slow but continuous on fresh detritus. In contrast, scallops lose weight on aged detritus (C/N ratio 12/1) and on protein rich fish food (C/N ratio 6/1).

REVIEW

Research on the feeding and digestion of estuarine organisms must emphasize those properties of the food chain which distinguish estuaries from other marine systems. As Ferguson and Murdoch (6) and Odum (23) have recently pointed out, estuarine food chains differ from those of the open sea in two important ways. First, the greater part of the primary production in estuaries is carried out by vascular plants (*Spartina, Zostera*, etc.) and macroalgae rather than by phytoplankton. Secondly, zooplankton is quantitatively less important in estuaries, while animals which utilize detritus and/or benthic microflora are the more important herbivores in the system. Understanding the origin, composition, and fate of the organic detritus in estuaries continues to be a major problem. Even the definition of the word "detritus" has been confusing (2). As used here detrital carbon in a sample is operationally defined as the concentration of total particulate carbon less the estimated concentration of living carbon as based upon the concentration of adenosine triphosphate (ATP) in the same sample (6).

Of the various feeding classifications to which the larger animals of the estuary can be assigned, there is a preponderance of suspension and deposit feeding types. Since both groups tend to be non-selective in the type of material ingested, investigations of what they are digesting and assimilating is central to understanding their energy and material transfers. Since organic detritus, microflora, and microfauna form the basis of the food web in estuaries, an understanding of the formation and transfer of these materials must be combined with studies of digestion and assimilation.

Thayer et al. (30) found the following distributions of standing crops of living organic matter in a shallow North Carolina estuary: benthic microheterotrophs > benthic macroinvertebrates > benthic microautotrophs > nekton = aquatic microheterotrophs > aquatic autotrophs. Many interpretations of these relative abundances are possible. For example, from one point of view the benthic microheterotrophs might be considered the most abundant, living food supply for other animals. Alternately, it could be reasoned that the benthic microheterotroph biomass is greatest because they are not being eaten. Such opposing interpretations could equally apply to the standing crops of other types of living and dead organic matter. Knowledge concerning rates of transfer of organic material and specific information on the use of material by organisms is necessary to evaluate these potential food supplies. For example, physiological studies such as those by Sushchenya (29) on crustaceans and Winter (34) on molluscs can be used to establish energy requirements and efficiencies of food utilization, while ecological research, such as Odum's (23) mullet studies, becomes critical to an understanding of how feeding and digestion occur under natural conditions.

The origin of inorganic nutrients and organic matter in estuaries has been extensively investigated. Estuaries receive inorganic nutrients from freshwater inflow and these nutrients are used in primary production by phytoplankton, macroalgae and vascular plants. Within estuarine systems, salt marshes classically have been considered major consumers of inorganic nutrients and major exporters of particulate organic matter in the form of detritus and phytoplankton. Studies, such as those of Haines (9) in Georgia, emphasize that nutrients supplied by rivers are almost totally consumed by the estuarine system. Most nutrients appear to be quickly used as river water mixes with salt water in the upper estuary. The nutrients are then recycled, lost in sediments, or transported through the estuary in much lower concentrations. Evidence from marshes on Long Island Sound (20, 31, 35, 36) suggests that on an annual basis there is little net flux of inorganic nitrogen or phosphorus between the marsh and the sound. Seasonal periods of export were followed by seasonal periods of import. There was a net import of phytoplankton and a small net import of fixed carbon. The net import of fixed carbon was the result of a continual influx of particulate carbon with a loss of dissolved carbon. These results from Long Island are very different from the results from Georgia (22) where there was a considerable net transport of particulate carbon "downstream" in the marsh channels. Other studies, such as that by Hinson et al. (10), continue to support the concept that large amounts of particulate matter are being exported from marshes adjacent to estuarine waters. Based upon such work it appears that marshes can be sinks as well as sources, not only for nutrients, but also for particulate organic matter. Also, physical factors, such as tidal amplitude and flow, and storm events, may be extremely important when considering the fluxes of materials in particular marshes.

The usual interpretation of detrital carbon in suspension in estuaries is that it is the final stages of bacterial breakdown of vascular plant material produced by marshes and subtidal grass beds. It is thought that this detrital material is covered with microorganisms which serve as the food for larger animals eating the particles (for discussion see Darnell, 2, 3). These concepts of organic detritus formation are being modified in a number of different ways.

One of the most interesting ideas currently being investigated is that detritus exported from marshes is new particulate organic matter synthesized by bacteria in the process of their breakdown of dead vascular plant material (26). Barber (1) and Paerl (24) have shown the importance of bacteria in the formation of detrital particles from dissolved substrate. Jannash (13) has described, in offshore waters, particulate organic matter which he interprets as being recently formed, nonliving aggregations of dissolved organic compounds which could be colonized by bacteria. Woodwell et al. (36) observed a significant export of dissolved organic matter from marshes which, when considered with the preceding studies, suggests that at least some of the suspended estuarine detritus could have its origin in dissolved carbon from the marshes. Sottile (27), working in Georgia, concluded that if dissolved organic carbon was exported from the marshes, it was a seasonal flux directly related to active plant metabolism, and not a tidal flux due to microbial activity on the salt marsh sediments. Although the dissolved carbon to particulate carbon transformation has not been evaluated quantitatively in natural estuarine waters, such processes could be occurring.

My observations indicate that small pieces of *Spartina* (less than 200 microns) settle rapidly from suspension whether they are fresh or aged in seawater for six months. Therefore, detritus derived directly from the physical breakdown of vascular plant material would tend to remain in place on the bottom or, if transported, would remain close to the sediment-water interface. Such detritus would be carried into suspension only in places with strong currents.

The concept that detritus in suspension is rich in attached bacteria is also being closely examined. Wiebe and Pomeroy (32) found that most detrital particles in suspension in estuarine waters in Georgia did not contain recognizable bacteria. The particles appeared to be small pieces of *Spartina* and clay rather than recent aggregations of dissolved substances, suggesting that suspended detritus is directly derived from the breakdown of vascular plant material. Furthermore, they concluded that it was unlikely that bacteria constitute an important energy source for suspension feeders in the open ocean. Their work implied that, because of the low number of bacteria on particles, animals feeding on suspended particles in estuaries do not use bacteria as an important energy source. Hobbie and Daley (11) described an improved method for direct counting of bacteria which apparently reveals large numbers of very small bacteria which could be missed with other techniques. Using this method Ferguson and Rublee (7) found low numbers of bacteria associated with particles. However, the bacterial volumes measured indicate that 45% of the bacterial carbon is associated with particulate matter, suggesting that bacteria/particle interactions are very significant. Ferguson and Rublee (7) also found that only 7-15% of the living carbon (based upon ATP analysis) was of bacterial origin, thus supporting the conclusion of Wiebe and Pomeroy (32) that bacteria are probably a minor energy source for suspension feeders.

In contrast to the living organic matter in suspension, it appears that the organic matter in deposits is relatively rich in bacteria and microfauna. Thayer et al. (30) found that, for the estuary as a whole, there was 1.7 to 20.0 times more biomass of suspended phytoplankton than heterotrophic organisms, while the heterotrophs in deposits had a biomass 11 to 49 times that of the autotrophs. These results suggest that, in terms of living carbon, deposit feeding organisms have a diet rich in bacteria and microfauna while suspension feeders have a diet rich in phytoplankton. However, in the diets of both, the amount of non-living carbon far outweighs the amount of living carbon.

The idea that microbial biomass on detritus is an important food source is very prevalent in the literature. Russell-Hunter (25) emphasized the importance of carbon to nitrogen (C/N) ratios in assessing the potential food value of a substance. A ratio of 17/1 is considered optimum for most animals. Odum and de la Cruz (22) showed that the C/N ratio of *Spartina* detritus dropped from approximately 45/1 to 11/1 (25) as microbial populations built up on the material. They suggest that this protein enrichment would make the particles a better food source than fresh *Spartina* for detritus feeding animals. Fisher (5), working in a North Carolina estuary, found that the total suspended material consisted of 80% inorganic (ash) and 20% organic matter. Of the organic carbon, approximately 25% was phytoplankton with the remainder being detritus and microfauna. The C/N ratio of the organic material was 11/1. Phytoplankton has a C/N ratio of 6/1 to 7/1 (25), hence the detritus and microfauna alone have a C/N ratio of approximately 12/1. The food value of this detritus and microfauna relative to the food value of phytoplankton for suspension feeders is unknown.

Recent studies on the decomposition of vascular plant material and the production of detritus have emphasized the importance of invertebrates in mechanically breaking down the material into successively smaller pieces, and the use of microorganisms, not the detritus per se, as a food source. Newell (21) found that two deposit feeding molluscs ingested detritus particles but digested only the microorganisms. Odum (23) found essentially the same phenomenon working with mullet. Fenchel (4) found that an amphipod, feeding on detrital particles, digested only the microorganisms. In addition, the amphipod reduced the particle size of the material, resulting in a subsequent increase in the surface area and microbial activity. Lopez (15) found that an amphipod, feeding on Spartina litter, increased the mineralization rate of the material by feeding on the microbial population. However, in contrast to other work, he suggested that much of the digested material was microbial debris rather than living cells. A complete understanding of feeding and digestion of any detritus-ingesting animal requires knowledge not only of which materials are ingested, but, more importantly, of which materials are actually digested and assimilated, and the amount of energy derived from these.

EVIDENCE FROM A SUSPENSION FEEDER

Measuring the growth rate of young animals on a variety of diets is one method of assessing the relative importance of various potential foods. The bay scallop, Argopecten irradians, has been used in studies (17, 18) which suggest that growth rates are directly correlated with phytoplankton concentration. Kirby-Smith (16) estimated that the total energy needs of scallops could be supplied by the phytoplankton even though the algae contributed only 20% of the total particulate carbon in suspension. The following experiment was designed to determine the relative food value to scallops of Spartina detritus and phytoplankton.

The experimental apparatus consisted of a set of four feed tanks connected with four experimental tanks containing bay scallops. The feed tanks were 64 liter polyethylene containers. The tanks were cleaned and refilled each day. The first feed tank was filled with seawater which had been passed through a 5 micron polyethylene filter bag. The second feed tank contained a phytoplankton culture which was obtained by enriching 19 liters of unfiltered seawater from the Beaufort Channel, N.C. to F/10 (8) with nitrate, phosphate, and silicate. The cultures were then grown in glass tanks placed in continuous, strong, fluorescent light for two to three days, at which time the cultures usually had between 30 and 100 $\mu g/l$ chlorophyll a. The cultures, consisting of populations of diatoms and small green algae, were then diluted in the feed tank to 64 liters with filtered seawater. The third feed tank contained 3.8 liters of a culture of aged Spartina detritus diluted to 64 liters with filtered seawater. The detritus was made by collecting blades of *Spartina alterniflora*, drying the material at 70°C, and grinding it to a powder with a particle size of 200 microns or less. The detritus culture consisted of one hundred grams of this powder mixed with 100 liters of unfiltered seawater and placed in a large plastic container in complete darkness at room temperature. The material was continuously stirred and allowed to age for six months before the start of the experiment. The fourth feed tank contained three grams of the dried *Spartina* powder suspended in 64 liters of filtered seawater. The concentration of aged and fresh detritus in the third and fourth feed tank was set to ensure equal concentrations of particulate organic carbon in the respective experimental tanks.

The experimental tanks consisted of four glass aquaria of approximately 19 liters volume maintained at $22 \pm 1°C$. Material from the feed tanks was pumped into the experimental tanks at 2.4 liters per hour. Three bay scallops, one to two centimeters in shell depth, were placed on a plastic screen and suspended off the bottom of each experimental tank. The shell depth of each scallop was measured four times during the 18-day experiment. At the end of the experiment the animals were sacrificed and the wet tissue weight was determined. A previously determined regression (16) of tissue weight to shell depth was used to estimate the expected weight of each animal. The actual weight was then expressed as a percentage of the expected tissue weight.

During the course of the experiment the particulate organic carbon (POC) of each experimental tank was measured (19) approximately 30 times. Chlorophyll *a* (fluorometric technique as given by Strickland and Parsons, 28) was measured once a day in the experimental tank which was being fed the phytoplankton culture.

On the 17th day of the experiment, samples of water from the feed tanks were analyzed for particulate nitrogen (Coleman Model 29 Nitrogen Analyzer), particulate ATP (12), and dissolved carbon (Beckman Model 915 Analyzer). Particulate samples were collected on precombusted glass fiber filters. The POC samples could not be analyzed, therefore the POC was estimated based on the C/N ratios obtained several weeks later using samples from an identical experimental setup. The ATP analysis was done by R. L. Ferguson of the National Marine Fisheries Service, Beaufort, North Carolina. Living carbon in the sample was estimated by multiplying the grams of ATP by 250. The percent living carbon was estimated by dividing the ATP carbon by the total particulate carbon and multiplying by 100.

The results of the growth experiments are shown in Fig. 1. The animals in filtered seawater and aged detritus had almost no shell growth past the first week. During this first week, the slight growth recorded can be explained as the lag effect while the animals are adjusting to experimental conditions (18). The aged detritus does not represent organic matter which would be found in nature. However, the material was easily recognizable as vascular plant detritus when viewed under the microscope; it had a C/N ratio of 12/1, and appeared to be similar to the natural detritus seen by Wiebe and Pomeroy (32). The tissue weight of these two groups was approximately 15% less than expected, suggesting that both filtered seawater and aged detritus have no nutritive value for bay scallops. Growth of the shell of scallops on a diet of phytoplankton was rapid and averaged approximately 0.43 millimeters per day. The tissue weight of the scallops in phytoplankton was approximately 25% above expected, indicating the high nutritive value of this diet. The growth of the shell of scallops on a diet of fresh detritus was slow but continuous, averaging 0.14 millimeters per day, or approximately one third of the rate in phytoplankton. The tissue weight of these animals was approximately the same as expected, suggesting an intermediate nutritive value for this diet. When the scallops were sacrificed at the end of the experiments, all except those in filtered seawater had stomachs and intestines full of material, indicating that they had been ingesting their respective diets. Fecal and pseudofecal formation was observed in the three tanks with

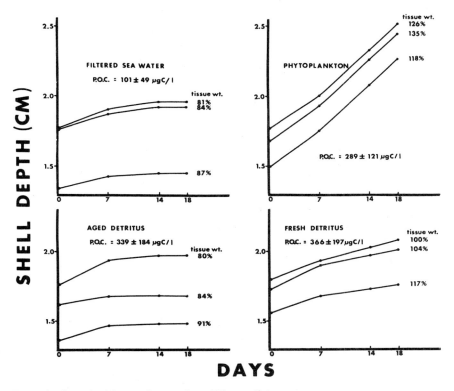

Figure 1. Growth of bay scallops on four different diets.

food, suggesting that the animals had been well supplied with suspended particulate matter.

The average POC for each experimental tank is presented in Fig. 1. The standard deviation indicates the magnitude of variability in the concentration of material in suspension around the animals during the 18-day experiments. The POC of the water in the three tanks which had food added includes the background POC of the filtered seawater. The experimental tank with phytoplankton had a mean chlorophyll *a* concentration of 2.6 micrograms per liter, slightly lower than levels found in the natural estuarine environment. However, the growth rate of the animals (0.43 cm per day, 22°C) was approximately the same as has been observed in scallops kept under field conditions, with the same temperature and chlorophyll *a* conditions (Kirby-Smith, unpublished).

The data on the relative composition of the diets prior to being supplied to the scallops is presented in Table 1. The scallops consumed all three food types. Since the total carbon content of the water in the three experimental tanks (Fig. 2) which had food was similar, the differences in the growth rates of the scallops is most easily explained by food quality rather than by quantity.

Table 1. Analysis of feed tanks on day 17 of the experiment.

Sample (Feed tanks)	Particulate nitrogen μg/l	Atomic C/N ratio	Particulate carbon μg/l[2]	Living carbon[1] ATP μg/l	Percent live carbon	Dissolved carbon μg/l
Filtered seawater	2.4	25	69.8	62.0	89	4200
Phytoplankton	114	11	1460	1990	136	3300
Aged detritus	341	12	4760	384	8	4200
Fresh detritus	552	34	21900	260	1.2	5000

[1] Ack — R.L. Ferguson, N.M.F.S., Beaufort, N.C.
[2] Estimated on C/N

Scallops did not grow on the aged detritus but did show some growth on the fresh detritus. If scallops were dependent upon the percent of living carbon for growth, then both fresh detritus and aged detritus would have to be considered poor food material. However, there was some growth on the fresh detritus, which suggests that a significant amount of digestible material is lost in the aging process. The C/N ratios in Table 1 would suggest that the phytoplankton (C/N = 11) and the aged detritus (C/N = 12) had a high protein content while the fresh detritus (C/N = 34) was relatively low in protein. From these data (Fig. 1, Table 1) it would appear that the C/N ratios, and presumably the protein content, do not necessarily indicate the nutritive value of a food resource for scallops. Burle and Kirby-Smith (in preparation) fed scallops on diets of suspended high protein trout ration (less than 200 microns in size, C/N ratio = 6) in an experiment similar to that described here. The scallops fed on this high protein material had no shell growth and lost weight even though they actively ingested the material and passed it through their digestive system. Winter (33) found similar results with bivalves using a large variety of different food items. It had been expected that such a high protein food source should be easily digested and assimilated.

ESTUARINE FEEDING AND DIGESTION—A FOCUS

Continuing research on the origin, composition, and rates of formation of detritus and microalgae will provide information which is basic to understanding estuarine food chains. The specific origin of detritus is important to studies which seek to explain pathways of the transfer of materials. However, regardless of its origin, the fate of detritus is poorly understood, especially in relation to non-specific deposit and suspension feeders which ingest large quantities of the material.

Based upon the data presented and reviewed here, one major question concerning estuarine food chains is the relative significance of microautotrophs, microheterotrophs, and detritus in the diets of suspension and deposit feeding organisms. Work with bay scallops would suggest that the microautotrophs (phytoplankton) in suspension are the only major energy sources for suspension feeders. However, this hypothesis needs to be tested for other typically estuarine species. The data for deposit and detritus feeding organisms are less clear. Odum (23) suggested that mullet might be obtaining a major portion of their energy from microautotrophs in certain habitats. However, most work with deposit feeders has not attempted to separate microautotrophs from microheterotrophs as potential food items. Using measurements of chlorophyll a, ATP, and direct counts of microbes, together with measurements of particulate carbon and particulate nitrogen, it is now possible to roughly estimate the standing crop of microautotrophs, microheterotrophs, and detritus in both deposits and in suspension (30, 11). Use of these techniques should provide answers to some of the problems presented here.

One of the most useful assays of the relative importance of various food items in a diet is the measurement of growth rates in a system in which the diet is artificially manipulated. One such approach would be the selective elimination of single, potential food items in natural deposits and suspensions. The results of these experiments would allow classification of estuarine animals on the basis of digestion and assimilation and thus clarify some of the pathways and rates of energy transfer in estuarine systems.

ACKNOWLEDGMENTS

The author would like to thank E. Barber, R. Barber, L. Barling and L. Garrigan for constructive review of the manuscript, R. Ferguson for ATP analysis, and L. Garrigan for laboratory assistance.

478   W. W. KIRBY-SMITH

LITERATURE CITED

1. Barber, R.T. 1966. Interaction of bubbles and bacteria in the formation of organic aggregates in seawater. Nature 211:257-258.
2. Darnell, R.M. 1967. The organic detritus program, p. 374-375. *In* G.H. Lauff (ed.), Estuaries. AAAS Publ. 83, Washington, D.C.
3. _____. 1967. Organic detritus in relation to the estuarine ecosystem, p. 376-382. *In* G.H. Lauff (ed.) Estuaries. AAAS Publ. 83, Washington, D.C.
4. Fenchel, T. 1970. Studies on the decomposition of organic detritus derived from the turtle grass *Thalassia testudinum*. Limnol. Oceanogr. 15:14-20.
5. Fisher, T.R., Jr. 1975. The quantity of estuarine seston as food for suspension feeders. Abstract, 38th Annual Meeting Amer. Soc. of Limnol. Oceanogr., Dalhousie Univ., Halifax, N.S.
6. Ferguson, R.L., and M.B. Murdoch. 1973. Microbial biomass in the Newport River Estuary, North Carolina. Proc. 2nd International Estuarine Research Conference. Myrtle Beach, South Carolina.
7. _____, and P. Rublee. 1976. Contribution of bacteria to standing crop of coastal plankton. Limnol. Oceanogr. 21(1): 141-145.
8. Guillard, R.R.L., and J.H. Ryther. 1962. Studies of marine planktonic diatoms. I. *Cyclotella nana* Hustedt and *Detonula confervacea* (Cleve) Gran. Can. J. Microbiol. 8:229-239.
9. Haines, E.B. 1974. Processes affecting production in Georgia coastal waters. Ph.D. Dissertation, Duke University, Durham, North Carolina.
10. Hinson, M.O., Jr., N.E. Armstrong, and J.C. Nelson. 1975. Nutrient exchange and export from a Texas marsh system. Abstract, 38th Annual Meeting, Amer. Soc. Limnol. and Oceanogr., Dalhousie Univ., Halifax, N.S.
11. Hobbie, J.E., and R.J. Daley. 1975. Improved methods for direct counts of bacteria. Abstract, 38th Annual Meeting, Amer. Soc. of Limnol. Oceanogr., Dalhousie Univ., Halifax, N.S.
12. Holm-Hansen, O. 1973. Determination of total microbial biomass by measurement of adenosine triphosphate, p. 73-89. *In* L.H. Stevenson and R.R. Colwell (eds.), Estuarine Microbial Ecology. Univ. South Carolina Press. Columbia, S.C.
13. Jannash, J.W. 1973. Bacterial content of particulate matter in offshore surface waters. Limnol. Oceanogr. 18(2):340-342.
14. Jorgensen, C.B. 1966. The Biology of Suspension Feeding. Pergamon Press, New York. 357 p.
15. Lopez, G.R. 1975. The role of detritus-ingesting invertebrates in the decomposition of *Spartina alterniflora* litter. Abstract, 38th Annual Meeting, Amer. Soc. Limnol. and Oceanogr., Dalhousie Univ., Halifax, N.S.
16. Kirby-Smith, W.W. 1970. Growth of the scallops, *Argopecten irradians concentricus* (Say) and *Argopecten gibbons* (Linne), as influenced by food and temperature. Ph.D. Thesis, Duke University, Durham, N.C. 126 p.
17. _____. 1972. Growth of the bay scallop; the influence of experimental water currents. Exp. Mar. Biol. Ecol. 8:7-18.
18. _____, and R.T. Barber. 1974. Suspension-feeding aquaculture systems: effects of phytoplankton concentration and temperature on the growth of the bay scallop. Aquaculture 3:135-145.
19. Menzel, D.W., and R.F. Vaccaro. 1964. The measurement of dissolved and particulate carbon in seawater. Limnol. Oceanogr. 9:138-142.

20. Moll, R.A. 1975. Production and consumption of phytoplankton in a salt marsh ecosystem. Abstract, 38th Annual Meeting, Amer. Soc. Limnol. and Oceanogr., Dalhousie Univ., Halifax, N.S.
21. Newell, R. 1965. The role of detritus in the nutrition of two marine deposit feeders, the prosobranch *Hydrobia ulvae* and the bivalve *Macoma balthica*. Proc. Zool. Soc. London, 144:25-45.
22. Odum, E.P., and A.A. de la Cruz. 1967. Particulate organic detritus in a Georgia salt marsh — estuarine ecosystem, p. 333-388. *In* C.H. Lauff (ed.), Estuaries, AAAS Pub. 83, Washington, D.C.
23. Odum, W.E. 1970. Utilization of the direct grazing and plant detritus food chains by the striped mullet *Mugil cephalus*, p. 222-240. *In* J.H. Steele (ed.), Marine Food Chains. Univ. Calif. Press, Berkeley.
24. Paerl, H.W. 1974. Bacterial uptake of dissolved organic matter in relation to detrital aggregation in marine and freshwater systems. Limnol. Oceanogr. 19(6):966-972.
25. Russell-Hunter, W.D. 1970. Aquatic Productivity. MacMillan Company, New York. 306 p.
26. Schultz, D.M., and J.G. Quinn. 1973. Fatty acid composition of organic detritus from *Spartina alterniflora*. Estuar. Coast. Mar. Sci. 1:177-190.
27. Sottile, W.S. 1974. Studies of Microbial Production and Utilization of Dissolved Organic Carbon in a Georgia Salt Marsh — Estuarine Ecosystem. Ph.D. Thesis, Univ. Georgia, Athens, Georgia, 153 p.
28. Strickland, J.D.H., and T.R. Parsons. 1972. A Practical Handbook of Seawater Analysis (2nd ed). Bull. 167 Fish. Res. Bd. Can. 310 p.
29. Sushchenya, L.M. 1970. Food rations, metabolism and growth of crustaceans, p. 127-141. *In* J.H. Steele (ed.), Marine Food Chains, Univ. Calif. Press, Berkeley.
30. Thayer, G.W., R.L. Ferguson, and M.A. Kjelson. 1974. Pools of organic matter, carbon, nitrogen, and energy in the Newport River Estuary. Ann. Rep. Atl. Estuarine Res. Center., Nat. Mar. Fish. Service, Beaufort, N.C. p. 245-252.
31. Whitney, D.E., G.M. Woodwell, and C.A.S. Hall. 1975. The flax pond ecosystem study: Exchanges of inorganic forms of nitrogen between a Long Island saltmarsh and Long Island Sound. Abstract, 38th Annual Meeting, Amer. Soc. Limnol. Oceanogr., Dalhousie Univ., Halifax, N.S.
32. Wiebe, W.J., and L.R. Pomeroy. 1972. Microorganisms and their association with aggregates and detritus in the sea: A microscopic study. Mem. Ist, Ital. Idrobiol., 29 Suppl: 325-352.
33. Winter, J.E. 1974. Growth in *Mytilus edulis* using different types of food. Ber. Deutsch. Wissenschaft. Komm. Meeresforsch. 23(4):360-375.
34. Winter, J. 1970. Filter feeding and food utilization in *Artica islandica* L. and *Modiolus modiolus* L. at different food concentrations, p. 196-206. *In* J.H. Steele (ed.), Marine Food Chains, Univ. Calif. Press, Berkeley.
35. Woodwell, G.M., and D.E. Whitney. 1975. The Flax Pond ecosystem study: Phosphate and total phosphorus exchanges between a Long Island salt marsh and Long Island Sound. Abstract, 38th Annual Meeting, Amer. Soc. Limnol. Oceanogr., Dalhousie Univ., Halifax, N.S.
36. Woodwell, G.M., D.E. Whitney, C.A.S. Hall, and R.A. Houghton. 1975. The flax pond ecosystem study: an evaluation of the ecosystem production equations for a salt marsh. Abstract, 38th Annual Meeting, Amer. Soc. Limnol. Oceanogr., Dalhousie Univ., Halifax, N.S.

CYCLING OF POLLUTANTS

Convened by:
Thomas W. Duke
U.S. Environmental Protection Agency
Environmental Research Laboratory
Gulf Breeze, Florida 32561

Estuaries continue to receive pollutants such as oil, heavy metals, pesticides and other toxic organics. It is fitting, therefore, that one session of this meeting be devoted to the impact of these pollutants on this productive environment. Because of their location, estuaries are susceptible to industrial, municipal, agricultural and similar wastes, transmitted through freshwater streams, and other pollutants derived from development of off-shore oil fields and wastes disposed in oceans. It is impossible to discuss all of the important pollutants which enter estuaries. However, for purposes of this session, we will discuss the impact of oil, heavy metals, and pesticides on ecosystems and on biological systems ranging from micro-organisms to fishes.

Studies designed to determine the impact of oil on the estuarine environment are especially important with the increased interest in development and transport of off-shore oil. Of particular interest is knowledge concerning the effect of oil on estuarine microbial populations and the effect of the microbial populations on oil. Most marshes include a high percentage of cellulolytic bacteria, and these bacteria are important in the breakdown or metabolism of oil. The concept of seeding certain species of bacteria or yeast is also of concern at this time.

The effect of metals on estuarine organisms and their environment continues to be investigated. The role of seagrass meadows in coastal ecosystems is just beginning to be documented. Before the impact of metals on these and other primary producers can be assessed, much baseline data must be developed.

Although there appears to be a decline in the level of residues of "hard" organochlorine pesticides, such as DDT, in marine organisms, even low levels of residue may affect the organisms and estuarine systems in which they occur. The distribution of the pesticides, various pathways of transfer and bioaccumulation are known in many instances, yet the ultimate effects of the pesticide on organisms and their environment are relatively unknown. Even less is understood about the synergistic, antagonistic and additive effects of metals, pesticides and toxic organics. The combined toxicities of methoxychlor, cadmium, and polychlorinated biphenyls are discussed in this session.

MICROBIOLOGICAL CYCLING OF OIL IN ESTUARINE MARSHLANDS

D.G. Ahearn, S.A. Crow, and N.H. Berner
Georgia State University
Atlanta, Georgia 30303
S.P. Meyers
Louisiana State University
Baton Rouge, Louisiana 70803

ABSTRACT: Indigenous microflora of sediments of *Spartina* marshes of the Louisiana coast include a high percentage of cellulolytic bacteria and ascosporogenous yeasts (*Pichia* and *Kluyveromyces*). At sites either accidentally or experimentally inundated with crude oil, the proportion of hydrocarbon-utilizing bacteria and yeasts increased. The marsh sediments contained low populations of hydrocarbonoclastic fungi with few strains showing significant oil-emulsifying properties. In culture, representative microorganisms readily utilized alkanes from C12 to C18. Oil utilization by a representative bacterium increased significantly as incubation temperatures were raised from 10 to 30 C, whereas yeasts showed peak activity at 20 C. Oil utilization by test microorganisms was negligible at 5 C.

Seeding of experimentally oiled marsh plots with a mixed culture of *Candida maltosa* and *C. lipolytica* demonstrated that these species survived in the oiled area without spreading to adjacent sites. In culture these strains gave significant emulsification of crude oil and utilized up to 90% of selected hydrocarbons.

INTRODUCTION

Recently, Hood et al. (8) reported that pollutant crude oil alters the bacterial populations of estuarine marshes. Long term effects of these alterations are unknown. Catastrophic oil pollution of coastal marshes in the United States has not been a major environmental concern, but recent political and economic occurrences may soon change this.

The Mid-East embargo on crude exports and the escalation of prices for foreign crude in 1974-75 accentuated the dependency of the United States on oil imports. This dependency had been developing since the late 1940's as the nation turned from coal to a cheaper and cleaner petroleum industrialization and from an exporter to a net importer of crude. By 1974 the U.S. was consuming about 18 million barrels of oil per day with less than two-thirds of it from domestic production (5).

The decline in the world's economy and the lesser demand for crude, brought on by escalating crude prices, has hindered the shipping of oil, the development of superports, and the search for new oil. These factors in turn have inculcated the nation with an awareness that alternate sources of energy will need to be developed. Considerable interest is being expressed in solar, geothermal, and atomic energy, but the prime source of power undoubtedly will be our vast coal reserves. The shift from the "petroleum age" was sparked prematurely by world politics, but limited resources made the change inevitable for the current century. Nevertheless, oil-based industries will require full development of domestic reserves for the practical transition to alternate sources of power. The production, piping and shipping of oil in coastal waters will be of major importance until after 2000. The oil processing and producing centers of the Gulf of Mexico coast will flourish with the advent of the projected superports. Therefore, the problems and potential of oil pollution for the Gulf of Mexico will be increased, not diminished, with the current energy crisis.

Much of the Gulf coast, particularily the south Louisiana coast, is a productive nursery ground for marine life. The Barataria Bay region of Louisiana is characterized by estuarine marshes dominated by *Spartina alterniflora*. Productivity of these coastal marshes has been estimated to be as great as 10 tons dry organic matter/acre/year (3). The rhizosphere of the *Spartina* is rich in microbes active in the transformation of the plant debris. Inbalance in this microbial ecosystem could result in detrimental modifications of the productivity of the entire Gulf.

For the past several years we have been studying the effects of crude oil and the subsequent addition of hydrocarbonoclastic fungi on the microbial ecology of the *Spartina* habitat. This report examines the effects of Louisiana crude oil and a mixed culture of yeasts on the indigenous microbial flora.

MATERIALS AND METHODS

Environmental monitoring. 1) Oil retention plots and control marsh were located near the entrance to Airplane Lake (Grande Isle, La.). Oiled areas were established in November 1970 and treated with 250 ml of Louisiana crude oil each month for from 9 to 24 months. An extensive description of the microbiology of this area has been made (7, 10, 11). 2) Martigan Point is a small island NW of Grande Isle in the Barataria Bay system. Parts of this island were heavily oiled following an oil pipeline break during October 1972. Samples were collected by the methods described by Meyers et al. (10). Microbiological studies involved the enumeration of total heterotrophic bacteria, total yeasts, hydrocarbonoclastic bacteria, and hydrocarbonoclastic fungi (yeasts and filamentous fungi).

Total heterotrophic bacteria were determined by spread technique with Marine Agar 2216 (Difco). Yeast numbers were determined on M12 agar acidified to pH 4.5 with sterile 10% lactic acid (10). Hydrocarbonoclastic microorganisms were determined by a modified MPN method (6). Following enumeration procedures, selected microorganisms were subcultured from enumeration media for further studies.

Physiological studies. Isolates of microorganisms from oil enriched sites, control marsh, and two yeasts, *C. lipolytica* and *C. maltosa*, from the culture collection of Georgia State University were tested for activity on crude oil and other hydrocarbons. Hydrocarbon assimilation was examined in several cultures at 26 C. Oil assimilation was tested by visual observation of growth in (Difco) yeast nitrogen base (YNB) with 2% Louisiana crude oil. Inococulum was prepared from cells grown in YNB broth with 0.01% glucose for 24-48 hours to deplete endogenous carbohydrates. Cultures were incubated on a roller drum (40-50 rpm). Growth was compared to uninoculated controls and recorded on a scale of 0-3 (0, no growth; 3, maximal growth compared to a glucose control). Selected isolates were studied for the metabolism of various and pure alkanes. Bacterial cultures were grown in a 1.0% yeast extract-sea water broth for 24 hours prior to inoculation. Medium containing 2% crude oil, 0.05% NH_4Cl, 0.05% K_2HPO_4, 1.0% Na_2HPO_4 in 75% sea water was inoculated with 1.0 ml of cell suspension. Cultures were extracted with hexane. The extract was fractionated by column chromatography on an alumina/silica gel column (2/15) (1 cm d × 30 cm). Fractions classed as alkane, aromatic, and asphaltene were eluted with 50 ml portions of hexane, benzene and 1/1 $CHCl_3$-MeOH respectively. All microbial cultures were compared to uninoculated controls incubated and extracted under the same conditions. Gravimetric determinations were made on the residual of both aromatic and asphaltic fractions, and the alkane fraction was divided into 2 portions, one analyzed by gas chromatography (GLC), the other by gravimetric methods.

Temperature studies. Organisms were examined for assimilation of crude oil at various temperatures from 6-30 C. Activity was assessed by growth, BOD, column chromatography with gravimetric analysis, and GLC analyses of the alkane fractions. Oxygen consumption was determined by a modification of the technique of Tool (13) with a Hach manometric apparatus (Model 2173). Inocula for respiration studies consisted of approximately $4×10^7$-$6×10^7$ cells growth in a YNB with 0.5% glucose for 24 hours. Respiration flasks contained 1% hydrocarbon and 0.01% yeast extract in sea water. Oxygen consumption was determined at 6, 20, and 30 C after incubation for 72 h. Controls containing hydrocarbon and yeast extract, inoculum and yeast extract, hydrocarbon only, inoculum only, or yeast extract only were included. Net oxygen consumption (mg/1) with 1.0% crude oil (v/v) is reported with corrections for both endogenous respiration and autoxidation.

Survival of seed cultures. Seed culture studies were conducted in estuarine waters and marsh with *C. lipolytica* and *C. maltosa* (syn. *C. subtropicalis*, 9). Oil retention devices were fabricated from cylindrical 189 liter vessels with lateral holes bored below the water line to allow tidal flushing. Containers were placed in estuarine water and 100 ml of crude oil was added to the container.

Additional survival tests were conducted in previously established oil plots. Cultures were added to marsh plots treated with 100 ml of oil. Following introduction of cultures, oil was added periodically over an eleven-month period. Samples were collected from the containers at monthly intervals over a one-year period, and yeast were enumerated as previously described (10, 11).

Hydrocarbon utilization by *C. maltosa* and *C. lipolytica* was studied with hexadecane (2% v/v) as the sole carbon source in YNB broth. Cultures were incubated on a roller drum (75 rpm) at 25 C for 2 weeks. Samples for gas chromatographic analysis were extracted with petroleum ether (2:1). Analysis was done with a flame ionization detector Varian Aerograph 2600 with 0.64 cm × 1.8 m glass column with 2% SE-30 on Gas. Chrom Q 100/120.

RESULTS

Indigenous microorganisms. Densities of proteolytic, cellulolytic, hydrocarbonoclastic, and total heterotrophic bacteria in oiled and non-oiled marsh sites are presented in Table 1. With the addition of crude oil, cellulolytic bacteria

Table 1. Populations of heterotrophic bacteria in oiled and non-oiled marsh sediments, 1973.

	Concentration of bacteria (log 10/g wet sediment)					
	Jan	Feb	Mar	May	June	July
Airplane Lake Site						
Control						
total	7.72	7.93	8.36	7.99	8.60	7.60
cellulolytic	4.97	6.34	6.15	5.85	4.36	6.20
proteolytic	5.78	6.57	6.45	6.59	6.30	6.45
hydrocarbonoclastic	5.52	4.78	5.36	4.89	3.90	4.23
Oil Plot						
total	7.84	8.11	8.32	7.96	8.18	7.43
cellulolytic	4.22	5.63	5.49	5.04	2.85	5.20
proteolytic	6.18	6.69	6.48	6.39	6.81	6.08
hydrocarbonoclastic	6.52	7.20	6.66	6.50	7.36	5.60
Martigan Point,						
Oil Spill Site						
Control						
total	–	–	–	7.81	8.70	7.95
cellulolytic	–	–	–	5.45	5.38	4.70
proteolytic	–	–	–	5.60	5.49	5.40
hydrocarbonoclastic	–	–	–	4.84	4.95	4.70
Station No. 1						
total	8.15	8.87	–	9.65	8.30	7.68
cellulolytic	4.98	7.34	–	5.98	4.41	3.85
proteolytic	6.25	7.00	–	6.58	5.28	6.15
hydrocarbonoclastic	6.60	7.73	–	7.04	5.53	6.96

appeared to decrease in number, whereas densities of hydrocarbonoclasts increased. A similar trend was noted for bacterial populations at the site of the Martigan Point oil spill.

Yeast populations in marsh sediments were enumerated before and 2 months after the addition of Louisiana crude oil. In response to oil, total cell populations doubled, and there was a shift in the microflora from species of *Pichia* and *Kluyveromyces* to species of *Rhodotorula* and *Trichosporon* (Table 2). We reported results similar to these in continuing analyses of the effects of oil on marshland yeast populations (1, 12).

Microorganisms selected from the predominant microflora of the oiled sediments were examined for their ability to grow with Louisiana crude oil as a carbon source (Table 3). Of the repersentative microorganisms, thirteen of the bacteria and three of the yeasts produced some emulsification, but only one bacterium and one yeast gave significant growth with marked emulsification of oil. The bacterium, *Corynebacterium* sp. MS-228, by gravimetric analysis of oil fractions showed preferential utilization of alkanes with maximal degradation near 30 C (Table 4). Some utilization of the aromatic fraction was evident but data for asphaltics were questionable. Gas chromatographic analyses of the alkane fraction with calculations of peak area ratios to pristane and to tetracosane suggested that C19-C23 alkanes were used preferentially. The yeast *Candida* sp. MS-309 gave little evidence of utilization by gravimetric analysis. Gas chromatographic analysis however indicated that C11-C13 alkanes accumulated in the culture system (Fig. 1).

Table 2. Effect of Louisiana crude oil on yeast populations in coastal marsh sediments.

Genera	Percent of total population	
	Before	After[a]
Pichia/Kluyveromyces	85-100	10-83
Rhodotorula/Rhodosporidium	< 10	5-50
Cryptococcus	< 15	< 10
Trichosporon/P. ohmeri	< 10	12-35
mean population[b]	7150	14772

[a]determined two months after oil addition, analysis of 33 samples in triplicate.
[b]colony forming units per mg wet sediment.

Table 3. Hydrocarbonoclastic activity of representative estuarine microorganisms.[a]

Total Isolates Examined		Growth on Crude			Emulsification
		Good	Weak	Negligible	
Bacteria	41	1	20	20	1
Yeast	27	3	7	17	1
Fungi	12	0	4	8	0

[a]Seven days growth, shaken culture in basal salts with 2.0% Louisiana crude oil.

Table 4. Comparison of weight change[a] of crude oil inoculated with culture MS-228 at 10, 20, and 30 C.

Fraction	10 C	20 C	30 C
Total	-0.4%[b]	-1.9%	-8.4%
Alkane	-0.1%	-2.6%	-9.5%
Aromatic	-2.8%	-1.9%	-4.8%
Asphaltic	+7.0%	+2.2%	+1.4%

[a]Compared to uninoculated controls for each temperature.
[b]Weight loss designated by (-); weight gain designated by (+).

Seed cultures. Two hydrocarbonoclastic yeasts *C. lipolytica* 37-1 isolated from a frankfurter and *C. maltosa* R-42 obtained from a freshwater holding pond of an asphalt refinery were compared with an isolate of *P. ohmeri* from the marsh sediments for growth on oil. The culture of *Pichia* was selected because of its active growth on crude with oil emulsification. The two non-marsh cultures showed greater growth, emulsification, and oxygen uptake on crude oil (Table 5) than the marsh yeast. In cultures with pure hydrocarbons as substrates, *C. lipolytica* and *C. maltosa* utilized n-alkanes from C9-C18; *C. maltosa* also showed some growth on octane. Gas chromatographic analyses demonstrated that both yeasts consumed up to 90% of hexadecane in a culture system within seven days. The persistence of these two yeasts after their introduction into the oiled marsh plots is shown in Table 6. Both yeasts were recovered from the oiled plots for up to seven months, but they did not spread into adjacent control areas. There was little change in the species composition of the indigenous fungal flora.

Table 5.   Oxygen consumption by yeasts after 72 hours growth on Louisiana crude oil.

Species	Temperature		
	6 C	20 C	30 C
Candida lipolytica	40[a]	312	285
Candida maltosa	25	268	245
Pichia ohmeri	0	185	—

[a]mg oxygen/1, average of 3 repeat tests; corrected for endogenous and uninoculated autoxidation controls.

DISCUSSION

The relative increase of hydrocarbonoclastic bacteria and the shift in the type of predominant yeast clearly demonstrate the effects of short term oil enrichment on marsh sediments. A reduction in the numbers of cellulolytic bacteria also appears to be coupled with short term intrusion by hydrocarbon pollutants. The development of only a restricted number of organisms capable of extensive

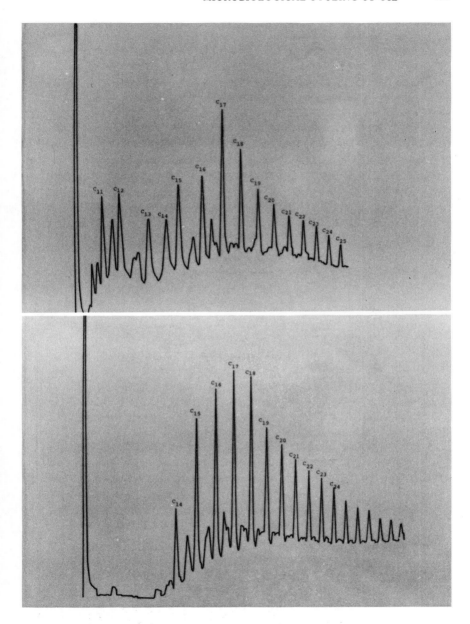

Figure 1. GLC analysis of alkane fraction of Louisiana crude oil incubated at 30 C for 4 days, uninoculated control (below) and after growth of *Candida* sp. MS 309 (above).

Table 6. Fungi isolated from oiled sediments Barataria Bay, Louisiana

| | Prior to Inoculation | Time after inoculation of seed cultures | | | | | | | |
| | | 2 months | | 3 months | | 7 months | | 11 months | |
		oiled	control[b]	oiled	control	oiled	control	oiled	control
Seed cultures:									
Candida maltosa R42	0[a]	+	0[c]	+	0	+	0	0	0
C. lipolytica 37-1	0	+	0	+	0	+	0	0	0
Indigenous Fungi:									
Yeast-like[d]									
Rhodotorula-Rhodosporidium sp.	+	+	+	+	+	+	+	+	+
Cryptococcus sp.	+	+	+	+	+	+	+	+	+
Kluyveromyces sp.	+	+	+	+	+	+	+	+	+
Pichia sp.	+	+	+	+	+	+	+	+	+
Trichosporon sp.	+	+	+	+	0	0	+	+	+
Aureobasidium sp.	+	0	+	0	+	0	+	0	+
Filamentous									
Cladosporium sp.	+	0	+	+	+	0	+	+	+
Penicillium sp.	+	0	+	+	+	0	+	0	+
Cephalosporium sp.	+	+	+	+	+	+	+	+	+
Fusarium sp.	+	+	0	+	+	+	+	+	0
Alternaria sp.	+	0	+	0	0	0	0	0	0
Trichoderma viride	0	0	0	0	0	0	0	0	+
Aspergillus sp.	0	0	0	0	+	0	0	+	0

[a] indicates not isolated, + present.
[b] control plots, periodically oiled, no inoculum.
[c] seed cultures introduced in non-oiled sediments were not recovered after 72 hours.
[d] major species; C. albidus, C. laurentii, K. drosophilarum, P. spartinae and T. cutaneum.

utilization of crude oil suggests that microbial cycling of oil, particularly the asphaltic fraction in a non-chronically oil polluted environment, will be slow. Hood et al. (8), reported extremely high ratios of hydrocarbonoclasts to total heterotrophs in chronically polluted marsh areas. The biodegradation of oil by *Corynebacterium* MS-228 increased with temperature to 30 C, whereas the yeasts *C. lipolytica* and *C. maltosa* gave their highest activity at 20 C. Laboratory data in this area can be extended only cautiously to field conditions, since water temperatures may differ dramatically from temperatures of the microorganism-oil milieu under differing environmental conditions. The laboratory experiments also were conducted with a suitable nitrogen supplement. Gibbs et al. (4) stressed the need for continuous replenishment of nitrogen for in situ oil degradation. The advantages and disadvantages of fertilization as well as seeding of oil for degradation in bunkers have been discussed in detail elsewhere (2).

Although little weight loss of oil was observed in some pure culture systems a considerable alteration of the chemical nature of the oil occurred. This type of activity is possible as an essential mechanism in the environmental removal of oil by a heterogenous microbial population. The microbial seeding of oiled areas would seem to be a means of hastening the biodegradation of oil in instances of cataclysmic pollution in sites not previously subjected to hydrocarbon pollution.

ACKNOWLEDGEMENTS

This research was supported in part by Office of Naval Research contract ONR N000-14-71-C-0145 and grant R803141-01 from the Environmental Protection Agency.

LITERATURE CITED

1. Ahearn, D.G., and S.P. Meyers. 1972. The role of fungi in the decomposition of hydrocarbons in the marine environment, p. 12-19. *In* A.H. Walters and E.H. Hueck van Plas (eds.), Biodeterioration of Materials. Vol. 2. John Wiley and Sons. New York.
2. _____, and _____, (eds.). 1973. Microbial Degradation of Oil Pollutants. Center for Wetland Resources, Louisiana State University Publication. No. LSU-SG-73-01.
3. Eleuterius, L.H., and S.P. Meyers. 1974. *Claviceps purpurea* on *Spartina* in coastal marshes. Mycologia 66:978-986.
4. Gibbs, C.F., K.B. Pugh, and A.R. Andrews. 1975. Quantitative studies on marine biodegradation of oil, II. Effect of temperature. Proc. R. Soc. Lond. B. 188:83-94.
5. Grove, N. 1974. Oil the dwindling treasure. Nat. Geograph. 145:792-825.
6. Gunkel, W. 1967. Experimentell-okologische Untersuchungen über die limitierenden Faktoren des midrobiellen Olablanes in marinen Milieu. Helgolander Wiss. Meeresunters. 15:210-225.
7. Hood, M.A., and A.R. Colmer. 1970. A comparison of three media in determination of bacterial content of sediments of Barataria Bay. Louisiana State University Coastal Studies Bull. 5:125-133.

8. _____, W.S. Bishop, Jr., F.W. Bishop, S.P. Meyers, and T. Whelan III. 1975. Microbial indicators of oil-rich salt marsh sediments. Appl. Microbiol. 30:982-987.

9. Meyer, S.A., K. Anderson, R.E. Brown, M. Th. Smith, D. Yarrow, G. Mitchell, and D.G. Ahearn. 1975. Physiological and DNA Characterization of *Candida maltosa* a hydrocarbon-utilizing yeast. Arch. Microbiol. 104:225-231.

10. Meyers, S.P., M.E. Nicholson, J. Rhee, P. Miles, and D.G. Ahearn. 1970. Mycological studies in Barataria Bay, Louisiana, and biodegradation of oyster grass, *Spartina alterniflora*. Louisiana State University Coastal Studies Bull. 5:111-124.

11. _____, D.G. Ahearn, and P.C. Miles. 1971. Characterization of yeasts in Barataria Bay. Louisiana State University Coastal Studies Bull. 6:7-15.

12. _____, _____, S. Crow, and N. Berner. 1973. The impact of oil on microbial marshland ecosystems, p. 221-228. *In* D.G. Ahearn and S.P. Meyers (eds.), The Microbial Degradation of Oil Pollutants. Center for Wetland Resources, Louisiana State University Publication. No. LSU-SG-73-01.

13. Tool, H.R. 1967. Manometric measurement of the biochemical oxygen demand. Water and Sewage Works. 114:211-218.

TRACE METAL CYCLES IN SEAGRASS COMMUNITIES

Warren Pulich[1]
University of Texas Marine Science Laboratory
Port Aransas, Texas 78373

Steve Barnes
Department of Chemistry
Texas A&I University at Corpus Christi
Corpus Christi, Texas 78411

Patrick Parker
University of Texas
Marine Science Laboratory
Port Aransas, Texas 78373

ABSTRACT: Processes which make essential trace metals, as well as other nutrients, readily available may account for the distribution of and patterns of succession in seagrass meadows. Complete inventories of Fe, Mn, Zn, and Cu were compared for the five species of Texas seagrasses from study areas in Redfish Bay and Northern Laguna Madre. Seasonal fluctuations in an area were minimal, while major areal differences between plants in Laguna Madre and Redfish Bay were detected. The seagrasses from Laguna Madre contained more Mn and less Fe (*Halodule* leaf avg. Fe/Mn ratio = 0.37, avg. Mn = 250 ppm) than Redfish Bay plants (*Halodule* leaf avg. Fe/Mn ratio = 4.0, avg. Mn = 90 ppm), although sediments from Laguna Madre had 50% lower concentrations of leachable (i.e. "available") Fe and Mn than Redfish Bay sediments. Measurements showed that seagrass detritus remained high in the 4 trace metals; thus unless the detritus is recycled at its source of production, large amounts of metals will be removed from the ecosystem. We suggest that recycling of detritus and gradual release of nutrients into the water column in the closed Laguna Madre system may be more advantageous to seagrass growth than increased levels of trace metals in the sediments of a more open system such as Redfish Bay.

---

[1] Author to whom correspondence should be addressed.

INTRODUCTION

The submergent marine flowering plants known as seagrasses are important members of coastal marine communities due to their roles as: 1) prodigious primary producers; 2) sediment stabilizers; 3) substrata for attachment of epiphytic organisms; 4) nursery grounds for marine animals; and 5) agents in biogeochemical cycles, eg. nitrogen and phosphorus.

This last function stems from their capacity to both absorb materials from and secrete materials into the sediments (via the roots) and the water column (via the leaves). When human activities contribute pollutants and toxic wastes to coastal waters, seagrasses might also absorb and accumulate these foreign materials. Thus pollutants would be concentrated and cycled through a seagrass-based food chain.

While heavy metals such as Pb, Hg and Cd are widely recognized as pollutants, even the so-called physiological "trace metals" required by all living cells (eg. Mn, Cu) become toxic when the concentration is too high. Therefore, these trace nutrients are potential pollutants. In order to substantiate the occurrence of trace metal pollution, "baseline" data must be accumulated from pristine areas prior to influx of trace metals.

Two main problems are associated with measurement of "baseline" levels of trace metals in plants (as well as animals): seasonal and geographic variations. Work by Walsh and Grow (17) has shown the degree of seasonal variation in trace metal composition of *Thalassia* and *Ruppia* from the same sampling site in Florida. Geographic variation in trace metal content of seagrasses could indicate an effect on seagrass production and growth by so-called "edaphic" factors. While this type of trace metal control of plant distribution has long been recognized for terrestrial plant systems (4, 6), and even some marsh plant systems (12), the idea has not been considered in relation to seagrasses.

In an attempt to provide "baseline" data on the trace metal content of seagrasses in the above frame of reference, we have examined the relationship between selected trace metals in seagrass plants and the physicochemical properties of sediments where the plants grow. The threefold objectives of this study were: (1) to determine baseline levels of Fe, Mn, Zn, and Cu in the Texas species of seagrasses from one locality (Redfish Bay); (2) to compare Fe, Mn, Zn, and Cu levels in Laguna Madre seagrasses and sediments with those from Redfish Bay; and (3) to delineate some of the steps in biogeochemical cycling of these trace metals in seagrass-dominated estuaries.

METHODS AND MATERIALS

Sampling sites were chosen for regular collecting in Redfish Bay and Northern Laguna Madre[2] along the S. Texas coast (Fig. 1). Three sites in L. Madre where

---

[2] Northern Laguna Madre refers to the Laguna Madre from Kennedy Causeway at Corpus Christi, Texas, south to Baffin Bay, Texas. Hereafter it will be referred to as L. Madre.

Ruppia, Halodule, and *Halophila* occur together corresponded to three similar sites in Redfish Bay. *Thalassia* and *Syringodium* were collected only at sites in Redfish Bay. Normally, sods of grass were dug up with a shovel and sediment washed free. In some cases, a special polyvinylchloride coring device was used to take a cylindrical core of sediment and grass, 20 cm diam. x ca. 30 cm long. The

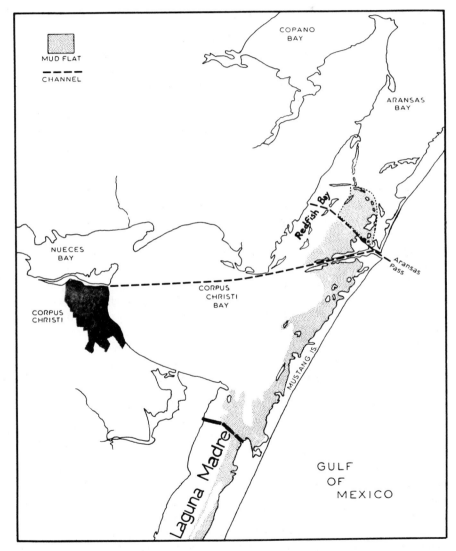

Figure 1. Map of the S. Texas coast, denoting northern Laguna Madre and Redfish Bay near Corpus Christi, Texas.

sediment and grass samples were both saved in this case; sediment was sampled from the root zone at a depth of 5-12 cm.

Grass samples were returned to the laboratory suspended in sea water in plastic bags. Within 24 hrs, the grass was hand-cleaned thoroughly of epiphytes, while submerged in sea water. The cleaned grass was next rinsed in tap water (3 times) and deionized water (2 times) in less than 3 mins. Then it was dried at 85 C in an oven for 2-4 days and dessicated over P_2O_5 until analysis. The sediments were similarly dried prior to analysis.

Plant material was analyzed for Fe, Mn, Zn, and Cu with a Perkin-Elmer 303 atomic absorption spectrophotometer after digestion with conc. red fuming nitric acid at ca. 100 C. Digested samples were dissolved in 3N HCl and filtered. Matrix effects were checked and found not to influence the determinations. A certified NBS standard (Orchard Leaves standard reference material #1571) was run along with all seagrass samples to check the accuracy of the method. Based on this standard, the precision of the analyses was: Fe, ± 20%; Mn, ± 5%; Zn, ± 10%; and Cu, ± 8%.

Leachable metals in sediments were extracted by the method of Chester and Hughes (2) with a hydroxylamine—acetic acid mixture. Trace metals were then analyzed as for plant material.

Photosynthesis of *Halophila* was measured on plants found growing at ca. 76 cm depth (at high tide), actual depths 79 cm L. Madre, 73 cm Redfish Bay. One shoot (stem and 6-7 leaves) was exposed to 20 μcuries $NaH^{14}CO_3$ (sp. act. 55.3 μc/μmole) in 50 ml filter-sterilized sea water containing 0.016 M Tris buffer pH 8.3. Samples were incubated in 100 ml sealed bottles at depths of 38 and 76 cm in the water column. Samples were taken by cutting a leaf off the stem, washing in fresh sea water for 2x 30 secs, blotting dry, and finally fuming over conc. HCl for 30 secs. Leaf samples were then dried at 40-50 C for a minimum of three days. ^{14}C assimilated by the whole leaves was counted using a Nuclear Measurements Corp. (Indianapolis) gas flow proportional counter (Model PC-4).

Particle size analysis of sediments from 5-12 cm depth was performed according to Folk (5) as modified by R. Harwood of the MSL (pers. commun.). The sizes of the various particle fractions were (in μm): clay, 0.06 - 3.9; silt, 3.9 - 62.5; sand > 62.5. Wet sediments were used for this determination.

RESULTS

Grain Size Analysis. Particle size distributions of root zone sediments from seagrass areas are presented in Fig. 2. TH-3, RS, and RH-1 are typical of *Thalassia, Syringodium,* and *Halodule - Halophila* substrates, respectively, in Redfish Bay. Differences between species are readily apparent. The shell content appears of major importance to *Thalassia.* All *Thalassia* beds examined (3) had at least 50% shell. Other species grow regularly in muddy sediments with much less shell; the amount of clay-sized sediment is often high. The pH of sediment interstitial water was always 6.5-7.0.

When compared with Redfish Bay sediments, L. Madre sediments (LM-1, -2, -3) appear much lower in shell material and have increased amounts of silt and clay. L. Madre sediments can therefore be described as more consolidated and of finer grain size than Redfish Bay. The high clay makes the L. Madre substrate much more compactable, while the shell and sand in Redfish Bay make these sediments firmer and more solid. The lack of visible shell could be related to the absence of naturally-occurring *Thalassia* in L. Madre. The pH of L. Madre sediment water was the same as Redfish Bay. Salinities averaged slightly higher in L. Madre sediment pore water than Redfish Bay (32°/oo L. Madre vs. 26°/oo Redfish, during winter and spring, 1974-1975).

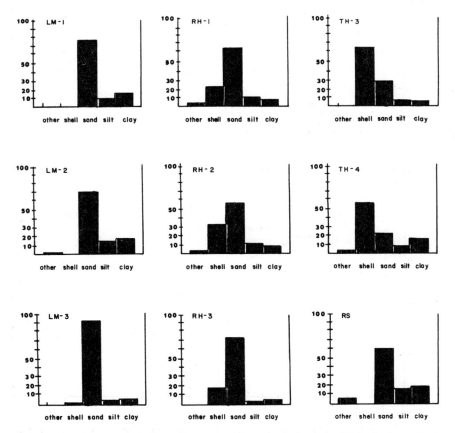

Figure 2. Grain size analysis for sediments from seagrass study areas. LM-1, -2, -3 are sites in Laguna Madre vegetated by *Halodule, Ruppia,* and *Halophila.* RH-1, -2, -3 are corresponding sites in Redfish Bay. TH-3 and TH-4 are sites in Redfish Bay where *Thalassia* occurs. RS is a site in Redfish Bay where *Syringodium* occurs.

Seasonal Variations in Trace Metals. A limited amount of data on seasonal variation in Fe, Mn, Zn, and Cu of *Halodule* was obtained (Fig. 3) for Redfish Bay and L. Madre populations. Fig. 3 shows that the leaf contents of Fe, Mn,

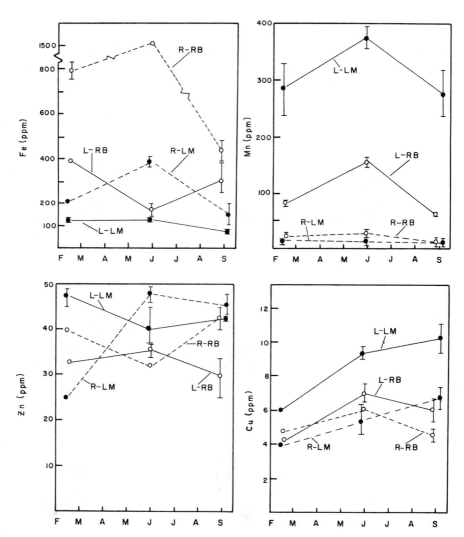

Figure 3. Seasonal variation in Fe, Mn, Zn, and Cu content (ppm dry wt.) of *Halodule wrightii* from study area LM-1 in Laguna Madre (= LM) and study area RH-1 in Redfish Bay (= RB). Open circles are for Redfish Bay plants; closed circles are for Laguna Madre plants. Solid lines are for leaves (= L); dashed lines are for roots (= R).

and Cu, as well as root contents of Fe, of plants in the two bay systems vary considerably over the course of the growing season. But *the trend* in variation for these particular metals at both localities is generally similar. Thus, for example, leaf Mn is low in late winter, rises to a peak in the late spring and then drops back to a lower level by late summer. The Zn content of plants did not vary in a predictable manner, nor did the root levels of Mn and Cu. From this, the validity of comparison of plants of different species or plants from different areas can be inferred if collected at the same time of year.

Trace Metal Content of Plants. The Fe, Mn, Zn, and Cu contents of the four Texas seagrasses are compared in Table 1. Several general observations emerge. (1) In all species except for *Syringodium*, Fe is more abundant in the roots than leaves. (2) The opposite pattern is followed by Mn, Mn being more abundant in leaves than roots in all species. (3) Zn distribution appears to parallel Fe somewhat, with *Syringodium* again presenting the opposite pattern from the other seagrasses. (4) Except for *Halophila*, which had much lower levels of Cu than other seagrasses and was not determined, Cu was higher in leaves than roots at this season. The Cu data reveals a potentially interesting difference between *Halophila* and the other species with respect to its ability to accumulate Cu or requirements for this element.

Fe/Mn Ratios in Plants. Some hint of a disparity between L. Madre and Redfish Bay populations of *Halodule* emerges from Fig. 3. However direct comparison of Fe/Mn ratios for *Halodule* leaves from Redfish Bay and L. Madre

Table 1.   Average Fe, Mn, Zn and Cu content of seagrasses collected during peak of 1974 summer season in Redfish Bay, Texas. *Halodule* and *Halophila* collected on 28 August and *Syringodium* on 1 August, all from same site. *Thalassia* collected 17 July from another site approximately 2 miles away. Values in μg/g dry wt. Values are averages of 2-3 determinations done on a single sample. Leaves = plant parts above ground, including stem. Roots = plant parts below ground, including rhizome.

	Halodule	Halophila	Syringodium	Thalassia
Iron				
Leaves	300 ± 38	1560 ± 189	776 ± 85	72 ± 13
Roots	480 ± 64	7200 ± 362	702 ± 200	230 ± 50
Manganese				
Leaves	53 ± 3	112 ± 8	184 ± 3	97 ± 10
Roots	10 ± 1	106 ± 8	33 ± 4	63 ± 20
Zinc				
Leaves	34 ± 2	20 ± 2	16 ± 2	18 ± 2
Roots	40 ± 4	27 ± 3	10 ± 2	37 ± 4
Copper				
Leaves	6.0 ± 1	<2.0	6.5 ± 2	5.0 ± 1
Roots	4.5 ± 1	<2.0	3.0 ± 1	3.7 ± 1

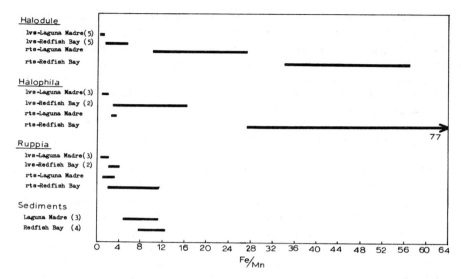

Figure 4. Comparison of Fe/Mn ratios of *Halodule wrightii*, *Halophila engelmanii*, and *Ruppia maritima* from Laguna Madre and Redfish Bay. Numbers in parentheses indicate number of samples involved. Leaf (= lvs) and root (= rts) tissues done separately. Sediment values are for LM-1, -2, -3 sites in Laguna Madre and RH-1, -2, -3, and RS sites in Redfish Bay.

regardless of season reveals a curious correlation (Fig. 4): L. Madre plants contain much lower levels of Fe and correspondingly higher levels of Mn. The resulting Fe/Mn ratio is much lower (less than 1.0) for L. Madre plants compared to Redfish Bay plants (greater than 1.0). This relationship exists independent of the season and the sampling sites within the two estuarine water systems. Moreover, the Fe/Mn phenomenon is exhibited by the other two submergent grasses common to both L. Madre and Redfish Bay, *Halophila* and *Ruppia*, though there are less data for these latter species. Roots of L. Madre plants are also distinguishable from roots of Redfish Bay plants on the basis of Fe/Mn ratios.

The iron content of these submerged grasses is reflected by an interesting morphological feature. We have often observed that seagrasses from Redfish Bay have a readily visible coating of reddish-colored material at the base of the shoots, in the area of the sediment-water interface. Much of the time this "rusty" material appears to coat the leaf sheath. These rusty portions of the leaves contain very high percentages of Fe (up to 2%) when analyzed (Table 2a), although Mn, Zn, and Cu are still at ppm levels. This would seem to indicate a preferential build-up or deposit of oxidized Fe. It is mainly observed in certain Redfish Bay areas. Possibly, this is caused either by the plants themselves

Table 2.   Fe, Mn, Zn and Cu content (μg/g dry wt) of (a) "rust" material from Redfish Bay plants and (b) seagrass detritus collected during winter 1974-75.

(a) "Rust" material. One sample of each species was analyzed; samples were collected in February, 1975.

	Halodule	Ruppia	Syringodium
Fe	20,500	17,100	4,300
Mn	77	818	205
Zn	51	61	25
Cu	8.1	8.2	9.0

(b) Seagrass detritus. One sample of "green" and one sample of "dried" detritus were analyzed. Both types of detritus were collected in January, 1975 from Redfish Bay.

	Halodule		Syringodium		Thalassia	
Fe	green	180	green	60	green	74
	dried	550			dried	580
Mn	green	122	green	68	green	61
	dried	228			dried	260
Zn	green	30	green	33	green	34
	dried	38			dried	46
Cu	green	7.4	green	7.2	green	2.2
	dried	7.8			dried	9.0

(oxygen diffusing out) or from microbial activity in the top portions of the sediments.

Fe and Mn in Sediments. Two reservoirs of trace metals were examined in an effort to determine the significance of the Fe/Mn levels in plants. Leachable Fe and Mn in bay sediments were considered the immediate source of trace metals. Upon extraction and measurement of this sediment fraction, a paradoxical pattern was revealed (Table 3). L. Madre sediments were lower in both "available" Fe (by about 1/3) and Mn (by 1/2) than Redfish Bay sediments. Zn was similar for both locations (between 5 and 10 ppm); Cu was below detectable limits ($<$ 5 ppm). Fe/Mn ratios in both sediments were similar (L. Madre 4.7 - 10.8; Redfish Bay 7.4-12.7).

Trace Metals in Seagrass Detritus. Seagrass detritus also proved to be high in trace metals (Table 2b). Green detritus is plant tissue that still has chlorophyll and is metabolically active to some extent. After about a week's time, it becomes necrotic and turns brown; at this stage it is denoted as dried detritus. When the quantity of detritus produced from seagrass is considered, then the quantities of trace metals contained therein appear sizeable indeed. Detritus is produced constantly throughout the growing season as leaves grow, senesce, and

Table 3. Leachable Fe and Mn content[1] of sediments from seagrass study areas. Sediments were from a depth of 5-12 cm. Extraction was by hydroxylamine—acetic acid.

Location	Station	Fe ppm	Mn ppm	Fe/Mn
Laguna Madre	35[2]	199	18.3	10.8
Laguna Madre	L.M. #2	76	16.1	4.7
Laguna Madre	Coburn[2]	120	12.9	9.3
Redfish Bay	Hd - II	418	33.8	12.4
Redfish Bay	Hd - Ia[2]	460	36.0	12.7
Redfish Bay	Hd - RI[2]	108	14.7	7.4
Redfish Bay	Syr - I	390	43.0	9.1
Redfish Bay	Th - F & F	200	9.6	20.8
Redfish Bay	Th - C.C.B.	282	33.1	8.5

[1] The average Fe for the 3 L. Madre stations was 132 ppm; average Mn was 15.8 ppm. This compares with the avg. Fe from the first 4 Redfish Bay stations of 344 ppm and an avg. Mn of 32 ppm.

[2] Values for these stations are the average of 4 replicate determinations on a single sample of sediment. All other stations represent the result of a single determination.

are shed. Wave action, water currents, and epiphyte growth hasten leaf loss during summer months. Towards the end of summer, seasonal die-off ensues and this continues on into the fall. This sink of trace metals would be available to the detrital food chain and could function as a source of trace (as well as other) nutrients to the ecosystem after microbial decomposition.

A factor affecting the cycling of grass detritus in the above manner is transport out of the ecosystem. The L. Madre system consists of an almost entirely land-locked lagoon; only the Gulf Intra-coastal Waterway provides access for exchange of sea water by tides or wind-driven currents (see Fig. 1). Thus most seagrass detritus (and other detritus) will remain where it is produced. Conversely, Redfish Bay is connected by natural passes and ship channels to surrounding bays and the Gulf of Mexico. Detritus can readily be carried out from the bay by tides and storms. In fact, Redfish Bay is subject to much more tidal flushing than L. Madre (14, 15). As a result it is more likely that detritus is transported out of Redfish Bay, whereas the converse would be true for L. Madre. Export of seagrass detritus is further aided by the fact that the green, freshly-shed, leaves can float on the water surface for up to a week, before they die and sink to the bottom.

Variation in Standing Crops. A striking variation in standing crops exists between seagrasses of L. Madre and Redfish Bay. L. Madre contains vast meadows of *Halodule* and the salt-tolerant aquatic plant, *Ruppia*, with less, but still substantial, amounts of *Halophila.* Though Redfish Bay also has good stands of these three grasses, plus *Thalassia* and *Syringodium*, the standing crop (biomass) of plants in L. Madre is much greater. Table 4 compares standing crop values of *Halodule* and *Halophila* from Redfish Bay with L. Madre. These values

Table 4.   Standing crop data and N content of seagrasses from Redfish Bay and N. Laguna Madre. Standing crop values obtained during August; N content is for early spring plants.

SPECIES	Standing Crop[1] (g dry wt/m^2) Redfish Bay	N. Laguna Madre	Productivity[2] g C/m^2/day	N. Content[3] (Leaves) % dry wt.
Thalassia	500 - 700	—	0.9 - 9.0	4.0
Halodule	100 - 200	400 - 500	-	Redfish 2.9 L. Madre 3.2
Halophila	15 - 30	50 - 100	-	Redfish 2.6
Syringodium	30 - 50	-	-	2.3

[1] *Thalassia* and *Halodule* standing crop data obtained by *Biology of Seagrass Ecosystems* course held at the Marine Science Inst., Port Aransas, Tex., Summer 1974. *Halophila* and *Syringodium* standing crop data by Pulich (unpublished).
[2] Productivity data from Odum and Hoskin (8).
[3] N content determined by Pulich (unpublished).

indicate the much denser, more luxuriant meadows which exist in L. Madre. Seagrass meadows in the open bays (higher wave-energy environments) usually show this less sparsely vegetated cover.

Photosynthetic Activity. The primary productivity of representative seagrasses from both areas was compared to determine if differences in photosynthetic metabolism were indirectly responsible for the differences in Fe/Mn ratios. If plants from L. Madre were capable of increased photosynthesis, this could indicate higher levels of Mn-containing enzymes in photosynthetic pathways. As determined by $H^{14}CO_3$ uptake at two different light intensities, *Halophila* plants growing in L. Madre had similar, but definitely not higher, photosynthetic rates than Redfish Bay plants (data not shown). Thus, photosynthetic metabolism of Redfish Bay seagrasses does not seem to be impaired compared to L. Madre seagrasses. Rather, some other factor accounts for the decreased standing crops of seagrasses in Redfish Bay.

DISCUSSION

L. Madre populations of seagrasses were readily distinguishable from Redfish Bay populations on the basis of standing crops and Fe/Mn ratios. The fact that three species showed this Fe/Mn characteristic argues against ecotypic differentiation of races. Rather it suggests a common response to an environmental condition on the part of all three species.

The nature of the environmental condition eliciting this response is not yet precisely defined, but several facts are pertinent. (a) The Fe/Mn ratios of the subsurface sediments do not satisfactorily account for Fe/Mn ratios of the plants. On the one hand, the higher levels of Fe in Redfish Bay sediments compared to L. Madre sediments correlate with the higher Fe contents of plants from Redfish Bay. On the other hand, even though Mn in Redfish Bay sediments is also higher than in L. Madre sediments, Mn is lower in Redfish Bay plants. Hence, the specific issue in question is the difference in Mn levels of the two populations of plants. (b) Fe inhibition of Mn accumulation (13) does not appear a strong possibility since Fe/Mn ratios of sediments from the two areas are almost identical. Moreover, it is tenuous to postulate that the 2- to 3-fold difference in Fe between L. Madre and Redfish Bay sediments would exert an antagonism to Mn accumulation when Fe in L. Madre is already very high. Based on the standing crop data, it is possible that plants in Redfish Bay are not growing as fast as in L. Madre. Therefore Fe could build up in Redfish Bay plants simply because Fe is readily taken up by plant roots, but leaf tissue is not produced fast enough to "dilute" this Fe reserve. (c) Some other environmental factor(s) besides Fe is believed to affect Mn accumulation by these three sea-grasses. Since photosynthesis of both populations appears normal, Redfish Bay plants would not appear Mn-deficient; plant tissue therefore is healthy. The standing crop data again support another possibility: that of a difference in the rate of supply of Mn, or the rate at which Mn becomes available to plants. On this basis, we postulate that local cycling of Mn (and possibly other nutrients) is a major environmental factor which differs between L. Madre and Redfish Bay.

Mn availability may be edaphically controlled. Each species of seagrass may have certain physiological requirements which are better satisfied in certain types of substrate. If the finer grain and low shell L. Madre sediments are indeed the preferred substrate for *Halodule* and *Halophila*, then the sediment properties may govern the availability of trace elements like Mn and Fe, making Mn more, and Fe less, available.

Another point of interest involves the shell ($CaCO_3$) fraction in *Thalassia* sediments from Redfish Bay. The $CaCO_3$ may promote N assimilation as proposed by Barker, et al. (1) or provide Ca for some other biochemical process. In regards to the latter possibility, it is noteworthy that *Thalassia* contains a slightly higher level of Ca on a dry wt basis than other species (over 1% compared to under 0.8% for others).

As suggested by Parker, et al. (11), cycling of trace elements must occur in seagrass-dominated ecosystems. Large quantities of trace elements are tied up in biomass (both live and detritus) at any one time, in some cases more than is available in underlying sediments. We believe that the present comparative study of L. Madre and Redfish Bay plants provides evidence specifically for the importance of Mn cycling: larger standing crops containing more Mn are found in an area with lower levels of available Mn in the subsurface sediments. Local cycling of trace minerals, a characteristic of mature ecosystems, has previously been documented for forests, deserts, and terrestrial grasslands (4), but not for rooted, aquatic plant communities.

Parker (10) studied the uptake of [59] Fe, [54] Mn, and [65] Zn by components of a *Thalassia*-dominated bottom community in Redfish Bay. His work showed that at the low levels present in the experiments, there was rapid disappearance of trace metals from the water column and that most of it accumulated in/on *Thalassia* leaves, rather than in the roots. Subsequently, much of the radioactivity appeared in the top 2 cm of the sediment, as though a diurnal cycle existed between plant leaves and sediments [see also Parker, et al. (11)]. These results demonstrate that seagrass leaves are capable of readily removing trace elements at low concentrations from the water column. Thus the supply of nutrients in the root zone (below 2 cm) would not be an absolute factor determining the nutrient status of these submerged angiosperms. Variations in the trace elements in the water column, and also the top 2 cm of the sediments, would be critical to seagrass growth.

The data of Parker, et al. (11) for *Halodule* (= *Diplanthera*) deserve comment. These workers measured an Fe/Mn ratio in whole plants from Redfish Bay of $735/1100 = 0.67$, which is well below the lowest value from our study for Redfish Bay of 1.00. Though their value may be in error (due particularly to algal epiphytes on leaves which are hard to clean), it is possible that conditions in Redfish Bay have changed over the 13 years since their study. As a result, Redfish Bay seagrasses may have adapted to a new environment, and in the process, Fe/Mn ratios have changed. This would be even more likely if the change in environmental conditions directly involved Mn in the sediment or water column. The present study demonstrates that some seagrasses (particularly *Halodule*) have a substantial amount of variability in Mn content from one habitat to another. The Fe/Mn ratio of such seagrasses may, in fact, represent a valid indicator of the status of nutrients and other growth conditions in an area.

An interesting implication of this study concerns the conservation of seagrass meadows along the Texas coast and possibly elsewhere. Seagrass production requires a delicate balance between absolute concentration of nutrient present and its relative availability. When natural disasters (eg., storms) or human pressures (eg., dredging) prevent nutrients from being recycled in grass meadows, then stabilization of the estuary bottom by seagrass beds may be endangered. Cycling of detritus in a local area (as unaesthetic as it is sometimes) is probably necessary for proper fertilization of *Halodule* grass meadows, and possibly other species of seagrass, because nutrients are supplied at the optimum rate and concentration for plant growth. Thus, extensive, dense meadows of *Halodule* may require enclosed bay areas with a minimum of channels.

ACKNOWLEDGMENTS

This research was partially supported by NSF grant GA 41648 to P. Parker. We are deeply indebted to Woei-lih Jeng and Gerald Pfeiffer for help with the field work portion of this study, to Ernest Guerrero for performing the grain size analyses, and to Ms. JoAnn Page and Lorene Christin for assistance in the preparation of the manuscript. Appreciation is also extended to the Marine Science Laboratory's crew of boat captains who participated in many of the field trips.

REFERENCES

1. Barker, A. V., R. J. Volk, and W. A. Jackson. 1966. Root environment acidity as a regulatory factor in ammonium assimilation by the bean plant. Plant Physiol. 41:1193-1199.
2. Chester, R., and M. J. Hughes. 1967. A chemical technique for the separation of ferromanganese minerals, carbonate minerals, and adsorbed trace elements from pelagic sediments. Chem. Geol. 2:249-262.
3. Conover, J. T. 1964. The ecology, seasonal periodicity, and distribution of benthic plants in some Texas lagoons. Bot. Mar. 7:4-41.
4. Epstein, E. 1972. Mineral Nutrition of Plants: Principles and Perspectives. John Wiley & Sons, Inc., New York. 412 p.
5. Folk, R. L. 1964. Petrology of Sedimentary Rocks. Hemphill's, Austin, Tex. 154 p.
6. Gerloff, G. C. 1963. Comparative mineral nutrition of plants. Ann. Rev. Plant Physiol. 14:107-124.
7. Kier, R. S., W. A. White, W. L. Fisher, D. Bell, B. C. Patton, and J. T. Woodman. 1974. Establishment of operational guidelines for Texas coastal zone management: Resource capability units II. Land resources of the Coastal Bend region, Texas. Final Report to National Science Foundation for Grant no. GI-34870X.
8. Odum, H. T., and C. M. Hoskin. 1958. Comparative studies of the metabolism of marine waters. Publ. Inst. Mar. Sci. Univ. Texas 5:16-46.
9. Oppenheimer, C. H., T. Isensee, W. B. Brogden, and D. Bowman. 1974. Establishment of operational guidelines for Texas coastal zone management: Biological uses criteria. Final Report to National Science Foundation for Grant no. GI-34870X.
10. Parker, P. L. 1966. Movement of radioisotopes in a marine bay: cobalt-60, iron-59, manganese-54, zinc-65, sodium-22. Publ. Inst. Mar. Sci. Univ. Texas 11:102-107.
11. _____, A. Gibbs, and R. Lawler. 1963. Cobalt, iron, and manganese in a Texas bay. Publ. Inst. Mar. Sci. Univ. Texas 10:28-32.
12. Pigott, C. D. 1969. Influence of mineral nutrition on the zonation of flowering plants in coastal saltmarshes, p. 25-35. In I. H. Rorison (ed.), Ecological Aspects of the Mineral Nutrition of Plants. Blackwell Scientific Publications, Oxford and Edinburgh.
13. Sanchez-Raya, A. J., A. Leal, M. Gomez-Ortega, and L. Recalde. 1974. Effect of iron on the absorption and translocation of manganese. Plant and Soil 41:429-434.
14. Smith, N. 1974. Intracoastal tides of Corpus Christi Bay. Cont. Mar. Sci. 18:205-219.
15. _____. 1975. A tide and circulation study of Upper Laguna Madre, May 1, 1974 to April 30, 1975. Final Report to National Park Service for contract CX700040146.
16. Thayer, G. W., D. A. Wolfe, and R. B. Williams. 1975. The impact of man on seagrass systems. Amer. Sci. 63:288-296.
17. Walsh, G. E., and T. E. Grow. 1973. Composition of *Thalassia testudinum* and *Ruppia maritima*. Quart. Jour. Florida Acad. Sci. 35:97-108.

DYNAMICS OF ORGANOCHLORINE PESTICIDES IN ESTUARINE

SYSTEMS: EFFECTS ON ESTUARINE BIOTA

Robert J. Livingston
Department of Biological Science
Florida State University
Tallahassee, Florida 32306

ABSTRACT: A literature review was made concerning the environmental impact of organochlorine pesticides on estuarine biota. Considerable laboratory information is available concerning the acute and chronic effects of these pesticides on aquatic organisms. Indications are that movement of such compounds through the environment is a complex process which involves numerous variables. Distribution of organochlorine pesticides is complicated by different attenuation mechanisms such as volatilization, microbial or chemical decomposition, biochemical and photochemical degradation, and bioconcentration. Various pathways of transfer and bioaccumulation have been shown. Despite a considerable effort to document the distribution of pesticide residues in estuaries in the United States, the ultimate fate of such chemicals in aquatic systems remains largely undetermined.

Recent interest has been shown in the chronic effects of pesticides on common estuarine organisms. This has included studies concerning embryological development and reproductive capacity of individual species. The data indicate maximal impact at some stage other than the adult, and that species-specific embryological response and yolk utilization are often involved in the reaction. Pesticide-induced behavioral anomalies have also been described. However, the ecological implications of such work remains unknown even though altered behavior could have a significant effect on the environmental impact of such toxicants. The few data involving synergistic relationships of different combinations of pollutants and modifying factors bring into focus the sheer complexity of the situation. With the exception of a few intensive field studies, relatively little is known concerning population and community response to chronic levels of pesticide residues in estuarine biota.

More long-term, integrated field and laboratory studies will be needed if the actual environmental effects of pesticides on estuarine ecosystems are to be determined.

INTRODUCTION

Considerable information exists concerning the effects of organochlorine pesticides on aquatic organisms. Various comprehensive reviews are available (30, 46, 55, 86, 99, 101). Such pesticides have been noted for their widespread distribution and persistence in the environment, an ability to concentrate in individual organisms and biomagnify from one trophic level to the next, and their toxicity to various forms of aquatic organisms. However, variations occur in stability and toxic activity from one compound to the next, so that no strict generalizations can be made regarding the group as a whole.

Butler (12) lists DDT (and its metabolites), dieldrin, endrin, toxaphene, and mirex as the most common polychlorinated compounds found in estuaries. Yet, relatively little comprehensive knowledge is available concerning the long-term effects of such compounds on estuarine systems *per se*. Odum (76) reviewed the vulnerability of estuarine biota to different forms of stress. The extreme complexity of a given estuarine environment together with the relatively high level of variability from one system to the next tend to obscure the potential effects of pesticide residues on estuarine populations and communities. This is especially true of sub-lethal concentrations which represent the overwhelming number of cases.

This review is primarily concerned with the distribution of organochlorine pesticides in estuarine systems, the determination of the environmental impact of organochlorine pesticides on estuarine biota, and the various criteria and limitations which must be considered in such an evaluation.

MOVEMENT AND DISTRIBUTION OF ORGANOCHLORINE PESTICIDES

Various authors have considered transfer mechanisms of pesticides in estuaries (7, 34, 42, 70, 79, 86, 92, 102, 103, 104, 105, 106). A graphic synopsis of such data is given in Fig. 1. Different factors determine pesticide transport in estuaries; a partial list would include pesticide type, time and area of application, local meteorological phenomena, biological activity, and physico-chemical characteristics of the drainage system. Aerial transfer of pesticides such as DDT has been reported (16, 93, 105); wind and rainfall patterns can influence the pesticide burden in various areas. Peterle (83) pointed out that pesticide translocation in soil is minimal with most of the transfer to aquatic areas being related to surface runoff and adsorption to (soil) particles. However, the mechanics of the adsorption-desorption rates are complex (7), and this involves individual components of suspended, particulate, and dissolved phases of pesticide concentration in runoff water. Thus, soil type, percent organics (98), time and form of application, rates of codistillation and evapotranspiration, and episodic rainfall

patterns all contribute to the movement of pesticides from upland areas into aquatic systems.

The physical and chemical properties of the individual pesticides affect the potential for adsorption to available organic and inorganic surfaces. Ultimately, siltation together with ionic flocculation of dissolved components cause the deposition of pesticides in estuaries. Such compounds, when adsorbed to detritus (77), in solution (103), and attached to particulate loads thus become available for concentration in estuarine food webs. Burnett (11), in a study of the distribution of DDT residues in *Emerita analoga*, found that organisms near sewer outfalls (contaminated with industrial pesticide wastes) contained 45 times more DDT than those near agricultural areas; bottom sediments acted as a reservoir of well over 100 metric tons of DDT. Olaffs et al. (79) postulated that actual quantities of persistent organochlorine pesticides in marine areas may be difficult to assess because of partitioning of such pesticides into sediments. Solubilization of pesticides with organic matter such as humic and fulvic acids, has been described (103) and could explain the relatively low rate of deposition of organochlorine pesticides in the sediments of river-dominated (highly organic) estuaries such as Apalachicola Bay in north Florida (60). Another complicating factor is the surface slick phenomenon (95) whereby organochlorine pesticides concentrate at surface convergences which then act as focal points for transfer and dissemination into marine biota. Evidently, pesticide concentration and movement can occur in various ways depending on the physico-chemical setting,

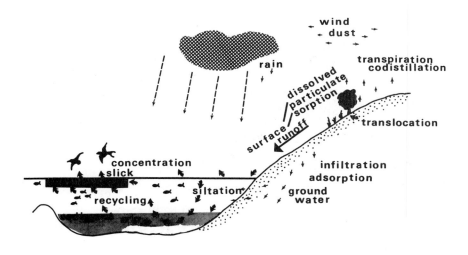

PESTICIDE MOVEMENT

Figure 1. Diagrammatic summary of potential modes of organochlorine pesticide transport in the environment.

the trophic structure, and the timed (seasonal) biotic fluctuations of a given system.

The extensive distribution of residues of organochlorine pesticides in estuaries indicates environmental persistence; however, various mechanisms of attenuation function to alter or degrade such chemicals. Processes such as volatization, microbial or chemical decomposition, biochemical and photochemical degradation, and bioconcentration account for the "disappearance" of such pesticides from water and sediments. Such functions are modified by salinity, temperature, pH, dissolved oxygen content, etc. For example, a composite scheme for potential environmental alteration of DDT is shown in Fig. 2. Patil et al. (81) found that biodegradation of various organochlorine pesticides in sea water was usually associated with biological agents such as microorganisms, plankton, and algae; such patterns resembled those also found in freshwater and terrestrial environments. In addition to such metabolic transformations, possible photolytic mechanisms have been proposed (DDE→PCB's) (70). When exposed to U.V. irradiation, compounds such as mirex with little or no known (aerobic) meta-

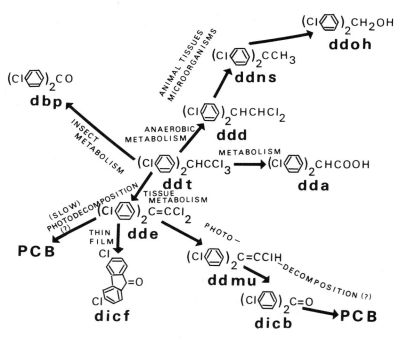

Figure 2. Composite projection of potential chemical reactions of p, p'-DDT in the environment (70, 81). This includes physical and biochemical reactions in various phases of transport. (Dicf = Dicofol or kelthane; PCB = polychlorinated biphenyls; dbp = p, p' - dichlorobenzophenone; dicb = dichlorobenzophenone; ddmu = 1-chloro-2, 2-bis (p-chlorophenyl) ethylene).

bolites (6) decompose into a series of photodegradation products (1,2) which can then accumulate in organisms (29).

By exposing simulated marsh communities to mirex bait held in sunlight, Cripe and Livingston (18) found that three photoproducts accumulated on the bait particles. One of these byproducts (the 8-monohydro isomer) was accumulated by oysters and one species of fish (*Fundulus similis*); thus, in addition to complex physico-chemical molecular alterations, species-specific feeding patterns were postulated as a factor in the differential accumulation of this compound. Considerably more work is needed to determine the potential for transformation and bioaccumulation of such metabolites and their cumulative impact on estuarine organisms.

BIOLOGICAL CONCENTRATION OF PESTICIDES

Pesticide bioaccumulation has been well documented (13, 24, 46, 71, 101); the concentration of organochlorine pesticides is assumed to be a function of their solubility characteristics and the mode of feeding and respiration of estuarine organisms. Direct concentration from ingestion of contaminated food (3), uptake from water passed over gill surfaces (28, 31, 40, 57), diffusion across the cuticle (23), and direct uptake from sediments (52) all contribute to such bioaccumulation. Often, biological concentration can result from combinations of mechanisms. Kobylinski and Livingston (52) showed that the hogchoker (*Trinectes maculatus*) absorbed pesticide residues from water contaminated by mirex-laden sediments as well as directly from the actual sediments. Stepwise concentration of pesticides from one trophic level to the next is an accepted mechanism for bioaccumulation of persistent pesticides (106); various studies have confirmed biological magnification experimentally (73, 85, 89). An alternate hypothesis proposed by Hamelink et al. (33) involves the control of the bioaccumulation process by various exchange equilibria based on the solubility characteristics of individual organochlorine pesticides. According to this, concentration potential varies with individual partition coefficients. Rice and Sikka (91) suggest a similar mechanism of DDT and dieldrin uptake by various species of marine algae. Of course, the interpretation of pesticide residue distribution is often complicated by varied distribution of fat content (46), seasonal periodicity of pesticide use, and potential changes in concentration factors (50). Smith and Cole (96) found that seasonal residue peaks in winter flounder (*Pseudopleuronectes americanus*) were associated with runoff conditions rather than application patterns. Thus, pesticide residue distribution is the result of various functions, and bioaccumulation of pesticides is probably related to species-specific patterns of feeding and movement.

DEVELOPMENT AND REPRODUCTION

A largely undeveloped yet important area of estuarine research concerns the effects of pesticides on the embryological and larval development of various species. Due to the complex life histories of estuarine and marine species, this is an extremely difficult question to answer. However, a species may be adversely affected even without a reduction in the spawning population (67). Work with various fishes indicates that non-polar contaminants such as DDT accumulate in eggs (90, 96); this has caused mortality to embryos and larvae at the yolk absorption stage. Adverse effects of DDT on the breeding potential of fishes such as *Cynoscion regalis, C. arenarius,* and *Pseudopleuronectes americanus* have been noted (49, 96, 97). Pesticide induced yolk sac fry mortality has been shown in lake trout (*Salvelinus namaycush*) when egg residues exceeded 2.9 ppm even though embryos and adults did not suffer acute effects (10). Field studies have shown adverse effects of DDT on embryos of rainbow trout (*Salmo gairdneri*) (43), eggs and fry of Atlantic salmon (*S. salar*) (67), eggs and fry of coho salmon (*Onchorhynchus kisutch*) (48), and the fertility of pike eggs (44). Laboratory studies have shown adverse effects of sublethal concentrations of organochlorine pesticides on reproduction, growth, and development of brook trout (*Salvelinus fontinalis*) (67, 72); such effects have also been indicated in the overall reproductive capacity of sailfin mollies (*Poecilia latipinna*) (57) and medaka (*Oryzias latipes*) (47). Koenig (53) noted that mirex and DDT acted synergistically in cyprinodontid larvae (*Adinia xenia*) even though there was no such effect with embryos or adults. In this instance, there was a delayed response that resembled the effect of mirex on certain estuarine crustaceans and fresh water crayfish (66).

Considerably less is known about the effects of such pesticides on the development of estuarine invertebrates. Bookhout et al. (9) found that relatively low levels of mirex affected survival, duration of developmental stages, and total time of development of the crab, *Rhithropanopeus harrisii.* They also showed that larvae of *Menippe mercenaria* were more sensitive (especially in the megalopal stage) to mirex than *Rhithropanopeus,* and that there were basic differences in concentration rates and developmental changes between the two species. These data indicate that the response of estuarine and coastal populations to organochlorine pesticides could be related to the cumulative effects of such compounds on specific (sensitive) developmental stages rather than the adults themselves, and that such effects would depend on species-specific patterns of embryological response and yolk utilization. Since estuaries serve as nursery areas for developmental stages of various organisms, this would have to be an important field of research.

BEHAVIOR

Behavioral changes caused by sublethal concentrations of DDT were discussed by Anderson (4). Such pesticides have been associated with aberrations in the CNS functions of various organisms. Anderson and Peterson (5) have shown dose-specific reactions to low levels of DDT; this involved the cold-block temperature response of a propeller tail reflex of brook trout. High and low doses resulted in respective increases and decreases of the cold-block temperatures. Effects of DDT on a conditioned learning response of fishes were reported (5) and qualified (45). Ogilvie and Anderson (78) showed that DDT caused changes in the temperature selection of Atlantic salmon; this was related to interference with normal acclimation functions. Peterson (84) found that various pesticides (DDT, DDD, DDE, methoxychlor, aldrin) altered the temperature selection behavior of juvenile Atlantic salmon. Hansen (36) showed that salinity preferences of mosquitofish (*Gambusia affinis*) were altered by DDT; affected fish selected higher salinities due to altered nerve sensitivity and/or changes in osmoregulatory ability. Effects of pesticides have also been analyzed with respect to other behavioral alterations such as vulnerability to predation (39) and locomotor responses (21, 22). Tolman (100) found that although the diurnal activity rhythms of *Adinia xenica* were not easily altered by mirex, there was an increase in the variability of individual activity patterns in exposed fish. Cairns et al. (15) developed an early warning system for industrial waste based on altered swimming and breathing movements of laboratory fish populations. Thus, different behavioral functions can be used as indicators of pollution.

Various studies have shown that aquatic organisms avoid pollutants (51). Hansen (35) showed that sheepshead minnows (*Cyprinodon variegatus*) avoided water containing DDT, endrin, dursban, and 2, 4-D although test fish actually preferred higher concentrations of DDT. Grass shrimp (*Palaemonetes pugio*) did not avoid such pesticides with the exception of the butoxyethanol ester of 2, 4-D (37). Susceptible and resistant mosquitofish avoided toxaphene, endrin, and parathion while DDT was avoided only by susceptible fish (56). The actual environmental significance of such experiments is open to question (24, 63). Avoidance could be detrimental if an estuarine population was prevented from reaching a spawning or breeding ground. However, an organism without an avoidance response such as the grass shrimp could be vulnerable to acute and chronic effects of exposure to such pollutants. Ultimately, the problem becomes one of habitat alteration and/or the prevention of resource utilization by a given population. One approach to this dilemma is an integrated laboratory-field analysis where long-term field data on species distribution in a gradient of a particular point-source of pollution is coupled with laboratory experiments concerning avoidance of the specific pollutant. Studies are presently being carried out at Florida State University concerning species reactions to pulp mill effluents and storm water runoff and the relationship of laboratory reactions and field response (58, 60, 62). There are strong indications that a well developed laboratory avoidance response to a given pollutant may have no relevance to the actual field distribution of a natural population. This is often a function of various complex physico-chemical and biological phenomena. Although the alteration of specific behavioral responses of aquatic organisms by pesticides remains a sensitive and potentially useful indicator of chronic impact, more work is needed concerning the environmental significance of such pesticide-induced reactions.

SYNERGISTIC INTERACTIONS AND MODIFYING EFFECTS

Relatively little information is available concerning the effects of mixtures of pesticides on estuarine organisms even though residue analysis has shown that pesticide burdens ordinarily involve more than one compound. One review of the literature (64) indicates that most of the emphasis in this field has been carried out with terrestrial invertebrates (notably insects) and mammals. These studies have shown induction and storage effects of pesticide interaction; however, relatively little is known concerning aquatic organisms. Although DDT, endrin, toxaphene, and methyl parathion had no additive effect on mosquitofish (26), a methoxychlor formulation (Metox) containing various compounds potentiated the effect of methoxychlor on *Daphnia magna* (14). Post and Garms (87) found that a solvent-emulsifier combination used with DDT was toxic to fishes. Mosser et al. (74) found varying synergistic and antagonistic responses of a marine diatom to mixtures of DDT and its metabolites. In this instance, counteraction rather than coprecipitation was found to be the synergistic mechanism. Koenig (53), working with the effects of mirex and DDT on the development of *Adinia xenica*, found differential synergistic effects which depended largely on the stage of development.

Other studies have concentrated on the interaction of organochlorine pesticides and various modifying factors. Reduced dissolved oxygen increased DDT toxicity to various freshwater fishes (59); Eisler (25) found that increased oxygen levels tended to decrease the toxicity of aldrin, malathion, and heptachlor to an estuarine fish although it increased the toxicity of phosdrin and methyl parathion. Various studies (25, 27, 41, 68) have shown that water temperature affects the toxicity of pesticides to various estuarine organisms. Koenig et al. (54) found that rapid reductions of temperature associated with cold fronts in north Florida combined with locally high DDT-R concentrations in blue crabs (*Callinectes sapidus*) to cause considerable mortality in the field.

Obviously, because of the overwhelming number of potential interactions among the various pesticides and modifying factors, no definitive summary of such effects is really possible. The available data indicate, however, that interacting factors can be instrumental to a comprehensive understanding of the impact of pesticides on estuarine organisms.

POPULATION AND COMMUNITY INTERACTIONS

Duke and Dumas (24) pointed to the importance of the "steady" (oscillatory) state in estuarine ecosystems. They considered that due to the complexity of such systems, several years of study are necessary for an evaluation of the effects of the pesticides on component systems. Little information is available. Again, integrated (laboratory-field) studies can be useful. Mosser et al. (75) showed that when DDT or PCB's were added to mixed cultures of sensitive marine diatoms and a resistant marine green alga, the normally dominant diatom was reduced in importance. Such pollutants could disrupt the species composition of marine phytoplankton communities. Model ecosystem studies (73, 94) have established the stability of dieldrin and DDT in aquatic systems. By placing such compounds in a simulated system composed of standard organisms, an analysis can be made concerning functions such as pesticide stability, bioconcentration, and species-specific reaction to the toxicant (18, 94).

Other forms of field experiments can be carried out. Harrington and Bidlingmayer (38) determined the acute effects of an aerial application of dieldrin to a marsh while Croker and Wilson (19) observed the short-term effects of a DDT application (0.2 pounds/acre) to a marsh. Reimold (88), in one of the few long-term studies of the effects of organochlorine pesticides on an estuary, found that reduction of toxaphene contamination was accompanied by increased species diversity of nektonic organisms. By the end of the study, such indices were actually higher in the previously affected areas than control (unaffected) systems. There are very few data, however, on the actual effects of pesticides on estuarine populations and communities.

SUMMARY AND CONCLUSIONS

A relatively rich body of scientific literature is available concerning the interactions of organochlorine pesticides and aquatic biota. Information exists concerning field residue analysis, acute and chronic (laboratory) toxicity, avian response (population biology, reproductive functions, etc.), and pesticide-induced resistance in natural, aquatic populations. Most of this literature has not been mentioned here since this review has emphasized relatively little known areas of pesticide research. Even with a comprehensive understanding of dose-dependent relationships and the use of appropriate application factors, there is still no direct or uniform mechanism to interpret such data in an ecological setting. In many cases, the laboratory results deal with standard adult organisms even though various studies have emphasized the susceptibility of developmental stages of estuarine and coastal species to organochlorine pesticides. In some cases, there is a more varied toxicological response within the ontogenetic history of a single species than could be found by an interspecific comparison.

The various molecular transmutations of organochlorine compounds in the environment add to the complexity of extrapolation from laboratory to field. This becomes even more confusing when synergistic interactions are taken into account. Species-specific bioaccumulation and magnification rates, variable trophic and behavioral relationships, and the broad array of micro-habitats encountered in any given estuarine system all tend to obscure an evaluation of population and community response to a specific pollutant. In addition, the temporal relationships of pulsed introductions of pesticides and the seasonally directed migrations of estuarine populations are not well understood even though such interactions are of critical importance with respect to the resilience of a given system.

Instead of discouraging field research, such complications only emphasize the need for data concerning the actual impact of a given level of pesticide pollution on the estuarine system as a whole. Knowledge of such estuarine functions is increasing, however. Long term fluctuations of estuarine assemblages have been described (20, 32, 60, 69, 80, 82) and some evaluations have been made concerning the impact of pollutants on estuarine populations and/or communities (8, 17, 61, 107). Once such study (60) noted a precipitous decline of DDT and PCB residues in Apalachicola Bay after the use of such compounds was sharply curtailed in the upland drainage system. A series of papers is presently being prepared to relate such decreases to long-term (4-5 year) trends of fish and invertebrate assemblages in the bay.

Although many of the more important organochlorine compounds have been restricted or banned from further use in the United States, we still need to answer certain questions concerning these and other persistent pesticides (101) since such chemicals are being used extensively in various areas of the world. Although there is considerable knowledge of the laboratory response of various estuarine organisms to certain organochlorine pesticides, more data are needed concerning *in situ* reactions at the population and community level before the picture is complete. This should include impact evaluation of trophic response, developmental reactions, and interactions at varying levels of complexity in different estuarine assemblages so that population tolerance and resilience of a given estuarine community can be assessed. The emphasis should thus be shifted to the integration of field and laboratory approaches so that the gap between residue analysis and ecological function can be narrowed.

LITERATURE CITED

1. Alley, E.G., B.R. Layton, and J.P. Minyard, Jr. 1974. Photoreduction of mirex in aliphatic amines. J. Agr. Food Chem. 22:727-729.
2. _____, D.A. Dollar, B.R. Layton, and J.P. Minyard, Jr. 1973. Photochemistry of mirex. J. Agr. Food Chem. 21:138-139.
3. Allison, D., B.J. Kallman, O.B. Cope, and C. VanValin. 1966. Some chronic effects of DDT on cutthroat trout. Fish Wildl. Serv. Res. Rept. No. 64:1-30.
4. Anderson, J.M. 1971. Sublethal effects and changes in ecosystems—Assessment of the effects of pollutants on physiology and behavior. Proc. Roy. Soc. Lond. B. 177:307-320.
5. _____, and M.R. Peterson. 1969. DDT: Sublethal effects on brook trout nervous system. Science 164:440-441.
6. Andrade, P.S.L., Jr., and W.B. Wheeler. 1974. Biodegradation of mirex by sewage sludge organisms. Bull. Env. Cont. Tox. 11:415-416.
7. Bailey, G.W., R.R. Swank, Jr., and H.P. Nicholson. 1974. Predicting pesticide runoff from agricultural land; a conceptual model. J. Environ. Qual. 3:95-102.
8. Bechtel, R.J., and B.J. Copeland. 1970. Fish species diversity indices as indicators of pollution in Galveston Bay, Texas. Contr. Mar. Sci. 15:103-132.

9. Bookhout, C.G., A.J. Wilson, Jr., T.W. Duke, and J.I. Lowe. 1972. Effects of mirex on the larval development of two crabs. Wat. Air Soil Pollut. 1:165-180.
10. Burdick, G.E., E.J. Harris, H.J. Dean, T.M. Walker, J. Skea, and D. Colby. 1964. The accumulation of DDT in lake trout and the effect on reproduction. Trans. Amer. Fish. Soc. 93:127-136.
11. Burnett, R. 1971. DDT residues: distribution of concentrations in *Emerita analoga* (Stimpson) along coastal California. Science 174:606-608.
12. Butler,   P.A.   1969. Monitoring pesticide pollution. Bioscience 19(10):889-896.
13. _____. 1973. Organochlorine residues in estuarine mollusks, 1965-72 National Pesticide Monitoring Program. Pest. Mon. J. 6:238-363.
14. Cabejszek, I., J. Maleszcwska, and J. Stanislawski. 1966. The effects of insecticides (aldrin, methoxychlor) on the physical-chemical properties of water and on aquatic organisms. Verh. Int. Verein. Theor. Angew. Limnol. 16:963-968.
15. Cairns, J., Jr., J.W. Hall, E.L. Morgan, R.E. Sparks, W.T. Waller, and G.F. Westlake. 1973. The development of an automated biological monitoring system for water quality. Virginia Wat. Res., Res. Cent. Bull. 59:1-50.
16. Cohen, J.M., and C. Pinkerton. 1966. Widespread translocation of pesticides by air transport and rain-out, p. 163-176. *In* Advances in Chemistry Series No. 60, Organic Pesticides in the Environment.
17. Copeland, B.J., and T.J. Bechtel. 1971. Species diversity and water quality in Galveston Bay, Texas. Wat. Air Soil Pollution 1:89-105.
18. Cripe, C.R., and R.J. Livingston. 1976. Dynamics of the pesticide mirex and its photoproducts in a simulated marsh system. Arch. Env. Cont. Tox. (in press).
19. Croker, R.A., and A.J. Wilson. 1970. Kinetics and effects of DDT in a tidal marsh ditch. Trans. Amer. Fish. Soc. 94:152-159.
20. Dahlberg, M.D., and E.P. Odum. 1970. Annual cycles of species occurrence, abundance, and diversity in Georgia estuarine fish populations. Am. Mid. Nat. 83:382-392.
21. Davy, F.B., H. Kleerekoper and P. Gensler. 1972. Effects of exposure to sublethal DDT on the locomotor behavior of the goldfish (*Carassius auratus*). J. Fish. Res. Bd. Canada 29(9): 1333-1336.
22. _____, H. Kleerekoper, and J.H. Matis. 1973. Effects of exposure to sublethal DDT on the exploratory behavior of the goldfish (*Carassius auratus*). Water Resour. Res. 9:900-905.
23. Derr, S.K., and M.J. Zabik. 1974. Bioactive compounds in the aquatic environment: Studies on the mode of uptake of DDE by the aquatic midge, *Chironomus tentans* (Diptera: Chironomidae). Arch. Env. Cont. Tox. 2:152-164.
24. Duke, T.W., and D.P. Dumas. 1974. Implications of pesticide residues in the coastal environment, p. 137-164. *In* F. J. Vernberg and W.B. Vernberg (eds.), Pollution and Physiology of Marine Organisms.
25. Eisler, R. 1970. Factors affecting pesticide-induced toxicity in an estuarine fish. Bur. Sport Fish. Wildl. Tech. Pap. 45: 19 p.
26. Ferguson, D.E., and C.R. Bingham. 1966. The effects of combinations of insecticides on susceptible and resistant mosquito fish. Bull. Env. Cont. Tox. 1:97-102.

518   R. J. LIVINGSTON

27. _____, D.D. Culley, W.D. Cotton, and R.P. Dodds. 1964. Resistance to chlorinated hydrocarbon insecticides in three species of fresh water fish. Bioscience 14:43-44.
28. _____, and C.P. Goodyear. 1967. The pathway of endrin entry in black bullheads, *Ictalurus melas*. Copeia 2(5):467-468.
29. Gibson, J.R., G.W. Ivie and H.W. Dorough. 1972. Fate of mirex and its major photodecomposition product in rats. J. Agr. Food Chem. 20(6):1246.
30. Gillett, J.W. 1970. The biological impact of pesticides in the environment. Env. Health Sci. Ser. No. 1; Oregon State Univ. Press, Corvallis. 210 p.
31. Grzenda, A.R., D.F. Paris, and W.J. Taylor. 1970. The uptake, metabolism and elimination of chlorinated residues by goldfish (*Carassius auratus*) fed a ^{14}C-DDT contaminated diet. Trans. Amer. Fish. Soc. 99:385-396.
32. Haedrich, R.L., and S.O. Haedrich. 1974. A seasonal survey of the fishes in the Mystic River, a polluted estuary in downtown Boston, Massachusetts. Est. Coast. Mar. Sci. 2:59-73.
33. Hamelink, J.L., R.C. Waybrant, and R.C. Ball. 1971. A proposal: Exchange equilibria control the degree chlorinated hydrocarbons are biologically magnified in lentic environments. Trans. Amer. Fish. Soc. 100:207-214.
34. Hansberry, R. 1970. The transport and accumulation of pesticides in environments and ecosystems, p. 1-27. *In* J.W. Gillett (ed.), The Biological Impact of Pesticides in the Environment.
35. Hansen, D.J. 1969. Avoidance of pesticides by untrained sheepshead minnows. Trans. Amer. Fish. Soc. 98:426-429.
36. _____. 1972. DDT and malathion: Effect on salinity selection by mosquitofish. Trans. Amer. Fish. Soc. 101:346-350.
37. _____, S.C. Schimmel and J.M. Keltner, Jr. 1973. Avoidance of pesticides by grass shrimp (*Palaemonetes pugio*). Bull. Env. Cont. Tox. 9:129-133.
38. Harrington, R.W., and W.L. Bidlingmayer. 1958. Effects of dieldrin on fishes and invertebrates of a salt marsh. J. Wildl. Mgmt. 22:76-82.
39. Hatfield, C.T., and J.M. Anderson. 1972. Effects of two insecticides on the vulnerability of Atlantic salmon (*Salmo salar*) parr to brook trout (*Salvelinus fontinalis*) predation. J. Fish. Res. Bd. Canada 29:27-29.
40. Holden, A.V. 1962. A study of the absorption of ^{14}C-labelled DDT from water by fish. Ann. Appl. Biol. 50:467-477.
41. Holland, H.T., and D.L. Coppage. 1970. Sensitivity to pesticides in three generations of sheepshead minnows. Bull. Env. Cont. Tox. 5:362-367.
42. Hom, W., R.W. Risebrough, A. Soutar, and D.R. Young. 1974. Deposition of DDE and polychlorinated biphenyls in dated sediments of the Santa Barbara Basin. Science 184:1197-1199.
43. Hopkins, C.L., S.R.B. Solly, and A.R. Ritchie. 1969. DDT in trout and its possible effect on reproduction potential. New Zealand J. Mar. Freshwater Res. 3:220-229.
44. Huisman, E.A., J.H. Koeman, and P. Wolff. 1971. An investigation into the influence of DDT and other chlorinated hydrocarbons on the fertility of the pike, p. 69-86. *In* Ann. Rept. of the organization for the improvement of the freshwater fishery.
45. Jackson, D.A., J.M. Anderson, and D.R. Gardner. 1970. Further investigations of the effect of DDT on learning in fish. Can. J. Zool. 48:577-580.
46. Johnson, D.W. 1968. Pesticides and fishes—a review of selected literature. Trans. Amer. Fish. Soc. 97:398-424.

47. Johnson, H.E. 1967. The effects of endrin on the reproduction of a fresh-water fish, *Oryzias latipes*. Ph. D. Dissertation, University of Washington. 136 p.

48. _____, and C. Pecor. 1969. Coho salmon mortality and DDT in Lake Michigan. Trans. 34th N. Amer. Wild. and Nat. Res. Conf., Wash. D.C. p. 159-166.

49. Joseph, E.B. 1972. The status of the sciaenid stocks of the middle Atlantic Coast. Chesapeake Sci. 13:87-100.

50. Kelso, J.R.M., and R. Frank. 1974. Organochlorine residues, mercury copper and cadmium in Yellow Perch, White Bass, and Smallmouth Bass, Long Point Bay, Lake Erie. Trans. Amer. Fish Soc. 103:577-581.

51. Kleerekoper, H., J.B. Waxman, and J. Matis. 1973. Interaction of tempera-ture and copper ions as orienting stimuli in the locomotor behavior of the goldfish (*Carassius auratus*). J. Fish. Res. Bd. Canada 30:725-728.

52. Kobylinski, G.J., and R.J. Livingston. 1975. Movement of mirex from sediment and uptake by the hogchoker, *Trinectes maculatus*. Bull. Env. Cont. Tox. 14:692-698.

53. Koenig, C.C. 1975. The effects of DDT and mirex alone and in combina-tion on the reproduction of a salt marsh cyprinodont fish, *Adinia xenica*. Ph. D. Dissertation, Florida State University, Tallahassee; 136 p.

54. _____, R.J. Livingston, and C.R. Cripe. 1976. Blue crab mortality: inter-action of temperature and DDT residues. Arch. Env. Cont. Tox. 4(1):119-128.

55. Kraybill, H.F. 1969. Biological effects of pesticides in mammalian systems. Ann. New York Acad. Sci. 160:1-422.

56. Kynard, B. 1974. Avoidance behavior of insecticide susceptible and resis-tant populations of mosquitofish to four insecticides. Trans. Amer. Fish. Soc. 103:557-561.

57. Lane, C.E., and R.J. Livingston. 1970. Some acute and chronic effects of dieldrin on the sailfin molly, (*Poecilia latipinna*). Trans. Amer. Fish. Soc. 99:489-495.

58. Lewis, F.G., III, and R.J. Livingston. 1976. Avoidance reactions of two species of marine fishes to kraft pulp mill effluent. J. Fish. Res. Bd. Canada, (in press).

59. Lincer, J.L., J.M. Solon, and J.H. Nair, III. 1970. DDT and endrin fish toxicity under static versus dynamic bioassay conditions. Trans. Amer. Fish. Soc. 99:13-19.

60. Livingston, R.J. 1974. Field and laboratory studies concerning the effects of various pollutants on estuarine and coastal organisms with application to the management of the Apalachicola Bay System (north Florida, U.S.A.). Florida Sea Grant Report #R/EM-1: 574 p.

61. _____. 1975. Impact of pulp mill effluents on estuarine and coastal fishes in Apalachee Bay, Florida, U.S.A. Mar. Biol. 32:19-48.

62. _____. 1976. Diurnal and seasonal fluctuations of estuarine organisms in a north Florida estuary. Est. Coastal Mar. Sci., (in press).

63. _____, C.R. Cripe, C.C. Koenig, F.G. Lewis, and B.D. DeGrove. 1974. A system for the determination of chronic effects of pollutants on the physi-ology and behavior of marine organisms. Florida Sea Grant Program Report 4:1-15.

64. _____, C.C. Koenig, J.L. Lincer, A. Michael, C. McAuliffe, R.J. Nadeau, R.E. Sparks, and B.E. Vaughan. 1974. Synergism and modifying effects:

Interacting factors in bioassay and field research, p. 225-304. *In* Marine bioassays workshop proceedings, Marine Technological Society, Washington, D.C.

65. Locke, D.O., and K. Havey. 1972. Effects of DDT upon salmon from Schoodic Lake, Maine. Trans. Amer. Fish Soc. 101:638-643.

66. Lowe, J.I., P.R. Parrish, A.J. Wilson, Jr., P.D. Wilson, and T.W. Duke, 1971. Effects of mirex on selected estuarine organisms. Trans. 36th North Amer. Wildl. Nat. Res. Conf., p. 171-186.

67. Macek, K.J. 1968. Reproduction in brook trout (*Salvelinus fontinalis*) fed sublethal concentrations of DDT. J. Fish. Res. Bd. Canada 25:1787-1796.

68. _____, C. Hutchinson, and O.B. Cope. 1969. Effects of temperature on the susceptibility of bluegills and rainbow trout to selected pesticides. Bull. Env. Cont. Tox. 3:174-183.

69. McErlean, A.J., and J.A. Mihursky. 1969. Species diversity-species abundance of fish populations: an examination of various methods. Proc. 22nd Conf. S.E. Coast Ass. Game Fish Comm. p. 367-372.

70. Maugh, T.H. 1973. DDT: An unrecognized source of polychlorinated biphenyls. Science 180:578-579.

71. Mehendale, H.M., L. Fishbein, M. Fields, and H.B. Matthews. 1972. Fate of mirex-^{14}C in the rat and plants. Bull. Env. Cont. Tox. 8:200-207.

72. Mehrle, P.M., and F.L. Mayer, Jr. 1975. Toxaphene effects on growth and development of brook trout (*Salvelinus fontinalis*). J. Fish. Res. Bd. Canada 32:609-613.

73. Metcalf, R.L., G.K. Sangha, and I.P. Kapoor. 1971. Model ecosystems for the evaluation of pesticide biodegradability and ecological magnification. Environ. Sci. Technol. 5:709-713.

74. Mosser, J.L., N.S. Fisher, and C.F. Wurster. 1972. Polychlorinated biphenyls and DDT alter species composition in mixed cultures of algae. Science 176:533-535.

75. _____, T.C. Teng, W.G. Walther, and C.F. Wurster. 1974. Interactions of PCB's, DDT, and DDE in a marine diatom. Bull. Env. Cont. Tox. 12:665-668.

76. Odum, W.E. 1970. Insidious alteration of the estuarine environment. Trans. Amer. Fish. Soc. 99:836-847.

77. _____, G.M. Woodwell, and C.F. Wurster. 1969. DDT residues absorbed from organic detritus by fiddler crabs. Science 164:576-577.

78. Ogilvie, D.M., and J.M. Anderson. 1965. Effect of DDT on temperature selection by young Atlantic salmon, *Salmo salar*. J. Fish. Res. Bd. Canada 22:503-512.

79. Oloffs, P.C., L.J. Albright, S.Y. Szeto, and J. Lau. 1973. Factors affecting the behavior of five chlorinated hydrocarbons in two natural waters and their sediments. J. Fish. Res. Bd. Canada 30:1619-1623.

80. Oviatt, C.A., and S.W. Nixon. 1973. The demersal fish of Narragansett Bay: An analysis of community structure, distribution, and abundance. Est. Coast. Mar. Sci. 1:361-378.

81. Patil, K.C., F. Matsumura, and G.M. Boush. 1972. Metabolic transformation of DDT, dieldrin, aldrin, and endrin by marine micro-organisms. Env. Sci. and Tech. 6(7):629-632.

82. Perret, W.S., and C.W. Callouet. 1974. Abundance and size of fishes taken by trawling in Vermillion Bay, Louisiana. Bull. Mar. Sci. 24:52-75.

83. Peterle, T.J. 1970. Translocation of pesticides in the environment, p. 11-16. *In* J. W. Gillett (ed.), The biological impact of pesticides in the environment. Oregon State University, Corvallis.
84. Peterson, R.H. 1973. Temperature selection of Atlantic salmon (*Salmo salar*) and brook trout (*Salvelinus fontinalis*) as influenced by various chlorinated hydrocarbons. J. Fish. Res. Bd. Canada 30:1091-1097.
85. Petrocelli, S.R., J.W. Anderson, and A.R. Hanks. 1975. Biomagnification of dieldrin residues by food-chain transfer from clams to blue crabs under controlled conditions. Bull. Env. Cont. Tox. 13:108-116.
86. Pimentel, D. 1972. Ecological impact of pesticides. Cornell Univ. Biol. Rept. 72(2):1-27.
87. Post, A., and R. Garmes. 1966. Die Empfindlichkeit einiger Tropsh Susswasserfiche gegenuber DDT und Baytex. Zeit. F. Ang. Zool. 53:487-494.
88. Reimold, R.J. 1974. Toxaphene interactions in estuarine ecosystems. Georgia Sea Grant Rept. 74-6: 80 p.
89. Reinert, R.E. 1972. Accumulation of dieldrin in an alga and the guppy. J. Fish. Res. Bd. Canada 29:1413-1418.
90. ____, and H.L. Bergman. 1974. Residues of DDT in lake trout (*Salvelinus namaycush*) and coho salmon (*Oncorhynchus kisutch*) from the great lakes. J. Fish. Res. Bd. Canada 31:191-199.
91. Rice, C.P., and H.C. Sikka. 1973. Fate of dieldrin in selected species of marine algae. Bull. Env. Cont. Tox. 9:116-123.
92. Risebrough, R.W. 1971. Chlorinated hydrocarbons, p. 259-286. *In* D.W. Hood (ed.), Impingement of Man on the Oceans. Wiley-Interscience, New York.
93. ____, R.J. Huggett, J.J. Griffin, and E.D. Goldberg. 1968. Pesticides: Transatlantic movements in the northeast trades. Science 159:1233-1236.
94. Sanborn, J.R., and C. Yu. 1973. The fate of dieldrin in a model ecosystem. Bull. Env. Cont. Tox. 10:340-346.
95. Seba, D.B., and E.F. Corcoran. 1969. Surface slicks as concentrators of pesticides in the marine environment. Pest. Mon. J. 3:190-193.
96. Smith, R.M., and C.F. Cole. 1972. Chlorinated hydrocarbon insecticide residues in winter flounder, *Pseudopleuronectes americanus*. J. Fish. Res. Bd. Canada 27:2374-2380.
97. ____, and C.F. Cole. 1973. Effects of egg concentration of DDT and dieldrin on development in winter flounder (*Pseudopleuronectes americanus*). J. Fish. Res. Bd. Canada 30:1894-1898.
98. Spencer, W.F., M.M. Cliath, W.J. Farmer, and R.A. Shepherd. 1974. Volatility of DDT residues in soil as affected by flooding and organic matter application. J. Environ. Quality 3:126-129.
99. Stickel, L.F. 1968. Organochlorine pesticides in the environment. Spec. Sci. Rept. Wildl. 119:150-183.
100. Tolman, A.J. 1974. Effect of chronic exposure to mirex on locomotor activity pattern in the diamond killifish. M.S. Thesis, Florida State University, Tallahassee. 90 p.
101. Walsh, Gerald E. 1972. Insecticides, herbicides, and polychlorinated biphenyls in estuaries. J. Wash. Acad. Sci. 62:122-139.
102. Weibel, S.R., R.B. Weidner, J.M. Cohen, and A.G. Christianson. 1966. Pesticides and other contaminants in rainfall and runoff. J. Amer. Water Works. Assoc. 58:1075-1084.

103. Wershaw, R.L., P.J. Burcar, and M.C. Goldberg. 1969. Interaction of pesticides with natural organic material. Env. Sci. Tech. 3:271-273.
104. Woodwell, G.M. 1967. Toxic substances and ecological cycles. Sci. Amer. 216:24-31.
105. _____, P.P. Craig, and H.A. Johnson. 1971. DDT in the biosphere: Where does it go? Science 174:1101-1108.
106. _____, C.F. Wurster, Jr., and P.A. Isaacson. 1967. DDT residues in an east coast estuary: A case of biological concentration of a persistent pesticide. Science 156:821-824.
107. Zimmerman, M.S., and R.J. Livingston. 1976. Effects of kraft mill effluents on benthic macrophyte assemblages in a shallow bay system (Apalachee Bay, north Florida, U.S.A.). Mar. Biol. 34:297-312.

METALS, PESTICIDES AND PCBs: TOXICITIES TO SHRIMP SINGLY

AND IN COMBINATION[1]

Del Wayne R. Nimmo and Lowell H. Bahner
U.S. Environmental Protection Agency
Environmental Research Laboratory
Sabine Island, Gulf Breeze, Florida 32561

ABSTRACT: The objective of this study was to assess potential deleterious effects of certain toxicants, singly and in combination, to penaeid shrimp. In nature, these shrimp are exposed to combinations of toxicants from industrial and municipal outfalls, from agricultural runoff or from dredge-and-fill operations.

The combined toxicities of methoxychlor and cadmium to penaeid shrimp, *Penaeus duorarum*, were either independent or additive, and varied with the method(s) of bioassay. Conclusions were based on the results of 10-, 25- and 30-day bioassays conducted with the toxicants added singly or in combination to flowing water of constant salinity and temperature.

Cadmium, but not methoxychlor, was accumulated by shrimp and methoxychlor appears to influence the processes of accumulation or loss of cadmium from tissues of shrimp.

INTRODUCTION

Water quality criteria or effluent guidelines for heavy metals usually do not take into account other toxicants which might exist in the effluent or receiving waters. However, it is well accepted that aquatic species are subjected to combinations of toxicants rather than to single toxicants in the environment. An example of this situation is the Southern California Bight, in which municipal wastewaters, storm runoff and aerial fallout are responsible for the occurrence of mercury, copper, DDT and other chlorinated hydrocarbons, such as PCBs in estuarine and marine waters (12).

[1] Contribution No. 271, Gulf Breeze Environmental Research Laboratory

523

Synergistic effects of toxic agents have been demonstrated in fresh water tests by Sprague (10), Sprague and Ramsay (11), and Cairns and Scheirer (3). Recently, Roales and Perlmutter (7, 8), found that combinations of methylmercury with copper or cygon (pesticide) and zinc appeared to have antagonistic (less than additive) effects to freshwater fish or fish embryos. In a saltwater system, Bahner and Nimmo (2) found that in short-term flow-through tests (48-hr and 96-hr), the combination of malathion-cadmium or of methoxychlor-cadmium appeared to be independent and additive. In this report, results of 10-, 25- and 30-day tests of the toxicity of Cd, methoxychlor and Aroclor® 1254 (a PCB), given singly and in combination to penaeid shrimp are discussed.

METHODS AND MATERIALS

All bioassays were conducted in flowing seawater at constant salinity and temperature (1). Sixty-five liters/hr of filtered water ($25\pm2°$ C and $20\pm2°/_{oo}$ salinity) were delivered to each 30 – 1 glass aquarium. Single toxicants were added to the water with metering pumps and combinations were obtained by simultaneously metering individual toxicants from separate syringes or flasks into the aquaria. Initially, shrimp were tested in a range of individual toxicants to determine the LC50, a mathematical expression referring to a calculated concentration of toxicant in which 50% of the experimental animals died within a prescribed time interval. Thereafter, combinations tested were conducted at or near the LC50's of the individual toxicants. These LC50's were calculated by probit analysis (5).

The procedures for collecting and acclimation of test animals are described in Bahner et al. and Nimmo et al. (1, 6).

Methoxychlor and Aroclor 1254 were analyzed by electron-capture gas chromatography, as outlined by Nimmo et al. (6). Cadmium in water and tissues of shrimp was analyzed by flameless atomic absorption spectroscopy, using the methods of Segar (9).

RESULTS AND DISCUSSION

Bioassays with single toxicants. Initially, bioassays were conducted with each toxicant singly, using different intervals of time, to determine LC50's. Earlier Bahner and Nimmo (2) reported the 96-hr acute toxicity of methoxychlor to be about 1000X more toxic to shrimp than was cadmium (Table 1). In 30-day tests, the toxicities were similar but lower (Table 1, Figs. 1 and 2). The 15-day LC50 for Aroclor 1254 was calculated from data reported by Nimmo et al. (6), and shown in Table 1.

®Registered trademark, Monsanto Co., St. Louis, Mo. Mention of commercial products does not constitute endorsement by the U.S. Environmental Protection Agency.

Bioassays with two toxicants. The 25-day combination test, conducted near the 30-day LC50 of each, indicated that the toxicities of methoxychlor and cadmium were additive (Fig. 3) and these results are consistent with those found in the 96-hr combination (2).

Table 1.   Summary of bioassays of single toxicants to pink shrimp, *Penaeus duorarum.*

Toxicant	LC50 (Measured Concentrations)	Length of Exposure
Cadmium	4.6 mg/ℓ	96-hr[1]
Cadmium	0.718 mg/ℓ	30-days
Methoxychlor	3.5 µg/ℓ	96-hr[1]
Methoxychlor	1.3 µg/ℓ	30-days
Aroclor 1254[2]	1.0 µg/ℓ	15-days

[1] Data from Bahner and Nimmo (2)
[2] Data from Nimmo et al., (6)

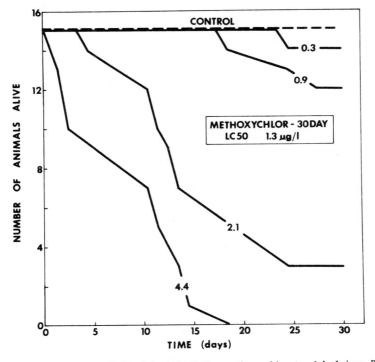

Figure 1. Acute toxicity (LC50) of the insecticide, methoxychlor, to pink shrimp, *Penaeus duorarum* during 30 day exposures.

Because toxicant concentrations vary in the environment, additivity at the LC50 concentration may not represent environmental effects in the field. Therefore, we attempted to distinguish the type of interaction between methoxychlor and cadmium using differing combinations of concentrations in 10-day tests (Fig. 4). Thus, the combinations of toxicants were prepared so as to have equal toxicity, but each was tested in increasing concentrations and counter to the other. The pairs of concentrations were arranged so that as the concentration of one toxicant increased, that of the other decreased. The combinations of concentrations were selected so that if the toxicities of methoxychlor and cadmium were additive, the sums of the toxicities of each pair of concentrations would be equal. Fig. 4 suggests that the two toxicants exert their effect independently; or, as Warren (12) has suggested there is "no interaction" (Fig. 5). Lack of interaction in this test does not invalidate the conclusion drawn from the previous

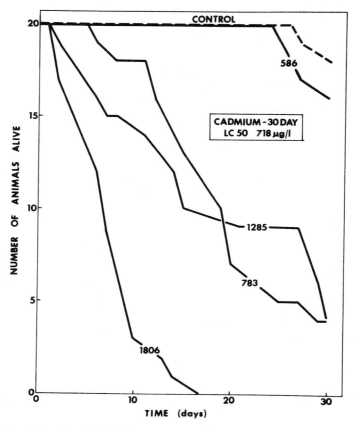

Figure 2. Acute toxicity (LC50) of cadmium to shrimp, *Penaeus duorarum* in 30 days.

test, which showed additivity, rather the difference is probably due to the relative concentration of each toxicant. Also, other researchers have shown the lack of interaction between two chemicals. The dosage-mortality curve shown in Fig. 4 is similar to that given for zebrafish embryos that were exposed to combinations of varying percentages of the 72-hr Tl_m concentrations for Cygon and zinc, and to that shown for the blue gouramis that were exposed for 96 hrs to combinations of copper and methylmercury (7, 8).

Bioassays with three toxicants. Polychlorinated biphenyls (PCBs) are industrial chemicals found in estuaries that possess properties similar to chlorinated

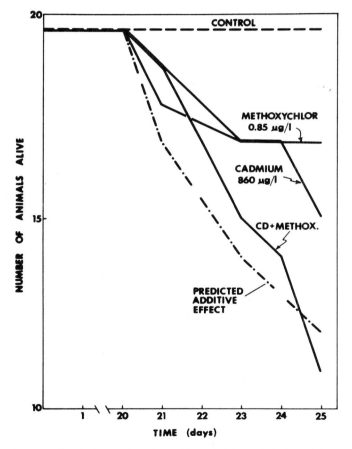

Figure 3. Toxicity of cadmium and methoxychlor administered singly and in combination to *Penaeus duorarum*. The predicted additive effect shown refers to the sum of the numbers of living shrimp in aquaria containing individual toxicants. (CD + Methox = 750 µg Cd/ℓ + 0.80 µg Methoxychlor/ℓ).

hydrocarbon pesticides (4). To determine the effect of one PCB (Aroclor 1254) in combination with cadmium and methoxychlor, we conducted bioassays using all possible single, 2-, and 3-way combinations of the toxicants. Cadmium and methoxychlor were administered at the 30-day LC50; Aroclor 1254, at the 15-day LC50. Again, the results showed that the toxicity of any combination was equal to the sum of each chemical tested singly (Fig. 6).

Accumulation of cadmium. After the exposures, cadmium, but not methoxychlor was detected in shrimp tissues. In the 25-day tests, with methoxychlor and cadmium singly, and in combination, there was no difference in cadmium accumulation in the muscle. There was less cadmium in the muscle of shrimp that had been exposed to methoxychlor alone than in controls (Table 2). Likewise, when cadmium was administered in combination with methoxychlor in the second test (Table 3), a statistically lower cadmium concentration was found when methoxychlor was present.

Pathology. Shrimp exposed to cadmium alone developed pronounced blackened areas on the gills, but this condition was reduced or absent when methoxychlor and cadmium were tested in combination.

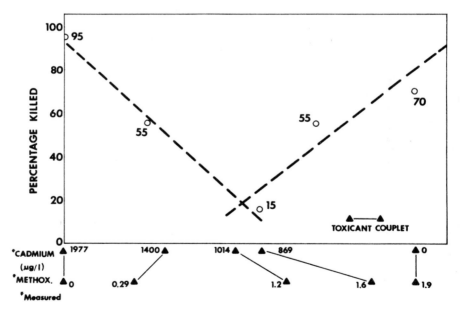

Figure 4. Dosage-mortality curves for toxicant combinations after 10 days. The toxicants, cadmium and methoxychlor were prepared and administered so as to have equal toxicity when tested separately. Along the abscissa we show each toxicant as a small triangle, and the combination in each aquarium, as a toxicant couplet.

Dover sole, *(Microstomus pacificus)* collected near the outfall of Palos Verdes Peninsula, California (an area in which the biota contained high DDT concentrations) did not contain as much arsenic, cadmium and selenium as specimens taken near Santa Catalina Island (13). Concentrations of these elements in sediments from the Palos Verdes outfall were greater by a factor of 15 for arsenic,

Table 2. Accumulation of cadmium in the pink shrimp when tested singly and in combination with methoxychlor.

Test Concentrations (μg/1) Nominal Measured		Percentage Mortality In 25 Days[1]	Cadmium in muscle mg/kg ± 2 S.E.M.
Control	—	0	0.4 ± 0.2
Cadmium			
1000	860	25	89.4 ± 17.4
Methoxychlor			
1.0	0.85	15	0.1 ± 0.0
Cadmium/Methoxychlor			
1000	730	55	94.1 ± 26.1
1.0	0.80		

[1] Bioassay conducted at 20 °/oo salinity and 25° C.

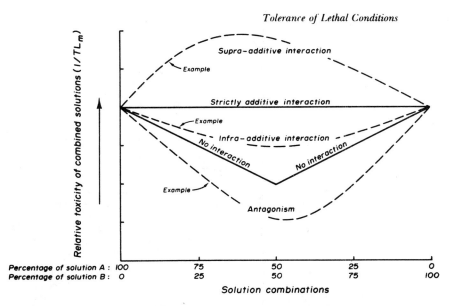

Figure 5. Possible kinds of interactions between two hypothetical toxicants tested in combination. After Warren (12).

160 for cadmium and 14 for selenium than those from Catalina Island. The apparent discrepancy in concentrations of metals in the tissues of biota may be due to the presence of organochlorine pesticides similar to DDT which influenced heavy metal accumulation.

CONCLUSIONS

No dramatic toxic interaction of the combination of methoxychlor and cadmium, or of the combination of methoxychlor-cadmium-PCB, to shrimp was

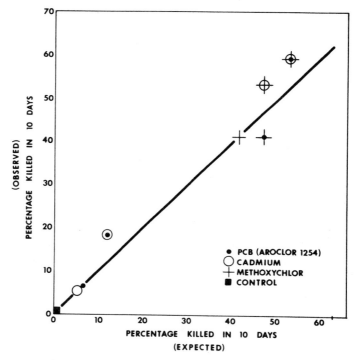

Figure 6. Comparison of toxicities of single vs. combined constituents to *Penaeus duorarum*. Measured concentrations in micrograms per liter ranged from 640-829 for cadmium, 0.9-1.0 for methoxychlor, 0.7-1.1 for Aroclor 1254. Each symbol represents the effect of a single toxicant, per aquarium; superimposed symbols represent the toxicants combined. Unity is denoted as the line constructed through the points equaling the observed percentage killed (ordinate) to that expected (abscissa) for each toxicant singly, and the origin. If any 2- or 3-way combination has elicited an exactly additive effect, the datum would fall on unity. If synergistic effects had been observed with any combination, the datum would lie to the left of unity; if antagonistic, to the right of unity. By constructing the Working-Hotelling confidence bands (regression analyses) on these data, they indicated that all toxicant combinations were probably additive ($\alpha = 0.05$).

evident. We have presented evidence that methoxychlor influences the accumulation in, or loss of cadmium from, the tissues of shrimp. The toxicities of the combinations, when compared to each toxicant tested singly, were independent and additive. Distinguishing between these relationships depends upon the method of assessment and the concentrations of each of the toxicants. We observed additivity, but no synergistic effects of the toxicants in the laboratory, although synergism might exist in the environment. Before allowable limits for effluents are established, therefore, background concentrations of all toxicants in receiving waters used in experimental systems must be known in order to properly evaluate the toxicity of the compound under study.

Table 3.   Accumulation of cadmium in tissues of shrimp in the presence of other chemicals.

Measured Concentrations in test media (μg/l)[1]			Cadmium in muscle mg/kg ± 2 S.E.M.	Concentration Factor
Cadmium[2]	PCB[3]	Methoxy.[4]		
640	ϕ	ϕ	15.58 ± 2.90	24
774	ϕ	1.1	9.90 ± 2.79	13
746	0.9	ϕ	13.99 ± 3.05	19
829	0.9	0.8	16.28 ± 5.44	20
Control			0.25 ± 0.1	0
ϕ	ϕ	1.0	0.25 ± 0.05	0
ϕ	0.7	ϕ	0.28 ± 0.1	0
ϕ	1.1	1.0	0.26 ± 0.07	0

[1] 10-day flowing water bioassays; 20° /oo, 25° C
[2] Cadmium chloride
[3] Aroclor 1254
[4] 1, 1, 1 - Trichloro-2, 2-bis-[p-methoxyphenyl] ethane

LITERATURE CITED

1. Bahner, L. H., C. D. Craft and D. R. Nimmo. 1975. A saltwater flow-through bioassay method with controlled temperature and salinity. Prog. Fish-Cult. 37:126-129.
2. _____, and D. R. Nimmo. In Press. Methods to assess effects of combinations of toxicants, salinity and temperature on estuarine animals. Presented at the 9th Annual Conference on Trace Substances in Environmental Health, 10-12 June 1975, Columbia, Mo.
3. Cairns, J. Jr., and A. Scheier. 1968. A comparison of the toxicity of some common industrial waste components tested individually and combined. Prog. Fish-Cult. 30:3-8.
4. Duke, T. W., J. I. Lowe, and A. J. Wilson, Jr. 1970. A polychlorinated biphenyl (Aroclor 1254®) in the water, sediment, and biota of Escambia Bay, Florida. Bull. Environ. Contam. Toxicol. 5:171-180.
5. Finney, D. J. 1971. Probit analysis. Cambridge University Press. 333p.

6. Nimmo, D. R., R. R. Blackman, A. J. Wilson, Jr. and J. Forester. 1971. Toxicity and distribution of Aroclor® 1254 in pink shrimp *Penaeus duorarum*. Mar. Biol. (Berlin) 11:191-197.

7. Roales, R. R., and A. Perlmutter. 1974. Toxicity of zinc and cygon, applied singly and jointly, to zebrafish embryos. Bull Environ. Contam. Toxicol. 12:475-480.

8. _____, and _____. 1974. Toxicity of methyl-mercury and copper, applied singly and jointly, to the blue gourami, *Trichogaster trichopterus*. Bull. Environ. Contam. Toxicol. 12:633-639.

9. Segar, D. A. 1971. The use of the heated graphite atomizer in marine sciences. Proc. 3rd International Congress of Atomic Absorption and Atomic Fluorescence Spectrometry. Adam Hilger, London. pp. 523-532.

10. Sprague, J. B. 1964. Lethal concentrations of copper and zinc for young Atlantic salmon. J. Fish. Res. Board Can. 21:17-26.

11. _____, and B. A. Ramsay. 1965. Lethal levels of mixed copper-zinc solutions for juvenile salmon. J. Fish. Res. Bd. Can. 22:425-432.

12. Warren, C. E. 1971. Biology and Water Pollution Control. W. B. Saunders Co. Philadelphia and London. 434 p.

13. Young, D. R., D. J. McDermott, T. C. Heesen and Tsu-Kai Jan. In Press. Presented at the American Chemical Society Symposium on Marine Chemistry in the Coastal Environment, 8-10 April 1975, Philadelphia, Pa.

INDEX

A

Abra alba, 436
Acanthaceae, 260
Acanthus ilicifolius, 260
Acrostichum aureum, 254, 263
Acta alba, 270
Acteocina canaliculata, 180
Adinia xenia, 512–513
Aedes aegypti, 288
Aegiceras corniculatum, 260
Aequipecten irradians, 438–440
Agonus cataphractus, 400
Agrostis stolonifera, 223
Aizoaceae, 263
Alkaline phosphatase, 84–87
Alligator mississippiensis, 227
Alosa pseudoharengus, 22
Alternanthera philoxeroides, 220
Alternaria sp., 490
Ambrosia artemisiifolia, 269
Ammonia, 119, 369–375
   excretion in crab, 372–376
   invertebrate excretion, 415–428
   origin and movement, 370–372
   relationship with blood pH, 369–370
   relationship with environmental salinity,
      369–370
   relationship with osmoregulation,
      372–374
Ammonia concentration, Long Island
      Sound, 33
Ampelisca abdita, 192
Amphiura
   chiajei, 192
   filiformis, 192
Anas fulvigula, 229
Anguilla
   anguilla, 400

   rostrata, 401
Apalachee Bay, 313, 315, 321, 328
Apalachicola Bay, 313–315, 328, 516
Aplysia californica, 455
Arenicola marina, 360, 414, 425, 455
Argopectens irradians, feeding on detritus,
      469, 473–477
Armandia brevis, 191
Ascaris, 352
Aspergillus sp., 490
Aster tripolium, 223
Asterias rubens, 351
Atriplex hastata, 223
Aureobasidium sp., 490
Avicennia, 254
   germinans, 255
   marina, 255
   officinalis, 255
Avicenniaceae, 255

B

Bacopa monnieri, 228
Bacteria
   activity in estuary, 381–389
   metabolism, 382
   removal by wetlands, 246–247
   survival in estuary, 386–389
   utilization of oil, 483–491
Batidaceae, 263
Batis maritima, 263
Benthic communities, 177–193
Beroë cucumis, 205
Behavior
   Callinectes sapidus, 315–318, 321–325,
      328–329
   effects of insect juvenile hormone mimic
      on, 279–288

533

A 6
B 7
C 8
D 9
E 0
F 1
G 2
H 3
I 4
J 5